Lecture Notes in Physics

Springer
Berlin
Heidelberg
New York
Barcelona
Hong Kong
London
Milan
Paris
Singapore
Tokyo

The Editorial Policy for Proceedings

The series Lecture Notes in Physics reports new developments in physical research and teaching – quickly, informally, and at a high level. The proceedings to be considered for publication in this series should be limited to only a few areas of research, and these should be closely related to each other. The contributions should be of a high standard and should avoid lengthy redraftings of papers already published or about to be published elsewhere. As a whole, the proceedings should aim for a balanced presentation of the theme of the conference including a description of the techniques used and enough motivation for a broad readership. It should not be assumed that the published proceedings must reflect the conference in its entirety. (A listing or abstracts of papers presented at the meeting but not included in the proceedings could be added as an appendix.)
When applying for publication in the series Lecture Notes in Physics the volume's editor(s) should submit sufficient material to enable the series editors and their referees to make a fairly accurate evaluation (e.g. a complete list of speakers and titles of papers to be presented and abstracts). If, based on this information, the proceedings are (tentatively) accepted, the volume's editor(s), whose name(s) will appear on the title pages, should select the papers suitable for publication and have them refereed (as for a journal) when appropriate. As a rule discussions will not be accepted. The series editors and Springer-Verlag will normally not interfere with the detailed editing except in fairly obvious cases or on technical matters.
Final acceptance is expressed by the series editor in charge, in consultation with Springer-Verlag only after receiving the complete manuscript. It might help to send a copy of the authors' manuscripts in advance to the editor in charge to discuss possible revisions with him. As a general rule, the series editor will confirm his tentative acceptance if the final manuscript corresponds to the original concept discussed, if the quality of the contribution meets the requirements of the series, and if the final size of the manuscript does not greatly exceed the number of pages originally agreed upon. The manuscript should be forwarded to Springer-Verlag shortly after the meeting. In cases of extreme delay (more than six months after the conference) the series editors will check once more the timeliness of the papers. Therefore, the volume's editor(s) should establish strict deadlines, or collect the articles during the conference and have them revised on the spot. If a delay is unavoidable, one should encourage the authors to update their contributions if appropriate. The editors of proceedings are strongly advised to inform contributors about these points at an early stage.
The final manuscript should contain a table of contents and an informative introduction accessible also to readers not particularly familiar with the topic of the conference. The contributions should be in English. The volume's editor(s) should check the contributions for the correct use of language. At Springer-Verlag only the prefaces will be checked by a copy-editor for language and style. Grave linguistic or technical shortcomings may lead to the rejection of contributions by the series editors. A conference report should not exceed a total of 500 pages. Keeping the size within this bound should be achieved by a stricter selection of articles and not by imposing an upper limit to the length of the individual papers. Editors receive jointly 30 complimentary copies of their book. They are entitled to purchase further copies of their book at a reduced rate. As a rule no reprints of individual contributions can be supplied. No royalty is paid on Lecture Notes in Physics volumes. Commitment to publish is made by letter of interest rather than by signing a formal contract. Springer-Verlag secures the copyright for each volume.

The Production Process

The books are hardbound, and the publisher will select quality paper appropriate to the needs of the author(s). Publication time is about ten weeks. More than twenty years of experience guarantee authors the best possible service. To reach the goal of rapid publication at a low price the technique of photographic reproduction from a camera-ready manuscript was chosen. This process shifts the main responsibility for the technical quality considerably from the publisher to the authors. We therefore urge all authors and editors of proceedings to observe very carefully the essentials for the preparation of camera-ready manuscripts, which we will supply on request. This applies especially to the quality of figures and halftones submitted for publication. In addition, it might be useful to look at some of the volumes already published. As a special service, we offer free of charge LaTeX and TeX macro packages to format the text according to Springer-Verlag's quality requirements. We strongly recommend that you make use of this offer, since the result will be a book of considerably improved technical quality. To avoid mistakes and time-consuming correspondence during the production period the conference editors should request special instructions from the publisher well before the beginning of the conference. Manuscripts not meeting the technical standard of the series will have to be returned for improvement.
For further information please contact Springer-Verlag, Physics Editorial Department II, Tiergartenstrasse 17, D-69121 Heidelberg, Germany

Series homepage – http://www.springer.de/phys/books/lnpp

Jerzy Kowalski-Glikman (Ed.)

Towards Quantum Gravity

Proceedings of the XXXV International
Winter School on Theoretical Physics
Held in Polanica, Poland, 2-11 February 1999

Springer

Editor

Jerzy Kowalski-Glikman
Institute of Theoretical Physics
University of Wrocław
Pl. Maxa Borna 9
50-204 Wrocław, Poland

Library of Congress Cataloging-in-Publication Data applied for

Die Deutsche Bibliothek - CIP-Einheitsaufnahme

Towards quantum gravity : proceedings of the XXXV International Winter School on Theoretical Physics, held in Polancia, Poland, 2 - 11 February 1999 / Jerzy Kowalski-Glikman (ed.). - Berlin ; Heidelberg ; New York ; Barcelona ; Hong Kong ; London ; Milan ; Paris ; Singapore ; Tokyo : Springer, 2000
(Lecture notes in physics ; Vol. 541)
ISBN 3-540-66910-8

ISSN 0075-8450
ISBN 3-540-66910-8 Springer-Verlag Berlin Heidelberg New York

Typesetting: Camera-ready by the authors/editor
Cover design: *design & production*, Heidelberg

Printed on acid-free paper
SPIN: 10720709 55/3144/du - 5 4 3 2 1 0

Preface

For almost forty years the Institute for Theoretical Physics of the University of Wrocław has organized winter schools devoted to current problems in theoretical physics. The XXXV International Winter School on Theoretical Physics, "From Cosmology to Quantum Gravity", was held in Polanica, a little town in southwest Poland, between 2nd and 11th February, 1999. The aim of the school was to gather together world-leading scientists working on the field of quantum gravity, along with a number of post-graduate students and young post-docs and to offer young scientists with diverse backgrounds in astrophysics and particle physics the opportunity to learn about recent developments in gravitational physics. The lectures covered macroscopic phenomena like relativistic binary star systems, gravitational waves, and black holes; and the quantum aspects, e.g., quantum space-time and the string theory approach.

This volume contains a collection of articles based on lectures presented during the School. They cover a wide spectrum of topics in classical relativity, quantum gravity, black hole physics and string theory. Unfortunately, some of the lecturers were not able to prepare their contributions, and for this reason I decided to entitle this volume "Towards Quantum Gravity", which better reflects its contents.

I would like to thank all the lecturers for the excellent lectures they gave and for the unique atmosphere they created during the School. Thanks are due to Professor Jan Willem van Holten and Professor Jerzy Lukierski for their help in organizing the school and preparing its scientific programme. Dobromila Nowak worked very hard, carrying out virtually all administrative duties alone. I would also like to thank the Institute for Theoretical Physics of the University of Wrocław, the University of Wrocław, the Foundation for Karpacz Winter Schools, and the Polish Committee for Scientific Research (KBN) for their financial support.

Wrocław, November 1999 *Jerzy Kowalski-Glikman*

Contents

List of Contributors

Giovanni Amelino-Camelia
Theory Division, CERN
1211 Geneva, Switzerland
Giovanni.Amelino-Camelia@cern.ch

Abhay Ashtekar
Center for Gravitational Physics
and Geometry
Department of Physics
The Pennsylvania State University
University Park, PA 16802, USA
ashtekar@phys.psu.edu

Brandon Carter
D.A.R.C., Observatoire de Paris
92 Meudon, France
Brandon.Carter@obspm.fr

Bernard de Wit
Institute for Theoretical Physics
Utrecht University
3508 TA Utrecht, The Netherlands
dewit@ruu.nl

Marcus Gaul
Max-Planck-Institut für Physik
(Werner-Heisenberg-Institut)
Föhringer Ring 6
80805 München, Germany; and
Sekt. Physik
Ludwig-Maximilians-Universität
Theresienstrasse 37
80333 München, Germany

Ivan Herger
Institute for Theoretical Physics
Utrecht University
3508 TA Utrecht, The Netherlands
Ivan.Herger@ruu.nl

Stuart Kauffman
Biosgroup LTD
317 Paseo de Peralta
Santa Fe, NM 87501-8943, USA
stu@biosgroup.com

Richard Kerner
Laboratoire de Gravitation
et Cosmologie Relativistes
Université Pierre-et-Marie-Curie -
CNRS ESA 7065, Tour 22
4-ème étage, Boîte 142
4, Place Jussieu, 75005 Paris, France
rk@ccr.jussieu.fr

Claus Kiefer
Fakultät Physik, Universität Freiburg
Hermann-Herder-Strasse 3
79104 Freiburg, Germany; and
Institute for Advanced Study Berlin
Wallotstrasse 19
14193 Berlin, Germany; kiefer@phyq6.physik.uni-freiburg.de

Jorma Louko
Max-Planck-Institut
für Gravitationsphysik
Am Mühlenberg 5
14476 Golm, Germany
louko@aei-potsdam.mpg.de

Luca Lusanna
Sezione INFN di Firenze
Largo E.Fermi 2
50125 Firenze, Italy
lusanna@fi.infn.it

Shahn Majid
Department of Applied Mathematics
and Theoretical Physics
Silver Street
Cambridge, CB3 9EW, England
Shahn.Majid@damtp.cam.ac.uk

Carlo Rovelli
Centre de Physique Théorique
CNRS Luminy
13288 Marseille, France; and
Department of Physics and Astronomy
University of Pittsburgh
Pittsburgh, PA 15260, USA
rovelli@science.unitn.it

Kostas Skenderis
Spinoza Institute
University of Utrecht
Leuvenlaan 4
3584 CE Utrecht, The Netherlands
skender@ruunat.phys.uu.nl

Lee Smolin
Center for Gravitational Physics
and Geometry
Department of Physics
The Pennsylvania State University
University Park, PA 16802, USA
smolin@phys.psu.edu

Jan Willem van Holten
NIKHEF, PO Box 41882
1008 DB Amsterdam
The Netherlands
t32@nikhef.nl

Are We at the Dawn of Quantum-Gravity Phenomenology?

Giovanni Amelino-Camelia[1]

Theory Division, CERN, CH-1211, Geneva, Switzerland**

Abstract. A handful of recent papers has been devoted to proposals of experiments capable of testing some candidate quantum-gravity phenomena. These lecture notes emphasize those aspects that are most relevant to the questions that inevitably come to mind when one is exposed for the first time to these research developments: How come theory and experiments are finally meeting in spite of all the gloomy forecasts that pervade traditional quantum-gravity reviews? Is this a case of theorists having put forward more and more speculative ideas until a point was reached at which conventional experiments could rule out the proposed phenomena? Or has there been such a remarkable improvement in experimental techniques and ideas that we are now capable of testing plausible candidate quantum-gravity phenomena? These questions are analysed rather carefully for the recent proposals of tests of space-time fuzziness using modern interferometers and tests of dispersion in the quantum-gravity vacuum using observations of gamma rays from distant astrophysical sources. I also briefly discuss other proposed quantum-gravity experiments, including those exploiting the properties of the neutral-kaon system for tests of quantum-gravity-induced decoherence and those using particle-physics accelerators for tests of models with large extra dimensions.

1 Introduction

Traditionally the lack of experimental input [1] has been the most important obstacle in the search for "quantum gravity", the new theory that should provide a unified description of gravitation and quantum mechanics. Recently there has been a small, but nonetheless encouraging, number of proposals [2–9] of experiments probing the nature of the interplay between gravitation and quantum mechanics. At the same time the "COW-type" experiments on quantum mechanics in a strong (classical) gravitational environment, initiated by Colella, Overhauser and Werner [10], have reached levels of sophistication [11] such that even gravitationally induced quantum phases due to local tides can be detected. In light of these developments there is now growing (although still understandably cautious) hope for data-driven insight into the structure of quantum gravity.

The primary objective of these lecture notes is the one of giving the reader an intuitive idea of how far quantum-gravity phenomenology has come. This is somewhat tricky. Traditionally experimental tests of quantum gravity were believed to be not better than a dream. The fact that now (some) theory and (some) experiments finally "meet" could have two very different explanations:

** Marie Curie Fellow (permanent address: Dipartimento di Fisica, Universitá di Roma "La Sapienza", Piazzale Moro 2, Roma, Italy it could be that experimental techniques and ideas have improved so much that now tests of plausible quantum-gravity effects are within reach, but it could also be that theorists have had enough time in their hands to come up with scenarios speculative enough to allow testing by conventional experimental techniques. I shall argue that experiments have indeed progressed to the point were some significant quantum-gravity tests are doable. I shall also clarify in which sense the traditional pessimism concerning quantum-gravity experiments was built upon the analysis of a very limited set of experimental ideas, with the significant omission of the possibility (which we now find to be within our capabilities) of experiments set up in such a way that very many of the very small quantum-gravity effects are somehow summed together. Some of the theoretical ideas that can be tested experimentally are of course quite speculative (decoherence, space-time foam, large extra dimensions, ...) but this is not so disappointing because it seems reasonable to expect that the new theory should host a large number of new conceptual/structural elements in order to be capable of reconciling the (apparent) incompatibility between gravitation and quantum mechanics. [An example of motivation for very new structures is discussed here in Section 10, which is a "theory addendum" reviewing some of the arguments [12] in support of the idea [13] that the mechanics on which quantum gravity is based might not be exactly the one of ordinary quantum mechanics, since it should accommodate a somewhat different (non-classical) concept of "measuring apparatus" and a somewhat different relationship between "system" and "measuring apparatus".]

The bulk of these notes gives brief reviews of the quantum-gravity experiments that can be done. The reader will be asked to forgive the fact that this review is not very balanced. The two proposals in which this author has been involved [5,7] are in fact discussed in greater detail, while for the experiments proposed in Refs. [2–4,8,9] I just give a very brief discussion with emphasis on the most important conceptual ingredients.

The students who attended the School might be surprised to find the material presented with a completely different strategy. While my lectures in Polanica were sharply divided in a first part on theory and a second part on experiments, here some of the theoretical intuition is presented while discussing the experiments. It appears to me that this strategy might be better suited for a written presentation. I also thought it might be useful to start with the conclusions, which are given in the next two sections. Section 4 reviews the proposal of using modern interferometers to set bounds on space-time fuzziness. In Section 5 I review the proposal of using data on GRBs (gamma-ray bursts) to investigate possible quantum-gravity induced *in vacuo* dispersion of electromagnetic radiation. In Section 6 I give brief reviews of other quantum-gravity experiments. In Section 7 I give a brief discussion of the mentioned "COW-type" experiments testing quantum mechanics in a strong classical gravity environment. Section 8 provides a "theory addendum" on various scenarios for bounds on the measurability of distances in quantum gravity and their possible relation to properties of the space-time foam. Section 9 provides a theory addendum on other works which are in one way or another related to (or relevant for) the content of these notes. Section 10 gives the mentioned theory addendum concerning ideas on a mechanics for quantum gravity that be not exactly of the type of ordinary quantum mechanics. Finally in Section 11 I give some comments on the outlook of quantum-gravity phenomenology, and I also emphasize the fact that, whether or not they turn out to be helpful for quantum gravity, most of the experiments considered in these notes are intrinsically significant as tests of quantum mechanics and/or tests of fundamental symmetries.

2 First the conclusions: what has this phenomenology achieved?

Let me start by giving an intuitive idea of how far quantum-gravity phenomenology has gone. Some of the views expressed in this section are supported by analyses which will be reviewed in the following sections. The crucial question is: Can we just test some wildly speculative ideas which have somehow surfaced in the quantum-gravity literature? Or can we test even some plausible candidate quantum-gravity phenomena?

Before answering these questions it is appropriate to comment on the general expectations we have for quantum gravity. It has been realized for some time now that by combining elements of gravitation with elements of quantum mechanics one is led to "interplay phenomena" with rather distinctive signatures, such as quantum fluctuations of space-time [14–16], and violations of Lorentz and/or CPT symmetries [17–23], but the relevant effects are expected to be very small (because of the smallness of the Planck length). Therefore in this "intuition-building" section the reader must expect from me a description of experiments with a remarkable sensitivity to the new phenomena.

Let me start from the possibility of quantum fluctuations of space-time. A prediction of nearly all approaches to the unification of gravitation and quantum mechanics is that at very short distances the sharp classical concept of space-time should give way to a somewhat "fuzzy" (or "foamy") picture, possibly involving virulent geometry fluctuations (sometimes depicted as wormholes and black holes popping in and out of the vacuum). Although the idea of space-time foam remains somewhat vague and it appears to have significantly different incarnations in different quantum-gravity approaches, a plausible expectation that emerges from this framework is that the distance between two bodies "immerged" in the space-time foam would be affected by (quantum) fluctuations. If urged to give a rough description of these fluctuations at present theorists can only guess that they would be of Planck length L_p ($L_p \sim 10^{-35}m$) magnitude and occurring at a frequency of roughly one per Planck time T_p ($T_p = L_p/c \sim 10^{-44}s$). One should therefore deem significant for space-time-foam research any experiment that monitors the distances between two bodies with enough sensitivity to test this type of fluctuations. This is exactly what was achieved by the analysis reported in Refs. [7,24], which was based on the observation that the most advanced modern interferometers (the ones normally used for detection of classical gravity waves) are exactly the natural instruments to study the fuzziness of distances. While I postpone to Section 4 a detailed discussion of these interferometry-based tests of fuzziness, let me emphasize already here that modern interferometers have achieved such a level of sensitivity that we are already in a position to **rule out** fluctuations in the distances of their test masses of the type discussed above, *i.e.* fluctuations of Planck-length magnitude occurring at a rate of one per each Planck time. This is perhaps the simplest way for the reader to picture intuitively the type of objectives already reached by quantum-gravity phenomenology.

Another very intuitive measure of the maturity of quantum-gravity phenomenology comes from the studies of *in vacuo* dispersion proposed in Ref. [5] (also see the more recent purely experimental analyses [25,26]). Deformed dispersion relations are a rather natural possibility for quantum gravity. For example, they emerge naturally in quantum gravity scenarios requiring a modification of Lorentz symmetry. Modifications of Lorentz symmetry could result from space-time discreteness (*e.g.* a discrete space accommodates a somewhat different concept of "rotation" with respect to the one of ordinary continuous spaces), a possibility extensively investigated in the quantum gravity literature (see, *e.g.*, Ref. [22]), and it would also naturally result from an "active" quantum-gravity vacuum of the type advocated by Wheeler and Hawking [14,15] (such a "vacuum" might physically label the space-time points, rendering possible the selection of a "preferred frame"). The specific structure of the deformation can differ significantly from model to model. Assuming that the deformation admits a series expansion at small energies E, and parametrizing the deformation in terms of an energy[1] scale E_{QG} (a scale characterizing the onset of quantum-gravity dispersion effects, often identified with the Planck energy $E_p = \hbar c/L_p \sim 10^{19} GeV$), for a massless particle one would expect to be able to approximate the deformed dispersion relation at low energies according to

$$c^2 \mathbf{p}^2 \simeq E^2 \left[1 + \xi \left(\frac{E}{E_{QG}} \right)^{\alpha} + O\left(\left(\frac{E}{E_{QG}} \right)^{\alpha+1} \right) \right] \qquad (1)$$

where c is the conventional speed-of-light constant. The scale E_{QG}, the power α and the sign ambiguity $\xi = \pm 1$ would be fixed in a given dynamical framework; for example, in some of the approaches based on dimensionful quantum deformations of Poincaré symmetries [21,27,28] one encounters a dispersion relation $c^2 \mathbf{p}^2 = E_{QG}^2 \left[1 - e^{E/E_{QG}} \right]^2$, which implies $\xi = \alpha = 1$. Because of the smallness of $1/E_{QG}$ it was traditionally believed that this effect could not be seriously tested experimentally (*i.e.* that for $E_{QG} \sim E_p$ experiments would only be sensitive to values of α much smaller than 1), but in Ref. [5] it was observed that recent progress in the phenomenology of GRBs [29] and other astrophysical phenomena should soon allow us to probe values of E_{QG} of the order of

[1] I parametrize deformations of dispersion relations in terms of an energy scale E_{QG}, which is implicitly assumed to be rather close to E_p, while I later parametrize the proposals for distance fuzziness with a length scale L_{QG}, which is implicitly assumed to be rather close to L_p. This is sometimes convenient in formulas, but it is of course somewhat redundant, since $E_p = \hbar c/L_p$.

(or even greater than) E_p for values of α as large as 1. As discussed later in these notes, $\alpha = 1$ appears to be the smallest value that can be obtained with plausible quantum-gravity arguments and several of these arguments actually point us toward the larger value $\alpha = 2$, which is still very far from present-day experimental capabilities. While of course it would be very important to achieve sensitivity to both the $\alpha = 1$ and the $\alpha = 2$ scenarios, the fact that we will soon test $\alpha = 1$ is a significant first step.

Another recently proposed quantum-gravity experiment concerns possible violations of CPT invariance. This is a rather general prediction of quantum-gravity approaches, which for example can be due to elements of nonlocality (locality is one of the hypotheses of the "CPT theorem") and/or elements of decoherence present in the approach. At least some level of non-locality is quite natural for quantum gravity as a theory with a natural length scale which might also host a "minimum length" [30–32,12,33]. Motivated by the structure of "Liouville strings" [19] (a non-critical string approach to quantum gravity which appears to admit a space-time foam picture) a phenomenological parametrization of quantum-gravity induced CPT violation in the neutral-kaon system has been proposed in Refs. [17,34]. (Other studies of the phenomenology of CPT violation can be found in Ref. [20,35].) In estimating the parameters that appear in this phenomenological model the crucial point is as usual the overall suppression given by some power of the Planck length. For the case in which the Planck length enters only linearly in the relevant formulas, experiments investigating the properties of neutral kaons are already setting significant bounds on the parameters of this phenomenological approach [2].

In summary, experiments are reaching significant sensitivity with respect to all of the frequently discussed features of quantum gravity that I mentioned at the beginning of this section: space-time fuzziness, violations of Lorentz invariance, and violations of CPT invariance. Other quantum-gravity experiments, which I shall discuss later in these notes, can probe other candidate quantum-gravity phenomena, giving additional breadth to quantum-gravity phenomenology.

Before closing this section there is one more answer I should give: how could this happen in spite of all the gloomy forecasts which one finds in most quantum-gravity review papers? The answer is actually simple. Those gloomy forecasts were based on the observation that under ordinary conditions the direct detection of a single quantum-gravity phenomenon would be well beyond our capabilities if the magnitude of the phenomenon is suppressed by the smallness of the Planck length. For example, in particle-physics contexts this is seen in the fact that the contribution from "gravitons" (the conjectured mediators of quantum-gravity interactions) to particle-physics processes with center-of-mass energy $\mathcal{E}$ is expected to be penalized by overall factors given by some power of the ratio $\mathcal{E}/(10^{19} GeV)$, which is an extremely small ratio even for an ideal particle accelerators ring built all around the Earth. However, small effects can become observable in special contexts and in particular one can always search for an experimental setup such that a very large number of the very small quantum-gravity contributions are effectively summed together. This later possibility is not unknown to the particle-physics community, since it has been exploited in the context of investigations of the particle-physics theories unifying the strong and electroweak interactions, were one encounters the phenomenon of proton decay. By finding ways to keep under observation very large numbers of protons, experimentalists have managed[2] to set highly significant bounds on proton decay [37], even though the proton-decay probability is penalized by the fourth power of the small ratio between the proton mass, which is of order $1 GeV$, and the mass of the vector bosons expected to mediate proton decay, which is conjectured to be of order $10^{16} GeV$. Just like proton-decay experiments are based on a simple way to put together very many of the small proton-decay effects[3] the experiments using modern interferometers to study space-time fuzziness and the experiments using GRBs to study violations of Lorentz invariance exploit simple ways to put together very many of the very small quantum-gravity effects. I shall explain this in detail in Sections 4 and 5.

3 Addendum to conclusions: any hints to theorists from experiments?

In the preceding section I have argued that quantum-gravity phenomenology, even being as it is in its infancy, is already starting to provide the first significant tests of plausible candidate quantum-gravity phenomena. It is of course just "scratching the surface" of whatever "volume" contains the full collection of experimental studies we might wish to perform, but we are finally getting started. Of course, a phenomenology programme is meant to provide input to the theorists working in the area, and therefore one measure of the achievements of a phenomenology programme is given by the impact it is having on theory studies. In the case of quantum-gravity experiments the flow of information from experiments to theory will take some time. The primary reason is that most quantum-gravity approaches have been guided (just because there was no alternative guidance from data) by various sorts of formal intuition for quantum gravity (which of course remain pure speculations as long as they are not confirmed by experiments). This is in particular true for the two most popular approaches to the unification of gravitation and quantum mechanics, *i.e.* "critical superstrings" [38,39] and "canonical/loop quantum gravity" [40]. Because of the type of intuition that went into them, it is not surprising that these "formal quantum gravity approaches" are proving extremely useful in providing us new ideas on how gravitation and quantum mechanics could resolve the apparent conflicts between their conceptual structures, but they are not giving us any ideas

[2] This author's familiarity [36] with the accomplishments of proton-decay experiments has certainly contributed to the moderate optimism for the outlook of quantum-gravity phenomenology which is found in these notes.

[3] For each of the protons being monitored the probability of decay is extremely small, but there is a significantly large probability that at least one of the many monitored protons decay.

on which experiments could give insight into the nature of quantum gravity. The hope that these formal approaches could eventually lead to new intuitions for the nature of space-time at very short distances has been realized only rather limitedly. In particular, it is still unclear if and how these formalisms host the mentioned scenarios for quantum fluctuations of space-time and violations of Lorentz and/or CPT symmetries. The nature of the quantum-gravity vacuum (in the sense discussed in the preceding section) appears to be still very far ahead in the critical superstring research programme and its analysis is only at a very preliminary stage within canonical/loop quantum gravity. In order for the experiments discussed in these notes to affect directly critical superstring research and research in canonical/loop quantum gravity it is necessary to make substantial progress in the analysis of the physical implications of these formalisms.

Still, in an indirect way the recent results of quantum-gravity phenomenology have already started to have an impact on theory work in these formal quantum gravity approaches. The fact that it is becoming clear that (at least a few) quantum-gravity experiments can be done has reenergized efforts to explore the physical implications of the formalisms. The best example of this way in which phenomenology can influence "pure theory" work is provided by Ref. [41], which was motivated by the results reported in Ref. [5] and showed that canonical/loop quantum gravity admits (under certain conditions, which in particular involve some parity breaking) the phenomenon of deformed dispersion relations, with deformation going linearly with the Planck length.

While the impact on theory work in the formal quantum gravity approaches is still quite limited, of course the new experiments are providing useful input for more intuitive/phenomelogical theoretical work on quantum gravity. For example, the analysis reported in Refs. [7,24], by ruling out the scheme of distance fluctuations of Planck length magnitude occurring at a rate of one per Planck time, has had significant impact [24,42] on the line of research which has been deriving intuitive pictures of properties of quantum space-time from analyses of measurability and uncertainty relations [12,43–45]. Similarly the "Liouville string" [19] inspired phenomenological approach to quantum gravity [34,46] has already received important input from the mentioned studies of the neutral-kaon system and will receive equally important input from the mentioned GRB experiments, once these experiments (in a few years) reach Planck-scale sensitivity.

It is possible that the availability of quantum-gravity experiments might also affect quantum-gravity theory in a more profound way: by leading to an increase in the amount of work devoted to intuitive phenomenological models. As mentioned the fact that until very recently no experiments were possible has caused most theoretical work on quantum gravity to be guided by formal intuition. Among all scientific fields quantum gravity is perhaps at present the one with the biggest unbalance between theoretical research devoted to formal aspects and theoretical research devoted to phenomenological aspects. In the next few years there could be an opportunity to render more balanced the theoretical effort on quantum gravity. This might happen not only because of the availability of an experimental programme but also because some of the formal approaches to quantum gravity have recently made such remarkable progress that they might soon be in a position to make the final leap toward physical predictions.

4 Interferometry and fuzzy space-time

In the preceding two sections I have described the conclusions which I believe to be supported by the present status of quantum-gravity phenomenology. Let me now start providing some support for those conclusions by reviewing my proposal [7,24] of using modern interferometers to set bounds on space-time fuzziness. I shall articulate this in subsections because some preliminaries are in order. Before going to the analysis of experimental data it is in fact necessary to give a proper (operative) definition of fuzzy distance and give a description of the type of stochastic properties one might expect of quantum-gravity-induced fluctuations of distances.

4.1 Operative definition of fuzzy distance

While nearly all approaches to the unification of gravity and quantum mechanics appear to lead to a somewhat fuzzy picture of space-time, within the various formalisms it is often difficult to characterize physically this fuzziness. Rather than starting from formalism, I shall advocate an operative definition of fuzzy space-time. More precisely for the time being I shall just consider the concept of fuzzy distance. I shall be guided by the expectation that at very short distances the sharp classical concept of distance should give way to a somewhat fuzzy distance. Since interferometers are ideally suited to monitor the distance between test masses, I choose as operative definition of quantum-gravity induced fuzziness one which is expressed in terms of quantum-gravity induced noise in the read-out of interferometers.

In order to properly discuss this proposed definition it will prove useful to briefly review some aspects of the physics of modern Michelson-type interferometers. These are schematically composed [47] of a (laser) light source, a beam splitter and two fully-reflecting mirrors placed at a distance L from the beam splitter in orthogonal directions. The light beam is decomposed by the beam splitter into a transmitted beam directed toward one of the mirrors and a reflected beam directed toward the other mirror; the beams are then reflected by the mirrors back toward the beam splitter, where [47] they are superposed[4]. The resulting interference pattern is extremely sensitive to changes in the positions of the mirrors relative to the beam splitter. The achievable sensitivity is

[4] Although all modern interferometers rely on the technique of folded interferometer's arms (the light beam bounces several times between the beam splitter and the mirrors before superposition), I shall just discuss the simpler "no-folding" conceptual setup. The readers familiar with the subject can easily realize that the observations here reported also apply to more realistic setups, although in some steps of the derivations the length L would have to be understood as the optical length (given by the actual length of the arms multiplied by the number of foldings).

so high that planned interferometers [48,49] with arm lengths L of 3 or 4 Km expect to detect gravity waves of amplitude h as low as $3 \cdot 10^{-22}$ at frequencies of about $100Hz$. This roughly means that these modern gravity-wave interferometers should monitor the (relative) positions of their test masses (the beam splitter and the mirrors) with an accuracy [50] of order $10^{-18}m$ and better.

In achieving this remarkable accuracy experimentalists must deal with classical physics displacement noise sources (*e.g.*, thermal and seismic effects induce fluctuations in the relative positions of the test masses) and displacement noise sources associated to effects of ordinary quantum mechanics (*e.g.*, the combined minimization of ***photon shot noise*** and ***radiation pressure noise*** leads to an irreducible noise source which has its root in ordinary quantum mechanics [47]). The operative definition of fuzzy distance which I advocate characterizes the corresponding quantum-gravity effects as an additional source of displacement noise. A theory in which the concept of distance is fundamentally fuzzy in this operative sense would be such that even in the idealized limit in which all classical-physics and ordinary-quantum-mechanics noise sources are completely eliminated the read-out of an interferometer would still be noisy as a result of quantum-gravity effects.

Upon adopting this operative definition of fuzzy distance, interferometers are of course the natural tools for experimental tests of proposed distance-fuzziness scenarios.

I am only properly discussing distance fuzziness although ideas on space-time foam would also motivate investigations of time fuzziness. It is not hard to modify the definition here advocated for distance fuzziness to describe time fuzziness by replacing the interferometer with some device that keeps track of the synchronization of a pair of clocks[5] I shall not pursue this matter further since I seem to understand[6] that sensitivity to time fluctuations is still significantly behind the type of sensitivity to distance fluctuations achievable with modern Michelson-type experiments.

4.2 Random-walk noise from random-walk models of quantum space-time fluctuations

As already mentioned in Section 2, it is plausible that a quantum space-time might involve in particular the fact that a distance D would be affected by fluctuations of magnitude $L_p \sim 10^{-35}m$ occurring at a rate of roughly one per each time interval of magnitude $t_p = L_p/c \sim 10^{-44}s$. Experiments monitoring the distance D between two bodies for a time T_{obs} (in the sense appropriate, *e.g.*,

[5] Actually, a realistic analysis of ordinary Michelson-type interferometers is likely to lead to a contribution from space-time foam to noise levels that is the sum (in some appropriate sense) of the effects due to distance fuzziness and time fuzziness (*e.g.* associated to the frequency/time measurements involved).

[6] This understanding is mostly based on recent conversations with G. Busca and P. Thomann who are involved in the next generation of ultra-precise clocks to be realized in microgravity (outer space) environments.

for an interferometer) could involve a total effect amounting to $n_{obs} \equiv T_{obs}/t_p$ randomly directed fluctuations of magnitude L_p. An elementary analysis allows to establish that in such a context the root-mean-square deviation σ_D would be proportional to $\sqrt{T_{obs}}$:

$$\sigma_D \sim \sqrt{c L_p T_{obs}} \,. \tag{2}$$

From the type of T_{obs}-dependence of Eq. (2) it follows [7] that the corresponding quantum fluctuations should have displacement amplitude spectral density $S(f)$ with the f^{-1} dependence[7] typical of "random walk noise" [51]:

$$S(f) = f^{-1} \sqrt{c L_p} \,. \tag{3}$$

In fact, there is a general connection between $\sigma \sim \sqrt{T_{obs}}$ and $S(f) \sim f^{-1}$, which follows [51] from the general relation

$$\sigma^2 = \int_{1/T_{obs}}^{f_{max}} [S(f)]^2 \, df \,, \tag{4}$$

valid for a frequency band limited from below only by the time of observation T_{obs}.

The displacement amplitude spectral density (3) provides a very useful characterization of the random-walk model of quantum space-time fluctuations prescribing fluctuations of magnitude L_p occurring at a rate of roughly one per each time interval of magnitude L_p/c. If somehow we have been assuming the wrong magnitude of distance fluctuations or the wrong rate (also see Subsection 4.4) but we have been correct in taking a random-walk model of quantum space-time fluctuations Eq. (3) should be replaced by

$$S(f) = f^{-1} \sqrt{c L_{QG}} \,, \tag{5}$$

where L_{QG} is the appropriate length scale that takes into account the correct values of magnitude and rate of the fluctuations.

If one wants to be open to the possibility that the nature of the stochastic processes associated to quantum space-time be not exactly (also see Section 8) the one of a random-walk model of quantum space-time fluctuations, then the f-dependence of the displacement amplitude spectral density could be different. This leads one to consider the more general parametrization

$$S(f) = f^{-\beta} \, c^{\beta - \frac{1}{2}} \, (\mathcal{L}_\beta)^{\frac{3}{2} - \beta} \,. \tag{6}$$

In this general parametrization the dimensionless quantity β carries the information on the nature of the underlying stochastic processes, while the length

[7] Of course, in light of the nature of the arguments used, one expects that an f^{-1} dependence of the quantum-gravity induced $S(f)$ could only be valid for frequencies f significantly smaller than the Planck frequency c/L_p and significantly larger than the inverse of the time scale over which, even ignoring the gravitational field generated by the devices, the classical geometry of the space-time region where the experiment is performed manifests significant curvature effects.

scale $\mathcal{L}_\beta$ carries the information on the magnitude and rate of the fluctuations. I am assigning an index β to $\mathcal{L}_\beta$ just in order to facilitate a concise description of experimental bounds; for example, if the fluctuations scenario with, say, $\beta = 0.6$ was ruled out down to values of the effective length scale of order, say, $10^{-27}m$ I would just write $\mathcal{L}_{\beta=0.6} < 10^{-27}m$. As I will discuss in Section 8, one might be interested in probing experimentally all values of β in the range $1/2 \leq \beta \leq 1$, with special interest in the cases $\beta = 1$ (the case of random-walk models whose effective length scale I denominated with $L_{QG} \equiv \mathcal{L}_{\beta=1}$), $\beta = 5/6$, and $\beta = 1/2$.

4.3 Comparison with gravity-wave interferometer data

Before discussing experimental bounds on $\mathcal{L}_\beta$ from gravity-wave interferometers, let us fully appreciate the significance of these bounds by getting some intuition on the actual magnitude of the quantum fluctuations I am discussing. One intuition-building observation is that even for the case $\beta = 1$, which among the cases I consider is the one with the most virulent space-time fluctuations, the fluctuations predicted are truly minute: the $\beta = 1$ relation (2) only predicts fluctuations with standard deviation of order $10^{-5}m$ on a time of observation as large as 10^{10} years (the size of the whole observable universe is about 10^{10} light years!!). In spite of the smallness of these effects, the precision [47] of modern interferometers (the ones whose primary objective is the detection of the classical-gravity phenomenon of gravity waves) is such that we can obtain significant information at least on the scenarios with values of β toward the high end of the interesting interval $1/2 \leq \beta \leq 1$, and in particular we can investigate quite sensitively the intuitive case of the random-walk model of space-time fluctuations. The operation of gravity-wave interferometers is based on the detection of minute changes in the positions of some test masses (relative to the position of a beam splitter). If these positions were affected by quantum fluctuations of the type discussed above, the operation of gravity-wave interferometers would effectively involve an additional source of noise due to quantum gravity.

This observation allows to set interesting bounds already using existing noise-level data obtained at the *Caltech 40-meter interferometer*, which has achieved displacement noise levels with amplitude spectral density lower than $10^{-18}m/\sqrt{Hz}$ for frequencies between 200 and 2000 Hz [50]. While this is still very far from the levels required in order to probe significantly the lowest values of β (for $\mathcal{L}_{\beta=1/2} \sim L_p$ and $f \sim 1000Hz$ the quantum-gravity noise induced in the $\beta = 1/2$ scenario is only of order $10^{-36}m/\sqrt{Hz}$), these sensitivity levels clearly rule out all values of L_{QG} (*i.e.* $\mathcal{L}_{\beta=1}$) down to the Planck length. Actually, even values of L_{QG} significantly smaller than the Planck length are inconsistent with the data reported in Ref. [50]; in particular, from the observed noise level of $3 \cdot 10^{-19}m/\sqrt{Hz}$ near 450 Hz, which is the best achieved at the *Caltech 40-meter interferometer*, one obtains [7] the bound $L_{QG} \leq 10^{-40}m$. As discussed above, the simplest random-walk model of distance fluctuations, the one with fluctuations of magnitude L_p occurring at a rate of one per each t_p time interval, would correspond to the prediction $L_{QG} \sim L_p \sim 10^{-35}m$ and it is therefore **ruled out by these data**. This experimental information implies that, if one was to insist on this type models, realistic random-walk models of quantum space-time fluctuations would have to be significantly less noisy (smaller prediction for L_{QG}) than the intuitive one which is now ruled out. Since, as I shall discuss, there are rather plausible scenarios for significantly less noisy random-walk models, it is important that experimentalists keep pushing forward the bound on L_{QG}. More stringent bounds on L_{QG} are within reach of the LIGO/VIRGO [48,49] generation of gravity-wave interferometers.[8]

In planning future experiments, possibly taylored to test these effects (unlike LIGO and VIRGO which were tailored around the properties needed for gravity-wave detection), it is important to observe that an experiment achieving displacement noise levels with amplitude spectral density S^* at frequency f^* will set a bound on $\mathcal{L}_\beta$ of order

$$\mathcal{L}_\beta < \left[S^* \, (f^*)^\beta \, c^{(1-2\beta)/2}\right]^{2/(3-2\beta)} , \tag{7}$$

which in particular for random-walk models takes the form

$$\mathcal{L}_\beta < \left[\frac{S^* \, f^*}{\sqrt{c}}\right]^2 . \tag{8}$$

The structure of Eq. (7) (and Eq. (8)) shows that there can be instances in which a very large interferometer (the ideal tool for low-frequency studies) might not be better than a smaller interferometer, if the smaller one achieves a very small value of S^*.

The formula (7) can also be used to describe as a function of β the bounds on $\mathcal{L}_\beta$ achieved by the data collected at the *Caltech 40-meter interferometer*. Using again the fact that a noise level of only $S^* \sim 3 \cdot 10^{-19} m/\sqrt{Hz}$ near $f^* \sim 450\ Hz$ was achieved [50], one obtains the bounds

$$[\mathcal{L}_\beta]_{caltech} < \left[\frac{3 \cdot 10^{-19} m}{\sqrt{Hz}} \, (450 Hz)^\beta \, c^{(1-2\beta)/2}\right]^{2/(3-2\beta)} . \tag{9}$$

Let me comment in particular on the case $\beta = 5/6$ which might deserve special attention because of its connection (which was derived in Refs. [7,24] and will be reviewed here in Section 8) with certain arguments for bounds on the measurability of distances in quantum gravity [24,45,43]. From Eq. (9) we

---

[8] Besides allowing an improvement on the bound on L_{QG} intended as a universal property of Nature, the LIGO/VIRGO generation of interferometers will also allow us to explore the idea that L_{QG} might be a scale that depends on the experimental context in such a way that larger interferometers pick up more of the space-time fluctuations. Based on the intuition coming from the Salecker-Wigner limit (here reviewed in Section 8), or just simply on phenomenological models in which distance fluctuations affect equally each L_p-long segment of a given distance, it would not be surprising if L_{QG} was a growing function of the length of the arms of the interferometer. This gives added significance to the step from the 40-meter arms of the existing Caltech interferometer to the few-Km arms of LIGO/VIRGO interferometers.

find that $\mathcal{L}_{\beta=5/6}$ is presently bound to the level $\mathcal{L}_{\beta=5/6} \leq 10^{-29}m$. This bound is remarkably stringent in absolute terms, but is still quite far from the range of values one ordinarily considers as likely candidates for length scales appearing in quantum gravity. A more significant bound on $\mathcal{L}_{\beta=5/6}$ should be obtained by the LIGO/VIRGO generation of gravity-wave interferometers. For example, it is plausible [48] that the "advanced phase" of LIGO achieve a displacement noise spectrum of less than $10^{-20}m/\sqrt{Hz}$ near 100 Hz and this would probe values of $\mathcal{L}_{\beta=5/6}$ as small as $10^{-34}m$.

In closing this subsection on interferometry data analysis relevant for space-time fuzziness scenarios, let me clarify how it happened that such small effects could be tested. As I already mentioned, one of the viable strategies for quantum-gravity experiments is the one finding ways to put together very many of the very small quantum-gravity effects. In these interferometric studies that I proposed in Ref. [7] one does indeed effectively sum up a large number of quantum space-time fluctuations. In a time of observation as long as the inverse of the typical gravity-wave interferometer frequency of operation an extremely large number of minute quantum fluctuations could affect the distance between the test masses. Although these fluctuations average out, they do leave traces in the interferometer. These traces grow with the time of observation: the standard deviation increases in correspondence of increases of the time of observation, while the amplitude spectral density of noise increases in correspondence of decreases of frequency (which again effectively means increases of the time of observation). From this point of view it is not surprising that plausible quantum-gravity scenarios ($1/2 \leq \beta \leq 1$) all involve higher noise at lower frequencies: the observation of lower frequencies requires longer times and is therefore affected by a larger number of quantum-gravity fluctuations.

4.4 Less noisy random-walk models of distance fluctuations?

The most significant result obtained in Refs. [7,24] and reviewed in the preceding subsection is that we can rule out the intuitive picture in which the distances between the test masses of the interferometer are affected by fluctuations of magnitude L_p occurring at a rate of one per each t_p time interval. Does this rule out completely the possibility of a random-walk model of distance fluctuations? or are we just learning that the most intuitive/naive example of such a model does not work, but there are other plausible random-walk models?

Without wanting to embark on a discussion of the plausibility of less noisy random-walk models, I shall nonetheless discuss some ideas which could lead to this noise reduction. Let me start by observing that certain studies of measurability of distances in quantum gravity (see Ref. [24] and the brief review of those arguments which is provided in parts of Section 8) can be interpreted as suggesting that L_{QG} might not be a universal length scale, *i.e.* it might depend on some specific properties of the experimental setup (particularly the energies/masses involved), and in some cases L_{QG} could be significantly smaller than L_p.

Another possibility one might want to consider [24] is the one in which the quantum properties of space-time are such that fluctuations of magnitude L_p

would occur with frequency somewhat lower than $1/t_p$. This might happen for various reasons, but a particularly intriguing possibility[9] is the one of theories whose fundamental objects are not pointlike, such as the popular string theories. For such theories it is plausible that fluctuations occurring at the Planck-distance level might have only a modest impact on extended fundamental objects characterized by a length scale significantly larger than the Planck length (*e.g.* in string theory the string size, or "length", might be an order of magnitude larger than the Planck length). This possibility is interesting in general for quantum-gravity theories with a hierarchy of length scales, such as certain "M-theory motivated" scenarios with an extra length scale associated to the compactification from 11 to 10 dimensions.

Yet another possibility for a random-walk model to cause less noise in interferometers could emerge if somehow the results of the schematic analysis adopted here and in Refs. [7,24] turned out to be significantly modified once we become capable of handling all of the details of a real interferometer. To clarify which type of details I have in mind let me mention as an example the fact that in my analysis the structure of the test masses was not taken into account in any way: they were essentially treated as point-like. It would not be too surprising if we eventually became able to construct theoretical models taking into account the interplay between the binding forces that keep together ("in one piece") a macroscopic test mass as well as some random-walk-type fundamental fluctuations of the space-time in which these macroscopic bodies "live". The interference pattern observed in the laboratory reflects the space-time fluctuations only filtered through their interplay with the mentioned binding forces of the macroscopic test masses. These open issues are certainly important and a lot of insight could be gained through their investigation, but there is also some confusion that might easily result[10] from simple-minded considerations (possibly guided by intuition developed using rudimentary table-top interferometers) concerning the macro-

[9] This possibility emerged in discussions with Gabriele Veneziano. In response to my comments on the possibility of fluctuations with frequency somewhat lower than $1/t_p$ Gabriele made the suggestion that extended fundamental objects might be less susceptible than point particles to very localized space-time fluctuations. It would be interesting to work out in some detail an example of dynamical model of strings in a fuzzy space-time.

[10] In particular, these and other elements of confusion are responsible for the incorrect conclusions on the Salecker-Wigner measurability limit which were drawn in the very recent Ref. [52]. The analysis reported in Ref. [52] relies on assumptions which are unjustified in the context of the Salecker-Wigner analysis (while they would be justified in the context of certain measurements using rudimentary table-top experimental setups). Contrary to the claim made in Ref. [52], the source of $\sqrt{T_{obs}}$ uncertainty considered by Salecker and Wigner cannot be truly eliminated; unsurprisingly, it can only be traded for another source of $\sqrt{T_{obs}}$ uncertainty. Some of the comments made in Ref. [52] also ignore the fact that, as already emphasized in Ref. [24] (and reviewed in Section 8 of these notes), only a relatively small subset of the quantum-gravity ideas that can be probed with modern interferometers is directly motivated by the Salecker-Wigner limit, while the bulk of the insight we can expect from such interfer-

scopic nature of the test masses used in modern interferometers. In closing this section let me try to offer a few relevant clarifications. I need to start by adding some comments on the stochastic processes I have been considering. In most physical contexts a series of random steps does not lead to $\sqrt{T_{obs}}$ dependence of σ because often the context is such that through the fluctuation-dissipation theorem the source of $\sqrt{T_{obs}}$ dependence is (partly) compensated (some sort of restoring effect). The hypothesis explored in these discussions of random-walk models of space-time fuzziness is that the type of underlying dynamics of quantum space-time be such that the fluctuation-dissipation theorem be satisfied without spoiling the $\sqrt{T_{obs}}$ dependence of σ. This is an intuition which apparently is shared by other authors; for example, the study reported in Ref. [53] (which followed by a few months Ref. [7], but clearly was the result of completely independent work) also models some implications of quantum space-time (the ones that affect clocks) with stochastic processes whose underlying dynamics does not produce any dissipation and therefore the "fluctuation contribution" to the T_{obs} dependence is left unmodified, although the fluctuation-dissipation theorem is fully taken into account. Since a mirror of an interferometer of LIGO/VIRGO type is in practice an extremity of a pendulum, another aspect that the reader might at first find counter-intuitive is that the $\sqrt{T_{obs}}$ dependence of σ, although coming in with a very small prefactor, for extremely large T_{obs} would seem to give values of σ too large to be consistent with the structure of a pendulum. This is a misleading intuition which originates from the experience with ordinary (non-quantum-gravity) analyses of the pendulum. In fact, the dynamics of an ordinary pendulum has one extremity "fixed" to a very heavy macroscopic and rigid body, while the other extremity is fixed to a much lighter (but, of course, still macroscopic) body. The usual stochastic processes considered in the study of the pendulum affect the heavier body in a totally negligible way, while they have strong impact on the dynamics of the lighter body. A pendulum analyzed according to a random-walk model of space-time fluctuations would be affected by stochastic processes which are of the same magnitude both for its heavier and its lighter extremity. [The bodies are fluctuating along with intrinsic space-time fluctuations, rather than fluctuating as a result of, say, collisions with air particles occurring in a conventional space-time.] In particular, in the directions orthogonal to the vertical axis the stochastic processes affect the position of the center of mass of the entire pendulum just as they would affect the position of the center of mass of any other body (the spring that connects the two extremities of the pendulum would not affect the motion of the overall center of mass of the pendulum).

5 Gamma-ray bursts and in-vacuo dispersion

Let me now discuss the proposal put forward in Ref. [5] (also see Ref. [54]), which exploits the recent confirmation that at least some gamma-ray bursters

ometric studies concerns the stochastic properties of "foamy" models of space-time, which are intrinsically interesting independently of the Salecker-Wigner limit.

are indeed at cosmological distances [55–58], making it possible for observations of these to provide interesting constraints on the fundamental laws of physics. In particular, such cosmological distances combine with the short time structure seen in emissions from some GRBs [59] to provide ideal features for tests of possible *in vacuo* dispersion of electromagnetic radiation from GRBs, of the type one might expect based on the intuitive quantum-gravity arguments reviewed in Section 2. As mentioned, a quantum-gravity-induced deformation of the dispersion relation for photons would naturally take the form $c^2\mathbf{p}^2 = E^2\,[1 + \mathcal{F}(E/E_{QG})]$, where E_{QG} is an effective quantum-gravity energy scale and $\mathcal{F}$ is a model-dependent function of the dimensionless ratio E/E_{QG}. In quantum-gravity scenarios in which the Hamiltonian equation of motion $\dot{x}_i = \partial H/\partial p_i$ is still valid (at least approximately valid; valid to an extent sufficient to justify the analysis that follows) such a deformed dispersion relation would lead to energy-dependent velocities for massless particles, with implications for the electromagnetic signals that we receive from astrophysical objects at large distances. At small energies $E \ll E_{QG}$, it is reasonable to expect that a series expansion of the dispersion relation should be applicable leading to the formula (1). For the case $\alpha = 1$, which is the most optimistic (largest quantum-gravity effect) among the cases discussed in the quantum-gravity literature, the formula (1) reduces to

$$c^2\mathbf{p}^2 \simeq E^2\left(1 + \xi\frac{E}{E_{QG}}\right) . \tag{10}$$

Correspondingly one would predict the energy-dependent velocity formula

$$v = \frac{\partial E}{\partial p} \sim c\left(1 - \xi\frac{E}{E_{QG}}\right) . \tag{11}$$

To elaborate a bit more than I did in Section 2 on the intuition that leads to this type of candidate quantum-gravity effect let me observe that [5] velocity dispersion such as described in (11) could result from a picture of the vacuum as a quantum-gravitational 'medium', which responds differently to the propagation of particles of different energies and hence velocities. This is analogous to propagation through a conventional medium, such as an electromagnetic plasma [60]. The gravitational 'medium' is generally believed to contain microscopic quantum fluctuations, such as the ones considered in the previous sections. These may [61] be somewhat analogous to the thermal fluctuations in a plasma, that occur on time scales of order $t \sim 1/T$, where T is the temperature. Since it is a much 'harder' phenomenon associated with new physics at an energy scale far beyond typical photon energies, any analogous quantum-gravity effect could be distinguished by its different energy dependence: the quantum-gravity effect would increase with energy, whereas conventional medium effects decrease with energy in the range of interest [60].

Also relevant for building some quantum-gravity intuition for this type of *in vacuo* dispersion and deformed velocity law is the observation [46,23] that this has implications for the measurability of distances in quantum gravity that fit well with the intuition emerging from heuristic analyses [12] based on a combination of arguments from ordinary quantum mechanics and general relativity.

[This connection between dispersion relations and measurability bounds will be here reviewed in Section 8.]

Notably, recent work [41] has provided evidence for the possibility that the popular canonical/loop quantum gravity [40] might be among the theoretical approaches that admit the phenomenon of deformed dispersion relations with the deformation going linearly with the Planck length ($L_p \sim 1/E_p$). Similarly, evidence for this type of dispersion relations has been found [46] in *Liouville* (non-critical) strings [19], whose development was partly motivated by an intuition concerning the "quantum-gravity vacuum" that is rather close to the one traditionally associated to the works of Wheeler [14] and Hawking [15]. Moreover, the phenomenon of deformed dispersion relations with the deformation going linearly with the Planck length fits rather naturally within certain approaches based on non-commutative geometry and deformed symmetries. In particular, there is growing evidence [23,27,28] for this phenomenon in theories living in the non-commutative Minkowski space-time proposed in Refs. [62,63,21], which involves a dimensionful (presumably Planck-length related) deformation parameter.

Equation (11) encodes a minute modification for most practical purposes, since E_{QG} is believed to be a very high scale, presumably of order the Planck scale $E_p \sim 10^{19}$ GeV. Nevertheless, such a deformation could be rather significant for even moderate-energy signals, if they travel over very long distances. According to (11) two signals respectively of energy E and $E + \Delta E$ emitted simultaneously from the same astrophysical source in traveling a distance L acquire a "relative time delay" $|\delta t|$ given by

$$|\delta t| \sim \frac{\Delta E}{E_{QG}} \frac{L}{c} . \tag{12}$$

Such a time delay can be observable if ΔE and L are large while the time scale over which the signal exhibits time structure is small. As mentioned, these are the respects in which GRBs offer particularly good prospects for such measurements. Typical photon energies in GRB emissions are in the range 0.1 – 100 MeV [59], and it is possible that the spectrum might in fact extend up to TeV energies [64]. Moreover, time structure down to the millisecond scale has been observed in the light curves [59], as is predicted in the most popular theoretical models [65] involving merging neutron stars or black holes, where the last stages occur on the time scales associated with grazing orbits. Similar time scales could also occur in models that identify GRBs with other cataclysmic stellar events such as failed supernovae Ib, young ultra-magnetized pulsars or the sudden deaths of massive stars [66]. We see from equations (11) and (12) that a signal with millisecond time structure in photons of energy around 10 MeV coming from a distance of order 10^{10} light years, which is well within the range of GRB observations and models, would be sensitive to E_{QG} of order the Planck scale.

In order to set a definite bound on E_{QG} it is necessary to measure L and to measure the time of arrival of different energy/wavelength components of a sharp peak within the burst. From Eq. (12) it follows that one could set a bound

$$E_{QG} > \Delta E \frac{L}{c|\tau|} \tag{13}$$

by establishing the times of arrival of the peak to be the same up to an uncertainty τ in two energy channels E and $E+\Delta E$. Unfortunately, at present we have data available only on a few GRBs for which the distance L has been determined (the first measurements of this type were obtained only in 1997), and these are the only GRBs which can be reliably used to set bounds on the new effect. Moreover, mostly because of the nature of the relevant experiments (particularly the BATSE detector on the Compton Gamma Ray Observatory), for a large majority of the GRBs on record only the portion of the burst with energies up to the MeV energy scale was observed, whereas higher energies would be helpful for the study of the phenomenon of quantum-gravity induced dispersion here considered (which increases linearly with energy). We expect significant improvements in these coming years. The number of GRBs with attached distance measurement should rapidly increase. A new generation of orbiting spectrometers, *e.g.* AMS [67] and GLAST [68], are being developed, whose potential sensitivities are very impressive. For example, assuming a E^{-2} energy spectrum, GLAST would expect to observe about 25 GRBs per year at photon energies exceeding 100 GeV, with time resolution of microseconds. AMS would observe a similar number at $E > 10$ GeV with time resolution below 100 nanoseconds.

While we wait for these new experiments, preliminary bounds can already be set with available data. Some of these bounds are "conditional" in the sense that they rely on the assumption that the relevant GRB originated at distances corresponding to redshift of $\mathcal{O}(1)$ (corresponding to a distance of ~ 3000 Mpc), which appears to be typical. Let me start by considering the "conditional" bound (first considered in Ref. [5]) which can be obtained from data on GRB920229. GRB920229 exhibited [69] micro-structure in its burst at energies up to $\sim$ 200 KeV. In Ref. [5] it was estimated conservatively that a detailed time-series analysis might reveal coincidences in different BATSE energy bands on a time-scale $\sim 10^{-2}$ s, which, assuming redshift of $\mathcal{O}(1)$ (the redshift of GRB920229 was not measured) would yield sensitivity to $E_{QG} \sim 10^{16}$ GeV (it would allow to set a bound $E_{QG} > 10^{16}$ GeV).

As observed in Ref. [54], a similar sensitivity might be obtainable with GRB980425, given its likely identification with the unusual supernova 1998bw [70]. This is known to have taken place at a redshift $z = 0.0083$ corresponding to a distance $D \sim 40$ Mpc (for a Hubble constant of 65 $\text{km}\,\text{sec}^{-1}\text{Mpc}^{-1}$) which is rather smaller than a typical GRB distance. However GRB980425 was observed by BeppoSAX at energies up to 1.8 MeV, which gains back an order of magnitude in the sensitivity. If a time-series analysis were to reveal structure at the $\Delta t \sim 10^{-3}$ s level, which is typical of many GRBs [71], it would yield the same sensitivity as GRB920229 (but in this case, in which a redshift measurement is available, one would have a definite bound, whereas GRB920229 only provides a "conditional" bound of the type discussed above).

Ref. [54] also observed that an interesting (although not very "robust") bound could be obtained using GRB920925c, which was observed by WATCH [72] and possibly in high-energy γ rays by the HEGRA/AIROBICC array above 20 TeV [73]. Several caveats are in order: taking into account the appropriate trial factor, the confidence level for the signal seen by HEGRA to be related to GRB920925c is only 99.7% ($\sim 2.7\sigma$), the reported directions differ by 9^0, and the redshift of the source is unknown. Nevertheless, the potential sensitivity is impressive. The events reported by HEGRA range up to $E \sim 200$ TeV, and the correlation with GRB920925c is within $\Delta t \sim 200$ s. Making the reasonably conservative assumption that GRB920925c occurred no closer than GRB980425, viz. $\sim$ 40Mpc, one finds a minimum sensitivity to $E_{QG} \sim 10^{19}$ GeV, modulo the caveats listed above. Even more spectacularly, several of the HEGRA GRB920925c candidate events occurred within $\Delta t \sim 1$ s, providing a potential sensitivity even two orders of magnitude higher.

As illustrated by this discussion, the GRBs have remarkable potential for the study of *in vacuo* dispersion, which will certainly lead to impressive bounds/tests as soon as improved experiments are put into operation, but at present the best GRB-based bounds are either "conditional" (example of GRB92022) or "not very robust" (example of GRB920925c). As a result, at present the best (reliable) bound has been extracted [74] using data from the Whipple telescope on a TeV γ-ray flare associated with the active galaxy Mrk 421. This object has a redshift of 0.03 corresponding to a distance of $\sim$ 100 Mpc. Four events with γ-ray energies above 2 TeV have been observed within a period of 280 s. These provide [74] a definite limit $E_{QG} > 4 \times 10^{16}$ GeV.

In passing let me mention that (as observed in Ref. [5,46]) pulsars and supernovae, which are among the other astrophysical phenomena that might at first sight appear well suited for the study of *in vacuo* dispersion, do not actually provide interesting sensitivities. Although pulsar signals have very well defined time structure, they are at relatively low energies and are presently observable over distances of at most 10^4 light years. If one takes an energy of order 1 eV and postulates generously a sensitivity to time delays as small as 1 μsec, one nevertheless reaches only an estimated sensitivity to $E_{QG} \sim 10^9$ GeV. With new experiments such as AXAF it may be possible to detect X-ray pulsars out to 10^6 light years, but this would at best push the sensitivity up to $E_{QG} \sim 10^{11}$ GeV. Concerning supernovae, it is important to take into account that neutrinos from Type II events similar to SN1987a, which should have energies up to about 100 MeV with a time structure that could extend down to milliseconds, are likely to be detectable at distances of up to about 10^5 light years, providing sensitivity to $E_{QG} \sim 10^{15}$ GeV, which is remarkable in absolute terms, but is still significantly far from the Planck scale and anyway cannot compete with the type of sensitivity achievable with GRBs.

It is rather amusing that GRBs can provide such a good laboratory for investigations of *in vacuo* dispersion in spite of the fact that the short-time structure of GRB signals is still not understood. The key point of the proposal in Ref. [5] is that sensitive tests can be performed through the serendipitous detection of short-scale time structure [69] at different energies in GRBs which are established to be at cosmological distances. Detailed features of burst time series enable (as already shown in several examples) the emission times in different energy ranges to be put into correspondence. Any time shift due to quantum-gravity would *increase* with the photon energy, and this characteristic dependence is separable from more conventional in-medium-physics effects, which *decrease* with energy. To distinguish any quantum-gravity induced shift from effects due to the source, one can use the fact that the quantum-gravity effect here considered is *linear* in the GRB distance.

This last remark applies to all values of α, but most of the observations and formulas in this section are only valid in the case $\alpha = 1$ (linear suppression). The generalization to cases with $\alpha \neq 1$ is however rather simple; for example, Eq. (13) takes the form (up to coefficients of order 1)

$$E_{QG} > \left[[(E + \Delta E)^\alpha - E^\alpha] \frac{L}{c|\tau|} \right]^{1/\alpha} . \tag{14}$$

Notice that here, because of the non-linearity, the right-hand side depends both on E and ΔE.

Before moving on to other experiments let me clarify what is the key ingredient of this experiment using observations of gamma rays from distant astrophysical sources that allowed to render observable the minute quantum-gravity effects. The ingredient is very similar to the one relevant for the studies of space-time fuzziness using modern interferometers, which I discussed in the preceding section; in fact, the gamma rays here considered are affected by a very large number of the minute quantum-gravity effects. Each of the dispersion-inducing quantum-gravity effect is very small, but the gamma rays emitted by distant astrophysical sources travel for a very long time before reaching us and can therefore be affected by an extremely large number of such effects.

6 Other quantum-gravity experiments

In this section I provide brief reviews of some other quantum-gravity experiments. The fact that the discussion here provided for these experiments is less detailed than the preceding discussions of the interferometry-based and GRB-based experiments is not to be interpreted as indicating that these experiments are somehow less significant: it is just that a detailed discussion of a couple of examples was sufficient to provide to the reader some general intuition on the strategy behind quantum-gravity experiments and it was natural for me to use as examples the ones I am most familiar with. For the experiments discussed in this section I shall just give a rough idea of the quantum-gravity scenarios that could be tested and of the experimental procedures which have been proposed.

6.1 Neutral kaons and CPT violation

One of the formalisms that has been proposed [17,2] for the evolution equations of particles in the space-time foam relies on a density-matrix picture. The foam is seen as providing a sort of environment inducing quantum decoherence even on isolated systems (*i.e.* systems which only interact with the foam). A given non-relativistic system (such as the neutral kaons studied by the CPLEAR collaboration at CERN) is described by a density matrix ρ that satisfies an evolution equation analogous to the one ordinarily used for the quantum mechanics of certain open systems:

$$\partial_t \rho = i[\rho, H] + \delta H \, \rho \tag{15}$$

where H is the ordinary Hamiltonian and δH, which has a non-commutator structure [2], represents the effects of the foam. δH is expected to be extremely small, suppressed by some power of the Planck length. The precise form of δH (which in particular would set the level of the new physics by establishing how many powers of the Planck length suppress the effect) has not yet been derived from some full-grown quantum gravity[11], and therefore phenomenological parametrizations have been introduced (see Refs. [17,75,20,35]). For the case in which the effects are only suppressed by one power of the Planck length (linear suppression) recent neutral-kaon experiments, such as the ones performed by CPLEAR, have set significant bounds [2] on the associated CPT-violation effects and forthcoming experiments are likely to push these bounds even further.

Like the interferometry-based and the GRB-based experiments, these experiments (which have the added merit of having started the recent wave of quantum-gravity proposals) also appear to provide significant quantum-gravity tests. As mentioned, the effect of quantum-gravity induced decoherence certainly qualifies as a traditional quantum-gravity subject, and the level of sensitivity reached by the neutral-kaon studies is certainly significant (as in the case of *in vacuo* dispersion and GRBs, one would like to be able to explore also the case of a quadratic Planck-length suppression, but it is nonetheless very significant that we have at least reached the capability to test the case of linear suppression). Also in this case it is natural to ask: how come we could manage this? What strategy allowed this neutral-kaon studies to evade the traditional gloomy forecasts for quantum-gravity phenomenology? While, as discussed above, in the interferometry-based and the GRB-based experiments the crucial element in the experimental proposal is the possibility to put together many quantum gravity effects, in the case of the neutral-kaon experiments the crucial element in the experimental proposal is provided by the very delicate balance of scales that characterizes the neutral-kaon system. In particular, it just happens to be true that the dimensionless ratio setting the order of magnitude of quantum-gravity effects in the linear suppression scenario, which is $c^2 M_{L,S}/E_p \sim 2 \cdot 10^{-19}$, is not much smaller than some of the dimensionless ratios characterizing the neutral-kaon system, notably the ratio $|M_L - M_S|/M_{L,S} \sim 7 \cdot 10^{-15}$ and the ratio $|\Gamma_L - \Gamma_S|/M_{L,S} \sim 1.4 \cdot 10^{-14}$. This renders possible for the quantum-gravity effects to provide observably large corrections to the physics of neutral kaons.

[11] Within the quantum-gravity approach here reviewed in Subsection 10.2, which only attempts to model certain aspects of quantum gravity, such a direct calculation might soon be performed.

6.2 Interferometry and string cosmology

Up to this point I have only reviewed experiments probing foamy properties of space-time in the sense of Wheeler and Hawking. A different type of quantum-gravity effect which might produce a signature strong enough for experimental testing has been discussed in the context of studies of a cosmology based on critical superstrings [76]. While for a description of this cosmology and of its physical signatures I must only refer the reader to the recent reviews in Ref. [77], I want to briefly discuss here the basic ingredients of the proposal [3] of interferometry-based tests of the cosmic gravitational wave background predicted by string cosmology.

In string cosmology the universe starts from a state of very small curvature, then goes through a long phase of dilaton-driven inflation reaching nearly Plankian energy density, and then eventually reaches the standard radiation-dominated cosmological evolution [76,77]. The period of nearly Plankian energy density plays a crucial role in rendering the quantum-gravity effects observable. In fact, this example based on string cosmology is quite different from the experiments I discussed earlier in these lectures also because it does not involve small quantum-gravity effects which are somehow amplified (in the sense for example of the amplification provided when many effects are somehow put together). The string cosmology involves a period in which the quantum-gravity effects are actually quite large. As clarified in Refs. [76,77] planned interferometers such as LIGO might be able to detect the faint residual traces of these strong effects occurred in a far past.

As mentioned, the quantum-gravity effects that, within string cosmology, leave a trace in the gravity-wave background are not of the type that requires an active Wheeler-Hawking foam. The relevant quantum-gravity effects live in the more familiar vacuum which we are used to encounter in the context of ordinary gauge theory. (Actually, for the purposes of the analyses reported in Refs. [76,77] quantum gravity could be seen as an ordinary gauge theory, although with unusual gauge-field content.) In the case of the Wheeler-Hawking foam one is tempted to visualize the vacuum as reboiling with (virtual) worm-holes and black-holes. Instead the effects relevant for the gravity-wave background predicted by string cosmology are more conventional field-theory-type fluctuations, although carrying gravitational degrees of freedom, like the graviton. Also from this point of view the experimental proposal discussed in Refs. [76,77] probes a set of candidate quantum-gravity phenomena which is complementary to the ones I have reviewed earlier in these notes.

6.3 Matter interferometry and primary state diffusion

The studies reported in Ref. [4] (and references therein) have considered how certain effectively stochastic properties of space-time would affect the evolution of quantum-mechanical states. The stochastic properties there considered are different from the ones discussed here in Sections 2, 3 and 4, but were introduced within a similar viewpoint, *i.e.* stochastic processes as effective description of quantum space-time processes. The implications of these stochastic properties for the evolution of quantum-mechanical states were modeled via the formalism of "primary state diffusion", but only rather crude models turned out to be treatable.

The approach proposed in Ref. [4] actually puts together some of the unknowns of space-time foam and the specific properties of "primary state diffusion". The structure of the predicted effects cannot be simply discussed in terms of elementary properties of space-time foam and a simple interpretation in terms of symmetry deformations does not appear to be possible. Those effects appear to be the net result of the whole formalism that goes into the approach. Moreover, as also emphasized by the authors, the crudeness of the models is such that all conclusions are to be considered as tentative at best. Still, the analysis reported in Ref. [4] is very significant as an independent indication of a mechanism, based on matter-interferometry experiments, that could unveil Planck-length-suppressed effects.

6.4 Colliders and large extra dimensions

It was recently suggested [78,79] that the characteristic quantum-gravity length scale might be given by a length scale L_D much larger than the Planck length in theories with large extra dimensions. It appears plausible that there exist models that are consistent with presently-available experimental data and have L_D as large as the $(TeV)^{-1}$ scale and (some of) the extra dimensions as large as a millimiter [79]. In such models the smallness of the Planck length is seen as the result of the fact that the strength of gravitation in the ordinary 3+1 space-time dimensions would be proportional to the square-root of the inverse of the large volume of the external compactified space multiplied by an appropriate (according to dimensional analysis) power of L_D.

Several studies have been motivated by the proposal put forward in Ref. [79], but only a small percentage of these studies considered the implications for quantum-gravity scenarios. Among these studies the ones reported in Refs. [8,9] are particularly significant for the objectives of these lectures, since they illustrate another completely different strategy for quantum-gravity experiments. It is there observed that within the realm of the ordinary 3+1 dimensional space-time an important consequence of the existence of large extra dimensions would be the presence of a tower of Kaluza-Klein modes associated to the gravitons. The weakness of the coupling between gravitons and other particles can be compensated by the large number of these Kaluza-Klein modes when the experimental energy resolution is much larger than the mass splitting between the modes, which for a small number of very large extra dimensions can be a weak requirement (*e.g.* for 6 millimiter-wide extra dimensions [79,8] the mass splitting is of a few MeV). This can lead to observably large [8,9] effects at planned particle-physics colliders, particularly CERN's LHC.

In a sense, the experimental proposal put forward in Refs. [8,9] is another example of experiment in which the smallness of quantum gravity effects is compensated by putting together a large number of such effects (putting together the contributions of all of the Kaluza-Klein modes).

Concerning the quantum-gravity aspects of the models with large extra dimensions proposed in Ref. [79], it is important to realize that, as emphasized in Ref. [24], if anything like the space-time foam here described in Sections 2, 3, 4 and 5 was present in such models the effective reduction of the quantum-gravity scale would naturally lead to foamy effects that are too large for consistency with available experimental data. Preliminary estimates based solely on dimensional considerations appear to suggest that [24] linear suppression by the reduced quantum-gravity scale would certainly be ruled out and even quadratic suppression might not be sufficient for consistency with available data. These arguments should lead to rather stringent bounds on space-time foam especially in those models in which some of the large extra dimensions are accessible to non-gravitational particles (see, *e.g.*, Ref. [80]), and should have interesting (although smaller) implications also for the popular scenario in which only the gravitational degrees of freedom have access to the large extra dimensions. Of course, a final verdict must await detailed calculations analysing space-time foam in these models with large extra dimensions. The first examples of this type of computations are given by the very recent studies in Refs. [81,82], which considered the implications of foam-induced light-cone deformation for certain examples of models with large extra dimensions.

7 Classical-space-time-induced quantum phases in matter interferometry

While of course the quantum properties of space-time are the most exciting effects we expect of quantum gravity, and probably the ones which will prove most useful in gaining insight into the fundamental structure of the theory, it is important to investigate experimentally all aspects of the interplay between gravitation and quantum mechanics. Among these experiments the ones that could be expected to provide fewer surprises (and less insight into the structure of quantum gravity) are the ones concerning the interplay between strong-but-classical gravitational fields and quantum matter fields. However, this is not necessarily true as I shall try to clarify within this section's brief review of the experiment performed nearly a quarter of a century ago by Colella, Overhauser and Werner [10], which, to my knowledge, was the first experiment probing some aspect of the interplay between gravitation and quantum mechanics. That experiment has been followed by several modifications and refinements (often labeled "COW experiments" from the initials of the scientists involved in the first experiment) all probing the same basic physics, *i.e.* the validity of the Schrödinger equation

$$\left[-\left(\frac{\hbar^2}{2\,M_I}\right)\boldsymbol{\nabla}^2 + M_G\,\phi(\boldsymbol{r})\right]\psi(t,\boldsymbol{r}) = i\,\hbar\,\frac{\partial\,\psi(t,\boldsymbol{r})}{\partial t} \tag{16}$$

for the description of the dynamics of matter (with wave function $\psi(t, \boldsymbol{r})$) in presence of the Earth's gravitational potential $\phi(\boldsymbol{r})$. [In (16) M_I and M_G denote the inertial and gravitational mass respectively.]

The COW experiments exploit the fact that the Earth's gravitational potential puts together the contribution of so many particles (all of those composing the Earth) that it ends up being large enough to introduce observable effects in rotating table-top interferometers. This was the first example of a physical context in which gravitation was shown to have an observable effect on a quantum-mechanical system in spite of the weakness of the gravitational force.

The fact that the original experiment performed by Colella, Overhauser and Werner obtained results in very good agreement [10] with Eq. (16) might seem to indicate that, as generally expected, experiments on the interplay between strong-but-classical gravitational fields and quantum matter fields should not lead to surprises and should not provide insight into the structure of quantum gravity. However, the confirmation of Eq. (16) does raise some sort of a puzzle with respect to the Equivalence Principle of general relativity; in fact, even for $M_I = M_G$ the mass does not cancel out in the quantum evolution equation (16). This is an observation that by now has also been emphasized in textbooks [83], but to my knowledge it has not been fully addressed even within the most popular quantum-gravity approaches, *i.e.* critical superstrings and canonical/loop quantum gravity. Which role should be played by the Equivalence Principle in quantum gravity? Which version/formulation of the Equivalence Principle should/could hold in quantum gravity?

Additional elements for consideration in quantum-gravity models will emerge if the small discrepancy between (16) and data reported in Ref. [84] (a refined COW experiment) is confirmed by future experiments. The subject of gravitationally induced quantum phases is also expanding in new directions [6,85], which are likely to provide additional insight.

8 Estimates of space-time fuzziness from measurability bounds

In the preceding Sections 4, 5, 6 and 7 I have discussed the experimental proposals that support the conclusions anticipated in Sections 2 and 3. This Section 8 and the following two sections each provide a "theoretical-physics addendum". In this section I discuss some arguments that appear to suggest properties of the space-time foam. These arguments are based on analyses of bounds on the measurability of distances in quantum gravity. The existence of measurability bounds has attracted the interest of several theorists, because these bounds appear to capture an important novel element of quantum gravity. In ordinary (non-gravitational) quantum mechanics there is no absolute limit on the accuracy of the measurement of a distance. [Ordinary quantum mechanics allows $\delta A = 0$ for any single observable A, since it only limits the combined measurability of pairs of conjugate observables.]

The quantum-gravity bound on the measurability of distances (whatever final form it actually takes in the correct theory) is of course intrinsically interesting, but here (as in previous works [7,24,12,86,13]) I shall be interested in the possibility that it might reflect properties of the space-time foam. This is of course not necessarily true: a bound on the measurability of distances is not necessarily associated to space-time fluctuations, but guided by the Wheeler-Hawking intuition on the nature of space-time one is tempted to interpret any measurability bound (which might be obtained with totally independent arguments) as an indicator of the type of irreducible fuzziness that characterizes space-time. One has on one hand some intuition about quantum gravity which involves stochastic fluctuations of distances and on the other hand some different arguments lead to intuition for absolute bounds on the measurability of distances; it is natural to explore the possibility that the two might be related, *i.e.* that the intrinsic stochastic fluctuations should limit the measurability just to the level suggested by the independent measurability arguments. Different arguments appear to lead to different measurability bounds, which in turn could provide different intuition for the stochastic properties of space-time foam.

8.1 Minimum-length noise

In many quantum-gravity approaches there appears to be a length scale L_{min}, often identified with the Planck length or the string length L_{string} (which, as mentioned, should be somewhat larger than the Planck length, plausibly in the neighborhood of $10^{-34}m$), which sets an absolute bound on the measurability of distances (a minimum uncertainty):

$$\delta D \geq L_{min} \; . \tag{17}$$

This property emerges in approaches based on canonical quantization of Einstein's gravity when analyzing certain gedanken experiments (see, *e.g.*, Refs. [30], [33] and references therein). In critical superstring theories, theories whose mechanics is still governed by the laws of ordinary quantum mechanics but with one-dimensional (rather than point-like) fundamental objects, a relation of type (17) follows from the stringy modification of Heisenberg's uncertainty principle [31]

$$\delta x \, \delta p \geq 1 + L_{string}^2 \, \delta p^2 \; . \tag{18}$$

In fact, whereas Heisenberg's uncertainty principle allows $\delta x = 0$ (for $\delta p \to \infty$), for all choices of δp the uncertainty relation (18) gives $\delta x \geq L_{string}$. The relation (18) is suggested by certain analyses of string scattering [31], but it might have to be modified when taking into account the non-perturbative solitonic structures of superstrings known as Dirichlet branes [38]. In particular, evidence has been found [87] in support of the possibility that "Dirichlet particles" (Dirichlet 0 branes) could probe the structure of space-time down to scales shorter than the string length. In any case, all evidence available on critical superstrings is consistent with a relation of type (17), although it is probably safe to say that some more work is still needed to firmly establish the string-theory value of L_{min}.

Having clarified that a relation of type (17) is a rather common prediction of theoretical work on quantum gravity, it is then natural to wonder whether such a relation is suggestive of stochastic distance fluctuations of a type that could significantly affect the noise levels of an interferometer. As mentioned relations such as (17) do not necessarily encode any fuzziness; for example, relation (17) could simply emerge from a theory based on a lattice of points with spacing L_{min} and equipped with a measurement theory consistent with (17). The concept of distance in such a theory would not necessarily be affected by the type of stochastic processes that lead to noise in an interferometer. However, if one does take as guidance the Wheeler-Hawking intuition on space-time foam it makes sense to assume that relation (17) might encode the net effect of some underlying physical processes of the type one would qualify as quantum space-time fluctuations. This (however preliminary) network of intuitions suggests that (17) could be the result of fuzziness for distances D of the type associated to stochastic fluctuations with root-mean-square deviation σ_D given by

$$\sigma_D \sim L_{min} \, . \tag{19}$$

The associated displacement amplitude spectral density $S_{min}(f)$ should roughly have a $1/\sqrt{f}$ behaviour

$$S_{min}(f) \sim \frac{L_{min}}{\sqrt{f}} \, , \tag{20}$$

which (using notation set up in Section 4) can be concisely described stating that $L_{min} \sim \mathcal{L}_{\beta=1/2}$. Eq. (20) can be justified using the general relation (4). Substituting the $S_{min}(f)$ of Eq. (20) for the $S(f)$ of Eq. (4) one obtains a σ that approximates the σ_D of Eq. (19) up to small (logarithmic) T_{obs}-dependent corrections. A more detailed description of the displacement amplitude spectral density associated to Eq. (19) can be found in Refs. [88,89]. For the objectives of these lectures the rough estimate (20) is sufficient since, if indeed $L_{min} \sim L_p$, from (20) one obtains $S_{min}(f) \sim 10^{-35} m/\sqrt{f}$, which is still very far from the sensitivity of even the most advanced modern interferometers, and therefore I shall not be concerned with corrections to Eq. (20).

8.2 Random-walk noise motivated by the analysis of a Salecker-Wigner gedanken experiment

Let me now consider a measurability bound which is encountered when taking into account the quantum properties of devices. It is well understood (see, *e.g.*, Refs. [12,13,90,44,45,32]) that the combination of the gravitational properties and the quantum properties of devices can have an important role in the analysis of the operative definition of gravitational observables. Since the analyses [30,33,31,87] that led to the proposal of Eq. (17) only treated the devices in a completely idealized manner (assuming that one could ignore any contribution to the uncertainty in the measurement of D due to the gravitational and quantum properties of devices), it is not surprising that analyses taking into account the gravitational and quantum properties of devices found more significant limitations to the measurability of distances.

Actually, by ignoring the way in which the gravitational properties and the quantum properties of devices combine in measurements of geometry-related physical properties of a system one misses some of the fundamental elements of novelty we should expect for the interplay of gravitation and quantum mechanics; in fact, one would be missing an element of novelty which is deeply associated to the Equivalence Principle. In measurements of physical properties which are not geometry-related one can safely resort to an idealized description of devices. For example, in the famous Bohr-Rosenfeld analysis [91] of the measurability of the electromagnetic field it was shown that the accuracy allowed by the formalism of ordinary quantum mechanics could only be achieved using idealized test particles with vanishing ratio between electric charge and inertial mass. Attempts to generalize the Bohr-Rosenfeld analysis to the study of gravitational fields (see, *e.g.*, Ref. [90]) are of course confronted with the fact that the ratio between gravitational "charge" (mass) and inertial mass is fixed by the Equivalence Principle. While ideal devices with vanishing ratio between electric charge and inertial mass can be considered at least in principle, devices with vanishing ratio between gravitational mass and inertial mass are not admissible in any (however formal) limit of the laws of gravitation. This observation provides one of the strongest elements in support of the idea [13] that the mechanics on which quantum gravity is based must not be exactly the one of ordinary quantum mechanics, since it should accommodate a somewhat different relationship between "system" and "measuring apparatus" and should not rely on the idealized "measuring apparatus" which plays such a central role in the mechanics laws of ordinary quantum mechanics (see, *e.g.*, the "Copenhagen interpretation").

In trying to develop some intuition for the type of fuzziness that could affect the concept of distance in quantum gravity, it might be useful to consider the way in which the interplay between the gravitational and the quantum properties of devices affects the measurability of distances. In Refs. [12,13] I have argued[12] that a natural starting point for this type of analysis is provided by the procedure for the measurement of distances which was discussed in influential work by Salecker and Wigner [92]. These authors "measured" (in the "*gedanken*" sense) the distance D between two bodies by exchanging a light signal between them. The measurement procedure requires *attaching*[13] a light-gun (*i.e.* a de-

[12] I shall comment later in these notes on the measurability analysis reported in Ref. [45], which also took as starting point the mentioned work by Salecker and Wigner, but advocated a different viewpoint and reached different conclusions.

[13] Of course, for consistency with causality, in such contexts one assumes devices to be "attached non-rigidly," and, in particular, the relative position and velocity of their centers of mass continue to satisfy the standard uncertainty relations of quantum mechanics.

vice capable of sending a light signal when triggered), a detector and a clock to one of the two bodies and *attaching* a mirror to the other body. By measuring the time T_{obs} (time of observation) needed by the light signal for a two-way journey between the bodies one also obtains a measurement of the distance D. For example, in flat space and neglecting quantum effects one simply finds that $D = cT_{obs}/2$. Within this setup it is easy to realize that the interplay between the gravitational and the quantum properties of devices leads to an irreducible contribution to the uncertainty δD. In order to see this it is sufficient to consider the contribution to δD coming from the uncertainties that affect the motion of the center of mass of the system composed by the light-gun, the detector and the clock. Denoting with x^* and v^* the position and the velocity of the center of mass of this composite device relative to the position of the body to which it is *attached*, and assuming that the experimentalists prepare this device in a state characterised by uncertainties δx^* and δv^*, one easily finds [92,13]

$$\delta D \geq \delta x^* + T_{obs}\delta v^* \geq \delta x^* + \left(\frac{1}{M_b} + \frac{1}{M_d}\right)\frac{\hbar T_{obs}}{2\,\delta x^*} \geq \sqrt{\frac{\hbar T_{obs}}{2}\frac{1}{M_d}}\,, \qquad (21)$$

where M_b is the mass of the body, M_d is the total mass of the device composed of the light-gun, the detector, and the clock, and I also used the fact that Heisenberg's *Uncertainty Principle* implies $\delta x^* \delta v^* \geq (1/M_b + 1/M_d)\hbar/2$. [The *reduced mass* $(1/M_b + 1/M_d)^{-1}$ is relevant for the relative motion.] Clearly, from (21) it follows that in order to reduce the contribution to the uncertainty coming from the quantum properties of the devices it is necessary to take the formal "classical-device limit," *i.e.* the limit[14] of infinitely large M_d.

Up to this point I have not yet taken into account the gravitational properties of the devices and in fact the "classical-device limit" encountered above is fully consistent with the laws of ordinary quantum mechanics. From a physical/phenomenological and conceptual viewpoint it is well understood that the formalism of quantum mechanics is only appropriate for the description of the results of measurements performed by classical devices. It is therefore not surprising that the classical-device (infinite-mass) limit turns out to be required in order to match the prediction $min\delta D = 0$ of ordinary quantum mechanics.

If one also takes into account the gravitational properties of the devices, a conflict with ordinary quantum mechanics immediately arises because the

[14] A rigorous definition of a "classical device" is beyond the scope of these notes. However, it should be emphasized that the experimental setups being here considered require the devices to be accurately positioned during the time needed for the measurement, and therefore an ideal/classical device should be infinitely massive so that the experimentalists can prepare it in a state with $\delta x\, \delta v \sim \hbar/M \sim 0$. It is the fact that the infinite-mass limit is not accessible in a gravitational context that forces one to consider only "non-classical devices." This observation is not inconsistent with conventional analyses of decoherence for macroscopic systems; in fact, in appropriate environments, the behavior of a macroscopic device will still be "closer to classical" than the behavior of a microscopic device, although the limit in which a device has exactly classical behavior is no longer accessible.

classical-device (infinite-mass) limit is in principle inadmissible for measurements concerning gravitational effects.[15] As the devices get more and more massive they increasingly disturb the gravitational/geometrical observables, and well before reaching the infinite-mass limit the procedures for the measurement of gravitational observables cannot be meaningfully performed [12,13,45]. In the Salecker-Wigner measurement procedure the limit $M_d \to \infty$ is not admissible when gravitational interactions are taken into account. At the very least the value of M_d is limited by the requirement that the apparatus should not turn into a black hole (which would not allow the exchange of signals required by the measurement procedure).

These observations render unavoidable the $\sqrt{T_{obs}}$-dependence of Eq. (21). It is important to realize that this $\sqrt{T_{obs}}$-dependence of the bound of the measurability of distances comes simply from combining elements of quantum mechanics with elements of classical gravity. As it stands it is not to be interpreted as a quantum-gravity effect. However, as clarified in the opening of this section, if one is interested in modeling properties of the space-time foam it is natural to explore the possibility that the foam be such that distances be affected by stochastic fluctuations with this typical $\sqrt{T_{obs}}$-dependence. The logic is here the one of observing that stochastic fluctuations associated to the foam would anyway lead to some form of dependence on T_{obs} and in guessing the specific form of this dependence the measurability analysis reviewed in this subsection can be seen as providing motivation for a $\sqrt{T_{obs}}$-dependence. From this point of view the measurability analysis reviewed in this subsection provides additional motivation for the study of random-walk-type models of distance fuzziness, whose fundamental stochastic fluctuations are characterized (as already discussed in Section 4) by root-mean-square deviation σ_D given by[16]

$$\sigma_D \sim \sqrt{L_{QG}\, c T_{obs}} \tag{22}$$

[15] This conflict between the infinite-mass classical-device limit (which is implicit in the applications of the formalism of ordinary quantum mechanics to the description of the outcome of experiments) and the nature of gravitational interactions has not been addressed within any of the most popular quantum gravity approaches, including critical superstrings [38,39] and canonical/loop quantum gravity [40]. In a sense somewhat similar to the one appropriate for Hawking's work on black holes [93], this "classical-device paradox" appears to provide an obstruction [13] for the use of the ordinary formalism of quantum mechanics for a description of quantum gravity.

[16] As discussed in Refs. [12,13,24], this form of σ_D also implies that in quantum gravity any measurement that monitors a distance D for a time T_{obs} is affected by an uncertainty $\delta D \geq \sqrt{L_{QG}\, c T_{obs}}$. This must be seen as a minimum uncertainty that takes only into account the quantum and gravitational properties of the measuring apparatus. Of course, an even tighter bound can emerge when taking into account also the quantum and gravitational properties of the system under observation. According to the estimates provided in Refs. [30,33] the contribution to the uncertainty coming from the system is of the type $\delta D \geq L_p$, so that the total contribution (summing the system and the apparatus contributions) might be of the type $\delta D \geq L_p + \sqrt{L_{QG}\, c T_{obs}}$.

and by displacement amplitude spectral density $S(f)$ given by

$$S(f) = f^{-1} \sqrt{L_{QG}\, c} \,. \tag{23}$$

Here the scale L_{QG} plays exactly the same role as in Section 4 (in particular $L_{QG} \equiv \mathcal{L}_{\beta=1}$ in the sense of Section 4). However, seeing L_{QG} as the result of Planck-length fluctuations occurring at a rate of one per Planck time immediately leads us to $L_{QG} \sim L_p$, whereas the different intuition which has gone into the emergence of L_{QG} in this subsection leaves room for different predictions. As already emphasized, by mixing elements of quantum mechanics and elements of gravitation one can only conclude that there must be some $\sqrt{T_{obs}}$-dependent irreducible contribution to the uncertainty in the measurement of distances. One can then guess that space-time foam might reflect this $\sqrt{T_{obs}}$-dependence and one can parametrize our ignorance by introducing L_{QG} in the formula $\sqrt{L_{QG}\, c T_{obs}}$. Within such an argument the estimate $L_{QG} \sim L_p$ could only be motivated on dimensional grounds (L_p is the only length scale available), but there is no direct estimate of L_{QG} within the argument advocated in this subsection. We only have (in the specific sense intended above) a lower limit on L_{QG} which is set by the bare analysis using straightforward combination of elements of ordinary quantum mechanics and elements of ordinary gravity. As seen above, this lower limit on L_{QG} is set by the minimum allowed value of $1/M_d$. Our intuition for L_{QG} might benefit from trying to establish this minimum allowed value of $1/M_d$. As mentioned, a conservative (possibly very conservative) estimate of this minimum value can be obtained by enforcing that M_d be at least sufficiently small to avoid black hole formation. In leading order (*e.g.*, assuming corresponding spherical symmetries) this amounts to the requirement that $M_d < \hbar S_d/(c L_p^2)$, where the length S_d characterizes the size of the region of space where the matter distribution associated to M_d is localized. This observation implies

$$\frac{1}{M_d} > \frac{c L_p^2}{\hbar} \frac{1}{S_d} \,, \tag{24}$$

which in turn suggests [12] that $L_{QG} \sim min[L_p^2/S_d]$:

$$\delta D \geq min \sqrt{\frac{1}{S_d} \frac{L_p^2\, c T_{obs}}{2}} \,. \tag{25}$$

Of course, this estimate is very preliminary since a full quantum gravity would be needed here; in particular, the way in which black holes were handled in my argument might have missed important properties which would become clear only once we have the correct theory. However, it is nevertheless striking to observe that the guess $L_{QG} \sim L_p$ appears to be very high with respect to the lower limit on L_{QG} which we are finding from this estimate; in fact, $L_{QG} \sim L_p$ would correspond to the maximum admissible value of S_d being of order L_p. Since my analysis only holds for devices that can be treated as approximately rigid[17] and

[17] The fact that I have included only one contribution from the quantum properties of the devices, the one associated to the quantum properties of the motion of the center

any non-rigidity could introduce additional contributions to the uncertainties, it is reasonable to assume that $max[S_d]$ be some small length (small enough that any non-rigidity would negligibly affect the measurement procedure), but it appears unlikely that $max[S_d] \sim L_p$. This observation might provide some encouragement for values of L_{QG} smaller than L_p, which after all is the only way to obtain random-walk models consistent with the data analysis reported in Refs. [7,24].

Later in this section I will consider a particular class of estimates for $max[S_d]$: if the correct quantum gravity is such that something like (25) holds but with $max[S_d]$ that depends on δD, one would have a different T_{obs}-dependence (and corresponding f-dependence), as I shall show in one example.

8.3 Random-walk noise motivated by linear deformation of dispersion relation

Besides the analysis of the Salecker-Wigner measurement procedure also the mentioned possibility of quantum-gravity-induced deformation of dispersion relations [5,46,41,21,27] would be consistent with the idea of random-walk distance fuzziness. The sense in which this is true is clarified by the arguments that follow.

Let me start by going back to the general relation (already discussed in Section 2):

$$c^2\mathbf{p}^2 \simeq E^2\left[1+\xi\left(\frac{E}{E_{QG}}\right)^{\alpha}\right] . \tag{26}$$

Scenarios (26) with $\alpha = 1$ are consistent with random-walk noise, in the sense that an experiment involving as a device (as a probe) a massless particle satisfying the dispersion relation (26) with $\alpha = 1$ would be naturally affected by a device-induced uncertainty that grows with $\sqrt{T_{obs}}$. From the deformed dispersion

of mass, implicitly relies on the assumption that the devices and the bodies can be treated as approximately rigid. Any non-rigidity of the devices could introduce additional contributions to the uncertainty in the measurement of D. This is particularly clear in the case of detector screens and mirrors, whose shape plays a central role in data analysis. Uncertainties in the shape (the relative position of different small parts) of a detector screen or of a mirror would lead to uncertainties in the measured quantity. For large devices some level of non-rigidity appears to be inevitable in quantum gravity. Causality alone (without any quantum mechanics) forbids rigid attachment of two bodies (*e.g.*, two small parts of a device), but is still consistent with rigid motion (bodies are not really attached but because of fine-tuned initial conditions their relative position and relative orientation are constants of motion). When Heisenberg's *Uncertainty Principle* is introduced rigid motion becomes possible only for bodies of infinite mass, whose trajectories can still be deterministic because of $\delta x\, \delta v \sim \hbar/M \sim 0$. Rigid devices are still available in ordinary quantum mechanics but they are peculiar devices, with infinite mass. When both gravitation and quantum mechanics are introduced rigid devices are no longer available since the infinite-mass limit is then inconsistent with the nature of gravitational devices.

relation (26) one is led to energy-dependent velocities [24]

$$v \simeq c \left[1 - \left(\frac{1+\alpha}{2} \right) \xi \left(\frac{E}{E_{QG}} \right)^{\alpha} \right] , \tag{27}$$

and consequently when a time T_{obs} has lapsed from the moment in which the observer (experimentalist) set off the measurement procedure the uncertainty in the position of the massless probe is given by

$$\delta x \simeq c\,\delta t + \delta v\, T_{obs} \simeq c\,\delta t + \frac{1+\alpha}{2}\,\alpha\,\frac{E^{\alpha-1}\,\delta E}{E_{QG}^{\alpha}} c\, T_{obs} , \tag{28}$$

where δt is the quantum uncertainty in the time of emission of the probe, δv is the quantum uncertainty in the velocity of the probe, δE is the quantum uncertainty in the energy of the probe, and I used the relation between δv and δE that follows from (27). Since the quantum uncertainty in the time of emission of a particle and the quantum uncertainty in its energy are related[18] by $\delta t\, \delta E \geq \hbar$, Eq. (28) can be turned into an absolute bound on the uncertainty in the position of the massless probe when a time T_{obs} has lapsed from the moment in which the observer set off the measurement procedure [24]

$$\delta x \geq c \frac{\hbar}{\delta E} + \frac{1+\alpha}{2}\,\alpha\,\frac{E^{\alpha-1}\,\delta E}{E_{QG}^{\alpha}} T_{obs} \geq \sqrt{\left(\frac{\alpha+\alpha^2}{2}\right)\left(\frac{E}{E_{QG}}\right)^{\alpha-1} \frac{c^2 \hbar T_{obs}}{E_{QG}}} . \tag{29}$$

For $\alpha = 1$ the E-dependence on the right-hand side of Eq. (29) disappears and one is led again to a δx of the type $(constant) \cdot \sqrt{T_{obs}}$:

$$\delta x \geq \sqrt{\frac{c^2 \hbar T_{obs}}{E_{QG}}} . \tag{30}$$

When massless probes are used in the measurement of a distance D, as in the Salecker-Wigner measurement procedure, the uncertainty (30) in the position of the probe translates directly into an uncertainty on D:

$$\delta D \geq \sqrt{\frac{c^2 \hbar T_{obs}}{E_{QG}}} . \tag{31}$$

This was already observed in Refs. [46,23,27] which considered the implications of deformed dispersion relations (26) with $\alpha = 1$ for the Salecker-Wigner measurement procedure.

---

[18] It is well understood that the $\delta t\, \delta E \geq \hbar$ relation is valid only in a weaker sense than, say, Heisenberg's Uncertainty Principle $\delta x\, \delta p \geq \hbar$. This has its roots in the fact that the time appearing in quantum-mechanics equations is just a parameter (not an operator), and in general there is no self-adjoint operator canonically conjugate to the total energy, if the energy spectrum is bounded from below [94,53]. However, the $\delta t\, \delta E \geq \hbar$ relation does relate δt intended as quantum uncertainty in the time of emission of a particle and δE intended as quantum uncertainty in the energy of that same particle.

Since deformed dispersion relations (26) with $\alpha = 1$ have led us to the same measurability bound already encountered both in the analysis of the Salecker-Wigner measurement procedure and the analysis of simple-minded random-walk models of quantum space-time fluctuations, if we assume again that such measurability bounds emerge in a full quantum gravity as a result of corresponding quantum fluctuations (fuzziness), we are led once again to random-walk noise:

$$\sigma_D \sim \sqrt{\frac{c^2 \hbar T_{obs}}{E_{QG}}} \,. \tag{32}$$

8.4 Noise motivated by quadratic deformation of dispersion relation

In the preceding subsection I observed that quantum-gravity deformed dispersion relations (26) with $\alpha = 1$ can also motivate random-walk noise $\sigma_D \sim (constant) \cdot \sqrt{T_{obs}}$. If we use the same line of reasoning that connects a measurability bound to a scenario for fuzziness when $\alpha \neq 1$ we appear to find $\sigma_D \sim \mathcal{G}(E/E_{QG}) \cdot \sqrt{T_{obs}}$, where $\mathcal{G}(E/E_{QG})$ is a (α-dependent) function of E/E_{QG}. However, in these cases with $\alpha \neq 1$ clearly the connection between measurability bound and fuzzy-distance scenario cannot be this simple; in fact, the energy of the probe E which naturally plays a role in the context of the derivation of the measurability bound does not have a natural counter-part in the context of the conjectured fuzzy-distance scenario.

In order to preserve the conjectured connection between measurability bounds and fuzzy-distance scenarios one can be tempted to envision that if $\alpha \neq 1$ the interferometer noise levels induced by space-time fuzziness might be of the type [24]

$$\sigma_D \sim \sqrt{\left(\frac{\alpha + \alpha^2}{2}\right) \left(\frac{E^*}{E_{QG}}\right)^{\alpha - 1} \frac{c^2 \hbar T_{obs}}{E_{QG}}} \,, \tag{33}$$

where E^* is some energy scale characterizing the physical context under consideration. [For example, at the intuitive level one might conjecture that E^* could characterize some sort of energy density associated with quantum fluctuations of space-time or an energy scale associated with the masses of the devices used in the measurement process.]

Since $\alpha \geq 1$ in all Quantum-Gravity approaches believed to support deformed dispersion relations it appears likely that the factor $(E^*/E_{QG})^{\alpha-1}$ would suppress the random-walk noise effect in all contexts with $E^* < E_{QG}$. Besides the case $\alpha = 1$ (linear deformation) also the case $\alpha = 2$ (quadratic deformation) deserves special interest since it can emerge quite naturally in quantum-gravity approaches (see, *e.g.*, Ref. [22]).

8.5 Noise with $f^{-5/6}$ amplitude spectral density

In Subsection 8.2 a bound on the measurability of distances based on the Salecker-Wigner procedure was used as motivation for experimental tests of interferometer noise of random-walk type, with f^{-1} amplitude spectral density and $\sqrt{T_{obs}}$ root-mean-square deviation. In this subsection I shall pursue further the observation that the relevant measurability bound could be derived by simply insisting that the devices do not turn into black holes. That observation allowed to derive Eq. (25), which expresses the minimum uncertainty δD on the measurement of a distance D (*i.e.* the measurability bound for D) as proportional to $\sqrt{T_{obs}}$ and $\sqrt{1/S_d}$. Within that derivation the minimum uncertainty is obtained in correspondence of $max[S_d]$, the maximum value of S_d consistent with the structure of the measurement procedure. I was therefore led to consider how large S_d could be while still allowing to disregard any non-rigidity in the quantum motion of the device (which would introduce additional contributions to the uncertainties). Something suggestive of the random-walk noise scenario emerged by simply assuming that $max[S_d]$ be independent of the accuracy δD that the observer would wish to achieve. However, as mentioned, the same physical intuition that motivates some of the fuzzy space-time scenarios here considered also suggests that quantum gravity might require a novel measurement theory, possibly involving a new type of relation between system and measuring apparatus. Based on this intuition, it seems reasonable to contemplate the possibility that $max[S_d]$ might actually depend on δD.

It is such a scenario that I want to consider in this subsection. In particular I want to consider the case $max[S_d] \sim \delta D$, which, besides being simple, has the plausible property that it allows only small devices if the uncertainty to be achieved is small, while it would allow correspondingly larger devices if the observer was content with a larger uncertainty. This is also consistent with the idea that elements of non-rigidity in the quantum motion of extended devices could be neglected if anyway the measurement is not aiming for great accuracy, while they might even lead to the most significant contributions to the uncertainty if all other sources of uncertainty are very small. [Salecker and Wigner [92] would also argue that "large" devices are not suitable for very accurate space-time measurements (they end up being "in the way" of the measurement procedure) while they might be admissible if space-time is being probed rather softly.]

In this scenario with $max[S_d] \sim \delta D$, Eq. (25) takes the form

$$\delta D \geq \sqrt{\frac{1}{S_d} \frac{L_p^2 \, c \, T_{obs}}{2}} \geq \sqrt{\frac{L_p^2 \, c \, T_{obs}}{2 \, \delta D}} \,, \tag{34}$$

which actually gives

$$\delta D \geq \left(\frac{1}{2} L_p^2 \, c \, T_{obs} \right)^{1/3} \,. \tag{35}$$

As done with the other measurability bounds, I have proposed [7,24] to take Eq. (35) as motivation for the investigation of a corresponding fuzziness scenario characterised by

$$\sigma_D \sim \left(\tilde{L}_{QG}^2 \, c \, T_{obs} \right)^{1/3} \,. \tag{36}$$

Notice that in this equation I replaced L_p with a generic length scale $\tilde{L}_{QG}$, since it is possible that the heuristic argument leading to Eq. (36) might have captured the qualitative structure of the phenomenon while providing an incorrect estimate of the relevant length scale. Also notice that Eq. (35) has the same form as the relations emerged in other measurability analyses [45,43], even though those analyses adopted a very different viewpoint (and even the physical role of the elements of Eq. (35) was different, as explained in the next section).

As observed in Refs. [7,24] the $T_{obs}^{1/3}$ dependence of σ_D is associated with displacement amplitude spectral density with $f^{-5/6}$ behaviour:

$$\mathcal{S}(f) = f^{-5/6}(\tilde{L}_{QG}^2\, c)^{1/3} \ . \tag{37}$$

Therefore the measurability analyses discussed in this subsection provides motivation for the investigation of the case $\beta = 5/6$ (using again the notation set up in Section 4).

9 Relations with other quantum gravity approaches

In this section I comment on the connections and the differences between some of the ideas which I reviewed in these notes and other quantum-gravity ideas.

9.1 Canonical Quantum Gravity

One of the most popular quantum-gravity approaches is the one in which the ordinary canonical formalism of quantum mechanics is applied to (some formulation of) Einstein's Gravity. Especially in light of the fact that [13] some of the observations reviewed in the previous sections suggest that quantum gravity should require a new mechanics, not exactly given by ordinary quantum mechanics, it is very interesting that some of the phenomena considered in the previous sections have also emerged in studies of canonical quantum gravity.

As mentioned, the most direct connection was found in the study reported in Ref. [41], which was motivated by Ref. [5]. In fact, Ref. [41] shows that the popular canonical/loop quantum gravity [40] admits the phenomenon of deformed dispersion relations, with the deformation going linearly with the Planck length.

Concerning the bounds on the measurability of distances it is probably fair to say that the situation in canonical/loop quantum gravity is not yet clear because the present formulations do not appear to lead to a compelling candidate "length operator." This author would like to interpret the problems associated with the length operator as an indication that perhaps something unexpected might actually emerge in canonical/loop quantum gravity as a length operator, possibly something with properties fitting the intuition of some of the scenarios for fuzzy distances which I reviewed. Actually, the random-walk space-time fuzziness model might have a (somewhat weak, but intriguing) connection with "quantum mechanics applied to gravity" at least to the level seen by comparison with the scenario discussed in Ref. [95], which was motivated by the intuition that is emerging from investigations of canonical/loop quantum gravity. The "moves" of Ref. [95] share many of the properties of the "random steps" of the random-walk models.

9.2 Critical and non-critical string theories

Unfortunately, in the popular quantum-gravity approach based on critical superstrings[19] not many results have been derived concerning directly the quantum properties of space-time. Perhaps the most noticeable such results are the ones on limitations on the measurability of distances emerged in the scattering analyses reported in Refs. [31,87], which I already mentioned.

A rather different picture is emerging (through the difficulties of this rich formalism) in *Liouville* (non-critical) strings [19], whose development was partly motivated by intuition concerning the quantum-gravity vacuum that is rather close to the one traditionally associated to the mentioned works of Wheeler and Hawking. Evidence has been found [46] in *Liouville* strings supporting the validity of deformed dispersion relations, with the deformation going linearly with the Planck/string length. In the sense clarified in Section 8.3 this approach might also host a bound on the measurability of distances which grows with $\sqrt{T_{obs}}$.

9.3 Other types of measurement analyses

Because of the lack of experimental input, it is not surprising that many authors have been seeking some intuition on quantum gravity by formal analyses of the ways in which the interplay between gravitation and quantum mechanics could affect measurement procedures. A large portion of these analyses produced a "$min[\delta D]$" with D denoting a distance; however, the same type of notation was used for structures defined in significantly different ways. Also different meanings have been given by different authors to the statement "absolute bound on the measurability of an observable." Quite important for the topics here discussed are the differences (which might not be totally transparent as a result of this unfortunate choice of overlapping notations) between the approach advocated in Refs. [7,12,13,24] and the approaches advocated in Refs. [92,44,45,43]. In Refs. [7,12,13,24] "$min[\delta D]$" denotes an absolute limitation on the measurability of a distance D. The studies [92,44,43] analyzed the interplay of gravity and quantum mechanics in defining a net of time-like geodesics, and in those studies "$min[\delta D]$" characterizes the maximum "tightness" achievable for the net of time-like geodesics. Moreover, in Refs. [92,44,45,43] it was required that the measurement procedure should not affect/modify the geometric observable being measured, and "absolute bounds on the measurability" were obtained in this

[19] As already mentioned the mechanics of critical superstrings is just an ordinary quantum mechanics. All of the new structures emerging in this exciting formalism are the result of applying ordinary quantum mechanics to the dynamics of extended fundamental objects, rather than point-like objects (particles).

specific sense. Instead, in Refs. [12,13,24] it was envisioned that the observable which is being measured might depend also on the devices (the underlying view is that observables in quantum gravity would always be, in a sense, shared properties of "system" and "apparatus"), and it was only required that the nature of the devices be consistent with the various stages of the measurement procedure (for example if a device turned into a black-hole some of the exchanges of signals needed for the measurement would be impossible). The measurability bounds of Refs. [12,13,24] are therefore to be intended from this more fundamental perspective, and this is crucial for the possibility that these measurability bounds be associated to a fundamental quantum-gravity mechanism for "fuzziness" (quantum fluctuations of space-time). The analyses reported in Refs. [92,44,45,43] did not include any reference to fuzzy space-times of the type operatively defined in terms of stochastic processes, as reviewed in Section 4.

The more fundamental nature of the bounds obtained in Refs. [12,13,24] is also crucial for the arguments suggesting that quantum gravity might require a new mechanics, not exactly given by ordinary quantum mechanics. The analyses reported in Refs. [92,44,45,43] did not include any reference to this possibility.

Having clarified that there is a "double difference" (different "min" and different "δD") between the meaning of $min[\delta D]$ adopted in Refs. [7,12,13,24] and the meaning of $min[\delta D]$ adopted in Refs. [92,44,45,43], it is however important to notice that the studies reported in Refs. [44,45,43] were among the first studies which showed how in some aspects of measurement analysis the Planck length might appear together with other length scales in the problem. For example, a quantum gravity effect naturally involving something of length-squared dimensions might not necessarily go like L_p^2: in some cases it could go like ΛL_p, with Λ some other length scale in the problem.

Interestingly, the analysis of the interplay of gravity and quantum mechanics in defining a net of time-like geodesics reported in Ref. [44] concluded that the maximum "tightness" achievable for the geodesics would be characterized by $\sqrt{L_p^2 R^{-1} s}$, where R is the radius of the (spherically symmetric) clocks whose world lines define the network of geodesics, and s is the characteristic distance scale over which one is intending to define such a network. The $\sqrt{L_p^2 R^{-1} s}$ maximum tightness discussed in Ref. [44] is formally analogous to Eq. (25), but, as clarified above, this "maximum tightness" was defined in a very different ("doubly different") way, and therefore the two proposals have completely different physical implications. Actually, in Ref. [44] it was also stated that for a single geodesic distance (which might be closer to the type of distance measurability analysis reported in Refs. [12,13,24]) one could achieve accuracy significantly better than the formula $\sqrt{L_p^2 R^{-1} s}$, which was interpreted in Ref. [44] as a direct result of the structure of a network of geodesics.

Relations of the type $min[\delta D] \sim (L_p^2 D)^{(1/3)}$, which are formally analogous to Eq. (35), were encountered in the analysis of maximum tightness achievable for a geodesics network reported in Ref. [43] and in the analysis of measurability of distances reported in Ref. [45]. Although once again the definitions of "min"

and "δD" used in these studies are completely different from the ones relevant for the "$min[\delta D]$" of Eq. (35).

10 Quantum gravity, no strings attached

Some of the arguments reviewed in these lecture notes appear to suggest that quantum gravity might require a mechanics not exactly of the type of ordinary quantum mechanics. In particular, the new mechanics might have to accommodate a somewhat different relationship (in a sense, "more democratic") between "system" and "measuring apparatus", and should take into account the fact that the limit in which the apparatus behaves classically is not accessible once gravitation is turned on. The fact that the most popular quantum-gravity approaches, including critical superstrings and canonical/loop quantum gravity, are based on ordinary quantum mechanics but seem inconsistent with the correspondence between formalism and measurability bounds of the type sought and found in non-gravitational quantum mechanics (through the work of Bohr, Rosenfeld, Landau, Peierls, Einstein, Salecker, Wigner and many others), represents, in this author's humble opinion, one of the outstanding problems of these approaches. Still, it is of great importance for quantum-gravity research that these approaches continue to be pursued very aggressively: they might eventually encounter along their development unforeseeable answers to these questions or else, as they are "pushed to the limit", they might turn out to fail in a way that provides insight on the correct theory. However, the observations pointing us toward deviations from ordinary quantum-mechanics could provide motivation for the parallel development of alternative quantum-gravity approaches. But how could we envision quantum gravity with no strings (or"canonical loops") attached? More properly, how can we devise a new mechanics when we have no direct experimental data on its structure? Classical mechanics was abandoned for quantum mechanics only after a relatively long period of analysis of physical problems such as black-body spectrum and photoelectric effect which contained very relevant information. We don't seem to have any such insightful physical problem. At best we might have identified the type of conceptual shortfall which Mach had discussed with respect to Newtonian gravity. It is amusing to notice that the analogy with Machian conceptual analyses might actually be quite proper, since at the beginning of this century we were invited to renounce to the comfort of the reference to "fixed stars" and now that we are reaching the end of this century we might be forced to renounce to the comfort of an idealized classical measuring apparatus.

Our task is that much harder in light of the fact that (unless something like large extra dimensions is verified in Nature) we must make a gigantic leap from the energy scales we presently understand to Planckian energy scales. While of course we must all hope someone clever enough can come up with the correct recipe for this gigantic jump, one less optimistic strategy that might be worth pursuing is the one of trying to come up with some effective theory useful for the description of new space-time-related phenomena occurring in an energy-scale range extending from somewhere not much above presently achievable energies up to somewhere safely below the Planck scale. These theories might provide guidance to experimentalists, and in turn (if confirmed by experiments) might provide a useful intermediate step toward the Planck scale. For those who are not certain that we can make a lucky guess of the whole giant step toward the Planck scale[20] this strategy might provide a possibility to eventually get to the Planck regime only after a (long and painful) series of intermediate steps. Some of the ideas discussed in the previous sections can be seen as examples of this strategy. In this section I collect additional relevant material.

10.1 A low-energy effective theory of quantum gravity

While the primary emphasis has been on experimental tests of quantum-gravity-motivated candidate phenomena, some of the arguments (which are based on Refs. [12,13,24]) reviewed in these lecture notes can be seen as attempts to indentify some of the properties that one could demand of a theory suitable for a first stage of partial unification of gravitation and quantum mechanics. This first stage of partial unification would be a low-energy effective theory capturing only some rough features of quantum gravity. In particular, as discussed in Refs. [23,13,24], it is plausible that the most significant implications of quantum gravity for low-energy (large-distance) physics might be associated with the structure of the non-trivial "quantum-gravity vacuum". A satisfactory picture of this vacuum is not available at present, and therefore we must generically characterize it as the appropriate new concept that in quantum gravity takes the place of the ordinary concept of "empty space"; however, it is plausible that some of the arguments by Wheeler, Hawking and others, attempting to develop an intuitive description of the quantum-gravity vacuum, might have captured at least some of its actual properties. Therefore the experimental investigations of space-time foam discussed in some of the preceding sections could be quite relevant for the search of a theory describing a first stage of partial unification of gravitation and quantum mechanics.

Other possible elements for the search of such a theory come from studies suggesting that this unification might require a new (non-classical) concept of measuring apparatus and a new relationship between measuring apparatus and system. I have reviewed some of the relevant arguments [12,13] through the discussion of the Salecker-Wigner setup for the measurement of distances, which manifested the problems associated with the infinite-mass classical-device limit. As mentioned, a similar conclusion was already drawn in the context of attempts (see, *e.g.*, Ref. [90]) to generalize to the study of the measurability of gravitational fields the famous Bohr-Rosenfeld analysis [91] of the measurability

[20] Understandably, some are rendered prudent by the realization that the ratio between the Planck length and the length scales which will be probed by LHC and LIGO is actually somewhat smaller than, say, the ratio between the typical lengths characterizing the size of small insects and the distance between the planet Pluto and the Sun.

of the electromagnetic field. It seems reasonable to explore the possibility that already the first stage of partial unification of gravitation and quantum mechanics might require a new mechanics. A (related) plausible feature of the correct effective low-energy theory of quantum-gravity is (some form of) a novel bound on the measurability of distances. This appears to be an inevitable consequence of relinquishing the idealized methods of measurement analysis that rely on the artifacts of the infinite-mass classical-device limit. If indeed one of these novel measurability bounds holds in the physical world, and if indeed the structure of the quantum-gravity vacuum is non-trivial and involves space-time fuzziness, it appears also plausible that this two features be related, *i.e.* that the fuzziness of space-time would be ultimately responsible for the measurability bounds. It is also plausible [23,13] that an effective large-distance description of some aspects of quantum gravity might involve quantum symmetries and noncommutative geometry (while at the Planck scale even more novel geometric structures might be required).

The intuition emerging from these considerations on a novel relationship between measuring apparatus and system and by a Wheeler-Hawking picture of the quantum-gravity vacuum has not yet been implemented in a fully-developed new formalism describing the first stage of partial unification of gravitation and quantum mechanics, but one can use this emerging intuition for rough estimates of certain candidate quantum-gravity effects. Some of the theoretical estimates that I reviewed in the preceding sections, particularly the ones on distance fuzziness, can be seen as examples of this.

Besides the possibility of direct experimental tests (such as some of the ones here reviewed), studies of low-energy effective quantum-gravity models might provide a perspective on quantum gravity that is complementary with respect to the one emerging from approaches based on proposals for a one-step full unification of gravitation and quantum mechanics. On one side of this complementarity there are the attempts to find a low-energy effective quantum gravity which are necessarily driven by intuition based on direct extrapolation from known physical regimes; they are therefore rather close to the phenomelogical realm but they are confronted by huge difficulties when trying to incorporate this physical intuition within a completely new formalism. On the other side there are the attempts of one-step full unification of gravitation and quantum mechanics, which usually start from some intuition concerning the appropriate formalism (*e.g.*, canonical/loop quantum gravity or critical superstrings) but are confronted by huge difficulties when trying to "come down" to the level of phenomenological predictions. These complementary perspectives might meet at the mid-way point leading to new insight in quantum gravity physics. One instance in which this mid-way-point meeting has already been successful is provided by the mentioned results reported in Ref. [41], where the candidate phenomenon of quantum-gravity induced deformed dispersion relations, which had been proposed within phenomenological analyses [46,23,5] of the type needed for the search of a low-energy theory of quantum gravity, was shown to be consistent with the structure of canonical/loop quantum gravity.

10.2 Theories on non-commutative Minkowski space-time

At various points in these notes there is a more or less explicit reference to deformed symmetries and noncommutative space-times[21]. Just in the previous subsection I have recalled the conjecture [23,13] that an effective large-distance description of some aspects of quantum gravity might involve quantum symmetries and noncommutative geometry. The type of *in vacuo* dispersion which can be tested [5] using observations of gamma rays from distant astrophysical sources is naturally encoded within a consistent deformation of Poincaré symmetries [23,27,28].

A useful structure (at least for toy-model purposes, but perhaps even more than that) appears to be the noncommutative (so-called "κ") Minkowski space-time [62,63,21]

$$[x^i, t] = i\lambda\, x^i, \quad [x^i, x^j] = 0 \tag{38}$$

where $i, j = 1, 2, 3$ and λ (commonly denoted[22] by $1/\kappa$) is a free length scale. This simple noncommutative space-time could be taken as a basis for an effective description of phenomena associated to a nontrivial foamy quantum-gravity vacuum[23]. When probed very softly such a space would appear as an ordinary Minkowski space-time[24], but probes of sufficiently high energy would be affected by the properties of the quantum-gravity foam and one could attempt to model (at least some aspects of) the corresponding dynamics using a noncommutative Minkowski space-time. In light of this physical motivation it is natural to assume that λ be related to the Planck length.

The so-called κ-deformed Poincaré quantum group [99] acts covariantly [63] on the κ-Minkowski space-time (38). The dispersion relation for massless spin-0 particles

$$\lambda^{-2}\left(e^{\lambda E} + e^{-\lambda E} - 2\right) - \boldsymbol{k}^2 e^{-\lambda E} = 0\,, \tag{39}$$

which at low energies describes a deformation that is linearly suppressed by λ (and therefore, if indeed $\lambda \sim L_p$, is of the type discussed in Section 5),

[21] The general idea of some form of connection between Planck-scale physics and quantum groups (with their associated noncommutative geometry) is of course not new, see *e.g.*, Refs. [96–98,?,21,100–103]. Moreover, some support for noncommutativity of space-time has also been found within measurability analyses [32,23].

[22] As for the notations L_{QG} and E_{QG}, this author is partly responsible [28] for the redundant convention of using the notation λ when the reader is invited to visualize a length scale and going back to the κ notation when instead it might be natural for the reader to visualize a length scale.

[23] In particular, within one particular attempt to model space-time foam, the one of *Liouville* non-critical strings [19], the time "coordinate" appears [104] to have properties that might be suggestive of a κ-Minkowski space-time.

[24] Generalizations would of course be necessary for a description of how the quantum-gravity foam affects spaces which are curved (non-Minkowski) at the classical level, and even for spaces which are Minkowski at the classical level a full quantum gravity of course would predict phenomena which could not be simply encoded in noncommutativity of Minkowski space.

emerges [21,27,28] as the appropriate Casimir of the κ-deformed Poincaré group. Rigorous support for the interpretation of (39) as a *bona fide* dispersion relation characterizing the propagation of waves in the κ-Minkowski space-time was recently provided in Ref. [28].

In Ref. [28] it was also observed that, using the quantum group Fourier transform which was worked out for our particular algebra in Ref. [105], there might be a rather simple approach to the definition of a field theory on the κ-Minkowski space-time. In fact, through the quantum group Fourier transform it is possible to rewrite structures living on noncommutative space-time as structures living on a classical (but nonAbelian) "energy-momentum" space. If one is content to evaluate everything in energy-momentum space, this observation gives the opportunity to by-pass all problems directly associated with the non-commutativity of space-time. While waiting for a compelling space-time formulation of field theories on noncommutative geometries to emerge, it seems reasonable to restrict all considerations to the energy-momentum space. This approach does not work for *any* noncommutative space-time but for all those where the space-time coordinate algebra is the enveloping algebra of a Lie algebra, with the Lie algebra generators regarded 'up side down' as noncommuting coordinates [106].[25]

Within this viewpoint a field theory is not naturally described in terms of a Lagrangian, but rather it is characterized directly in terms of Feynman diagrams. In principle, according to this proposal a given ordinary field theory can be "deformed" into a counterpart living in a suitable noncommutative space-time not by fancy quantum-group methods but simply by the appropriate modification of the momentum-space Feynman rules to those appropriate for a nonAbelian group. Additional considerations can be found in Ref. [28], but, in order to give at least one example of how this nonAbelian deformation could be applied, let me observe here that the natural propagator of a massless spin-0 particle on κ-Minkowski space-time should be given in energy-momentum space by the inverse of the operator in the dispersion relation (39), i.e. in place of $D = (\omega^2 - \boldsymbol{k}^2 - m^2)^{-1}$ one would take

$$D_\lambda = \left(\lambda^{-2}(e^{\lambda\omega} + e^{-\lambda\omega} - 2) - e^{-\lambda\omega}\boldsymbol{k}^2\right)^{-1} . \tag{40}$$

As discussed in Ref. [28] the elements of this approach to field theory appear to lead naturally to a deformation of CPT symmetries, which would first show up in experiments as a violation of ordinary CPT invariance. The development of realistic field theories of this type might therefore provide us a single formalism in which both *in vacuo* dispersion and violations of ordinary CPT invariance could be computed explicitly (rather than being expressed in terms of unknown parameters), connecting all of the aspects of these candidate quantum-gravity phenomena to the value of $\lambda \equiv 1/\kappa$. One possible "added bonus" of this approach could be associated to the fact that also loop integration must be appropriately deformed, and it appears plausible [28] that (as in other quantum-group based

[25] Another (partly related, but different) κ-Minkowski motivated proposal for field theory was recently put forward in Ref. [107]. I thank J. Lukierski for bringing this paper to my attention.

approaches [98]) the deformation might render ultraviolet finite some classes of diagrams which would ordinarily be affected by ultraviolet divergences.

11 Conservative motivation and other closing remarks

Since this paper started off with the conclusions, readers might not be too surprised of the fact that I devote most of the closing remarks to some additional motivation. These remarks had to be postponed until the very end also because in reviewing the experiments it would have been unreasonable to take a conservative viewpoint: those who are so inclined should find in the present lecture notes encouragement for unlimited excitement. However, before closing I must take a step back and emphasize those reasons of interest in this emerging phenomenology which can be shared even by those readers who are approaching all this from a conservative viewpoint.

In reviewing these quantum-gravity experiments I have not concealed my (however moderate) optimism regarding the prospects for data-driven advances in quantum-gravity research. I have reminded the reader of the support one finds in the quantum-gravity literature for the type of phenomena which we can now start to test, particularly distance fuzziness and violations of Lorentz and/or CPT symmetries and I have also emphasized that it is thanks to recent advances in experimental techniques and ideas that these phenomena can be tested (see, for example, the role played by the remarkable sensitivities recently achieved with modern interferometers in the experimental proposal reviewed in Section 4 and the role played by very recent break-throughs in GRB phenomenology in the experimental proposal reviewed in Section 5). But now let me emphasize that even from a conservative viewpoint these experiments are extremely significant, especially those that provide tests of quantum mechanics and tests of fundamental symmetries. One would not ordinarily need to stress this, but since these lectures are primarily addressed to young physics students let me observe that of course this type of tests is crucial for a sound development of our science. Even if there was no theoretical argument casting doubts on them, we could not possibly take for granted (extrapolating *ad infinitum*) ingredients of our understanding of Nature as crucial as its mechanics laws and its symmetry structure. We should test quantum mechanics and fundamental symmetries anyway, we might as well do it along the directions which appear to be favoured by some quantum-gravity ideas.

A somewhat related observation can be made concerning the fact that most of these experiments actually test only one of the two main branches of quantum-gravity proposals: the proposals in which (in one or another fashion) quantum decoherence is present. There is in fact a connection (whose careful discussion I postpone to future publications) between decoherence and the type of violations of Lorentz and CPT symmetries and the type of power-law dependence on T_{obs} of distance fuzziness here considered. The portion of our community which finds appealing the arguments supporting the decoherence-inducing Wheeler-Hawking space-time foam (and certain views on the so-called "black-hole information paradox") finds in these recent developments in quantum-gravity phenomenology an opportunity for direct tests of some of its intuition. The rest of our community has developed an orthogonal intuition concerning the quantum-gravity realm, in which there is no place for quantum decoherence. Even this second group might be looking forward to the outcome of experiments on quantum decoherence, since the results are going to put under serious test the alternative approach. Moreover, the fact that we are finally at least at the point of testing decoherence-involving quantum-gravity approaches (something which was also supposed to be impossible) should be seen as encouragement for the hope that even other quantum-gravity approaches will eventually be tested experimentally.

Even though there is of course no guarantee that this new phenomenology will be able to uncover important elements of the structure of quantum gravity, the fact that such a phenomenological programme exists suffices to make a legitimate (empirical) science of quantum gravity, a subject often derided as a safe heaven for theorists wanting to speculate freely without any risk of being proven wrong by experiments. As emphasized in Refs. [85,108] (and even in the non-technical press [109]) this can be an important turning point in the development of the field. Concerning the future of quantum-gravity phenomenology let me summarize my expectations in the form of a response to the question posed by the title of these notes: I believe that we are indeed at the dawn of quantum-gravity phenomenology, but the forecasts call for an extremely long and cloudy day with only a few rare moments of sunshine. Especially for those of us motivated by theoretical arguments suggesting that at the end of the road there should be a wonderful revolution of our understanding of Nature (perhaps a revolution of even greater magnitude than the one undergone during the first years of this 20th century), it is crucial to profit fully from the few glimpses of the road ahead which quantum-gravity phenomenology will provide.

Acknowledgements

First of all I would like to thank the organizers of this XXXV Winter School, especially for their role in creating a very comfortable informal atmosphere, which facilitated exchanges of ideas among the participants. My understanding of Refs. [2], [3] and [8] benefited from conversations with R. Brustein, M. Gasperini, G.F. Giudice, N.E. Mavromatos, G. Veneziano and J.D. Wells, which I very greatfully acknowlegde. Still on the "theory side" I am grateful to several colleagues who provided encouragement and stimulating feed-back, particularly D. Ahluwalia, A. Ashtekar, J. Ellis, J. Lukierski, C. Rovelli, S. Sarkar, L. Smolin and J. Stachel. On the "experiment side" I would like to thank F. Barone, E. Bloom, J. Faist, R. Flaminio, L. Gammaitoni, T. Huffman, L. Marrucci, M. Punturo and J. Scargle for informative conversations (some were nearly tutorials) on various aspects of interferometry and gamma-ray-burst experiments.

References

1. C.J. Isham, *Structural issues in quantum gravity*, in *Proceedings of General relativity and gravitation* (Florence 1995).
2. J. Ellis, J. Lopez, N. Mavromatos, D. Nanopoulos and CPLEAR Collaboration, Phys. Lett. B364 (1995) 239.
3. R. Brustein, M. Gasperini, M. Giovannini and G. Veneziano, Phys. Lett. B361 (1995) 45.
4. I.C. Percival and W.T. Strunz, quant-ph/9607011, Proc. R. Soc. A453 (1997) 431.
5. G. Amelino-Camelia, J. Ellis, N.E. Mavromatos, D.V. Nanopoulos and S. Sarkar, Nature 393 (1998) 763.
6. D.V. Ahluwalia, Mod. Phys. Lett. A13 (1998) 1393.
7. G. Amelino-Camelia, gr-qc/9808029, Nature 398 (1999) 216.
8. G.F. Giudice, R. Rattazzi and J.D. Wells, Nucl. Phys. B544 (1999) 3.
9. E.A. Mirabelli, M. Perelstein and M.E. Peskin, Phys. Rev. Lett. 82 (1999) 2236.
10. R. Colella, A.W. Overhauser and S.A. Werner, Phys. Rev. Lett. 34 (1975) 1472.
11. Young, B., Kasevich, M., and Chu, S. *Atom Interferometry* (Academic Press, 1997).
12. G. Amelino-Camelia, Mod. Phys. Lett. A9 (1994) 3415; *ibid.* A11 (1996) 1411.
13. G. Amelino-Camelia, Mod. Phys. Lett. A13 (1998) 1319.
14. J.A. Wheeler, in *Relativity, groups and topology*, ed. B.S. and C.M. De Witt (Gordon and Breach, New York, 1963).
15. S.W. Hawking, Phys. Rev. D18 (1978) 1747; S.W. Hawking, D.N. Page and C.N. Pope, Nucl. Phys. B170 (1980) 283.
16. A. Ashtekar, C. Rovelli and L. Smolin, Phys. Rev. Lett. 69 (1992) 237.
17. J. Ellis, J.S. Hagelin, D.V. Nanopoulos and M. Srednicki, Nucl. Phys. B241 (1984) 381.
18. D.N. Page, Gen. Rel. Grav. 14 (1982) 299; L. Alvarez-Gaume and C. Gomez, Commun. Math. Phys. 89 (1983) 235.
19. J. Ellis, N. Mavromatos and D.V. Nanopoulos, Phys. Lett. B293 (1992) 37.
20. P. Huet and M.E. Peskin, Nucl. Phys. B434 (1995) 3.
21. J. Lukierski, A. Nowicki and H. Ruegg, Ann. Phys. 243 (1995) 90.
22. G. 't Hooft, Class. Quant. Grav. 13 (1996) 1023.
23. G. Amelino-Camelia, Phys. Lett. B392 (1997) 283.
24. G. Amelino-Camelia, gr-qc/9903080.
25. S.D. Biller *et al.*, gr-qc/9810044.
26. B.E. Schaefer, astro-ph/9810479.
27. G. Amelino-Camelia, J. Lukierski and A. Nowicki, hep-th/9706031, Phys. Atom. Nucl. 61 (1998) 1811-1815; gr-qc/9903066, Int. J. Mod. Phys. A (in press).
28. G. Amelino-Camelia and S. Majid, hep-th/9907110.
29. See R. Wijers, Nature 393 (1998) 13, and references therein.
30. T. Padmanabhan, Class. Quantum Grav. 4 (1987) L107.
31. G. Veneziano, Europhys. Lett. 2 (1986) 199; D.J. Gross and P.F. Mende, Nucl. Phys. B303 (1988) 407; D. Amati, M. Ciafaloni, G. Veneziano, Phys. Lett. B216 (1989) 41; K. Konishi, G. Paffuti, P. Provero, Phys. Lett. B234 (1990) 276; T. Yoneya, Mod. Phys. Lett. A4 (1989) 1587.
32. D.V. Ahluwalia, Phys. Lett. B339 (1994) 301.
33. See, *e.g.*, L.J. Garay, Int. J. Mod. Phys. A10 (1995) 145.
34. J. Ellis, J. Lopez, N.E. Mavromatos and D.V. Nanopoulos, Phys. Rev. D53 (1996) 3846.
35. F. Benatti and R. Floreanini, Nucl. Phys. B488 (1997) 335.

36. G. Amelino-Camelia, *SO(10) grandunification model with proton lifetime of the order of* 10^{33} *years*, (Laurea thesis, Facoltá di Fisica dell'Universitá di Napoli, 1990).
37. K.S. Hirata et al. (KAMIOKANDE-II Collaboration), Phys. Lett. B220 (1989) 308.
38. J. Polchinski, *Superstring Theory and Beyond*, (Cambridge University Press, Cambridge, 1998).
39. M.B. Green, J.H. Schwarz and E. Witten, *Superstring theory* (Cambridge Univ. Press, Cambridge, 1987)
40. A. Ashtekar, Phys. Rev. Lett. 57 (1986) 2244; C. Rovelli and L. Smolin, Phys. Rev. Lett. 61 (1988) 1155.
41. R. Gambini and J. Pullin, Phys. Rev. D59 (1999) 124021.
42. Y.J. Ng and H. van Dam, gr-qc/9906003.
43. F. Karolyhazy, Il Nuovo Cimento A42 (1966) 390.
44. L. Diosi and B. Lukacs, Phys. Lett. A142 (1989) 331.
45. Y.J. Ng and H. Van Dam, Mod. Phys. Lett. A9 (1994) 335.
46. G. Amelino-Camelia, J. Ellis, N.E. Mavromatos and D.V. Nanopoulos, Int. J. Mod. Phys. A12 (1997) 607.
47. P.R. Saulson, *Fundamentals of interferometric gravitational wave detectors* (World Scientific, Singapore, 1994).
48. A. Abramovici *et al.*, Science 256 (1992) 325.
49. C. Bradaschia *et al.*, Nucl. Instrum. Meth. A289 (1990) 518.
50. A. Abramovici *et al.*, Phys. Lett. A218 (1996) 157.
51. V. Radeka, IEEE Trans. Nucl. Sci. NS16 (1969) 17; Ann. Rev. Nucl. Part. Sci. 38 (1988) 217.
52. R.J. Adler, I.M. Nemenman, J.M. Overduin and D.I. Santiago, gr-qc/9909017.
53. I.L. Egusquiza, L.J. Garay and J.M. Raya, quant-ph/9811009.
54. G. Amelino-Camelia, J. Ellis, N.E. Mavromatos, D.V. Nanopoulos and S. Sarkar, astro-ph/9810483.
55. J. van Paradis *et al.*, Nature 386 (1997) 686-689.
56. P.J. Groot *et al.*, IAU Circ. (1997) 6660.
57. M.L. Metzeger *et al.*, Nature 387 (1997) 878-880.
58. M.L. Metzeger *et al.*, IAU Circ. (1997) 6676.
59. G.J. Fishman and C.A. Meegan, Annu. Rev. Astron. Astrophys. 33 (1995) 415-458.
60. J.I. Latorre, P. Pascual and R. Tarrach, Nucl. Phys. B437 (1995) 60-82.
61. L.J. Garay, Phys. Rev. Lett. 80 (1998) 2508.
62. S. Zakrzewski, Journ. Phys. A27 (1994) 2075.
63. S. Majid and H. Ruegg, Phys. Lett. B334 (1994) 348.
64. M.G. Baring, astro-ph/9711256, in *Towards a Major Atmospheric Cerenkov Detector, Proc. Kruger National Park TeV Workshop* (ed de Jager O.C.) (Wesprint, Potchefstroom, in press).
65. M.J. Rees astro-ph/9701162, *Proc. 18th Texas Symp. on Relativistic Astrophysics* (eds Olinto A., Friemann J. and Schramm D.N.) (World Scientific, in press).
66. P. Mészáros astro-ph/9711354, in *Gamma-Ray Bursts, Proc. 4th Huntsville Symposium* (eds Meegan C., Preece R. and Koshut T.) (AIP, in press)
67. AMS Collaboration, S. Ahlen *et al.*, Nucl. Instrum. Meth. A350 (1994) 351-367.
68. GLAST Team, E.D. Bloom *et al.*, *Proc. Intern. Heidelberg Workshop on TeV Gamma-ray Astrophysics*, eds. H.J. Volk and F.A. Aharonian (Kluwer, 1996) pp.109-125.
69. J.D. Scargle, J. Norris and J. Bonnell, astro-ph/9712016.

70. S.R. Kulkarni *et al.*, Nature 395 (1998) 663; T.J. Galama *et al.*, Nature 395 (1998) 670.
71. K.C. Walker, B.E. Schaefer and E.E. Fenimore, astro-ph/9810271.
72. S.Y. Sazonov *et al.*, Astron. Astrophys. Suppl. 129 (1998) 1.
73. H. Krawczynski *et al.*, astro-ph/9611044, Proc. Intern. School of Cosmic-Ray Astrophysics, Erice, 1996 (World Scientific, in press).
74. P.J. Boyle *et al.*, astro-ph/9706132, Proc. 25th Intern. Cosmic ray Conf., Durban, Vol.3, 61 (eds Potgieter M.S. *et al.*,) (Wesprint, Potchefstroom, 1998).
75. J. Ellis, J.L. Lopez, N.E. Mavromatos and D.V. Nanopoulos, Phys. Rev. D53 (1996) 3846.
76. G. Veneziano, Phys. Lett. B265 (1991) 287; M. Gasperini and G. Veneziano, Astropart. Phys. 1 (1993) 317.
77. R. Brustein, gr-qc/9810063; G. Veneziano, hep-th/9902097; M. Gasperini, hep-th/9907067; M. Maggiore, gr-qc/9909001.
78. I. Antoniadis, Phys. Lett. B246 (1990) 377; J. Lykken, Phys. Rev. D54 (1996) 3693; E. Witten, Nucl.Phys. B471 (1996) 135.
79. N. Arkani-Hamed, S. Dimopoulos and G. Dvali, Phys. Lett. B429 (1998) 263.
80. K.R. Dienes, E. Dudas and T. Gherghetta, Phys. Lett. B436 (1998) 55.
81. Hong-wei Yu and L.H. Ford, gr-qc/9907037
82. A. Campbell-Smith, J. Ellis, N.E. Mavromatos and D.V. Nanopoulos, hep-th/9907141.
83. J.J. Sakurai, *Modern Quantum Mechanics*, (Addison-Wesley, Reading, 1994) pp.126-129.
84. K.C. Littrell, B.E. Allman and S.A. Werner, Phys. Rev. A56 (1997) 1767.
85. D.V. Ahluwalia, Nature 398 (1999) 199.
86. G. Amelino-Camelia, gr-qc/9804063, Mod. Phys. Lett. A13 (1998) 1155; gr-qc/9808047, in *Proceedings of 7th International Colloquium on Quantum Groups and Integrable Systems*
87. D. Kabat and P. Pouliot, Phys. Rev. Lett. 77 (1996) 1004; M.R. Douglas, D. Kabat, P. Pouliot, S.H. Shenker, Nucl. Phys. B485 (1997) 85.
88. M.-T. Jaekel and S. Reynaud, Europhys. Lett. 13 (1990) 301.
89. M.-T. Jaekel and S. Reynaud, Phys. Lett. B185 (1994) 143.
90. P.G. Bergmann and G.J. Smith, Gen. Rel. Grav. 4 (1982) 1131.
91. N. Bohr and L. Rosenfeld, Kgl. Danske Videnskab S. Nat. Fys. Medd. 12 (1933) 1.
92. E.P. Wigner, Rev. Mod. Phys. 29 (1957) 255; H. Salecker and E.P. Wigner, Phys. Rev. 109 (1958) 571.
93. S.W. Hawking, Nature 248 (1974) 30.
94. W. Pauli, *Die allgemeinen Prinzipien der Wellenmechanik. Handbuch der Physik*, edited by S. Fluegge (Springer, 1958).
95. F. Markopoulou, gr-qc/9704013; F. Markopoulou and L. Smolin, Phys. Rev. D58 (1998) 084033
96. S. Majid, *Hopf algebras for physics at the Planck scale*, Class. Quantum Grav. 5 (1988) 1587.
97. S. Majid, *Non-commutative-geometric Groups by a Bicrossproduct Construction*, (PhD thesis, Harvard mathematical physics, 1988).
98. S. Majid, *On q-regularization*, Int. J. Mod. Phys. A5 (1990) 4689-4696.
99. J. Lukierski, A. Nowicki, H. Ruegg, and V.N. Tolstoy, Phys. Lett. B264 (1991) 331.
100. A. Nowicki, E. Sorace and M. Tarlini, Phys. Lett. B302 (1993) 419.
101. M. Maggiore, Phys. Lett. B319 (1993) 83-86.

102. S. Doplicher, K. Fredenhagen and J.E. Roberts. Phys. Lett. B331 (1994) 39; S. Doplicher, K. Fredenhagen and J.E. Roberts. Commun. Math. Phys. 172 (1995) 187; S. Doplicher, Annales Poincare Phys. Theor. 64 (1996) 543.
103. A. Kempf, J. Math. Phys. 35 (1994) 4483; A. Kempf, G. Mangano, and R.B. Mann, Phys. Rev. D52 (1995) 1108; A. Kempf and G. Mangano, Phys. Rev. D55 (1997) 7909.
104. G. Amelino-Camelia, J. Ellis, N.E. Mavromatos and D.V. Nanopoulos, Mod. Phys. Lett. A12 (1997) 2029.
105. S. Majid and R. Oeckl, *Twisting of quantum differentials and the Planck scale Hopf agebra*, preprint 1998, to appear Commun. Math. Phys.
106. S. Majid, *Duality principle and braided geometry*, Springer Lec. Notes in Physics 447 (1995) 125.
107. P. Kosinski, J. Lukierski and P. Maslanka, hep-th/9902037.
108. A. Ashtekar, gr-qc/9901023.
109. G. Musser, Scientific American, October 1998 issue; R. Matthews, New Scientist, 20 March 1999 issue; M. Brooks, New Scientist, 19 June 1999 issue.

Classical and Quantum Physics of Isolated Horizons: A Brief Overview

Abhay Ashtekar[1,2]

[1] Center for Gravitational Physics and Geometry
Department of Physics, The Pennsylvania State University
University Park, PA 16802, USA
[2] Institute for Theoretical Physics,
University of California, Santa Barbara, CA 93106, USA

Abstract. The arena normally used in black holes thermodynamics was recently generalized to incorporate a broad class of physically interesting situations. The key idea is to replace the notion of stationary event horizons by that of 'isolated horizons.' Unlike event horizons, isolated horizons can be located in a space-time *quasi-locally*. Furthermore, they need not be Killing horizons. In particular, a space-time representing a black hole which is itself in equilibrium, but whose exterior contains radiation, admits an isolated horizon. In spite of this generality, the zeroth and first laws of black hole mechanics extend to isolated horizons. Furthermore, by carrying out a systematic, non-perturbative quantization, one can explore the quantum geometry of isolated horizons and account for their entropy from statistical mechanical considerations. After a general introduction to black hole thermodynamics as a whole, these recent developments are briefly summarized.

1 Motivation

In the seventies, there was a flurry of activity in black hole physics which brought out an unexpected interplay between general relativity, quantum field theory and statistical mechanics [1–4]. That analysis was carried out only in the semi-classical approximation, i.e., either in the framework of Lorentzian quantum field theories in curved space-times or by keeping just the leading order, zero-loop terms in Euclidean quantum gravity. Nonetheless, since it brought together the three pillars of fundamental physics, it is widely believed that these results capture an essential aspect of the more fundamental description of Nature. For over twenty years, a concrete challenge to all candidate quantum theories of gravity has been to derive these results from first principles, without invoking semi-classical approximations.

Specifically, the early work is based on a somewhat ad-hoc mixture of classical and semi-classical ideas – reminiscent of the Bohr model of the atom – and generally ignored the quantum nature of the gravitational field itself. For example, statistical mechanical parameters were associated with macroscopic black holes as follows. The laws of black hole mechanics were first derived in the framework of *classical* general relativity, without any reference to the Planck's constant $\hbar$ [2]. It was then noted that they have a remarkable similarity with the laws of thermodynamics if one identifies a multiple of the surface gravity κ of the black hole with temperature and a corresponding multiple of the area $a_{\rm hor}$ of its horizon with entropy. However, simple dimensional considerations and thought experiments showed that the multiples must involve $\hbar$, making quantum considerations indispensable for a fundamental understanding of the relation between black hole mechanics and thermodynamics [1]. Subsequently, Hawking's investigation of (test) quantum fields propagating on a black hole geometry showed that black holes emit thermal radiation at temperature $T_{\rm rad} = \hbar\kappa/2\pi$ [3]. It therefore seemed natural to assume that black holes themselves are hot and their temperature $T_{\rm bh}$ is the same as $T_{\rm rad}$. The similarity between the two sets of laws then naturally suggested that one associate an entropy $S_{\rm bh} = a_{\rm hor}/4\hbar$ with a black hole of area $a_{\rm hor}$. While this procedure seems very reasonable, it does not provide a 'fundamental derivation' of the thermodynamic parameters $T_{\rm bh}$ and $S_{\rm bh}$. The challenge is to derive these formulas from first principles, i.e., by regarding large black holes as statistical mechanical systems in a suitable quantum gravity framework.

Recall the situation in familiar statistical mechanical systems such as a gas, a magnet or a black body. To calculate their thermodynamic parameters such as entropy, one has to first identify the elementary building blocks that constitute the system. For a gas, these are molecules; for a magnet, elementary spins; for the radiation field in a black body, photons. What are the analogous building blocks for black holes? They can not be gravitons because the underlying space-times were assumed to be stationary. Therefore, the elementary constituents must be non-perturbative in the field theoretic sense. Thus, to account for entropy from first principles within a candidate quantum gravity theory, one would have to: i) isolate these constituents; ii) show that, for large black holes, the number of quantum states of these constituents goes as the exponential of the area of the event horizon; and, iii) account for the Hawking radiation in terms of processes involving these constituents and matter quanta.

These are difficult tasks, particularly because the very first step –isolating the relevant constituents– requires new conceptual as well as mathematical inputs. Furthermore, in the semi-classical theory, thermodynamic properties have been associated not only with black holes but also with cosmological horizons. Therefore, ideally, the framework has to be sufficiently general to encompass these diverse situations. It is only recently, more than twenty years after the initial flurry of activity, that detailed proposals have emerged. The more well-known of these comes from string theory [27] where the relevant elementary constituents are associated with D-branes which lie outside the original perturbative sector of the theory. The purpose of this contribution is to summarize the ideas and results from another approach which emphasizes the quantum nature of geometry, using non-perturbative techniques from the very beginning. Here, the elementary constituents are the quantum excitations of geometry itself and the Hawking process now corresponds to the conversion of the quanta of geometry to quanta of matter. Although the two approaches seem to be strikingly different from one another, as I will indicate, in a certain sense they are complementary.

2 Key Issues

In the last section, I focussed on quantum issues. However, the status of *classical* black hole mechanics, which provided much of the inspiration in quantum considerations, has itself remained unsatisfactory in some ways. Therefore, in a systematic approach, one has to revisit the classical theory before embarking on quantization.

The zeroth and first laws of black hole mechanics refer to equilibrium situations and small departures therefrom. Therefore, in this context, it is natural to focus on isolated black holes. However, in standard treatments, these are generally represented by *stationary* solutions of field equations, i.e, solutions which admit a time-translation Killing vector field *everywhere*, not just in a small neighborhood of the black hole. While this simple idealization is a natural starting point, it seems to be overly restrictive. Physically, it should be sufficient to impose boundary conditions at the horizon which ensure *only the black hole itself is isolated.* That is, it should suffice to demand only that the intrinsic geometry of the horizon be time independent, whereas the geometry outside may be dynamical and admit gravitational and other radiation. Indeed, we adopt a similar viewpoint in ordinary thermodynamics; in the standard description of equilibrium configurations of systems such as a classical gas, one usually assumes that only the system under consideration is in equilibrium and stationary, not the whole world. For black holes, in realistic situations one is typically interested in the final stages of collapse where the black hole is formed and has 'settled down' or in situations in which an already formed black hole is isolated for the duration of the experiment (see figure 1). In such situations, there is likely to be gravitational radiation and non-stationary matter far away from the black hole, whence the space-time as a whole is not expected to be stationary. Surely, black hole mechanics should incorporate in such situations.

A second limitation of the standard framework lies in its dependence on *event* horizons which can only be constructed retroactively, after knowing the *complete* evolution of space-time. Consider for example, Figure 2 in which a spherical star of mass M undergoes a gravitational collapse. The singularity is hidden inside the null surface Δ_1 at $r = 2M$ which is foliated by a family of marginally trapped surfaces and would be a part of the event horizon if nothing further happens. Suppose instead, after a very long time, a thin spherical shell of mass δM collapses. Then Δ_1 would not be a part of the event horizon which would actually lie slightly outside Δ_1 and coincide with the surface $r = 2(M + \delta M)$ in distant future. On physical grounds, it seems unreasonable to exclude Δ_1 a priori from thermodynamical considerations. Surely one should be able to establish the standard laws of laws of mechanics not only for the event horizon but also for Δ_1.

Another example is provided by cosmological horizons in de Sitter space-time [4]. In this case, there are no singularities or black-hole event horizons. On the other hand, semi-classical considerations enable one to assign entropy and temperature to these horizons as well. This suggests the notion of event horizons is too restrictive for thermodynamical analogies. We will see that this

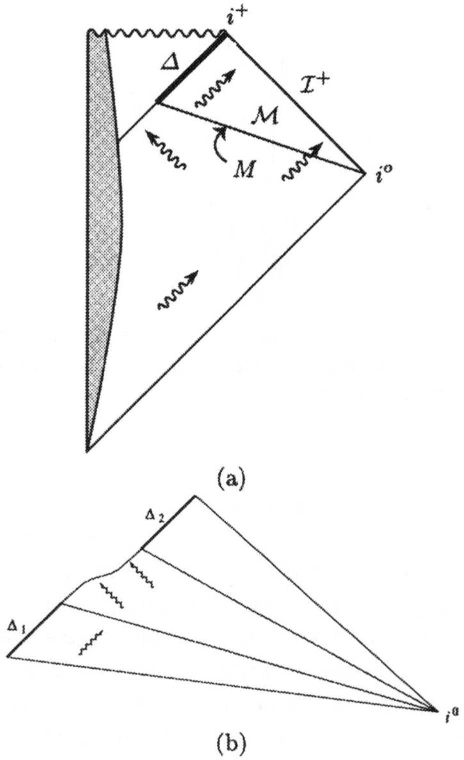

Fig. 1. (a) A typical gravitational collapse. The portion Δ of the horizon at late times is isolated. The space-time $\mathcal{M}$ of interest is the triangular region bounded by Δ, $\mathcal{I}^+$ and a partial Cauchy slice M. (b) Space-time diagram of a black hole which is initially in equilibrium, absorbs a small amount of radiation, and again settles down to equilibrium. Portions Δ_1 and Δ_2 of the horizon are isolated.

is indeed the case; as far as equilibrium properties are concerned, the notion of event horizons can be replaced by a more general, quasi-local notion of 'isolated horizons' for which the familiar laws continue to hold. The surface Δ_1 in figure 2 as well as the cosmological horizons in de Sitter space-times are examples of isolated horizons.

At first sight, it may appear that only a small extension of the standard framework, based on stationary event horizons, is needed to overcome the limitations discussed above. However, this is not the case. For example, in the stationary context, one identifies the black-hole mass with the ADM mass defined at spatial infinity. In the presence of radiation, this simple strategy is no longer viable since radiation fields well outside the horizon also contribute to the ADM mass. Hence, to formulate the first law, a new definition of the black hole mass is needed. Similarly, in the absence of a global Killing field, the notion of surface gravity has to be extended in a non-trivial fashion. Indeed, even if space-time happens to be static in a neighborhood of the horizon —already a

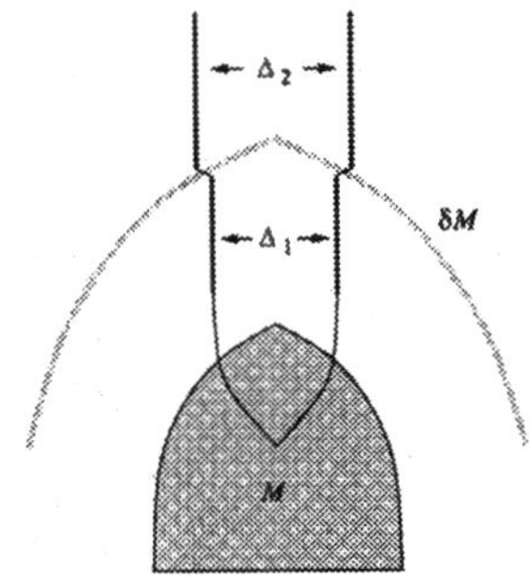

Fig. 2. A spherical star of mass M undergoes collapse. Later, a spherical shell of mass δM falls into the resulting black hole. While Δ_1 and Δ_2 are both isolated horizons, only Δ_2 is part of the event horizon.

stronger condition than contemplated above— the notion of surface gravity is ambiguous because the standard expression fails to be invariant under constant rescalings of the Killing field. When a *global* Killing field exists, the ambiguity is removed by requiring the Killing field be unit at *infinity*. Thus, contrary to intuitive expectation, the standard notion of the surface gravity of a stationary black hole refers not just to the structure at the horizon, but also to infinity. This 'normalization problem' in the definition of the surface gravity seems especially difficult in the case of cosmological horizons in (Lorentzian) space-times whose Cauchy surfaces are compact. Apart from these conceptual problems, a host of technical issues must also be resolved. In Einstein-Maxwell theory, the space of stationary black hole solutions is three dimensional whereas the space of solutions admitting isolated horizons is *infinite*-dimensional since these solutions admit radiation near infinity. As a result, new techniques have to be used and these involve some functional analytic subtleties.

This set of issues has a direct bearing on quantization as well. For, in a systematic approach, one would first extract an appropriate sector of the theory in which space-time geometries satisfy suitable conditions at interior boundaries representing horizons, then introduce a well-defined action principle tailored to these boundary conditions, and, finally, use the resulting Lagrangian or Hamiltonian frameworks as points of departure for constructing the quantum theory. If one insists on using *event* horizons, these steps are difficult to carry out because the resulting boundary conditions do not translate in to (quasi-)local restrictions on fields. Indeed, for event horizon boundaries, there is *no* action principle available in the literature. The restriction to *globally* stationary space-times causes additional difficulties. For, by no hair theorems, the space of stationary solutions admitting event horizons is finite dimensional and quantization of this 'mini-superspace' would ignore all field theoretic effects *by fiat*. Indeed, most treatments of black hole mechanics are based on differential geometric identities and field equations, and are not at all concerned with such issues related to quantization.

Thus, the first challenge is to find a new framework which achieves, in a single stroke, three goals: i) it overcomes the two limitations of black hole mechanics by finding a better substitute for stationary event horizons; ii) generalizes laws of black hole mechanics to the new, more physical paradigm; and, iii) leads to a well-defined action principle and Hamiltonian framework which can serve as springboards for quantization. The second challenge is then to: i) carry out quantization non-perturbatively; ii) obtain a quantum description of the horizon geometry; and, iii) account for the the horizon entropy statistical mechanically by counting the underlying micro-states. As discussed in the next section, these goals have been met for non-rotating isolated horizons.

3 Summary

In this section, I will sketch the main ideas and results on the classical and quantum physics of isolated horizons and provide a guide to the literature where details can be found.

3.1 Isolated horizons

The detailed boundary conditions defining non-rotating isolated horizons were introduced in [10,12]. Basically, an isolated horizon Δ is a null 3-surface, topologically $S^2 \times R$, foliated by a family of marginally trapped 2-spheres. Denote the normal direction field to Δ by $[\ell^a]$. Being null, it is also tangential to Δ. The boundary conditions require that it be expansion-free, so that the area of the marginally trapped surface remains constant 'in time'. Assuming that the matter fields under consideration satisfy a very weak 'energy condition' at Δ, the Raychaudhuri equation then implies that there is no flux of matter across Δ. More detailed analysis also shows that there is no flux of gravitational radiation. (More precisely, the Newman-Penrose curvature component Ψ_0 vanishes on Δ.) These properties capture the idea that the horizon is isolated. Denote the second null normal to the family of marginally trapped 2-spheres by $[n^a]$. There are additional conditions on the Newman-Penrose spin coefficients associated with $[n^a]$ which ensure that Δ is a *future* horizon with no rotation.

Event horizons of static black holes of the Einstein-Maxwell-Dilaton theory are particular examples of non-rotating isolated horizons. The cosmological horizons in de Sitter space-time provide other examples. However, there are many other examples as well; the space of solutions admitting isolated horizons is in fact *infinite* dimensional [14,12].

All conditions in the definition are *local* to Δ whence the isolated horizon can be located quasi-locally; unlike the event horizon, one does not have to know the *entire* space-time to determine whether or not a given null surface is an isolated horizon. Also, there may be gravitational or other radiation arbitrarily close to Δ. Therefore, in general, space-times admitting isolated horizons need not be stationary *even in a neighborhood of* Δ; isolated horizons need not be Killing horizons [14]. In spite of this generality, the intrinsic geometry, several of the curvature components and several components of the Maxwell field at any isolated horizon are the same as those at the event horizon of Reissner-Nordström space-times [12,10,13]. This similarity greatly simplifies the detailed analysis.

Finally, isolated horizons are special cases of Hayward's trapping horizons [18], the most important restriction being that the direction field $[\ell^a]$ is assumed to be expansion-free. Physically, as explained above, this restriction captures the idea that the horizon is 'isolated', i.e., we are dealing with an equilibrium situation. The restriction also gives rise to some mathematical simplifications which, in turn, make it possible to introduce a well-defined action principle and Hamiltonian framework. As we will see below, these structures play an essential role in the proof of the generalized first law and in passage to quantization.

3.2 Mechanics

Let me begin by placing the present work on mechanics of isolated horizons in the context of other treatments in the literature. The first treatments of the zeroth and first laws were given by Bardeen, Carter and Hawking [2] for black holes surrounded by rings of perfect fluid and this treatment was subsequently generalized to include other matter sources [5]. In all these works, one restricted oneself to globally stationary space-times admitting event horizons and considered transitions from one such space-time to a nearby one. Another approach, based on Noether charges, was introduced by Wald and collaborators [19,6]. Here, one again considers stationary event horizons but allows the variations to be arbitrary. Furthermore, this method is applicable not only for general relativity but for stationary black holes in a large class of theories. In both approaches, the surface gravity κ and the mass M of the hole were defined using the global Killing field and referred to structure at infinity.

The zeroth and first laws were generalized to arbitrary, non-rotating isolated horizons Δ in the Einstein-Maxwell theory in [11,12] and dilatonic couplings were incorporated in [13]. In this work, the surface gravity κ and the mass M_Δ of the isolated horizon refer only to structures *local* to Δ.[1] As mentioned in section 3.1, the space $\mathcal{IH}$ of solutions admitting isolated horizons is infinite dimensional and static solutions constitute only a finite dimensional sub-space $\mathcal{S}$ of $\mathcal{IH}$. Let us restrict ourselves to the non-rotating case for comparison. Then, in treatments based on the Bardeen-Carter-Hawking approach, one restricts oneself only to $\mathcal{S}$ and variations tangential to $\mathcal{S}$. In the Wald approach, one again restricts oneself to points of $\mathcal{S}$ but the variations need not be tangential to $\mathcal{S}$. In the present approach, on the other hand, the laws hold at *any* point of $\mathcal{IH}$ and *any* tangent

[1] In standard treatments, static solutions are parametrized by the ADM mass M, electric and magnetic charges Q and P, dilatonic charge D, cosmological constant Λ and the dilatonic coupling parameter α. Of these, M and D are defined *at infinity*. In the generalized context of isolated horizons, on the other hand, one must use parameters that are intrinsic to Δ. Apriori, it is not obvious that this can be done. It turns out that we can trade M with the area a_Δ of the horizon and D with the value ϕ_Δ of the dilaton field on Δ. Boundary conditions ensure that ϕ_Δ is a constant.

vector at that point. However, so far, our results pertain only to *non-rotating* horizons in a restricted class of theories.

The key ideas in the present work can be summarized as follows. It is clear from the setup that surface gravity should be related to the acceleration of $[\ell^a]$. Recall, however, the acceleration is not a property of a direction field but of a vector field. Therefore, to define surface gravity, we must pick out a specific vector field ℓ^a from the equivalence class $[\ell^a]$. Now, the shear, the twist, and the expansion of the direction field $[\ell^a]$ all vanish for *any* choice of normalization. Therefore, we can not use these fields to pick out a preferred ℓ^a. However, it turns out that the expansion $\Theta_{(n)}$ of n^a *is* sensitive to its normalization. Furthermore, in static solutions, $\Theta_{(n)}$ is determined entirely by the intrinsic parameters of the horizon. Therefore, it is natural to require that $\Theta_{(n)}$ be the same function of the parameters on *any* isolated horizon. Although it is not apriori obvious, the available rescaling freedom in fact suffices to meet this requirement on *any* isolated horizon. Furthermore, the condition *uniquely* picks out a vector field n^a from the equivalence class $[n^a]$. Having a preferred n^a at our disposal, using the standard normalization $\ell \cdot n = -1$ we can then select an ℓ^a from the equivalence class $[\ell^a]$ uniquely. Finally, we define surface gravity κ to be the acceleration of this 'properly normalized' ℓ^a; i.e., we set $\ell^a \nabla_a \ell^b = \kappa \ell^b$ On Δ.

By construction, κ, so defined, yields the 'correct' surface gravity in the six parameter family of static, dilatonic black-holes. However, the key question is: Do the zeroth and first laws hold for general isolated horizons? This is a key test of our strategy of defining κ in the general case. The answer is in the affirmative.

The zeroth law –constancy of κ on isolated horizons– is established as follows. First, our boundary conditions on $[\ell^a]$ and $[n^a]$ directly imply that κ is constant on each trapped 2-surface. Next, one can show that κ can be expressed in terms of the Weyl curvature component Ψ_2 and the expansion $\Theta_{(n)}$. Finally, the Bianchi identity $\nabla_{[a} R_{bc]de} = 0$, the form of the Ricci tensor component Φ_{11} dictated by our boundary conditions on the matter stress-energy, and our 'normalization condition' on $\Theta_{(n)}$ imply that κ is also constant along the integral curves of ℓ^a. Hence κ is constant on any isolated horizon. To summarize, even though our boundary conditions allow for the presence of radiation arbitrarily close to Δ, they successfully extract enough structure intrinsic to the horizons of static black holes to ensure the validity of the zeroth law. Our derivation brings out the fact that the zeroth law is really local to the horizon: Degrees of freedom of the isolated horizon 'decouple' from excitations present elsewhere in space-time.

To establish the first law, one must first introduce the notion of mass M_Δ of the isolated horizon. The idea is to define M_Δ using the Hamiltonian framework. For this, one needs a well-defined action principle. Fortunately, even though the boundary conditions were designed only to capture the notion of an isolated horizon in a quasi-local fashion, they turn out to be well-suited for the variational principle. However, just as one must add a suitable boundary term at infinity to the Einstein-Hilbert action to make it differentiable in the asymptotically flat context, we must now add another boundary term at Δ. Somewhat surprisingly, the new boundary term turns out to be the well-known Chern-Simons action (for the self-dual connection). This specific form is not important to classical considerations. However, it plays a key role in the quantization procedure. The boundary term at Δ is different from that at infinity. Therefore one can not simultaneously absorb both terms in the bulk integral using Stokes' theorem. Finally, to obtain a well-defined variational principle for the Maxwell part of the action, one needs a partial gauge fixing at Δ. One can follow a procedure similar to the one given above for fixing the rescaling freedom in n^a and ℓ^a. It turns out that, not only does this strategy make the Maxwell action differentiable, but it also uniquely fixes the scalar potential Φ at the horizon.

Having the action at one's disposal, one can pass to the Hamiltonian framework.[2] Now, it turns out that the symplectic structure has, in addition to the standard bulk term, a surface term at Δ. The surface term is inherited from the Chern-Simons term in the action and is therefore precisely the Chern-Simons symplectic structure with a specific coefficient (i.e., in the language of the Chern-Simons theory, a specific value of the 'level' k). The presence of a surface term in the symplectic structure is somewhat unusual; for example, the boundary term at infinity in the action does *not* induce a boundary term in the symplectic structure.

The Hamiltonian consists of a bulk integral and two surface integrals, one at infinity and one at Δ. The presence of two surface integrals is not surprising; for example one encounters it even in the absence of an internal boundary, if the space-times under consideration have two asymptotic regions. As usual, the bulk term is a linear combination of constraints and the boundary term at infinity is the ADM energy. Using several examples as motivation, we interpret the surface integral at horizon as the horizon mass M_Δ [12]. This interpretation is supported by the following result: If the isolated horizon extends to future time-like infinity i^+, under suitable assumptions one can show that M_Δ is equal to the future limit, along $\mathcal{I}^+$, of the Bondi mass. Finally, note that M_Δ is *not* a fundamental, independent attribute of the isolated horizon; it is a function of the area a_Δ and charges Q_Δ, P_Δ which are regarded as the fundamental parameters.

Thus, we can now assign to any isolated horizon, an area a_Δ, a surface gravity κ, an electric potential Φ and a mass M_Δ. The electric charge Q_Δ can be defined using the electro-magnetic and dilatonic fields field *at* Δ [13]. All quantities are defined in terms of the local structure at Δ. Therefore, one can now ask: if one moves from *any* space-time in $\mathcal{IH}$ to *any* nearby space-time through a variation δ, how do these quantities vary? An explicit calculation shows:

$$\delta M_\Delta = \frac{1}{8\pi G}\, \kappa \delta a_\Delta + \Phi \delta Q_\Delta \,.$$

Thus, the first law of black hole mechanics naturally generalizes to isolated horizons. (As usual, the magnetic charge can be incorporated via the standard duality rotation.) This result provides additional support for our strategy of defining κ, Φ and M_Δ.

[2] This passage turns out not to be as straightforward as one might have imagined because there are subtle differences between the variational principles that lead to the Lagrangian and Hamiltonian equations of motion. See [12].

In static space-times, the mass M_Δ of the isolated horizon coincides with the ADM mass M defined at infinity. In general, M_Δ is the difference between M and the 'radiative energy' of space-time. However, as in the static case, M_Δ continues to include the energy in the 'Coulombic' fields —i.e., the 'hair'— associated with the charges of the horizon, even though it is defined locally at Δ. This is a subtle property but absolutely essential if the first law is to hold in the form given above. To my knowledge, none of the quasi-local definitions of mass shares this property with M_Δ. Finally, isolated horizons provide an appropriate framework for discussing the 'physical process version' of the first law for processes in which the charge of the black hole changes. The standard strategy of using the ADM mass in place of M_Δ appears to run in to difficulties [12] and, as far as I am aware, this issue was never discussed in the literature in the usual context of context of static event horizons.

3.3 Quantum geometry in the bulk

In this sub-section, I will make a detour to introduce the basic ideas we need from quantum geometry. For simplicity, I will ignore the presence of boundaries and focus just on the structure in the bulk.

There is a common expectation that the continuum picture of space-time, used in macroscopic physics, would break down at the Planck scale. This expectation has been shown to be correct within a non-perturbative, background independent approach to quantum gravity (see [7] and references therein).[3] The approach is background independent in the sense that, at the fundamental level, there is neither a classical metric nor any other field to perturb around. One only has a bare manifold and *all* fields, whether they represent geometry or matter, are quantum mechanical from the beginning. Because of the subject matter now under consideration, I will focus on geometry.

Quantum mechanics of geometry has been developed systematically over the last three years and further exploration continues [7]. The emerging theory is expected to play the same role in quantum gravity that differential geometry plays in classical gravity. That is, quantum geometry is not tied to a specific gravitational theory. Rather, it provides a kinematic framework or a language to formulate dynamics in a large class of theories, including general relativity and supergravity. In this framework, the fundamental excitations of gravity/geometry are one-dimensional, rather like 'polymers' and the continuum picture arises only as an approximation involving coarse-graining on semi-classical states. The one dimensional excitations can be thought of as flux lines of area [21]. Roughly, each line assigns to a surface element it crosses one Planck unit of area. More

[3] The necessity of a non-perturbative approach is illustrated by the following simple example. The energy levels of a harmonic oscillator are discrete. However, it would be difficult to see this fundamental discreteness if one were to solve the problem perturbatively, starting from the Hamiltonian of a free particle. Similarly, if one *begins* with a continuum background geometry and then tries to incorporate the quantum effects perturbatively, it would be difficult to unravel discreteness in the spectra of geometric operators such as areas of surfaces or volumes of regions.

precisely, the area assigned to a surface is obtained by algebraic operations (involving group-representation theory) at points where the flux lines intersect the surface. As is usual in quantum mechanics, quantum states of geometry are represented by elements of a Hilbert space [20]. I will denote it by $\mathcal{H}_{\text{bulk}}$. The basic object for spatial Riemannian geometry continues to be the triad, but now represented by an operator(-valued distribution) on $\mathcal{H}_{\text{bulk}}$ [21]. All other geometric quantities —such as areas of surfaces and volumes of regions— are constructed from the triad and represented by self-adjoint operators on $\mathcal{H}_{\text{bulk}}$. The eigenvalues of all geometric operators are discrete; geometry is thus quantized in the same sense that the energy and angular momentum of the hydrogen atom are quantized [21].

There is however, one subtlety: there is a one-parameter ambiguity in this non-perturbative quantization [22]. The parameter is positive, labeled γ and called the Immirzi parameter. This ambiguity is similar to the θ ambiguity in the quantization of Yang-Mills theories. For all values of γ, one obtains the same classical theory, expressed in different canonical variables. However, quantization leads to a one-parameter family of *inequivalent* representations of the basic operator algebra. In particular, in the sector labeled by γ the spectra of the triad —and hence, all geometric— operators depend on γ through an overall multiplicative factor. Therefore, while the qualitative features of quantum geometry are the same in *all* γ sectors, the precise eigenvalues of geometric operators vary from one sector to another. The γ-dependence itself is simple —effectively, Newton's constant G is replaced by γG in the γ-sector. Nonetheless, to obtain unique predictions, it must be eliminated and this requires an additional input. Note however that since the ambiguity involves a single parameter, as with the θ ambiguity in QCD, one judiciously chosen experiment would suffice to eliminate it. Thus, for example, if we could measure the quantum of area , i.e., smallest non-zero value that area of any surface can have, we would know which value of γ is realized in Nature. Any further experiment would then be a test of the theory. Of course, it is not obvious how to devise a feasible experiment to measure the area quantum directly. However, we will see that it is possible to use black hole thermodynamics to introduce suitable thought experiments. One of them can determine the value of γ and the other can then serve as consistency checks.

3.4 Quantum geometry of horizon and entropy

Ideas introduced in the last three sub-sections were combined and further developed to systematically analyze the quantum geometry of isolated horizons and calculate their statistical mechanical entropy in [10,15,16]. (For earlier work, see [23,24].) In this discussion, one is interested in space-times with an isolated horizon with *fixed* values a_o, Q_o and ϕ_o of the intrinsic horizon parameters, the area, the electric charge, and the value of the dilaton field.

The presence of an isolated horizon Δ manifests itself in the classical theory through boundary conditions. As usual, we can use some of the boundary conditions to eliminate certain gauge degrees of freedom at Δ. The remaining, 'true' degree of freedom are coded in an Abelian connection V defined intrinsically on

Δ. V is constructed from the self-dual spin connection in the bulk. It is interesting to note that there are *no* surface degrees of freedom associated with matter: Given the intrinsic parameters of the horizon, boundary conditions imply that matter fields defined intrinsically on Δ can be completely expressed in terms of geometrical (i.e., gravitational) fields at Δ. One can also see this feature in the symplectic structure. While the gravitational symplectic structure acquires a surface term at Δ, matter symplectic structures do not. We will see that this fact provides a simple explanation of the fact that, among the set of intrinsic parameters natural to isolated horizons, entropy depends only on area.

Of particular interest to the present Hamiltonian approach is the pull-back of V to the 2-sphere S_Δ (orthogonal to ℓ^a and n^a) at which the space-like 3-surfaces M used in the phase space construction intersect Δ. (See figure 1(a).) This pull-back —which I will also denote by V for simplicity— is precisely the $U(1)$ spin-connection of the 2-sphere S_Δ. Not surprisingly, the Chern-Simons symplectic structure for the non-Abelian self-dual connection that I referred to in Section 3.2 can be re-expressed in terms of V. The result is unexpectedly simple [10]: the surface term in the total symplectic structure is now just the Chern-Simons symplectic structure for the *Abelian* connection V! The only remaining boundary condition relates the curvature $F = dV$ of V to the triad vectors. This condition is taken over as an operator equation. Thus, in the quantum theory, neither the intrinsic geometry nor the curvature of the horizon are frozen; neither is a classical field. Each is allowed to undergo quantum fluctuations but because of the operator equation relating them, they have to fluctuate in tandem.

To obtain the quantum description in presence of isolated horizons, therefore, one begins with a fiducial Hilbert space $\mathcal{H} = \mathcal{H}_{\text{bulk}} \otimes \mathcal{H}_{\text{surface}}$ where $\mathcal{H}_{\text{bulk}}$ is the Hilbert space associated with the bulk polymer geometry and $\mathcal{H}_{\text{surface}}$ is the Chern-Simons Hilbert space for the connection V.[4] The quantum boundary condition says that only those states in $\mathcal{H}$ are allowed for which there is a precise intertwining between the bulk and the surface parts. However, because the required intertwining is 'rigid', apriori it is not clear that the quantum boundary conditions would admit *any* solutions at all. For solutions to exist, there has to be a very delicate matching between certain quantities on $\mathcal{H}_{\text{bulk}}$ calculated from the bulk quantum geometry and certain quantities on $\mathcal{H}_{\text{surface}}$ calculated from the Chern-Simons theory. The precise numerical coefficients in the surface calculation depend on the numerical factor in front of the surface term in the symplectic structure (i.e., on the Chern-Simons level k) which is itself determined in the classical theory by the coefficient in front of the Einstein-Hilbert action and our classical boundary conditions. Thus, the existence of a coherent quantum theory of isolated horizons requires that the three corner stones —classical general relativity, quantum mechanics of geometry and Chern-Simons theory—

[4] In the classical theory, all fields are smooth, whence the value of any field in the bulk determines its value on Δ by continuity. In quantum theory, by contrast, the measure is concentrated on generalized fields which can be arbitrarily discontinuous, whence surface states are no longer determined by bulk states. A compatibility relation does exist but it is introduced by the quantum boundary condition. It ensures that the total state is invariant under the permissible internal rotations of triads.

be united harmoniously. Not only should the three conceptual frameworks fit together seamlessly but certain *numerical coefficients*, calculated independently within each framework, have to match delicately. Fortunately, these delicate constraints are met and there the quantum boundary conditions admit a sufficient number of solutions.

Because we have fixed the intrinsic horizon parameters, is is natural to construct a microcanoniocal ensemble from eigenstates of the corresponding operators with eigenvalues in the range $(q_o - \delta q, q_o + \delta q)$ where δq is very small compared to the fixed value q_o of the intrinsic parameters. Since there are no surface degrees of freedom associated with matter fields, let us focus on area, the only gravitational parameter available to us. Then, we only have to consider those states in $\mathcal{H}_{\mathrm{bulk}}$ whose polymer excitations intersect S_Δ in such a way that they endow it with an area in the range $(a_o - \delta a, a_o + \delta a)$ where δa is of the order of ℓ_{Pl}^2 (with ℓ_{Pl}, the Planck length). Denote by $\mathcal{P}$ the set of punctures that any one of these polymer states makes on S_Δ, each puncture being labeled by the eigenvalue of the area operator at that puncture. Given such a bulk state, the quantum boundary condition tells us that only those Chern-Simons surface states are allowed for which the curvature is concentrated at punctures and the range of allowed value of the curvature at each puncture is dictated by the area eigenvalue at that puncture. Thus, for each $\mathcal{P}$, the quantum boundary condition picks out a sub-space $\mathcal{H}^{\mathcal{P}}_{\mathrm{surface}}$ of the surface Hilbert space $\mathcal{H}_{\mathrm{surface}}$. Thus, the quantum geometry of the isolated horizon is effectively described by states in

$$\mathcal{H}^{\mathrm{phys}}_{\mathrm{surface}} = \bigoplus_{\mathcal{P}} \mathcal{H}^{\mathcal{P}}_{\mathrm{surface}}$$

as $\mathcal{P}$ runs over all possible punctures and area-labels at each puncture, compatible with the requirement that the total area assigned to S_Δ lie in the given range.

One can visualize this quantum geometry as follows. Given any one state in $\mathcal{H}^{\mathcal{P}}_{\mathrm{surface}}$, the connections V are flat everywhere except at the punctures and the holonomy around each puncture is fixed. Using the classical interpretation of V as the metric compatible spin connection on S_Δ we conclude that, in quantum theory, the intrinsic geometry of the horizon is flat except at the punctures. At each puncture, there is a deficit angle, whose value is determined by the holonomy of V around that puncture. Since each puncture corresponds to a polymer excitation in the bulk, polymer lines can be thought of as 'pulling' on the horizon, thereby producing deficit angles in an otherwise flat geometry (see figure 3). Each deficit angle is quantized and the angles add up to 2π as in a discretized model of a 2-sphere geometry. Thus, the quantum geometry of an isolated horizon is quite different from its smooth classical geometry. In addition, of course, each polymer line endows the horizon with a small amount of area and these area elements add up to provide the horizon with total area in the range $(a_0 - \delta a, a_o + \delta a)$. Thus, one can intuitively picture the quantum horizon as the surface of a large, water-filled balloon which is suspended with a very large number of wires, each exerting a small tug on the surface at the point of contact and giving rise to a 'conical singularity' in the geometry.

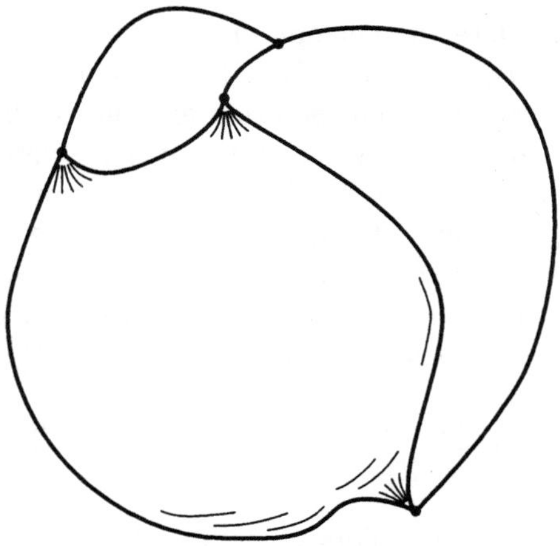

(a)

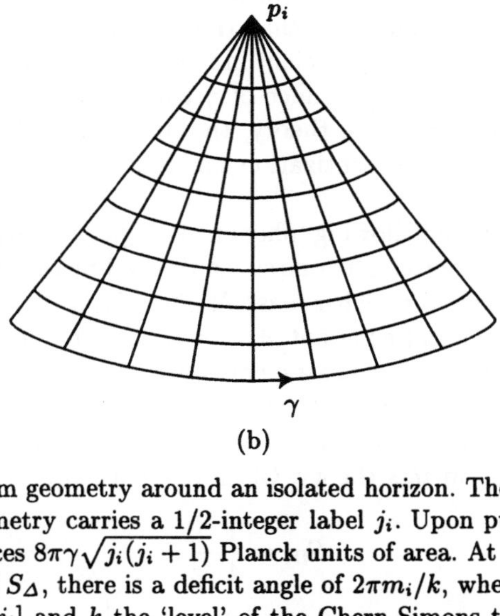

(b)

Fig. 3. (a) Quantum geometry around an isolated horizon. The i-th polymer excitation of the bulk geometry carries a 1/2-integer label j_i. Upon puncturing the horizon 2-sphere S_Δ, it induces $8\pi\gamma\sqrt{j_i(j_i+1)}$ Planck units of area. At each puncture, in the intrinsic geometry of S_Δ, there is a deficit angle of $2\pi m_i/k$, where m_i is a 1/2-integer in the interval $[-j_i, j_i]$ and k the 'level' of the Chern-Simons theory. (b) Magnified view of a puncture p_i. The holonomy of the $U(1)$ connection V around a loop γ surrounding any puncture p_i determines the deficit angle at p_i. Each deficit angle is quantized and they add up to 2π.

Finally, one can calculate the entropy of the quantum micro-canonical ensemble. We are not interested in the *full* Hilbert space since the 'bulk-part' includes, e.g., states of gravitational radiation and matter fields far away from Δ. Rather, we wish to consider only the states of the isolated horizon Δ itself. Therefore, we are led to trace over the 'bulk states' to construct a density matrix $\rho_{\rm IH}$ describing a maximum-entropy mixture of surface states for which the intrinsic parameters lie in the given range. The statistical mechanical entropy is then given by $S = -{\rm Tr}\, \rho_{\rm IH} \ln \rho_{\rm IH}$. As usual, the trace can be obtained simply by counting states, i.e., by computing the dimension $\mathcal{N}$ of $\mathcal{H}^{\rm phys}_{\rm surface}$. We have:

$$\mathcal{N} = \exp\left(\frac{\gamma_o}{\gamma}\frac{a_o}{4\ell_{\rm Pl}^2}\right) \qquad \text{where} \qquad \gamma_o = \frac{\ln 2}{\pi\sqrt{3}}$$

Thus, the number of micro-states does go exponentially as area. This is a non-trivial result. For example if, as in the early treatments, one ignores boundary conditions and the Chern-Simons term in the symplectic structure and does a simple minded counting, one finds that the exponent in $\mathcal{N}$ is proportional to $\sqrt{a_o}$. However, our numerical coefficient in front of the exponent depends on the Immirzi parameter γ. The appearance of γ can be traced back directly to the fact that, in the γ-sector of the theory, the area eigenvalues are proportional to γ. Thus, because of the quantization ambiguity, the γ-dependence of $\mathcal{N}$ is inevitable.

We can now adopt the following 'phenomenological' viewpoint. In the infinite dimensional space $\mathcal{IH}$, one can fix one space-time admitting isolated horizon, say the Schwarzschild space-time with mass $M_o >> M_{\rm Pl}$, (or, the de Sitter space-time with the cosmological constant $\Lambda_o << 1/\ell_{\rm Pl}^2$). For agreement with semi-classical considerations, in these cases, entropy should be given by $S = (a_o/4\ell_{\rm Pl}^2)$ which can happen only in the sector $\gamma = \gamma_o$ of the theory. The theory is now completely determined and we can go ahead and calculate the entropy of any other isolated horizon in *this* theory. Clearly, we obtain:

$$S_{\rm IH} = \frac{1}{4}\frac{a_o}{\ell_{\rm Pl}^2}$$

for *all* isolated horizons. Furthermore, in this γ-sector, the statistical mechanical temperature of any isolated horizon is given by Hawking's semi-classical value $\kappa\hbar/2\pi$ [8,23]. Thus, we can do one thought experiment —observe the temperature of a large black black hole from far away— to eliminate the Immirzi ambuguity and fix the theory. This theory then predicts the correct entropy and temperature for all isolated horizons in $\mathcal{IH}$ with $a_o >> \ell_{\rm Pl}^2$.

The technical reason behind this univerality is trivial. However, the conceptual argument is not because it is quite non-trivial that $\mathcal{N}$ depends only on the area and not on values of other charges. Furthermore, the space $\mathcal{IH}$ is infinite dimensional and it is not apriori obvious that one should be able to give a statistical mechanical account of entropy of *all* isolated horizons in one go. Indeed, values of fields such as Ψ_4 and ϕ_2 can be vary from one isolated horizon to another even when they have same intrinsic parameters. This freedom could well have introduced obstructions, making quantization and entropy calculation impossible. That this does not happen is related to but independent of the fact that this feature did not prevent us from extending the laws of mechanics from static event horizons to general isolated horizons.

4 Discussion

Perhaps the most pleasing aspect of this analysis is the existence of a single framework to encompass diverse ideas at the interface of general relativity, quantum theory and statistical mechanics. In the classical domain, this framework generalizes laws of black hole mechanics to physically more realistic situations. At the quantum level, it provides a detailed description of the quantum geometry of horizons and leads to a statistical mechanical calculation of entropy. In both domains, the notion of isolated horizons provides an unifying arena enabling us to handle different types of situations —e.g., black holes and cosmological horizons— in a single stroke. In the classical theory, the same line of reasoning allows one to establish the zeroth and first laws for *all* isolated horizons. Similarly, in the quantum theory, a single procedure leads one to quantum geometry and entropy of *all* isolated horizons. By contrast, in other approaches, fully quantum mechanical treatments seem to be available only for stationary black holes. Indeed, to my knowledge, even in the static case, a complete statistical mechanical calculation of the entropy of cosmological horizons has not been available. Finally, our extension of the standard *Killing* horizon framework sheds new light on a number of issues, particularly the notion of mass of associated to an horizon and the physical process version of the first law [12].

However, the framework presented here is far from being complete and provides promising avenues for future work. First, while some of the motivation behind our approach is similar to the considerations that led to the interesting series of papers by Brown and York, not much is known about the relation between the two frameworks. It would be interesting to explore this relation, and more generally, to relate the isolated horizon framework to the semi-classical ideas based on Euclidean gravity. Second, while the understanding of the micro-states of an isolated horizon is fairly deep by now, work on a quantum gravity derivation of the Hawking radiation has just begin [17]. Using general arguments based on Einstein's A and B coefficients [1] and the known micro-states of an isolated horizon, one can argue that the envelope of the line spectrum emitted by a black hole should be thermal. However, further work is necessary to make sure that the details are correct. As far as the zeroth and first laws and the entropy calculation are concerned, the obvious open problem is the extension to incorporate non-zero angular momentum. As indicated in [10,12], the extension of the classical theory should be relatively straightforward, although it may well pose some technical challenges. To incorporate rotation, only one condition (on spin-coefficients associated with n^a) in the present definition of non-rotating isolated horizon needs to be weakened. Work has already begun on this problem. The extension of the entropy calculation, on the other hand, may turn out to be trickier for it may well require a new technical insight. On a long range, the outstanding challenge is to obtain a deeper understanding of the Immirzi ambiguity and the associated issue of renormalization of Newton's constant. For any value of γ, one obtains the 'correct' classical limit. However, as far as black hole thermodynamics is concerned, it is only for $\gamma = \gamma_o$ that one seems to obtain agreement with quantum field theory in curved space-times. Is this value of γ robust? Can one make further semi-classical checks? A pre-requisite for this investigation is a better handle on the issue of semi-classical states. A major effort will soon be devoted to this issue.

Let me conclude with a comparison between the entropy calculation in this approach and those performed in string theory. First, there are some obvious differences. In the present approach, one begins with the sector of the classical theory containing space-times with isolated horizons and then proceeds with quantization. consequently, one can keep track of the physical, curved geometry. In particular, one can see that, as required by physical considerations, the degrees of freedom which account for entropy can interact with the physical exterior of the black hole. In string theory, by contrast, actual calculations are generally performed in flat space and non-renormalization arguments and/or duality conjectures are then invoked to argue that the results so obtained refer to macroscopic black holes. Therefore, relation to the curved space geometry and physical meaning of the degrees of freedom which account for entropy is rather obscure. More generally, lack of direct contact with physical space-time can also lead to practical difficulties while dealing with macroscopic situations. For example, in string theory, it may be difficult to account for the entropy normally associated with de Sitter horizons. On the other hand, in the study of genuinely quantum, Planck size black holes, this 'distance' from the curved space-time geometry may turn out to be a blessing, as classical curved geometry will not be an appropriate tool to discuss physics in these situations. In particular, a description which is far removed from space-time pictures may be better suited in the discussion of the last stages of Hawking evaporation and the associated issue of 'information loss'.

Another advantage of the string-theory approach is that entropy calculations have been carried out in a number of space-time dimensions. By contrast, so far the framework presented here is applicable only to four dimensions.[5] Also, our quantization procedure has an inherent ambiguity which trickles down to the entropy calculation. By contrast, calculations in string theory are free of this problem. On the other hand, almost all detailed calculations in string theory have been carried out only for (a sub-class of) extremal or near-extremal black holes. While these black holes are especially simple to deal with mathematically, unfortunately, they are not of direct relevance to astrophysics, i.e., to the physical world we live in. More recently, using the Maldecena conjecture, stringy calculations have been extended to non-extremal black holes with $R^2_{\rm Sch} >> 1/\Lambda$, where $R_{\rm Sch}$ is the Schwarzschild radius. However, the numerical coefficient in

[5] However, an extension of the underlying non-perturbative framework to higher dimensions was recently proposed by Freidel, Krasnov and Puzzio.

front of the entropy turns out to be incorrect and it is not yet clear whether inclusion of non-Abelian interactions, which are ignored in the current calculations, would restore the numerical coefficient to its correct value. Furthermore, it appears that a qualitatively new strategy may be needed to go beyond the $R^2_{\rm Sch} >> 1/\Lambda$ approximation. Finally, as in other results based on the Maldecena conjecture, the underlying boundary conditions at infinity are quite unphysical since the radius of the compactified dimensions is required to equal the cosmological radius. Hence the relevance of these mathematically striking results to our physical world remains unclear. In the current approach, by contrast, ordinary, astrophysical black holes in the physical, four space-time dimensions are included from the beginning.

In spite of this differences, there are some striking similarities. Our polymer excitations resemble stings. Our horizon looks like a 'gravitational 2-brane'. Our polymer excitations ending on the horizon, depicted in figure 3, closely resemble strings with end points on a membrane. As in string theory, our '2-brane' carries a natural gauge field. Furthermore, the horizon degrees of freedom arise from this gauge field. These similarities seem astonishing. However, a closer look brings out a number of differences as well. In particular, being horizon, our '2-brane' has a direct interpretation in terms of the curved *space-time geometry* and our $U(1)$ connection is the *gravitational* spin-connection on the horizon. Nonetheless, it may well be that, when quantum gravity is understood at a deeper level, it will reveal that the striking similarities are not accidental, i.e., that the two decriptions are in fact closely related.

Acknowledgments: The material presented in this report is based on joint work with John Baez, Chris Beetle, Alex Corichi, Steve Fairhurst and especially Kirill Krasnov. I am grateful to them for collaboration and countless discussions. Special thanks are due to Chris Beetle for his help with figures. I would like to thank Jerzy Lewandowski for sharing his numerous insights on the isolated horizon boundary conditions. I have also profited from comments made by Brandon Carter, Helmut Friedrich, Sean Hayward, Gary Horowitz, Ted Jacobson, Don Marolf, Jorge Pullin, Daniel Sudarsky, Istavan Racz, Oscar Reula, Carlo Rovelli, Thomas Thiemann and Robert Wald. Finally, I would like to thank the organizers of Mazatlan workshop on black holes, the Karpacz winter school, the Val Morin workshop on black holes, and the Montreal conference on general relativity and relativistic astrophysics, where many of these discussions took place. This work was supported in part by the NSF grants PHY94-07194, PHY95-14240, INT97-22514 and by the Eberly research funds of Penn State.

References

Classics

1. J.D. Bekenstein, Black holes and entropy, *Phys. Rev.* **D7** 2333-2346 (1973).
   J.D. Bekenstein, Generalized second law of thermodynamics in black hole physics, *Phys. Rev.* **D9** 3292-3300 (1974).
   J. D. Bekenstein and A. Meisels, Einstein A and B coefficients for a black hole, *Phys. Rev.* **D15**, 2775-2781 (1977).
2. J.W. Bardeen, B. Carter and S.W. Hawking, The four laws of black hole mechanics, *Commun. Math. Phys.* **31** 161-170 (1973).
   B. Carter, Black hole equilibrium states, In *Black Holes*, ed. B. DeWitt and C. De-Witt, Gordon and Breach, New York, 1973.
3. S.W. Hawking, Particle creation by black holes, *Commun. Math. Phys.* **43** 199-220 (1975).
4. G. Gibbons and S.W. Hawking, Cosmological event horizons, thermodynamics, and particle creation, *Phys. Rev.* **D15** 2738-2751 (1977).

Reviews

5. M. Heusler, *Black Hole Uniqueness Theorems*, Cambridge University Press, Cambridge, 1996.
6. R.M. Wald, *Quantum Field Theory in Curved Space-Time and Black Hole Thermodynamics*, University of Chicago Press, Chicago, 1994.
7. A. Ashtekar, Quantum mechanics of geometry, In *The Narlikar Festschrift*, ed. N. Dadhich and A. Kembhavi, 1999, also available as gr-qc/990123.
8. A. Ashtekar, K. Krasnov. Quantum geometry and black holes, In *Black holes, gravitational radiation and the Universe*, ed. B. Bhawal and B.R. Iyer, Kluwer, Dordrecht, 149-170 (1998); also available as gr-qc/9804039.
9. G. Horowitz, Quantum states of black holes, In *Black holes and relativistic stars*, ed. R.M. Wald, U. Chicago press, Chicago, 1998.

Main papers where details can be found

10. A. Ashtekar, A. Corichi and K. Krasnov, Isolated horizons: the classical phase space, *Adv. Theor. Math. Phys.* in press; also available as gr-qc/9905089.
11. A. Ashtekar, C. Beetle and S. Fairhurst, Isolated horizons: a generalization of black hole mechanics, *Class. Quantum Grav.* **16** L1-L7 (1999).
12. A. Ashtekar, C. Beetle and S. Fairhurst, Mechanics of isolated horizons, Submitted to *Class. Quantum Grav.* also available as gr-qc/9907068.
13. A. Ashtekar and A. Corichi, Laws governing dilatonic isolated horizons, Pre-print.
14. J. Lewandowski, Space-times admitting isolated horizons, Submitted to *Class. Quantum Grav.*, also available as gr-qc/9907058.
15. A. Ashtekar, J. Baez, A. Corichi, K. Krasnov, Quantum geometry and black hole entropy, *Phys. Rev. Lett.* **80** 904-907 (1998).
16. A. Ashtekar, J. Baez, K. Krasnov, Quantum geometry of isolated horizons horizons and black hole entropy, Pre-print.
17. K. Krasnov, Quantum geometry and thermal radiation form black Holes, *Class. Quantum. Grav.* **16**, 563 (1999).

Related references: Classical black holes

18. S. Hayward, General laws of black hole dynamics, *Phys. Rev.* **D49** 6467-6474 (1994).
S. Hayward, Spin coefficient form of the new laws of black hole dynamics, *Class. Quant. Grav.* **11** 3025-3036 (1994).
S. Hayward, Relativistic thermodynamics, gr-qc/9803007.
19. D. Sudarski and R.M. Wald, Extrema of mass, stationarity and staticity, and solutions to the Einstein-Yang-Mills equations, *Phys. Rev.* **D46** 1453-1474 (1992).
R.M. Wald, Black hole entropy is Noether charge, *Phys. Rev.* **D48** 3427-3431 (1993).

Related references: Quantum Geometry

20. A. Ashtekar and J. Lewandowski, Representation theory of analytic holonomy $C^\star$-algebras, In *Knots and quantum gravity*, ed. J. Baez, Oxford University Press, Oxford, 1994.
A. Ashtekar and J. Lewandowski, Projective techniques and functional integration for gauge theories, *J. Math. Phys.* **36**, 2170-2191 (1995).
J. Baez, Diffeomorphism invariant generalized measures on the space of connections modulo gauge transformations, In *The Proceedings of the conference on quantum topology*, ed. D. Yetter, World Scientific, Singapore, 1994.
J. Baez, Generalized measures in gauge theory, *Lett. Math. Phys.* **31**, 213-223 (1994).
D. Marolf and J. Mourão, On the support of the Ashtekar-Lewandowski measure, *Commun. Math. Phys.* **170**, 583-605 (1995).
C. Rovelli and L. Smolin, Spin networks and quantum gravity, *Phys. Rev.* **D52**, 5743-5759 (1995).
J. Baez, Spin-networks in gauge theory, *Adv. Math.* **117**, 253-272 (1996).
J. Baez, Spin networks in non-perturbative quantum gravity, In *The Interface of Knots and Physics*, ed. L. Kauffman, American Mathematical Society, Providence, 1996, pp. 167-203.
21. A. Ashtekar and J. Lewandowski, Differential geometry on the space of connections using projective techniques, *J. Geo. Phys.* **17**, 191-230 (1995).
C. Rovelli and L. Smolin, Area and volume in quantum gravity, *Nucl. Phys.* **B442**,
A. Ashtekar and J. Lewandowski, Quantum theory of geometry I: Area operators, *Class. Quant. Grav.* **14**, A55-A81 (1997).
K. Krasnov, On the constant that fixes the area spectrum in canonical quantum gravity, *Class. Quant. Grav.* **15**, L1-L4 (1998).
22. F. Barbero, From Euclidean to Lorentzian general relativity: the real way, *Phys. Rev.* **D54**, 1492-1499 (1996).
G. Immirzi, Quantum Gravity and Regge Calculus, *Nucl. Phys. Proc. Suppl.* **57**, 65-72 (1997).
R. Gambini, O. Obregón and J. Pullin, Yang-Mills analogs of the Immirzi ambiguity, *Phys. Rev.* **D59**, 047505 (1998).
C. Rovelli, T. Thiemann, The Immirzi parameter in quantum general relativity, *Phys. Rev.* **D57**, 1009-1014 (1998).

Related references: Quantum black holes

23. K. Krasnov, Geometrical entropy from loop quantum gravity, *Phys. Rev.* **D55** 3505-3513 (1997).
K. Krasnov, On statistical mechanics of Schwarzschild black holes, *Gen. Rel. Grav.* **30**, No. 1, 53-68 (1998).

24. C. Rovelli, Black hole entropy from loop quantum gravity, *Phys. Rev. Lett.* **14** 3288-3291 (1996).
C. Rovelli, Loop quantum gravity and black hole physics, *Helv. Phys. Acta.* **69** 582-611 (1996).
M. Barreira, M. Carfora, C. Rovelli, Physics with non-perturbative quantum gravity: radiation from a quantum black hole, *Gen. Rel. and Grav.* **28**, 1293-1299 (1996).
25. R.K. Kaul and P. Majumdar, Quantum black hole entropy, *Phys. Lett.* **B439** 267-270, (1998).
26. J. Wheeler, It from bits and quantum gravity, In *Sakharov Memorial Lectures on Physics*, ed. L. Keldysh and V. Feinberg, Nova Science, New York, 1992, Vol. 2.
27. A. Strominger and C. Vafa, Microscopic Origin of the Bekenstein-Hawking Entropy, *Phys. Lett.* **B379**, 99-104, (1996)
J. Maldacena and A. Strominger, Statistical entropy of four-dimensional extremal black holes, *Phys. Rev. Lett.* **77** 428-429 (1996).

Old and New Processes of Vorton Formation

Brandon Carter[1]

D.A.R.C., Observatoire de Paris 92 Meudon, France

Abstract. Among the likely consequences of cosmic string formation, one of the most important possibilities is the formation of equilibrium configurations, known as vortons, for current carrying loops. This article provides a concise review of available quantitative estimates of the vorton population that would be produced in various cosmic string scenarios. Attention is drawn to previously unconsidered mechanisms that might give rise to much more prolific vorton formation that has been envisaged hitherto.

This review is an updated version of a previous very brief overview[1] of the theory of vortons, meaning equilibrium states of cosmic string loops, and of the cosmological processes by which they can be produced in various scenarios. The main innovation here is to draw attention to the possibility of greatly enhanced vorton formation in cases for which the cosmic string current is of the strictly chiral type [2] that arises naturally in certain kinds of supersymmetric field theory.

It is rather generally accepted[3] that among the conceivable varieties of local topological defects of the vacuum that might have been generated at early phase transitions, the *vortex type* defects describable on a macrosopic scale as *cosmic strings* are the kind that is most likely to actually occur – at least in the post inflationary epoch – because the other main categories, namely walls and local monopoles, would produce a catastrophic cosmological mass excess. Even a single wall stretching accross a Hubble radius would by itself be too much, while in the case of monopoles it is their collective density that would be too high unless the relevant phase transition occurred at an energy far below that of the G.U.T. level, a possibility that is commonly neglected on the grounds that no monopole formation occurs in the usual models for the transitions in the relevant range, of which the most important is that of electroweak symmetry breaking.

The case of cosmic strings is different. One reason is that – although they are not produced in the standard electroweak model – strings are indeed produced at the electroweak level in many of the commonly considered (e.g. supersymmetric) alternative models. A more commonly quoted reason why the case of strings is different, even if they were formed at the G.U.T level, is that – while it may have an important effect in the short run as a seed for galaxy formation – such a string cannot be cosmologically dangerous just by itself, while a distribution of cosmic strings is also cosmologically harmless because (unlike "local" as opposed to "global" monopoles) they will ultimately radiate away all their energy and disappear. However while this latter consideration is indeed valid in the case of ordinary Goto-Nambu type strings, it was pointed out by Davis and Shellard[4] that it need not apply to "superconducting" current-carrying strings of the kind originally introduced by Witten[5]. This is because the occurrence of stable currents allows loops of string to be stabilized in states known as "vortons", so that they cease to radiate.

The way this happens is that the current, whether timelike or spacelike, breaks the Lorentz invariance along the string worldsheet [6–9], thereby leading to the possibility of rotation, with velocity v say. The centrifugal effect of this rotation, may then compensate the string tension T in such a way as to produce an equilibrium configuration, i.e. what is known as a *vorton*, in which

$$T = v^2 U \,, \tag{1}$$

where U is the energy per unit length in the corotating rest frame[10,11]. Such a vorton state will be stable, at least classically, if it minimises the energy for given values of the pair of conserved quantities characterising the current in the loop, namely the phase winding number N say, and the corresponding particle number Z say, whose product determines the mass M of the ensuing vorton state according to a rough order of magnitude formula of the form

$$M \approx |NZ|^{1/2} m_{\rm x} \tag{2}$$

where $m_{\rm x}$ is the relevant Kibble mass, whose square is the zero current limit value of both T and U. If the current is electromagnetically coupled, with charge coupling constant e, then there will be a corresponding vorton charge $Q = Ze$.

Whereas the collective energy density of a distribution of non-conducting cosmic strings will decay in a similar manner to that of a radiation gas, in contrast for a distribution of relic vortons the energy density will scale like that of ordinary matter. Thus, depending on when and how efficiently they were formed, and on how stable they are in the long run, such vortons might eventually come to dominate the density of the universe. It has been rigorously established[12–14] that circular vorton configurations of this kind will commonly (though not always) be stable in the dynamic sense at the classical level, but very little is known so far about non-circular configurations or about the question of stability against quantum tunnelling effects, one of the difficulties being that the latter is likely to be sensitively model dependent.

In the earliest crude quantitative estimates[4,15] of the likely properties of a cosmological vorton distribution produced in this way, it was assumed not only that the Witten current was stable against leakage by tunnelling, but also that the mass scale m_σ characterising the relevant carrier field was of the same order of magnitude as the Kibble mass scale m_x characterising the string itself, which will normally be given approximately by the mass of the Higgs field responsible for the relevant vacuum symmetry breaking. The most significant development in the more detailed investigations carried out more recently[16,17] was the extension to cases in which m_σ is considerably smaller than m_x. A rather extreme example that immediately comes to mind is that for which m_x is postulated to be at the G.U.T. level, while m_σ is at the electroweak level in which case it was found that the resulting vorton density would be far too low to be cosmologically significant.

The simplest scenarios are those for which (unlike the example just quoted) the relation

$$\frac{\sqrt{m_\sigma}}{m_{\rm x}} \gtrsim 1 \tag{3}$$

is satisfied in dimensionless Planck units as a rough order of magnitude inequality. In this case the current condensation would have ocurred during the regime in which (as pointed out by Kibble[18] in the early years of cosmic string theory) the dynamics was dominated by friction damping. Under these circumstances, acording to the standard picture[3], the string distribution will consist of wiggles and loops of which the most numerous will be the shortest, characterised by a length scale ξ say below which smaller scale structure will have been smoothed out by friction damping. The number density n of these smallest and most numerous loops will be given by the (dimensionally obvious) formula

$$n \approx \frac{1}{\xi^3}\,, \tag{4}$$

in which the smoothing length scale ξ itself is given by

$$\xi \approx \sqrt{t\tau}\,, \tag{5}$$

where τ is the relevant friction damping timescale and t is the cosmological time, which, using Planck units, will be expressible in terms of the cosmological temperature Θ by

$$t \approx \frac{1}{\Theta^2}\,, \tag{6}$$

in the radiation dominated epoch under consideration. According to the usual description of the friction dominated epoch [19,3], the relevant damping timescale will be given by

$$\tau \approx \frac{m_{\rm x}^2}{\Theta^3}\,, \tag{7}$$

from which it can be seen that the smoothing lengthscale ξ that characterises the smallest and most numerous string loops will be given roughly by the well known formula

$$\xi \approx \frac{m_{\rm x}}{\Theta^{5/2}}\,. \tag{8}$$

At the time of condensation of the current carrier field on the strings, when the temperature reaches a value $\Theta \approx m_\sigma$, the corresponding thermal fluctuation wavelength λ will be given by

$$\lambda \approx \frac{1}{m_\sigma}\,. \tag{9}$$

Taken around the circumference, of order ξ, of a typical small string loop, the number of such fluctuation wavengths will be of order ξ/λ. In the cases considered previously [16,17] it was assumed that the fluctuations would be randomly orientated and would therefore tend to cancel each other out so that, by the usual kind of random walk process the net particle and winding numbers taken around the loop as a whole would be expected to be of the order of the square root of this number of wavelengths, i.e. one would typically obtain

$$N \approx Z \approx \sqrt{\frac{\xi}{\lambda}}. \tag{10}$$

However a new point to which I would like to draw attention here is that the random walk cancellation effect will not apply in case for which the current is of strictly chiral type so that the string dynamics is of the kind whose special integrability properties have recently been pointed out [2]. This case arises [5] when the string current is attributable to (necessarily uncharged) fermionic zero modes moving in an exlusively rightwards (or exclusively leftwards) direction. In such a case, the possibility of cancellation between left moving and right moving fluctuations does not arise, so that (as in the ordinary kind of diode rectifier circuit used for converting alternating current to direct curent) there is an effective filter ensuring that the fluctuations induced on the string will all have the same orientation. In such a case only one of the quantum numbers in the formula (2) will be independent, i.e. they will be restricted by a relation of the form $N = Z$, and their expected value will be of the order of the total number of fluctuation wavelengths round the loop (not just the square root thereof as in the random walk case). In such a strictly chiral case the formula (2) should therefore be evaluated using an estimate of the form

$$N = Z \approx \frac{\xi}{\lambda}, \tag{11}$$

instead of (10)

Whereas even smaller loops will have been entirely destroyed by the friction damping process, those that are present at the time of the current condensation can survive as vortons, whose number density will be reduced in inverse proportion to the comoving volume, i.e. proportionally to Θ^3, relative to the initial number density value given by (4) when $\Theta \approx m_\sigma$. Thus (assuming the current on each string is strictly conserved during the subsequent evolution) when the cosmological temperature has fallen to a lower value $\Theta \ll m_\sigma$, the expected number density n of the vortons will be given as a constant fraction of the corresponding number density $\approx \Theta^3$ of black body photons by the rough order of magnitude formula

$$\frac{n}{\Theta^3} \approx \left(\frac{\sqrt{m_\sigma}}{m_x}\right)^3 m_\sigma^3 . \tag{12}$$

In the previously considered cases [16,17], for which the random walk formula (10) applies, the typical value of the quantum numbers of vortons in the resulting population will be given very roughly by

$$N^2 \approx Z^2 \approx \frac{m_{\rm x}}{m_\sigma^{3/2}} \,. \tag{13}$$

According to (2), this implies a typical vorton mass given by

$$M \approx \Big(\frac{m_x}{\sqrt{m_\sigma}}\Big)^{3/2} , \tag{14}$$

which, in view of (3), will never exceed the Planck mass. It follows in this case that, in order to avoid producing a cosmological mass excess, the value of m_σ in this formula should not exceed a limit that works out to be of the order of 10^{-9}, and the limit is even be smaller, $m_\sigma \ll 10^{-11}$, when the two scales m_σ and m_x are comparable.

The new point to which I wish to draw attention here is that for the strictly chiral case, as characterised by (11) instead of (10), the formula (2) for the vorton mass gives a typical value

$$M \approx \frac{m_{\rm x}^2}{m_\sigma^{3/2}} , \tag{15}$$

which is greater than what is given by the usual formula (14) by a factor $m_{\rm x}^{1/2} m_\sigma^{-3/4}$. Although the vorton to photon number density ratio (12) will not be affected, the corresponding mass density $\rho = Mn$ of the vorton distribution will be augmented by the same factor $m_{\rm x}^{1/2} m_\sigma^{-3/4}$. This augmentation factor will be expressible simply as $m_\sigma^{-1/4}$ when the two scales m_σ and m_x are comparable, in which case the requirement that a cosmological mass excess should be avoided leads to the rather severe limit $m_\sigma \lesssim 10^{-14}$. This mass limit works out to be of the order of a hundred TeV, which is within the range that is commonly envisaged for the electroweak symmetry breaking transition.

The foregoing conclusion can be construed as meaning that if strictly chiral current carrying strings were formed (within the framework of some generalised, presumably supersymmetric, version of the Standard electroweak model) during the electroweak symmetry breaking phase transition, then the ensuing vorton population might conceivably constitute a significant fraction of the cosmological dark matter distribution in the universe. Although, according to (12), the number density of such chiral vortons would be rather low, their typical mass, as given according to (15) by $M \approx \sqrt{m_\sigma}$ would be rather large, about 10^{-7} in Planck units, which works out as about 10^9 TeV.

An alternative kind of scenario that naturally comes to mind is that in which the cosmic strings themselves were formed at an energy scale $m_{\rm x}$ in the GUT range (of the order of 10^{-3} in Planck units) but in which the current did not condense on the string until the thermal energy scale had dropped to a value m_σ that was nearer the electroweak value (below the order of 10^{-14} in Planck units). Since this very much lower condensation temperature would be outside the friction dominated range characterised by (3), the reasonning summarised above would not be applicable. Preliminary evaluations of the (relatively inefficient) vorton production that would arise from current condensation after the end of the friction dominated period are already available [17] for the usual random walk case, but analogous estimates for aumentation that might arise in the strictly chiral case have not yet been carried out. The reason why it is not so easy to evaluate the consequences of current condensation after the end of the friction dominated epoch (when radiation damping becomes the main dissipation mechanism) is that most of the loops present at the time of the current condensation would have been be too small to give vortons stable against quantum decay processes, a requirement which imposes a lower limit

$$M \gtrsim \frac{m_{\rm x}^2}{m_\sigma} \tag{16}$$

on the mass of a viable vorton. This condition is satisfied automatically by the masses estimated in the manner described above for vortons formed by condensation during the friction dominated era characterised by (3). On the other hand when (3) is not satisfied – in which case the lower limit (16) will evidently exceed the Planck mass – then the majority of loops present at the time of the carrier condensation phase transition at the temperature $\Theta \approx m_\sigma$ will not acquire the rather large quantum number values that would be needed to make them ultimately viable as vortons. It is not at all easy to obtain firmly conclusive estimates of the small fraction that will satisfy this viability condition. However it should not be too difficult to carry out an adaptation to the strictly chiral case of the kind of tentative provisional estimates (based on simplifying assumptions whose confirmation will require much future work) that have already been provided [17] for the generic case of currents built up by the usual random walk process.

The possibility of strictly chiral current formation is not the only mechanism whereby vorton formation might conceivably be augmented relative to what was predicted on the basis [17] of the previous estimates, which took no account of electromagnetic effects. There cannot be any electromagnetic coupling in the strictly chiral case [2], and in other cases where electromagnetic coupling will be typically be present it has been shown [20] that it will usually have only a minor perturbing effect on the vorton equilibrium states. However it has recently been remarked [21] that even though the averaged "direct" current that is relevant for vorton formation may be small, the local "alternating" current can have a sufficiently large amplitude, I say, for its interaction with the surrounding black body radiation plasma to provide the dominant friction damping mechanism, with a damping time scale that instead of (7) will be given in rough order of magnitude by

$$\tau \approx \frac{m_{\rm x}^2}{I\Theta^2} \,. \tag{17}$$

As can be seen from (6), this means that instead of being restricted to the very early epoch when cosmological temperature was above Kibble limit value, i.e.

when $\Theta \gtrsim \sqrt{m_x}$, the period of friction domination can be extended indefinitely if the current amplitude satisfies

$$I \gtrsim m_x^2 , \tag{18}$$

a requirement that is easily compatible with Witten's [5] bosonic current saturation bound $I \lesssim em_x$ (where $e \simeq 1/\sqrt{137}$ is the electromagnetic charge coupling constant), and that is in most cases compatible even with the more severe limit $I \lesssim em_\sigma$ that applies in cases for which instead of arising as a bosonic condensate, the current is due to femionic zero modes. Such a tendency to prolongation of friction dominance will presumably delay the decay of small scale loop structure and so may plausibly be expected to augment the efficiency of vorton formation in cases when m_σ is below the limit given by (3), but a quantitative estimate of just how large this effect is likely to be will require a considerable amount of future work.

Despite the possibility that the effciency of vorton formation may have been underestimated by previous work, it still seems unlikely that vortons can constitute more than a small fraction of the missing matter in the universe. However this does not mean that vortons could not give rise to astrophysically interesting effects: in particular it has recently been suggested by Bonazzola and Peter[22] that they might account for otherwise inexplicable cosmic ray events.

The author is grateful to many colleagues, particularly Patrick Peter and Anne Davis, for helpful discussions on numerous occasions.

References

1. B. Carter, *Int. J. Theor. Phys.* **36** 2451-2458 (1997). [astro-ph/9712116]
2. B. Carter, P. Peter, "Dynamics and integrability property of the chiral string model" [hep-th/9905025]
3. E. P. S. Shellard & A. Vilenkin, *Cosmic strings and other topological defects*, Cambridge University Press (1994).
4. R. L. Davis, E. P. S. Shellard, *Phys. Rev.* **D38**, 4722 (1988); Nucl. Phys. B **323**, 209 (1989);
5. E. Witten, Nucl. Phys. B **249**, 557 (1985).
6. B. Carter, Phys. Lett. **B 224**, 61 (1989) and **B 228**, 466 (1989).
7. B. Carter, in *Formation and Interactions of Topological Defects* (NATO ASI **B349**), ed R. Brandenberger & A-C. Davis, *pp* 303-348 (Plenum, New York, 1995).
8. P. Peter, *Phys. Rev.* **D45**, 1091 (1992).
9. P. Peter, *Phys. Rev.* **D46**, 3335 (1992).
10. B. Carter, Phys. Lett. **238**, 166 (1990).
11. B. Carter, P. Peter, A. Gangui, *Phys Rev.* **D55**, 4647 (1997) [hep-ph/9609401].
12. B. Carter, X. Martin, Ann. Phys. **227**, 151 (1993).
13. X. Martin, *Phys. Rev.* **D50**, 749 (1994).
14. X. Martin, P. Peter, *Phys. Rev* **D51**, 4092 (1995).

15. B. Carter, Ann. N.Y. Acad.Sci., **647**, 758 (1991);
16. B. Carter, Proceedings of the XXXth Rencontres de Moriond, Villard-sur-Ollon, Switzerland, 1995, Edited by B. Guideroni and J. Tran Thanh Vân (Editions Frontières, Gif-sur-Yvette, 1995) [hep-ph/9511225].
17. R. Brandenberger, B. Carter, A-C. Davis & M. Trodden, *Phys.Rev.* **D54**, 6059 (1966) [hep-ph/9605382].
18. T. W. B. Kibble, J. Math. Phys. A **9**, 1387 (1976); Phys. Rep. **67**, 183 (1980).
19. T. Vachaspati, A. Vilenkin, *Phys Rev.* **D31**, 3035, (1985).
20. P. Peter, *Phys. Lett.* **B298**, 60 (1993).
21. K. Dimopoulos & A.-C. Davis, *Phys. Rev.* **D57**, 692 (1998). [hep-ph/9705302)
22. S. Bonazzola & P. Peter, *Astropart. Phys.* **7**, 161 (1997). [hep-ph/9701246]

Anti-de Sitter Supersymmetry

Bernard de Wit and Ivan Herger

Institute for Theoretical Physics, Utrecht University
3508 TA Utrecht, Netherlands

Abstract. We give a pedagogical introduction to certain aspects of supersymmetric field theories in anti-de Sitter space. Among them are the presence of masslike terms in massless wave equations, irreducible unitary representations and the phenomenon of multiplet shortening.

1 Introduction

Recently the study of field theory in anti-de Sitter space has received new impetus by the observation that the near-horizon geometry of black branes, which usually involves anti-de Sitter space as a factor, is related to a field theory associated with the massless modes of open strings that are attached to a certain number n of parallel Dirichlet branes, separated by small distances [1]. In certain cases there thus exists a connection between superconformal field theories in flat space, living on the boundary of an anti-de Sitter space-time, and gauged supergravity. The most striking example is that of $N = 4$ supersymmetric Yang-Mills theory in four space-time dimensions with gauge group $\mathrm{U}(n)$, and IIB supergravity or superstring theory compactified on the five-dimensional sphere.

In these lectures we intend to give a pedagogical introduction to field theories and supersymmetry in anti-de Sitter space. The subject is not new. Already in the thirties Dirac considered wave equations that are invariant under the anti-de Sitter group [2]. Later, in 1963, he discovered the 'remarkable representation' which is now known as the singleton [3]. Shortly afterwards there was a series of papers by Fronsdal and collaborators discussing the representations of the anti-de Sitter group [4]. Quantum field theory in anti-de Sitter space was studied, for instance in [5,6]. Many new developments were inspired by the discovery that gauged supergravity theories have ground states corresponding to anti-de Sitter spacetimes [7–16]. This led to a study of the stability of these ground states with respect to fluctuations of the scalar fields [17] as well as to an extended discussion of supermultiplets in anti-de Sitter space [17–22].

In these notes we will be able to cover only a few of these topics. We restrict ourselves to an introduction to supersymmetry in anti-de Sitter space and discuss the presence of the so-called masslike terms in wave equations for various fields in anti-de Sitter space. Then we will analyze the various irreducible representations of the anti-de Sitter isometry group, using a variety of techniques, and at the end we will consider the consequences for supermultiplets. We emphasize the issue of multiplet shortening for both multiplets of given spin and for supermultiplets.

2 Supersymmetry and anti-de Sitter space

Let us start with simple supergravity in an unspecified number of space-time dimensions. Two important terms in any supergravity Lagrangian are the Einstein Lagrangian of general relativity and the Rarita-Schwinger Lagrangian for the gravitino field(s),

$$\mathcal{L} = -\tfrac{1}{2}e\,R(\omega) - \tfrac{1}{2}e\bar{\psi}_\mu\,\Gamma^{\mu\nu\rho}D_\nu(\omega)\psi_\rho + \cdots\,, \tag{1}$$

where the covariant derivative on a spinor ψ reads

$$D_\mu(\omega)\psi = \left(\partial_\mu - \tfrac{1}{4}\omega_\mu{}^{ab}\,\Gamma_{ab}\right)\psi\,, \tag{2}$$

and $\omega_\mu{}^{ab}$ is the spin-connection field defined such that the torsion tensor (or a supercovariant version thereof) vanishes. The action corresponding the above Lagrangian is locally supersymmetric up to terms cubic in the gravitino field. The supersymmetry transformations contain the terms,

$$\delta e_\mu{}^a = \tfrac{1}{2}\bar{\epsilon}\,\Gamma^a\psi_\mu\,, \qquad \delta\psi_\mu = D_\mu(\omega)\epsilon\,. \tag{3}$$

Extending this Lagrangian to a fully supersymmetric one is not always possible. It may require additional fields and only when the dimension of space-time is less than twelve does one know solutions for interacting theories.

Let us now include a cosmological term into the above Lagrangian as well as a suitably chosen masslike term for the gravitino field,

$$\begin{aligned}\mathcal{L} = &-\tfrac{1}{2}e\,R(\omega) - \tfrac{1}{2}e\bar{\psi}_\mu\,\Gamma^{\mu\nu\rho}D_\nu(\omega)\psi_\rho\\ &+\tfrac{1}{4}g(d-2)e\,\bar{\psi}_\mu\Gamma^{\mu\nu}\psi_\nu + \tfrac{1}{2}g^2(d-1)(d-2)\,e + \cdots\,.\end{aligned} \tag{4}$$

As it turns out the corresponding action is still locally supersymmetric, up to terms that are cubic in the gravitino field, provided that we introduce an extra term to the transformation rules,

$$\delta e_\mu{}^a = \tfrac{1}{2}\bar{\epsilon}\,\Gamma^a\psi_\mu\,, \qquad \delta\psi_\mu = \left(D_\mu(\omega) + \tfrac{1}{2}g\Gamma_\mu\right)\epsilon\,. \tag{5}$$

This demonstrates that, a priori, supersymmetry does not forbid a cosmological term, but it must be of definite sign (at least, if the ground state is to preserve supersymmetry). For a discussion see [23,24] and references therein. Again, to construct a fully supersymmetric field theory is difficult and in this case there are even stronger restrictions on the number of space-time dimensions than in the case without a cosmological term. The Lagrangian (4) was first written down in [25] in four space-time dimensions and the correct interpretation of the masslike term was given in [26].

The Einstein equation corresponding to (4) reads (suppressing the gravitino field),

$$R_{\mu\nu} - \tfrac{1}{2}g_{\mu\nu}\,R + \tfrac{1}{2}g^2(d-1)(d-2)\,g_{\mu\nu} = 0\,, \tag{6}$$

which implies,

$$R_{\mu\nu} = g^2(d-1)\, g_{\mu\nu}\,, \qquad R = g^2 d(d-1)\,. \tag{7}$$

Hence we are dealing with a d-dimensional Einstein space. The maximally symmetric solution of this equation is an anti-de Sitter space, whose Riemann curvature equals

$$R_{\mu\nu}{}^{ab} = 2g^2\, e_\mu{}^{[a}\, e_\nu{}^{b]}\,. \tag{8}$$

This solution leaves all the supersymmetries intact just as flat Minkowski space does. One can verify this directly by considering the supersymmetry variation of the gravitino field and by requiring that it vanishes in the bosonic background. This happens for spinors $\epsilon(x)$ satisfying

$$\big(D_\mu(\omega) + \tfrac{1}{2} g\Gamma_\mu\big)\,\epsilon = 0\,. \tag{9}$$

Spinors satisfying this equation are called Killing spinors. Consequently also $(D_\mu(\omega) + \frac{1}{2}g\Gamma_\mu)(D_\nu(\omega) + \frac{1}{2}g\Gamma_\nu)\epsilon$ must vanish. Antisymmetrizing this expression in μ and ν then yields the integrability condition

$$\Big(-\tfrac{1}{4}R_{\mu\nu}{}^{ab}\,\Gamma_{ab} + \tfrac{1}{2}g^2\,\Gamma_{\mu\nu}\Big)\epsilon = 0\,, \tag{10}$$

which is precisely satisfied in anti-de Sitter space.

Because anti-de Sitter space is maximally symmetric, it has $\frac{1}{2}d(d+1)$ isometries which constitute the group $\mathrm{SO}(d-1,2)$. As we have just seen, anti-de Sitter space is consistent with supersymmetry. This is just as for flat Minkowski space, which has the same number of isometries but now corresponding to the Poincaré group, and which is also consistent with supersymmetry. The two cases are clearly related since flat space is obtained in the limit $g \to 0$. The algebra of the combined bosonic and fermionic symmetries will be called the anti-de Sitter superalgebra. Note again that the above derivation is based on an incomplete theory and in general one will need to introduce additional fields. The structure of the anti-de Sitter algebra changes drastically for dimensions $d > 7$ (see [27] and references cited therein). For $d \leq 7$ the bosonic subalgebra coincides with the anti-de Sitter algebra. There are N-extended versions, where we introduce N supersymmetry generators, each transforming as a spinor under the anti-de Sitter group. These N generators transform under a compact group, whose generators appear in the $\{Q, \bar{Q}\}$ anticommutator. For $d > 7$ the bosonic subalgebra can no longer be restricted to the anti-de Sitter algebra and the algebra corresponding to a compact group, but one needs extra bosonic generators that transform as high-rank antisymmetric tensors under the Lorentz group. In contrast to this, there exists an (N-extended) super-Poincaré algebra associated with flat Minkowski space of any dimension, whose bosonic generators correspond to the Poincaré group, possibly augmented with the generators of a compact group associated with rotations of the supercharges.

It is possible to describe anti-de Sitter space as a hypersurface embedded into a $(d+1)$-dimensional embedding space. Denoting the extra coordinate of the embedding space by Y^-, so that we have coordinates Y^A with $A = -,0,1,2,\ldots,d-1$, this hypersurface is defined by

$$-(Y^-)^2 - (Y^0)^2 + \mathbf{Y}^2 = \eta_{AB}\, Y^A Y^B = -g^{-2}\,. \tag{11}$$

Obviously, the hypersurface is invariant under linear transformations that leave the metric $\eta_{AB} = \mathrm{diag}\,(-,-,+,+,\ldots,+)$ invariant. These transformations constitute the group $\mathrm{SO}(d-1,2)$. The $\frac{1}{2}d(d+1)$ generators denoted by M_{AB} act on the embedding coordinates by

$$M_{AB} = Y_A \frac{\partial}{\partial Y^B} - Y_B \frac{\partial}{\partial Y^A}\,, \tag{12}$$

where we lower and raise indices by contracting with η_{AB} and its inverse η^{AB}. It is now easy to evaluate the commutation relations for the M_{AB},

$$[M_{AB}, M_{CD}] = \eta_{BC}\, M_{AD} - \eta_{AC}\, M_{BD} - \eta_{BD}\, M_{AC} + \eta_{AD}\, M_{BC}\,. \tag{13}$$

Anti-de Sitter space is a homogeneous space, which means that any two points on it can be related via an isometry. It has the topology of S^1 [time] $\times\ \mathbf{R}^{d-1}$. When unwrapping S^1 one finds the universal covering space denoted by CadS, which has the topology of $\mathbf{R}^d$. There are many ways to coordinatize anti-de Sitter space but we will try to avoid using specific coordinates.

On spinors, the anti-de Sitter algebra can be realized by the following combination of gamma matrices,

$$M_{AB} = \tfrac{1}{2}\Gamma_{AB} = \begin{cases} \tfrac{1}{2}\Gamma_{ab} \text{ for } A,B = a,b\,, \\ \tfrac{1}{2}\Gamma_a \ \text{ for } A = -\,, B = a \end{cases} \tag{14}$$

with $a,b = 0,1,\ldots,d-1$. Our gamma matrices satisfy the Clifford property $\{\Gamma^a\,,\,\Gamma^b\} = 2\,\eta^{ab}\,\mathbf{1}$, where $\eta^{ab} = \mathrm{diag}\,(-,+,\ldots,+)$.

The commutator of two supersymmetry transformations yields an infinitesimal general-coordinate transformation and a tangent-space Lorentz transformation. For example, we obtain for the vielbein,

$$\begin{aligned}[][\delta_1,\delta_2]\, e_\mu{}^a &= \tfrac{1}{2}\bar\epsilon_2\, \Gamma^a\, \delta_1\psi_\mu - \tfrac{1}{2}\bar\epsilon_1\, \Gamma^a\, \delta_2\psi_\mu \\ &= D_\mu(\tfrac{1}{2}\bar\epsilon_2\, \Gamma^a\epsilon_1) + \tfrac{1}{2}g\,(\bar\epsilon_2\, \Gamma^{ab}\epsilon_1)\, e_{\mu b}\,. \end{aligned} \tag{15}$$

Again we remind the reader of the fact that we are dealing with an incomplete theory. For a complete theory the above result should hold uniformly on all the fields (possibly modulo field equations). As before we have ignored terms proportional to the gravitino field. In the anti-de Sitter background the vielbein is left invariant by the combination of symmetries on the right-hand side. Consequently the metric is invariant under these coordinate transformations and we have the so-called Killing equation,

$$\delta g_{\mu\nu} = D_\mu \xi_\nu + D_\nu \xi_\mu = 0\,, \tag{16}$$

where $\xi_\mu = \frac{1}{2}\bar\epsilon_2 \Gamma_\mu \epsilon_1$ is a Killing vector and where $\epsilon_{1,2}$ are Killing spinors. Since $D_\mu \xi_\nu = \frac{1}{2} g \bar\epsilon_2 \Gamma_{\mu\nu} \epsilon_1$, the right-hand side of (15) vanishes for this choice of supersymmetry parameters, and ξ^μ satisfies the Killing equation (16). As for all Killing vectors, higher derivatives can be decomposed into the Killing vector and its first derivative, e.g. $D_\mu(g\,\bar\epsilon_2 \Gamma_{\nu\rho}\epsilon_1) = -g^2\, g_{\mu[\rho}\xi_{\nu]}$. The Killing vector can be decomposed into the $\frac{1}{2}d(d+1)$ Killing vectors of the anti-de Sitter space.

For later use we record the anti-de Sitter superalgebra, which in addition to (13) contains the (anti-)commutation relations,

$$\begin{aligned}
\{Q_\alpha, \bar Q_\beta\} &= -\tfrac{1}{2}(\Gamma_{AB})_{\alpha\beta}\, M^{AB}\,,\\
[M_{AB}, \bar Q_\alpha] &= \tfrac{1}{2}(\bar Q\, \Gamma_{AB})_\alpha\,.
\end{aligned} \tag{17}$$

As we alluded to earlier this algebra changes its form when considering N supersymmetry generators, which rotate under the action of a compact group. The generators of this group will then also appear on the right-hand side of the $\{Q, \bar Q\}$ anticommutator. Beyond $d = 7$ there are extra bosonic charges associated with higher-rank Lorentz tensors. However, in these lectures, we will mainly be dealing with the case $N = 1$ and we will always assume that $d \leq 7$.

3 Anti-de Sitter supersymmetry and masslike terms

In flat Minkowski space we know that all fields belonging to a supermultiplet are subject to field equations with the same mass. This must be so because the momentum operators commute with the supersymmetry charges, so that P^2 is a Casimir operator. For supermultiplets in anti-de Sitter space this is not longer the case, so that masslike terms will not necessarily be the same for different fields belonging to the same multiplet. This phenomenon will be illustrated below in a specific example, namely a chiral supermultiplet in four spacetime dimensions. Further clarification will be given later in sections 4 and 7.

A chiral supermultiplet in four spacetime dimensions consists of a scalar field A, a pseudoscalar field B and a Majorana spinor field ψ. In anti-de Sitter space the supersymmetry transformations of the fields are proportional to a spinor parameter $\epsilon(x)$, which is a Killing spinor in the anti-de Sitter space, i.e. $\epsilon(x)$ must satisfy the Killing spinor equation (9). We allow for two constants a and b in the supersymmetry transformations, which we parametrize as follows,

$$\begin{aligned}
&\delta A = \tfrac{1}{4}\bar\epsilon\psi\,, \qquad \delta B = \tfrac{1}{4} i\bar\epsilon\gamma_5\psi\,,\\
&\delta\psi = \not{\mathrm{d}}(A + i\gamma_5 B)\epsilon - (a\,A + ib\,\gamma_5 B)\epsilon\,.
\end{aligned} \tag{18}$$

The coefficient of the first term in $\delta\psi$ has been chosen such as to ensure that $[\delta_1, \delta_2]$ yields the correct coordinate transformation $\xi^\mu D_\mu$ on the fields A and B. To determine the constants a and b and the field equations of the chiral multiplet, we consider the closure of the supersymmetry algebra on the spinor field. After some Fierz reordering we find

$$[\delta_1, \delta_2]\psi = \xi^\mu D_\mu \psi + \tfrac{1}{16}(a-b)\,\bar\epsilon_2\gamma^{ab}\epsilon_1\,\gamma_{ab}\psi - \tfrac{1}{2}\xi^\rho\gamma_\rho[\not{\mathrm{d}}\psi + \tfrac{1}{2}(a+b)\psi]\,. \tag{19}$$

We point out that derivatives acting on $\epsilon(x)$ occur in this calculation at an intermediate stage and should not be suppressed in view of (9). However, they produce terms proportional to g which turn out to cancel in the above commutator. Now we note that the right-hand side should constitute a coordinate transformation and a Lorentz transformation, possibly up to a field equation. Obviously, the coordinate transformation coincides with (15) but the correct Lorentz transformation is only reproduced provided that $a - b = 2g$. If we now denote the mass of the fermion by $m = \frac{1}{2}(a + b)$, so that the last term is just the Dirac equation with mass m, then we find

$$a = m + g\,, \qquad b = m - g\,. \tag{20}$$

Consequently, the supersymmetry transformation of the ψ equals

$$\delta\psi = \not{\mathrm{d}}(A + i\gamma_5 B)\epsilon - m(A + i\gamma_5 B)\epsilon - g(A - i\gamma_5 B)\,\epsilon\,, \tag{21}$$

and the fermionic field equation equals $(\not{\mathrm{d}} + m)\psi = 0$. The second term in (21), which is proportional to m, can be accounted for by adding an auxiliary field to the supermultiplet. The third term, which is proportional to g, can be understood as a compensating S-supersymmetry transformation associated with auxiliary fields in the supergravity sector (see, e.g., [28]). In order to construct the corresponding field equations for A and B, we consider the variation of the fermionic field equation. Again we have to take into account that derivatives on the supersymmetry parameter are not equal to zero. This yields the following second-order differential equations,

$$\begin{aligned}
[\Box_{\mathrm{adS}} + 2g^2 - m(m-g)]\,A &= 0\,,\\
[\Box_{\mathrm{adS}} + 2g^2 - m(m+g)]\,B &= 0\,,\\
[\Box_{\mathrm{adS}} + 3g^2 - m^2]\,\psi &= 0\,.
\end{aligned} \tag{22}$$

The last equation follows from the Dirac equation. Namely, one evaluates $(\not{\mathrm{d}} - m)(\not{\mathrm{d}} + m)\psi$, which gives rise to the wave operator $\Box_{\mathrm{adS}} + \frac{1}{2}[\not{\mathrm{d}}, \not{\mathrm{d}}] - m^2$. The commutator yields the Riemann curvature of the anti-de Sitter space. In an anti-de Sitter space of arbitrary dimension d this equation then reads,

$$[\Box_{\mathrm{adS}} + \tfrac{1}{4}d(d-1)g^2 - m^2]\psi = 0\,, \tag{23}$$

which, for $d = 4$ agrees with the last equation of (22). A striking feature of the above result is that the field equations (22) all have different mass terms, in spite of the fact that they belong to the same supermultiplet. Consequently, the role of mass is quite different in anti-de Sitter space as compared to flat Minkowski space. This will be elucidated later.

For future applications we also evaluate the Proca equation for a massive vector field,

$$D^\mu(\partial_\mu A_\nu - \partial_\nu A_\mu) - m^2\,A_\nu = 0\,. \tag{24}$$

This leads to $D^\mu A_\mu = 0$, so that the field equation reads $D^2 A_\nu - [D^\mu, D_\nu]A_\mu - m^2 A_\nu = 0$ or, in anti-de Sitter space,

$$[\Box_{\rm adS} + (d-1)g^2 - m^2]\, A_\mu = 0\,. \tag{25}$$

The g^2 term in the field equations for the scalar fields can be understood from the observation that the scalar D'Alembertian can be extended to a conformally invariant operator (see e.g. [28]),

$$\Box + \frac{1}{4}\frac{d-2}{d-1}R = \Box + \tfrac{1}{4}d(d-2)\,g^2\,, \tag{26}$$

which seems the obvious candidate for a massless wave operator for scalar fields. Indeed, for $d=4$, we do reproduce the g^2 dependence in the first two equations (22). Observe that the Dirac operator d is also conformally invariant and so is the wave equation associated with the Maxwell field.

4 The quadratic Casimir operator

To make contact between the masslike terms in the wave equations and the properties of the irreducible representations of the anti-de Sitter group, it is important that we establish a relation between the D'Alembertian in anti-de Sitter space and the quadratic Casimir operator $\mathcal{C}_2$ of the isometry group. We will use $\mathcal{C}_2$ later on in our discussion of the unitary irreducible representations of the anti-de Sitter algebra. In this section, we will use the $(d+1)$-dimensional flat embedding space, introduced in section 2, to obtain such a relation for the scalar D'Alembertian. In the embedding space, the latter is equal to to

$$\Box_{d+1} = \eta^{AB}\frac{\partial}{\partial Y^A}\frac{\partial}{\partial Y^B}\,. \tag{27}$$

Denoting $\partial_A = \partial/\partial Y^A$ and $Y^2 = \eta_{AB}Y^AY^B$, we straightforwardly derive an expression for the quadratic Casimir operator associated with the anti-de Sitter group $\mathrm{SO}(d-1,2)$,

$$\begin{aligned}\mathcal{C}_2 &= -\tfrac{1}{2}\,M^{AB}\,M_{AB}\\ &= -Y^A\,\partial^B(Y_A\partial_B - Y_B\partial_A)\\ &= -Y^2\,\Box_{d+1} + Y^A\partial_A\,(Y^B\partial_B + d - 1)\,.\end{aligned} \tag{28}$$

The group $\mathrm{SO}(d-1,2)$ has more Casimir operators but the others are of higher order in the generators and will not play a role in the following. We now introduce different coordinates. We express the Y^A in terms of coordinates X^A, where $X^\mu = x^\mu$ with $\mu = 0,1,\ldots,d-1$ and X^- is defined by

$$X^- = \rho = \sqrt{-\eta_{AB}\,Y^AY^B}\,. \tag{29}$$

Furthermore, we require the ρ-dependence to be such that

$$Y^A(X) = \rho\, y^A(x)\,, \tag{30}$$

so that $y^A(x)\, y^B(x)\, \eta_{AB} = -1$. With this choice of coordinates one readily derives the following relations ($\widehat{\partial}_A = \partial/\partial X^A$),

$$\begin{aligned} \widehat{\partial}_- Y^A &= \frac{1}{\rho} Y^A \,, & \widehat{\partial}_- Y^A\, \eta_{AB}\, Y^B &= -\rho \,, \\ \widehat{\partial}_\mu Y^A\, \eta_{AB}\, Y^B &= 0 \,, & \widehat{\partial}_\mu Y^A\, \eta_{AB}\, \widehat{\partial}_- Y^B &= 0 \,, \\ \widehat{\partial}_- = \frac{\partial}{\partial \rho} = \widehat{\partial}_- Y^A \frac{\partial}{\partial Y^A} &= \frac{1}{\rho} Y^A\, \partial_A \,. \end{aligned} \tag{31}$$

In the new coordinate system the metric is given by

$$\widehat{g}_{AB} = \widehat{\partial}_A Y^C\, \eta_{CD}\, \widehat{\partial}_B Y^D = \begin{pmatrix} \rho^2 g_{\mu\nu} & 0 \\ 0 & -1 \end{pmatrix} \tag{32}$$

where $g_{\mu\nu}$ is the induced metric on the d-dimensional anti-de Sitter space (with radius equal to unity). Note that $\hat{g} \equiv \det \hat{g}_{AB} = -\rho^{2d} \det g_{\mu\nu} = -\rho^{2d} g$.

The D'Alembertian of the embedding space in the new coordinates is equal to (observe that derivatives act on all quantities on the right)

$$\begin{aligned} \Box_{d+1} &= \frac{1}{\sqrt{\widehat{g}}}\, \widehat{\partial}_A\, \widehat{g}^{AB}\, \sqrt{\widehat{g}}\, \widehat{\partial}_B \\ &= \frac{1}{\sqrt{-g}}\, \frac{1}{\rho^d} \Big\{ \partial_-\, \widehat{g}^{--}\, \rho^d \sqrt{-g}\, \partial_- + \partial_\mu\, g^{\mu\nu}\, \rho^{d-2} \sqrt{-g}\, \partial_\nu \Big\} \\ &= -\frac{\partial^2}{\partial \rho^2} - \frac{d}{\rho} \frac{\partial}{\partial \rho} + \rho^{-2}\, \Box_{\mathrm{adS}} \,, \end{aligned} \tag{33}$$

where $\Box_{\mathrm{adS}}$ is the D'Alembertian for the anti-de Sitter space of unit radius. Combining this with the expression (28) for the Casimir operator, we find

$$\mathcal{C}_2 = \rho^2\, \Box_{d+1} + \rho \frac{\partial}{\partial \rho} \Big(\rho \frac{\partial}{\partial \rho} + d - 1 \Big) = \Box_{\mathrm{adS}} \,. \tag{34}$$

Hence the $\partial/\partial\rho$ terms cancel as expected and the Casimir operator is just equal to the normalized anti-de Sitter D'Alembertian with unit anti-de Sitter radius. Note that this result cannot be used for other than spinless fields.

Let us now return to the wave equation for massless scalars (26). According to this equation, massless $s = 0$ fields lead to representations whose Casimir operator is equal to

$$\mathcal{C}_2 = -\tfrac{1}{4} d(d-2) \,. \tag{35}$$

Indeed, later in these lectures we will see that the Casimir operator for a massless $s = 0$ representation in four spacetime dimensions is equal to -2.

5 Unitary representations of the anti-de Sitter algebra

In this section we discuss unitary representations of the anti-de Sitter algebra. For definiteness we will mainly look at the case of four spacetime dimensions. We refer to [4] for some of the original work, and to [19,20] where some of this work was reviewed. In order to underline the general features we start in d spacetime dimensions. Obviously, the group $\mathrm{SO}(d-2,2)$ is noncompact. This implies that unitary representations will be infinitely dimensional. The generators are then all anti-hermitean,

$$M_{AB}^{\dagger} = -M_{AB}\,. \tag{36}$$

Note that the covering group of $\mathrm{SO}(d-1,2)$ has the generators $\frac{1}{2}\Gamma_{\mu\nu}$ and $\frac{1}{2}\Gamma_{\mu}$. They act on spinors, which are finite-dimensional objects. These generators, however, have different hermiticity properties from the ones above.

The compact subgroup of the anti-de Sitter group is $\mathrm{SO}(2)\times\mathrm{SO}(d-1)$ corresponding to rotations of the compact anti-de Sitter time and spatial rotations. It is convenient to decompose the $\frac{1}{2}d(d+1)$ generators as follows. First, the generator M_{-0} is related to the energy operator when the radius of the anti-de Sitter space is taken to infinity. The eigenvalues of this generator, which is associated with motions along the circle, are quantized in integer units in order to have single-valued functions, unless one goes to the covering space CadS. So we define the energy operator H by

$$H = -iM_{-0}\,. \tag{37}$$

Obviously the generators of the spatial rotations are the operators M_{ab} with $a,b=1,\ldots,d-1$. Note that we have changed notation: here and henceforth the indices $a,b,\ldots$ refer only to spacelike indices. The remaining $2(d-1)$ generators M_{-a} and M_{0a} are combined into pairs of mutually conjugate operators,

$$M_a^{\pm} = -iM_{0a} \pm M_{-a}\,, \tag{38}$$

and we have $(M_a^{+})^{\dagger} = M_a^{-}$. The anti-de Sitter commutation relations now read

$$\begin{aligned}
[H, M_a^{\pm}] &= \pm M_a^{\pm}\,,\\
[M_a^{\pm}, M_b^{\pm}] &= 0\,,\\
[M_a^{+}, M_b^{-}] &= -2(H\,\delta_{ab} + M_{ab})\,.
\end{aligned} \tag{39}$$

Obviously, the $M_a^{\pm}$ play the role of raising and lowering operators: when applied to an eigenstate of H with eigenvalue E, application of $M_a^{\pm}$ yields a state with eigenvalue $E\pm 1$.

In this section we restrict ourselves to the bosonic case. Nevertheless, let us already briefly indicate how some of the other (anti-)commutators of the anti-de

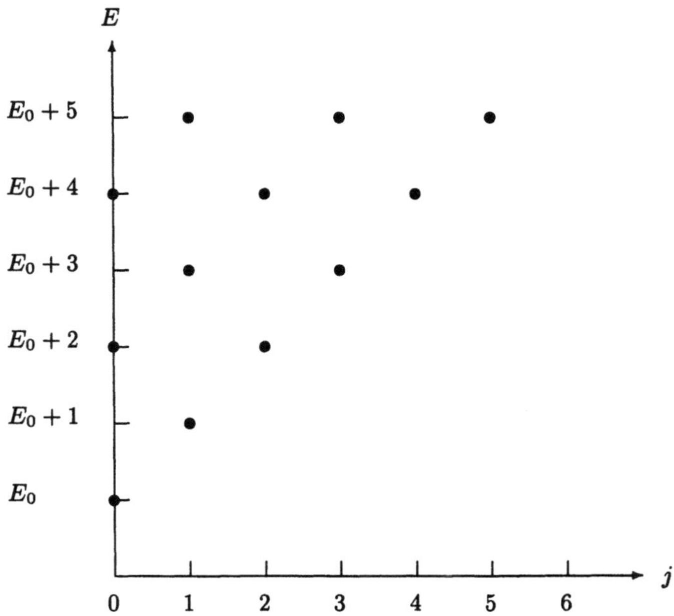

Fig. 1. States of the $s = 0$ representation in terms of the energy eigenvalues E and the angular momentum j. Each point has a $(2j + 1)$-fold degeneracy.

Sitter superalgebra decompose c.f. (17),

$$\begin{aligned}
\{Q_\alpha\,, Q^\dagger_\beta\} &= H\,\delta_{\alpha\beta} - \tfrac{1}{2}iM_{ab}\,(\Gamma^a\Gamma^b\Gamma^0)_{\alpha\beta} \\
&\quad +\tfrac{1}{2}(M^+_a\,\Gamma^a\,(1+i\Gamma^0) + M^-_a\,\Gamma^a\,(1-i\Gamma^0))_{\alpha\beta}\,, \\
[H\,, Q_\alpha] &= -\tfrac{1}{2}i(\Gamma^0\,Q)_\alpha\,, \\
[M^\pm_a\,, Q_\alpha] &= \mp\tfrac{1}{2}(\Gamma_a(1\mp i\Gamma^0)\,Q)_\alpha\,.
\end{aligned} \tag{40}$$

For the anti-de Sitter superalgebra, all the bosonic operators can be expressed as bilinears of the supercharges, so that in principle one could restrict oneself to fermionic operators only and employ the projections $(1 \pm i\Gamma^0)Q$ as the basic lowering and raising operators. However, this is not quite what we will be doing later in section 7.

Let us now assume that the spectrum of H is bounded from below,

$$H \geq E_0\,, \tag{41}$$

so that in mathematical terms we are considering lowest-weight irreducible unitary representations. The lowest eigenvalue E_0 is realized on states that we denote by $|E_0, s\rangle$, where E_0 is the eigenvalue of H and s indicates the value of the total angular momentum operator. Of course there are more quantum numbers, e.g. associated with the angular momentum operator directed along some axis (in $d = 4$ there are thus $2s + 1$ degenerate states), but this is not important for

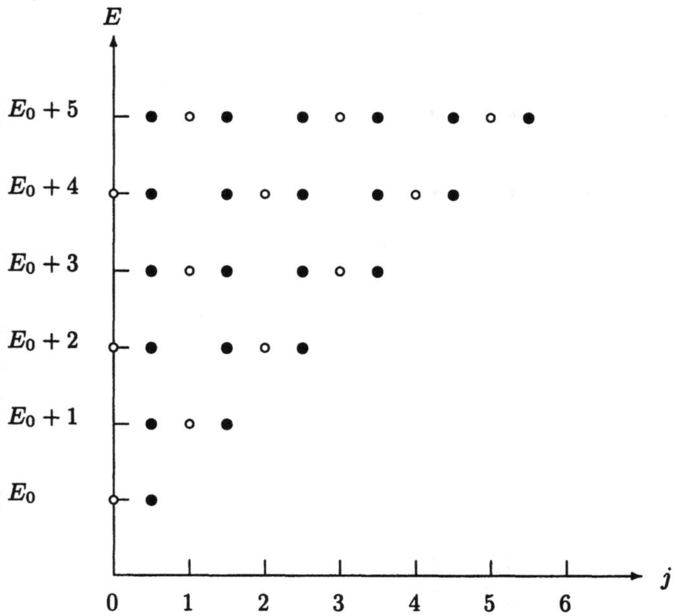

Fig. 2. States of the $s = \frac{1}{2}$ representation in terms of the energy eigenvalues E and the angular momentum j. Each point has a $(2j+1)$-fold degeneracy. The small circles denote the original $s = 0$ multiplet from which the spin-$\frac{1}{2}$ multiplet has been constructed by taking a direct product.

the construction and these quantum numbers are suppressed. Since states with $E < E_0$ should not appear, ground states are characterized by the condition,

$$M_a^- \left|E_0, s\right\rangle = 0\,. \tag{42}$$

The representation can now be constructed by acting with the raising operators on the vacuum state $|E_0, s\rangle$. To be precise, all states of energy $E = E_0 + n$ are constructed by an n-fold product of creation operators M_a^+ In this way one obtains states of higher eigenvalues E with higher spin. The simplest case is the one where the vacuum has no spin ($s = 0$). For given eigenvalue E, the highest spin state is given by the traceless symmetric product of $E - E_0$ operators M_a^+ on the ground state. These states are shown in Fig. 1.

Henceforth we specialize to the case $d = 4$ in order to keep the aspects related to spin simple. To obtain spin-$\frac{1}{2}$ is trivial; one simply takes the direct product with a spin-$\frac{1}{2}$ state. That implies that every point with spin j in Fig. 1 generates two points with spin $j \pm \frac{1}{2}$, with the exception of points associated with $j = 0$, which will simply move to $j = \frac{1}{2}$. The result of this is shown in Fig. 2.

Likewise one can take the direct product with a spin-1 state, but now the situation is more complicated as the resulting multiplet is not always irreducible. In principle, each point with spin j now generates three points, associated with j and $j \pm 1$, again with the exception of the $j = 0$ points, which simply move to $j = 1$. The result of this procedure is shown in Fig. 3.

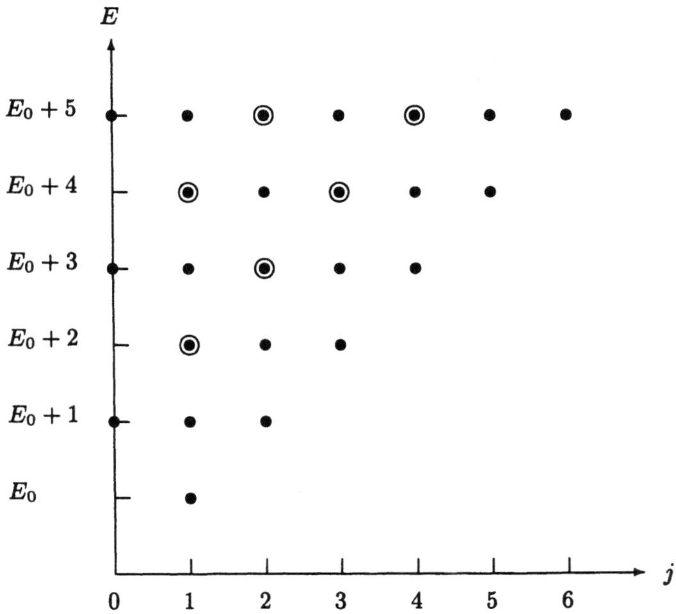

Fig. 3. States of the $s = 1$ representation in terms of the energy eigenvalues E and the angular momentum j. Observe that there are now points with double occupancy, indicated by the circle superimposed on the dots. These points could combine into an $s = 0$ multiplet with ground state $|E_0 + 1, s = 0\rangle$. This $s = 0$ multiplet becomes reducible and can be dropped when $E_0 = 2$, as is explained in the text. The remaining points then constitute a massless $s = 1$ multiplet, shown in Fig. 4.

Let us now turn to the quadratic Casimir operator, which for d spacetime dimensions can be written as

$$\begin{aligned} \mathcal{C}_2 &= -\tfrac{1}{2} M^{AB} M_{AB} \\ &= H^2 - \tfrac{1}{2}\{M_a^+, M_a^-\} - \tfrac{1}{2}(M_{ab})^2 \\ &= H(H-d+1) - \tfrac{1}{2}(M_{ab})^2 - M_a^+ M_a^- \,. \end{aligned} \tag{43}$$

Applying the last expression on the ground state $|E_0, s\rangle$ and assuming $d = 4$ we derive

$$\mathcal{C}_2 = E_0(E_0 - 3) + s(s+1)\,, \tag{44}$$

and, since $\mathcal{C}_2$ is a Casimir operator, this result holds for any state belonging to the corresponding irreducible representation. Note, that the angular momentum operator is given by $\boldsymbol{J}^2 = -\frac{1}{2}(M_{ab})^2$.

We can apply this result to an excited state (which is generically present in the spectrum) with $E = E_0 + 1$ and $j = s - 1$. Here, we assume that the ground state has $s \geq 1$. In that case we find

$$\begin{aligned} \mathcal{C}_2 &= (E_0+1)(E_0-2) + s(s-1) - \Big| M_a^- |E_0+1, s-1\rangle \Big|^2 \\ &= E_0(E_0-3) + s(s+1)\,, \end{aligned} \tag{45}$$

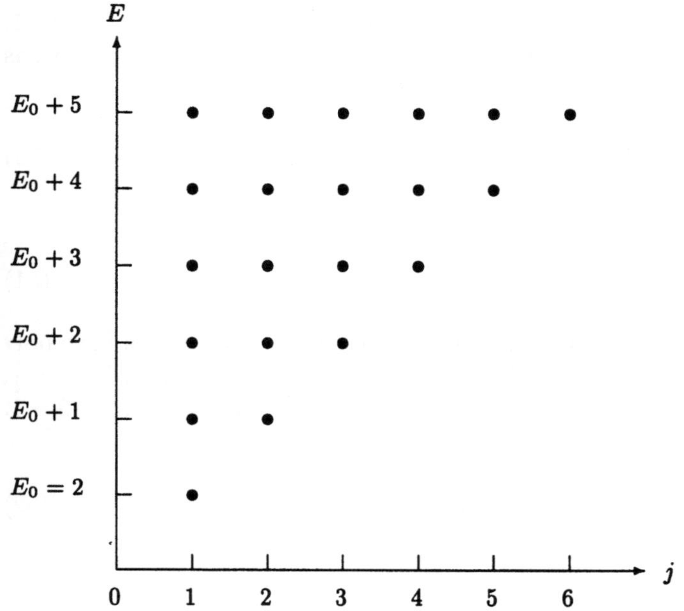

Fig. 4. States of the massless $s=1$ representation in terms of the energy eigenvalues E and the angular momentum j. Now E_0 is no longer arbitrary but it is fixed to $E_0=2$.

so that

$$E_0 - s - 1 = \tfrac{1}{2}\Big|M_a^- |E_0+1, s-1\rangle\Big|^2 . \tag{46}$$

This shows that $E_0 \geq s+1$ in order to have a unitary multiplet. When $E_0 = s+1$, however, the state $|E_0+1, s-1\rangle$ is itself a ground state, which decouples from the original multiplet, together with its corresponding excited states. This can be interpreted as the result of a gauge symmetry and therefore we call these multiplets *massless*. Hence massless multiplets with $s \geq 1$ are characterized by

$$E_0 = s+1 , \qquad \text{for} \quad s \geq 1 . \tag{47}$$

For these particular values the quadratic Casimir operator is

$$\mathcal{C}_2 = 2(s^2-1) . \tag{48}$$

Although this result is only derived for $s \geq 1$, it also applies to massless $s=0$ and $s=\frac{1}{2}$ representations, as we shall see later. Massless $s=0$ multiplets have either $E_0=1$ or $E_0=2$, while massless $s=\frac{1}{2}$ multiplets have $E_0=\frac{3}{2}$.

One can try and use the same argument again to see if there is a possibility that even more states decouple. Consider for instance a state with the same spin as the ground state, with energy E. In that case

$$E(E-3) = E_0(E_0-3) + \Big|M_a^- |E, s\rangle\Big|^2 . \tag{49}$$

For spin $s \geq 1$, this condition is always satisfied in view of the bound $E_0 \geq s+1$. But for $s = 0$, one can apply (49) for the first excited $s = 0$ state which has $E = E_0 + 2$. In that case one derives

$$2(2E_0 - 1) = \left| M_a^- |E_0 + 2, s = 0\rangle \right|^2 , \tag{50}$$

so that

$$E_0 \geq \tfrac{1}{2} . \tag{51}$$

For $E_0 = \frac{1}{2}$ we have the so-called singleton representation, where we have only one state for a given value of the spin. A similar result can be derived for $s = \frac{1}{2}$, where one can consider the first excited state with $s = \frac{1}{2}$, which has $E = E_0 + 1$. One then derives

$$2(E_0 - 1) = \left| M_a^- |E_0 + 1, s = \tfrac{1}{2}\rangle \right|^2 , \tag{52}$$

so that

$$E_0 \geq 1 . \tag{53}$$

For $E_0 = 1$ we have the spin-$\frac{1}{2}$ singleton representation, where again we are left with just one state for every spin value. The existence of these singleton representations was first noted by Dirac [3]. They are shown in Fig. 5. Both singletons have the same value of the Casimir operator,

$$\mathcal{C}_2 = -\tfrac{5}{4} . \tag{54}$$

From the above it is clear that we are dealing with the phenomenon of multiplet shortening for specific values of the energy and spin of the ground state. This can be understood more generally from the fact that the $[M_a^+, M_b^-]$ commutator acquires zero or negative eigenvalues for certain values of E_0 and s. We will return to this phenomenon in section 7 in the context of the anti-de Sitter superalgebra.

6 The oscillator construction

There exists a constructive procedure for determining the unitary irreducible representations of the anti-de Sitter algebra, which is known as the oscillator method. This method can be used for any number of dimensions and also for the supersymmetric extension of the anti-de Sitter algebra [29,30]. Here we will demonstrate it for the case of four spacetime dimensions.

Consider an even $n = 2p$ or an odd $n = 2p + 1$ number of bosonic harmonic oscillators, whose creation and annihilation operators transform as doublets under the compact subgroup $\mathrm{U}(1) \times \mathrm{SU}(2)$ of the covering group $\mathrm{Sp}(4) \cong \mathrm{SO}(3,2)$.

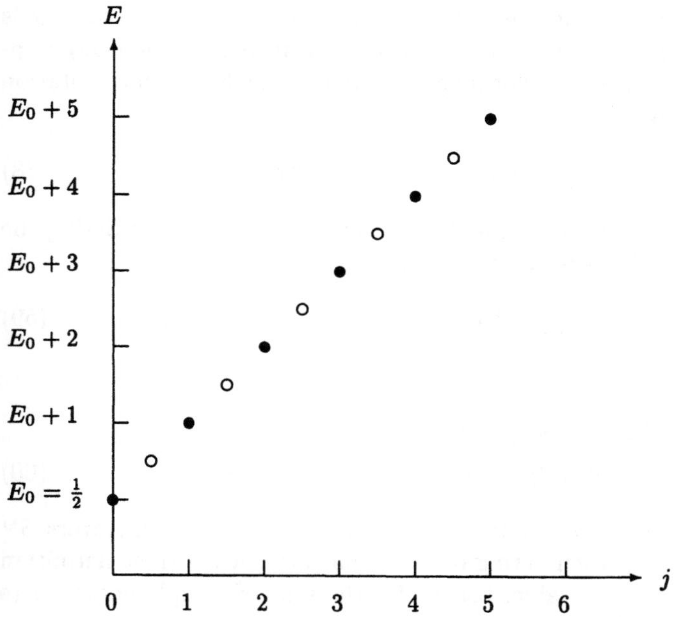

Fig. 5. The spin-0 and spin-$\frac{1}{2}$ singleton representations. The solid dots indicate the states of the $s=0$ singleton, the circles the states of the $s=\frac{1}{2}$ singleton. It is obvious that singletons contain much less degrees of freedom than a generic local field. The value of E_0, which denotes the spin-0 ground state energy, is equal to $E_0=\frac{1}{2}$. The $s=\frac{1}{2}$ singleton ground state has an energy equal to unity, as is explained in the text.

We introduce pairs of mutually commuting annihilation operators $a_i(r)$ and $b_i(r)$ labeled by $r=1,\ldots,p$ and an optional annihilation operator c_i when we wish to consider an odd number of oscillators. The indices i are the doublet indices associated with SU(2). The nonvanishing commutation relations are

$$\begin{aligned}
[a_i(r), a^j(s)] &= \delta_i{}^j\,\delta_{rs}\,,\\
[b_i(r), b^j(s)] &= \delta_i{}^j\,\delta_{rs}\,,\\
[c_i, c^j] &= \delta_i{}^j\,,
\end{aligned} \tag{55}$$

where the creation operators carry upper SU(2) indices and are defined by $a^i=(a_i)^\dagger$, $b^i=(b_i)^\dagger$ and $c^i=(c_i)^\dagger$. The generators of $\mathrm{U}(1)\times\mathrm{SU}(2)$ are then given by

$$U^i{}_j = a^i\cdot a_j + b_j\cdot b^i + \tfrac{1}{2}(c^i\,c_j + c_j\,c^i)\,, \tag{56}$$

where $a_i\cdot a^j$ stands for $\sum_r a_i(r)\,a^j(r)$. The U(1) generator will be denoted by $Q=\frac{1}{2}U^i{}_i$ and can be expressed as

$$\begin{aligned}
Q &= \tfrac{1}{2}(a^i\cdot a_i + b_i\cdot b^i + \tfrac{1}{2}c^ic_i + \tfrac{1}{2}c_ic^i)\\
&= \tfrac{1}{2}(a^i\cdot a_i + b^i\cdot b_i + c^ic_i + 2p + 1)\\
&= \tfrac{1}{2}(N+n)\,,
\end{aligned} \tag{57}$$

where N is the number operator for the oscillator states. Observe that Q is associated with the generator that we previously identified with the energy operator. The other generators, transforming according to the $\mathbf{3}+\bar{\mathbf{3}}$ representation of SU(2), are defined by

$$S^{ij} = (S_{ij})^\dagger = a^i \cdot b^j + a^j \cdot b^i + c^i c^j \,. \tag{58}$$

It is now easy to identify the raising and lowering operators by considering the commutation relations of Q with all the other operators,

$$[Q\,,U^i{}_j] = 0\,, \qquad [Q\,,S^{ij}] = S^{ij}, \qquad [Q\,,S_{ij}] = -S_{ij}\,. \tag{59}$$

Together with

$$\begin{aligned}
&[S^{ij},S^{kl}] = [S_{ij},S_{kl}] = 0\,,\\
&[S^{ij},S_{kl}] = \delta^i{}_k\,U^j{}_l + \delta^i{}_l\,U^j{}_k + \delta^j{}_k\,U^i{}_l + \delta^j{}_l\,U^i{}_k\,,
\end{aligned} \tag{60}$$

we recover all commutation relations of SO(3,2). Obviously, the operators S^{ij} raise the eigenvalue of Q, when acting on its eigenstates, while their hermitian conjugates S_{ij} lower the eigenvalue. Let us, for the sake of completeness, write down the commutation relations of Q with the oscillators,

$$[Q\,,a^i] = \tfrac{1}{2}a^i\,, \qquad [Q\,,a_i] = -\tfrac{1}{2}a_i\,. \tag{61}$$

We see that a^i raises the energy by half a unit whereas a_i lowers it by the same amount. The same relations hold of course true for the oscillators b^i and c^i. The ground state $|\Omega\rangle$ is then defined by

$$S_{ij}|\Omega\rangle = 0\,. \tag{62}$$

The representation is built by acting with an arbitrary product of raising operators S^{ij} on the ground state. Depending on the number of oscillators we have chosen certain states will be present whereas others will not. In this way the shortening of the multiplets will be achieved automatically. Experience has taught us that the oscillator construction is complete in the sense that it yields all unitary irreducible representations. However, it is not possible to describe the construction for arbitrary dimension, as every case has its own characteristic properties.

The obvious choice for $|\Omega\rangle$ is the vacuum state $|0\rangle$ of the oscillator algebra. However, there are other possibilities. For example, we can act on $|0\rangle$ by any number of different creation operators, i.e. $a^i(r_1)\,a^j(r_2)\,b^k(r_3)\cdots|0\rangle$, as long as we do not include a pair $a^i(r_1)\,b^j(r_2)$ with $r_1 = r_2$, unless it is anti-ssymmetrized in indices i and j. The reason is that S_{ij} consists of terms that are linear in both $a_i(r_1)$ and $b_j(r_2)$ annihilation operators with $r_1 = r_2$ and with symmetrized SU(2) indices. Let us now turn to a number of relevant examples in order to clarify the procedure.

Assume that we have a single harmonic oscillator (i.e. $n = 1$). Then there are two possible ground states. One is $|\Omega\rangle = |0\rangle$. In that case we have $E_0 =$

$Q = \frac{1}{2}$ and $s = 0$. The states take the form of products of *even* numbers of creation operators, i.e. $c^i\, c^j\, c^k \cdots |0\rangle$, which are symmetric in the SU(2) indices because the creation operators are mutually commuting. Obviously these states comprise states of spin 1, 2, 3, ... with multiplicity one. This is the $s = 0$ singleton representation. The spin-$\frac{1}{2}$ singleton follows from choosing the ground state $|\Omega\rangle = c^i|0\rangle$, which has $E_0 = Q = 1$ and $s = \frac{1}{2}$. The states are again generated by even product of creation operators which lead to states of spin $\frac{3}{2}$, $\frac{5}{2}$, ... with multiplicity one.

Let us now consider the case of two oscillators ($n = 2$). Here we distinguish the following ground states and corresponding irreducible representations:

- One obvious ground state is the oscillator ground state, $|\Omega\rangle = |0\rangle$. In that case we have $E_0 = Q = 1$ and $s = 0$. This is the massless $s = 0$ representation.
- Alternative ground states are $|\Omega\rangle = a^i|0\rangle$ or $|\Omega\rangle = b^i|0\rangle$. In that case the ground state has $E_0 = Q = \frac{3}{2}$ and $s = \frac{1}{2}$. This is the massless $s = \frac{1}{2}$ representation.
- Yet another option is to choose $|\Omega\rangle$ equal to m annihilation operators exclusively of the a-type or of the b-type, applied to $|0\rangle$. This ground state has $E_0 = Q = 1 + \frac{1}{2}m$ and $s = \frac{1}{2}m$. From the values of E_0 and s one deduces that these are precisely the massless spin-s representations.
- Finally one may choose $|\Omega\rangle = (a^i\, b^j - a^j\, b^i)|0\rangle$, which has $E_0 = Q = 2$ and $s = 0$. This is the second massless $s = 0$ representation.

To sum up, for a single oscillator one recovers the singleton representations and for two oscillators one obtains all massless representations. The excited states in a given representation are constructed by applying arbitrary products of an *even* number of creation operators on the ground state. For more than two oscillators, one obtains the massive representations. This pattern, sometimes with small variations, repeats itself for other than four spacetime dimensions.

7 The superalgebra OSp(1|4)

In this section we return to the anti-de Sitter superalgebra. We start from the (anti-)commutation relations already established in (39) and (40). For definiteness we discuss the case of four spacetime dimensions with a Majorana supercharge Q. This allows us to make contact with the material discussed in section 3. These anti-de Sitter multiplets were discussed in [17–20].

We choose conventions where the gamma matrices are given by

$$\Gamma^0 = \begin{pmatrix} -i\mathbf{1} & 0 \\ 0 & i\mathbf{1} \end{pmatrix}, \qquad \Gamma^a = \begin{pmatrix} -i\sigma^a & 0 \\ 0 & i\sigma^a \end{pmatrix}, \quad a = 1,2,3, \tag{63}$$

and write the Majorana spinor Q in the form

$$Q = \begin{pmatrix} q_\alpha \\ \varepsilon_{\alpha\beta}\, q^\beta \end{pmatrix}, \tag{64}$$

where $q^\alpha \equiv q^\dagger_\alpha$ and the indices $\alpha, \beta, \ldots$ are two-component spinor indices. We substitute these definitions into (40) and obtain

$$\begin{aligned}
[H\,, q_\alpha] &= -\tfrac{1}{2} q_\alpha\,, \\
[H\,, q^\alpha] &= \tfrac{1}{2} q^\alpha\,, \\
\{q_\alpha\,, q^\beta\} &= (H\,\mathbf{1} + \boldsymbol{J}\cdot\boldsymbol{\sigma})_\alpha{}^\beta\,, \\
\{q_\alpha\,, q_\beta\} &= M_a^-\,(\sigma^a\sigma^2)_{\alpha\beta}\,, \\
\{q^\alpha\,, q^\beta\} &= M_a^+\,(\sigma^2\sigma^a)^{\alpha\beta}\,,
\end{aligned} \tag{65}$$

where we have defined the angular momentum operator $J_a = -\frac{1}{2}\, i\, \varepsilon_{abc} M^{bc}$. We see that the operators q_α and q^α are lowering and raising operators, respectively. They change the energy of a state by half a unit.

In analogy to the bosonic case, we study unitary irreducible representations of the OSp(1|4) superalgebra. We assume that there exists a lowest-weight state $|E_0, s\rangle$, characterized by the fact that it is annihilated by the lowering operators q_α,

$$q_\alpha |E_0, s\rangle = 0\,. \tag{66}$$

In principle we can now choose a ground state and build the whole representation upon it by applying products of raising operators q^α. However, we only have to study the *antisymmetrized* products of the q^α, because the symmetric ones just yield products of the operators M_a^+ by virtue of (65). Products of the M_a^+ simply lead to the higher-energy states in the anti-de Sitter representations of given spin that we considered in section 5. By restricting ourselves to the antisymmetrized products of the q^α we thus restrict ourselves to the ground states upon which the full anti-de Sitter representations are build. These ground states are $|E_0, s\rangle$, $q^\alpha|E_0, s\rangle$ and $q^{[\alpha}q^{\beta]}|E_0, s\rangle$. Let us briefly discuss these representations for different s.

The $s = 0$ case is special since it contains less anti-de Sitter representations than the generic case. It includes the spinless states $|E_0, 0\rangle$ and $q^{[\alpha}q^{\beta]}|E_0, 0\rangle$ with ground-state energies E_0 and $E_0 + 1$, respectively. There is one spin-$\frac{1}{2}$ pair of ground states $q^\alpha|E_0, 0\rangle$, with energy $E_0 + \frac{1}{2}$. As we will see below, these states correspond exactly to the scalar field A, the pseudo-scalar field B and the spinor field ψ of the chiral supermultiplet, that we studied in section 3.

For $s \geq \frac{1}{2}$ we are in the generic situation. We obtain the ground states $|E_0, s\rangle$ and $q^{[\alpha}q^{\beta]}|E_0, s\rangle$ which have both spin s and which have energies E_0 and $E_0 + 1$, respectively. There are two more (degenerate) ground states, $q^\alpha|E_0, s\rangle$, both with energy $E_0 + \frac{1}{2}$, which decompose into the ground states with spin $j = s - \frac{1}{2}$ and $j = s + \frac{1}{2}$.

As in the purely bosonic case of section 5, there can be situations in which states decouple so that we are dealing with multiplet shortening associated with gauge invariance in the corresponding field theory. The corresponding multiplets are then again called massless. We now discuss this in a general way analogous to the way in which one discusses BPS multiplets in flat space. Namely, we consider the matrix elements of the operator $q_\alpha\, q^\beta$ between the $(2s+1)$-degenerate ground states $|E_0, s\rangle$,

$$\begin{aligned}\langle E_0, s|\, q_\alpha q^\beta\, |E_0, s\rangle &= \langle E_0, s|\{q_\alpha\,, q^\beta\}|E_0, s\rangle \\ &= \langle E_0, s|(E_0\,\mathbf{1} + \boldsymbol{J}\cdot\boldsymbol{\sigma})_\alpha{}^\beta|E_0, s\rangle\,. \end{aligned} \tag{67}$$

This expression constitutes an hermitean matrix in both the quantum numbers of the degenerate groundstate and in the indices α and β, so that it is $(4s+2)$-by-$(4s+2)$. Because we assume that the representation is unitary, this matrix must be positive definite, as one can verify by inserting a complete set of intermediate states between the operators q_α and q^β in the matrix element on the left-hand side. Obviously, the right-hand side is manifestly hermitean as well, but in order to be positive definite the eigenvalue E_0 of H must be big enough to compensate for possible negative eigenvalues of $\boldsymbol{J}\cdot\boldsymbol{\sigma}$, where the latter is again regarded as a $(4s+2)$-by-$(4s+2)$ matrix. To determine its eigenvalues, we note that $\boldsymbol{J}\cdot\boldsymbol{\sigma}$ satisfies the following identity,

$$(\boldsymbol{J}\cdot\boldsymbol{\sigma})^2 + (\boldsymbol{J}\cdot\boldsymbol{\sigma}) = s(s+1)\mathbf{1}\,, \tag{68}$$

as follows by straightforward calculation. This shows that $\boldsymbol{J}\cdot\boldsymbol{\sigma}$ has only two (degenerate) eigenvalues (assuming $s \neq 0$, so that the above equation is not trivially satisfied), namely s and $-(s+1)$. Hence in order for (67) to be positive definite, E_0 must satisfy the inequality

$$E_0 \geq s+1\,, \quad \text{for } s \geq \tfrac{1}{2}\,, \tag{69}$$

If the bound is saturated, i.e. if $E_0 = s+1$, the expression on the right-hand side of (67) has zero eigenvalues so that there are zero-norm states in the multiplet which decouple. In that case we must be dealing with a massless multiplet. As an example we mention the case $s = \frac{1}{2}, E_0 = \frac{3}{2}$, which corresponds to the massless vector supermultiplet in four spacetime dimensions. Observe that we have multiplet shortening here without the presence of central charges.

One can also use the oscillator method discussed in the previous section to construct the irreducible representations. This is, for instance, done in [21,22].

Armed with these results we return to the masslike terms of section 3 for the chiral supermultiplet. The ground-state energy for anti-de Sitter multiplets corresponding to the scalar field A, the pseudo-scalar field B and the Majorana spinor field ψ, are equal to E_0, E_0+1 and $E_0+\frac{1}{2}$, respectively. The Casimir operator therefore takes the values

$$\begin{aligned}\mathcal{C}_2(A) &= E_0(E_0-3)\,,\\ \mathcal{C}_2(B) &= (E_0+1)(E_0-2)\,,\\ \mathcal{C}_2(\psi) &= (E_0+\tfrac{1}{2})(E_0-\tfrac{5}{2}) + \tfrac{3}{4}\,.\end{aligned} \tag{70}$$

For massless anti-de Sitter multiplets, we know that the quadratic Casimir operator is given by (48), so we present the value for $\mathcal{C}_2 - 2(s^2-1)$ for the three multiplets, i.e

$$\begin{aligned} C_2(A) + 2 &= (E_0 - 1)(E_0 - 2)\,, \\ C_2(B) + 2 &= E_0(E_0 - 1)\,, \\ C_2(\psi) + \tfrac{3}{2} &= (E_0 - 1)^2\,. \end{aligned} \tag{71}$$

The terms on the right-hand side are not present for massless fields and we should therefore identify them somehow with the common mass parameter. Comparison with the field equations (22) shows for $g = 1$ that we obtain the correct contributions provided we make the identification $E_0 = m + 1$. Observe that we could have made a slightly different identification here; the above result remains the same under the interchange of A and B combined with a change of sign in m (the latter is accompanied by a chiral redefinition of ψ).

Outside the context of supersymmetry, we could simply assign independent mass terms with a mass parameter μ for each of the fields, by equating $C_2 - 2(s^2 - 1)$ to μ^2. In this way we obtain

$$E_0(E_0 - 3) - (s + 1)(s - 2) = \mu^2\,, \tag{72}$$

which leads to

$$E_0 = \tfrac{3}{2} \pm \sqrt{(s - \tfrac{1}{2})^2 + \mu^2}\,. \tag{73}$$

For $s \geq \frac{1}{2}$ we must choose the plus sign in (73) in order to satisfy the unitarity bound $E_0 \geq s + 1$. For $s = 0$ both signs are acceptable as long as $\mu^2 \leq \frac{3}{4}$. Observe, however, that μ^2 can be negative but remains subject to the condition $\mu^2 \geq -(s - \frac{1}{2})^2$ in order that E_0 remains real. For $s = 0$, this is precisely the bound of Breitenlohner and Freedman for the stability of the anti-de Sitter background against small fluctuations of the scalar fields [17].

We can also compare $C_2 - 2(s^2 - 1)$ to the conformal wave operator for the corresponding spin. This shows that (again with unit anti-de Sitter radius), $C_2 = \Box_{\mathrm{adS}} + \delta_{\mathrm{s}}$, where δ_{s} is a real number depending on the spin of the field. Comparison with the field equations of section 3 shows that δ_{s} equals 0, $\frac{3}{2}$ and 3, for $s = 0, \frac{1}{2}$ and 1, respectively.

In the case of N-extended supersymmetry the supercharges transform under an SO(N) group and we are dealing with the so-called OSp(N|4) algebras. Their representations can be constructed by the methods discussed in these lectures. However, the generators of SO(N) will now also appear on the right-hand side of the anticommutator of the two supercharges, thus leading to new possibilities for multiplet shortening. For an explicit discussion of this we refer the reader to [19].

8 Conclusions

In these lectures we discussed the irreducible representations of the anti-de Sitter algebra and its superextension. Most of our discussion was restricted to four spacetime dimensions, but in principle the same methods can be used for anti-de Sitter spacetimes of arbitrary dimension.

For higher-extended supergravity, the only way to generate a cosmological constant is by elevating a subgroup of the rigid invariances that act on the gravitini to a local group. This then leads to a cosmological constant, or to a potential with possibly a variety of extrema, and corresponding masslike terms which are quadratic and linear in the gauge coupling constant, respectively. So the relative strength of the anti-de Sitter and the gauge group generators on the right-hand side of the $\{Q, \bar{Q}\}$ anticommutator is not arbitrary and because of that maximal multiplet shortening can take place so that the theory can realize a supermultiplet of massless states that contains the graviton and the gravitini. Of course, this is all under the assumption that the ground state is supersymmetric. But these topics are outside the scope of these lectures and will be reviewed elsewhere [31].

We thank M. Günaydin for valuable comments. IH is supported by the Swiss National Science Foundation through the graduate fellowship 83EU-053229. This work is also supported by the European Commission TMR programme ERBFMRX-CT96-0045.

References

1. J. Maldacena, The large-N limit of superconformal field theories and supergravity, Adv. Theor. Math. Phys. **2** (1998) 231, `hep-th/9711200`.
2. P.A.M. Dirac, The electron wave equation in de Sitter space, Ann. Math. **36** (1935) 657.
3. P.A.M. Dirac, A remarkable representation of the $3+2$ de Sitter group, J. Math. Phys. **4** (1963) 901.
4. C. Fronsdal, Elementary particles in a curved space, Rev. Mod. Phys. **37** (1965) 221; Elementary particles in a curved space II, Phys. Rev. **D10** (1974) 589; C. Fronsdal and R.B. Haugen, Elementary particles in a curved space III, Phys. Rev. **D12** (1975) 3810; C. Fronsdal, Elementary particles in a curved space IV, Phys. Rev. **D12** (1975) 3819.
5. S.J. Avis, C.J. Isham and D. Storey, Quantum field theory in anti-de Sitter space-time, Phys. Rev. **D18** (1978) 3565.
6. D.W. Düsedau and D.Z. Freedman, Lehmann spectral representation for anti-de Sitter quantum field theory, Phys. Rev. **D33** (1986) 389.
7. D.Z. Freedman, Supergravity with axial-gauge invariance, Phys. Rev. **D15** (1977) 1173.
8. D.Z. Freedman and A. Das, Gauge internal symmetry in extended supergravity, Nucl. Phys. **B120** (1977) 221.
9. D.Z. Freedman and J.H. Schwarz, N=4 supergravity theory with local SU(2) × SU(2) invariance, Nucl. Phys. **B137** (1978) 333.
10. B. de Wit and H. Nicolai, Extended supergravity with local SO(5) invariance, Nucl. Phys. **B188** (1981) 98.
11. B. de Wit and H. Nicolai, $N = 8$ supergravity with local SO(8) × SU(8) invariance, Phys. Lett. **108B** (1982) 285; $N = 8$ supergravity, Nucl. Phys. **B208** (1982) 323.

12. S.J. Gates and B. Zwiebach, Gauged $N = 4$ supergravity with a new scalar potential, Phys. Lett. **123B** (1983) 200.
13. M. Pernici, K. Pilch and P. van Nieuwenhuizen, Gauged maximally extended supergravity in seven dimension, Phys. Lett. **143B** (1984) 103.
14. C.M. Hull, More gaugings of $N = 8$ supergravity, Phys. Lett. **148B** (1984) 297.
15. F. Gianni, M. Pernici and P. van Nieuwenhuizen, Gauged $N = 4$ supergravity in six dimension, Phys. Rev. **D30** (1984) 1680.
16. M. Günaydin, L.J. Romans and N.P. Warner, Compact and non-compact gauged supergravity theories in five dimensions, Nucl. Phys. **B272** (1986) 598.
17. P. Breitenlohner and D.Z. Freedman, Stability in gauged extended supergravity, Ann. Phys. **144** (1982) 249.
18. W. Heidenreich, All linear unitary irreducible representations of de Sitter supersymmetry with positive energy, Phys. Lett. **110B** (1982) 461.
19. D.Z. Freedman and H. Nicolai, Multiplet shortening in $\mathrm{OSp}(N|4)$, Nucl. Phys. **B237** (1984) 342.
20. H. Nicolai, Representations of supersymmetry in anti-de Sitter space, in Supersymmetry and Supergravity '84, Proceedings of the Trieste Spring School, eds. B. de Wit, P. Fayet, P. van Nieuwenhuizen (World Scientific, 1984).
21. M. Günaydin and N.P. Warner, Unitary supermultiplets of $\mathrm{OSp}(8|4,\mathbf{R})$ and the spectrum of the S^7 compactification of eleven-dimensional supergravity, Nucl. Phys. **B272** (1986) 99.
22. M. Günaydin, P. van Nieuwenhuizen and N.P. Warner, General construction of the unitary representations of anti-de Sitter superalgebras and the spectrum of the S^4 compactification of eleven-dimensional supergravity, Nucl. Phys. **B255** (1985) 63.
23. S. Ferrara, Algebraic properties of extended supergravity in de Sitter space, Phys. Lett. **69B** (1977) 481.
24. B. de Wit and A. Zwartkruis, $\mathrm{SU}(2,2|1,1)$ supergravity and $N = 2$ supersymmetry with arbitrary cosmological constant, Class. Quantum Grav. **4** (1987) L59.
25. P.K. Townsend, Cosmological constant in supergravity, Phys. Rev. **D15** (1977) 2802.
26. S. Deser and B. Zumino, Broken supersymmetry and supergravity, Phys. Rev. Lett. **38** (1977) 1433.
27. W. Nahm, Supersymmetries and their representations, Nucl. Phys. **B135** (1978) 149.
28. B. de Wit, Multiplet calculus, in Supersymmetry and Supergravity '82, Proceedings of the Trieste School, eds. S. Ferrara, J.G. Taylor and P. Van Nieuwenhuizen (World Scientific, 1983).
29. M. Günaydin and C. Saclioglu, Oscillator like unitary representations of noncompact groups with a Jordan structure and the noncompact groups of supergravity, Commun. Math. Phys. **91** (1982) 159.
30. M. Günaydin, Oscillator like unitary representations of noncompact groups and supergroups and extended supergravity theories, in Int. Colloq. on Group Theoretical Methods in Physics, Istanbul 1982, eds. M. Serdaroglu and E. Inönü, Lect. Notes in Physics **180** (Springer, 1983).
31. B. de Wit, Gauged supergravity, lectures at the Trieste Spring School 1999, to be published in the proceedings.

Combinatorial Dynamics and Time in Quantum Gravity

Stuart Kauffman[1] and Lee Smolin[2]

[1] Biosgroup LTD
317 Paseo de Peralta,
Santa Fe, New Mexico, 87501-8943
[2] Center for Gravitational Physics and Geometry
Department of Physics
The Pennsylvania State University
University Park, PA, USA 16802

Abstract. We describe the application of methods from the study of discrete dynamical systems to the study of histories of evolving spin networks. These have been found to describe the small scale structure of quantum general relativity and extensions of them have been conjectured to give background independent formulations of string theory. We explain why the the usual equilibrium second order critical phenomena may not be relevant for the problem of the continuum limit of such theories, and why the relevant critical phenomena analogue to the problem of the continuum limit is instead non-equilibrium critical phenomena such as directed percolation. The fact that such non-equilibrium critical phenomena may be self-organized implies the possibility that the classical limit of quantum theories of gravity may exist without fine tuning of parameters. We note that dynamical theories of the kind described here may be formulated so as not to employ the notion of a fixed configuration space, and so avoid problems of constructibility of configuration spaces based on taking the quotient by the diffeomorphism group. In such a theory time plays a necessarily fundamental role.

1 Introduction

The idea that space and time are fundamentally discrete is very old and has often reappeared in the history of the search for a quantum theory of gravity[1] However, it is only recently that concrete results from attempts to construct a quantum theory have gravity have been found which suggest very strongly that such a theory must be based on a discrete structure. These results come from the quantization of general relativity[3,4], string theory[5] and the thermodynamics of black holes[6–8]. (For reviews see[9–12].)

If space and time are discrete, then the study of the dynamics of the spacetime may benefit from our understanding of other discrete dynamical systems such as cellular automata[14], froths[15] and binary networks[16]. The importance of this may be seen once it is appreciated that a key problem in any discrete theory of quantum gravity must be the recovery of continuous space time and the fields that live on it as an approximation in an appropriate continuum limit.

[1] see, for example [1,2].

This continuum limit, which will be also related to the classical limit of the theory, (because the physical cutoff l_{Planck} which marks the transition between the discrete and continuous picture is proportional to $\hbar$) is then a problem in critical phenomena[13]. As one doesn't want the existence of classical spacetime to rest on some fine tunings of parameters, this must presumably be some kind of spontaneous, or self-organized critical phenomena[17][2].

However, there is a key element which distinguishes quantum gravity from other kinds of quantum and statistical systems. This is that the causal structure is dynamical. As a result, the usual second order equilibrium critical phenomena may not be relevant for the continuum limit of quantum theories of gravity, as its connection to quantum field theory relies on rotation from a Euclidean to Lorentzian metric and this is not well defined when the fluctuating degrees of freedom are the metric (or causal structure.) Instead, the relevant statistical physics analogue to the problem of the classical limit will be non-equilibrium critical phenomena[18]. To see why, let us consider the issue of critical behavior for a discrete dynamical systems whose only attribute is causal structure. Consider a set $\mathcal{P}$ of N events, such that for any two of them p and q one may have either $p > q$, (meaning p is to the causal future of q), or $q > p$, or neither, but never both. This gives the set $\mathcal{P}$ the structure of a partially ordered set, or poset. In addition, if one assumes that there are no time like loops and that the poset is locally finite (which means that there are only a finite number of events in the intersection of the future of any event and the past of any other) one has what is called a *causal set.* One may then invent an action which depend on the causal relations and then study the quantum statistical physics of such a set, in the limit of large N.

This program has been pursued by physicists interested in using it as a model of quantum gravity, particularly by Myers, Sorkin[19], 'tHooft[20] and collaborators. This is motivated by the fact that the events of any Lorentzian spacetime form a poset, where $p < q$ is the causal relation arising from the lightcone structure of the metric. In fact, if the causal structure is given, the spactime metric is determined up to an overall function.

Sorkin and collaborators have conjectured that the causal structure is sufficient to define a satisfactory quantum theory of spacetime[19]. However, there is reason to believe that this may not be the case, and that additional structure associated with what may loosely be called the properties of space, must be introduced. One reason for this is that the models where the degrees of freedom are only causal structure do not seem, at least so far, to have yielded the kinds of results necessary to answer the key questions about the emergence of the classical limit.

As a result, recently, Markopoulou proposed adding structure to poset models of spacetime taken from results in other approaches to quantum gravity [21]. Her idea has been to combine the discrete causal structure of poset construction with descriptions of a discrete quantum spatial geometry which has emerged

[2] Indeed this is a general problem for particle physics, brought on by the hierarchy problem, which is the existence of several widely separated scales.

from the study of non-perturbative quantum gravity. These descriptions are usually expressed in terms of spin networks, which are graphs whose edges are labeled with half-integers, $1/2, 1, 3/2, \ldots$ which represent quantum mechanical spins. Originally invented by Penrose[1], more recently they have been shown to represent faithfully a basis of exact non-perturbative states of the quantum gravitational field[3,4]. Extensions of the spin network states have also been constructed that are relevant for supergravity[29] and other extensions have been proposed in the context of a conjectured background independent formulation of string theory[25–27]

To show how the discrete causal structure of posets may be fitted to a discrete description of both spacetime and spatial geometry we may need to describe the structure of a causal set $\mathcal{P}$ in more detail. The *Alexandrov neighborhood* of two events p and q, $A(p,q)$, consist of all x such that $p < x < q$. 't Hooft has proposed that the number of events in $A(p,q)$ should be a measure of its volume, in Planck units. If the poset is taken by events picked randomly from a Lorentzian manifold, using the measure given by the volume element, there is then exactly enough information in the poset to reconstruct the metric, in the limit of an infinite number of events. Using the Alexandrov neighborhoods of a poset, we may then construct a discrete model of a spacetime geometry. When the theory has a good classical limit that should approximate a continuous spacetime geometry.

In classical general relativity it is possible to define an infinite number of spatial slices, which have defined on them three dimensional Reimannian geometries. There are an infinite number of ways to slice a spacetime into a sequence of spatial slices, each of which may be associated with surfaces of simultaneity defined by a family of observers and clocks moving in the spacetime. Because the choice of how to slice spacetime into a series of spatial geometries is arbitrary time in general relativity is referred to as being "many-fingered".

A completely analogous notion of spatial geometry can be defined strictly in terms of a poset. To do this we consider a set of events $\Sigma \subset \mathcal{P}$ which consists of events y_i such that no two of them are causally related (i.e. neither $y_i < y_j$ or $y_j < y_i$ for all pairs in Σ.) These may be called "spacelike related". If no event of $\mathcal{P}$ may be added to Σ preserving the condition of no causal relations it is a maximal set of spacelike related events. Such sets are called *antichains* or *discrete spacelike slices* of $\mathcal{P}$. The basic idea of [21] is then to endow the antichains of causal sets with the properties of discrete quantum geometries represented by spin networks. The result gives a notion of a quantum spacetime, which is discrete but which has many of the attributes of continuous spacetime, including causal structure, spacelike slices and many-fingered time. As described in [21] discrete sets having these properties can be constructed by beginning with a spin network and then altering it by a series of local moves.

The purpose of this paper is to raise several key issues involved in the study of the continuum limits in this kind of formulation of quantum gravity. It is written for statistical physicists, relativists and quantum field theorists. Our intention in writing it is mainly to point the attention of people in these fields to the existence of a class of problems in which methods used to study non-equilibrium critical phenomena may play an important role in studies of quantum gravity.

In the next section we describe the basic structure of a causally evolving spin network, in language we hope is accessible to statistical physicists. We do not give any details about how these structures are related to general relativity or its quantization, these may be found elsewhere[3,4,22,23,10,9,11]. Section 3 and 4 then discuss the problem of the classical limit of this theory In section 5 some structures are defined on the set of quantum states of the theory, which are then used in sections 6 and 7, in the context of a simplified model, to argue for the existence of a classical limit that may reproduce general relativity. Section 8 then introduces a new question, which is how the dynamics of the theory is to be chosen. We suggest that it may be reasonable for the dynamics to evolve as the spacetime does, leading to the classical limit as a kind of self-organized critical phenomena.

Finally, in section 9 we discuss a new issue concerning the problem of time in quantum cosmology, which concerns the question of whether the physical configuration space of the theory can be constructed by any finite procedure.

2 Combinatorial descriptions of quantum spacetime

There are actually several closely related versions of the spin network description of quantum spatial geometry[24–26]. As our interest here is on the analysis of their dynamics, we will consider only one kind of model, which is the easiest to visualize. This is associated with combinatorial triangulations[21][3].

We describe first the quantum geometry of space, then how these evolve to make combinatorial spacetimes.

2.1 Combinatorial description of spatial geometry

A combinatorial m-simplex is a set of m points, $e_1, ...e_m$ called the vertices, together with all the subsets of those points. Those subsets with two elements, $e_{12} = \{e_1, e_2\}$... are called edges, those with three $e_{123} = \{e_1, e_2, e_3\}$... faces and so on. A combinatorial tetrahedron is a combinatorial 4 simplex.

A three dimensional simplicial psuedomanifold, T, consists of a set of N combinatorial tetrahedra joined such that each face is in exactly two tetrahedra. Many such psuedomanifolds define manifolds, in which case the neighborhoods of the edges and nodes are homeomorphic to the neighborhoods of edges and nodes in triangulations of Euclidean three space. These are constraints on the construction of the psuedomanifold, which are called the manifold conditions. When they are not satisfied, we have a more general structure of a psuedomanifold. Many psuedomanifolds can be constructed from manifolds by identifying two or more edges or nodes.

The sets on which the manifold conditions fail to be satisfied constitute defects in the topology defined by the combinatorial triangulation. Under suitable

[3] Its exact relationship to the spin network states which arise in canonical quantum gravity is complicated, due to some subtleties which need not concern us here. These are discussed in [28].

choices of the evolution rules these defects propagate in time, forming extended objects, with dimension up to two less than the dimension of the spacetime. When the discrete spacetime has a dynamics, as we will describe below, laws of motion for the extended objects are induced. It is very interesting that string theory in its present form has in it extended objects of various dimensions; the relationship between those "branes" and the defects in psuedomanifolds is under investigation[28].

A psuedomanifold may be labeled by attaching suitable labels to the faces and tetrahedra. For quantum gravity it is useful to consider labels that come from the representation theory of some algebra G, which may be a Lie algebra, a quantum Lie algebra, a supersymmetry algebra, or something more general. Such algebras are characterized by a set of representations, $i, j, k...$ and by product rules for decomposing products of representations, $j \otimes k = \sum_l f^l_{jk} l$, where the f^l_{jk} are integers. Each such algebra has associated to it linear vector spaces V_{ijkl}, which consists of the linear maps $\mu : i \otimes j \otimes k \otimes l \rightarrow 1$, where 1 is the one dimensional identity representation. It is then usual to label a model of quantum gravity with algebra G by associating a representation k with each face and an intertwinor $\mu \in V_{ijkl}$ to each tetrahedra, where i, j, k, l label its four faces. The pseudomanifold T, together with a set of labels is denoted Γ and called a labeled pseudomanifold.

It is particularly convenient to work with a quantum group at a root of unity, as the label sets in these cases are finite. In canonical quantum gravity, the quantum deformation is related to the cosmological constant[30,31].

To each labeled pseudomanifold Γ we associate a basis state $|\Gamma >$ of a quantum theory of gravity. The set of such states spans the state space of the theory, $\mathcal{H}$, whose inner product is chosen so that the topologically distinct $|\Gamma >$'s comprise an orthonormal basis.

Each labeled pseudomanifold is also dual to a *spin network*, which is a combinatorial graph constructed by drawing an edge going through each face and joining the four edges that enter every tetrahedra at a vertex[1,4]. The edges are then labeled by representations and the nodes by intertwinors[4].

If one wants a simpler model one may simply declare all labels to be identical and leave them out. These are called "frozen models[28]". Frozen models are like the dynamical triangulation models of Euclidean quantum gravity, except that there are different kinds of simplices, corresponding to causal ordering. We may also consider "partly frozen" models in which the spins on the faces are all equal, but the intertwinors are allowed to vary over a set of allowed values.

One of the results of the canonical quantization of general relativity is a geometrical interpretation for the spins and intertwinors of spin networks. Given

[4] Note that the pseudomanifolds have more information than the spin networks, for a given spin network may come from several combinatorial triangulations. The spin network structure may be extended so as to code this additional information, for example by extending the edges into tubes as in [25,26]. For simplicity in this paper we stick to psuedomanifolds. In some papers these are also called "dual spin networks"[21].

the correspondence of labeled triangulations to spin networks, this interpretation may be applied directly to the simplices of the labeled spin networks. Doing this, we find that each face f_{abc} of the combinatorial triangulation has an area, which is related to the spin j_{abc} on the face by the formula[3],

$$A_{abc} = l^2_{Planck}\sqrt{j_{abc}(j_{abc}+1)} \tag{1}$$

There are also quanta of volume associated with the combinatorial tetrahedras of the combinatorial triangulations. This correspondence is more complicated, and is motivated as well from canonical quantum gravity. Associated with the finite dimensional space of intertwinors $\mathcal{H}_{j_\alpha}$ at each node, where the spins of the 4 incident edges are fixed to be j_α, is a volume operator $\mathcal{V}_{j_\alpha}$[10,3]. These operators are constructed in canonical quantum gravity[10,3] and shown to be hermitian[32]. They are also finite and diffeomorphism invariant, when constructed through an appropriate regularization procedure[10,3]. Their spectra have been computed[32], yielding a set of eigenvalues $\{v^I_{j_\alpha}\}$ and eigenstates $|v^I_{j_\alpha} > \in \mathcal{H}_{j_\alpha}$. These eigenvalues are given, in units of l^3_{Planck} by certain combinatorial expressions found in [32]. Thus, a combinatorial triangulation represents a quantum geometry where the faces have areas and the tetrahedra volumes, which depend on the labelings in the way we have described.

2.2 Causal evolution of quantum geometries

We now follow the proposal of [21] and construct combinatorial quantum spacetimes by applying a set of evolution rules to the states we have just described. A basis state $|\Gamma_0 > \in \mathcal{H}$ may evolve to one of a finite number of possible successor states $|\Gamma^I_0 >$. Each $|\Gamma^I_0 >$ is derived from $|\Gamma_0 >$ by application of one of four possible moves, called Pachner moves[]. These moves modify the state $|\Gamma_0 >$ in a local region involving one to four adjacent tetrahedra.

Consider any subset of Γ consisting of n adjacent tetrahedra, where n is between 1 and 4, which make up n out of the 5 tetrahedra of a four-simplex S_4. Then there is an evolution rule by which those n tetrahedra are removed, and replaced by the other $5-n$ tetrahedra in the S_4. This is called a Pachner move. The different possible moves are called $n \to (5-n)$ moves (Thus, there are $1 \to 4$, $2 \to 3$, etc. moves. The new tetrahedra must be labeled, by new representations j and intertwiners k. For each move there are 15 labels involved, 10 representations on the faces and 5 intertwinors on the tetrahedra. This is because the labels involved in the move are exactly those of the four simplex S_4. For each n there is then an amplitude $\mathcal{A}_{n\to 5-n}$ that is a function of the 15 labels. A choice of these amplitudes for all possible labels, for the four cases $1 \to 4, ..., 4 \to 1$, then constitutes a choice of the dynamics of the theory.

The application of one of the possible Pachner moves to Γ_0, together with a choice of the possible labelings on the new faces and tetrahedra the move creates, results in a new labeled pseudomanifold state Γ_1. This differs from Γ_0 just in a region which consisted of between 1 and 4 adjacent tetrahedra. The process may be continued a finite number of times N, to yield successor labeled pseudomanifold states $\Gamma_2, ...\Gamma_N$.

Any particular set of N moves beginning with a state Γ_0 and ending with a state Γ_N defines a four dimensional combinatorial structure, which we will call a *history*, $\mathcal{M}$ from Γ_0 to Γ_N. Each history consists of N combinatorial four simplices. The boundary of $\mathcal{M}$, is a set of tetrahedra which fall into two connected sets so that $\partial\mathcal{M} = \Gamma_0 \cup \Gamma_1$. All tetrahedra not in the boundary of $\mathcal{M}$ are contained in exactly two four simplices of $\mathcal{M}$.

Each history $\mathcal{M}$ is a causal set, whose structure is determined as follows. The tetrahedra of each four simplex, S_4 of $\mathcal{M}$ are divided into two sets, which are called the past and the future set. This is possible because each four simplex contains tetrahedra in two states Γ_i and Γ_{i+1} for some i between 0 and N. Those in Γ_i were in the group that were wiped out by the Pachner move, which were replaced by those in Γ_{i+1}. Those that were wiped out are called the past set of that four simplex, the new ones, those in Γ_{i+1} are called the future set. With the exception of those in the boundary, every tetrahedron is in the future set of one four simplex and the past set of another.

The causal structure of $\mathcal{M}$ is then defined as follows. The tetrahedra of $\mathcal{M}$ make up a causal set defined as follows. Given two tetrahedra T_1 and T_2 in $\mathcal{M}$, we say T_2 is to the future of T_1 (written $T_2 > T_1$) iff there is a sequence of causal steps that begin on T_1 and end on T_2. A causal step is a step from a tetrahedron which is an element of the past set of some four simplex, S_4 to any tetrahedron which is an element of the future set of the same four simplex. By construction, there are no closed causal loops, so the partial ordering gives a causal set.

Each history $\mathcal{M}$ may also be foliated by a number of spacelike slices Γ. These are the anitchains that we defined in section 1

Each Γ_i in the original construction of $\mathcal{M}$ constitutes a spacelike slice of $\mathcal{M}$. But there are also many other spacelike slices in $\mathcal{M}$ that are not one of the Γ_i. In fact, given any spacelike slice Γ in $\mathcal{M}$ there are a large, but finite, number of slices which are differ from it by the application of one Pachner move. Because of this, there is in this formulation a discrete analogue of the many fingered time of the canonical picture of general relativity.

2.3 How the dynamics are specified

We have now defined quantum spatial geometry and quantum spacetime histories, both completely combinatorially. To turn this structure into a physical theory we must invent some dynamics. Although it is not the only possible starting point (and we will discuss another in section 8) it is best to begin by being conservative and using the standard notion of the path integral. We then assign to each history $\mathcal{M}$ an amplitude $\mathcal{A}[\mathcal{M}]$ given by

$$\mathcal{A}[\mathcal{M}] = \prod_i A[i] \tag{2}$$

where the product is over the moves, or equivalently the 4-simplices, labeled by i. $A[i]$ is the amplitude for that four simplex, which will be a function of its causal structure ($1 \rightarrow 4$ or the others) and the labels on its faces and tetrahedra. The dynamics is specified by giving the complex function $A[i]$, which depends on the possible causal structures and labels, a choice of such a function is equivalent to a choice of an action.

The amplitude for the transition from an initial state $|i>$ to a final state $|f>$, both in $\mathcal{H}$ is then given by

$$T[i,f] = \sum_{\mathcal{M}|\partial\mathcal{M}=|i>\cup|f>} \mathcal{A}[\mathcal{M}] \tag{3}$$

where the sum is over all histories from the given initial and final state.

The theory is then specified by giving the kinematics, which is the algebra from which the label set is chosen and the dynamics, which is the choice of functions $A[i]$. One important question, which we will now discuss, is whether there are choices that lead to theories that have a good classical limit.

3 The problem of the classical limit and its relationship to critical phenomena

Having defined the class of models we will study, we now turn to our main subject, which is the problem of the classical limit and its relation to problems in non-equilibrium critical phenomena. We begin by making the following observation: Suppose that the amplitudes of each move were real numbers of the form,

$$A[i] = e^{-S(i)} \tag{4}$$

Then the sum over histories can be considered to define a statistical system, whose partition function is of the form,

$$Z[i,f] = \sum_{\mathcal{M}|\partial\mathcal{M}=|i>\cup|f>} e^{-\sum_i S(i)} \tag{5}$$

Thus we have a statistical average over histories, each weighed by a probability, just as in non-equilibrium systems such as percolation problems. In fact, there is an exact relationship with directed percolation problems, as the following example shows.

In Figure (1) we show the setup of a $1+1$ directed percolation problem. The degrees of freedom are the arrows, each of which points to the future, which is upwards in the picture. The *value* or state of an arrow is whether it is on or off. A history, $\mathcal{M}$ of a directed percolation problem is a record of which arrows are on. One such history is shown in Figure (2).

In the simplest version of directed percolation, each arrow is turned on with a probability p. There is a critical probability p^* at which the percolation phase transition takes place. Below p^* the on arrows make up disconnected clusters of finite size, whereas for $p > p^*$ the on arrows almost always form a single

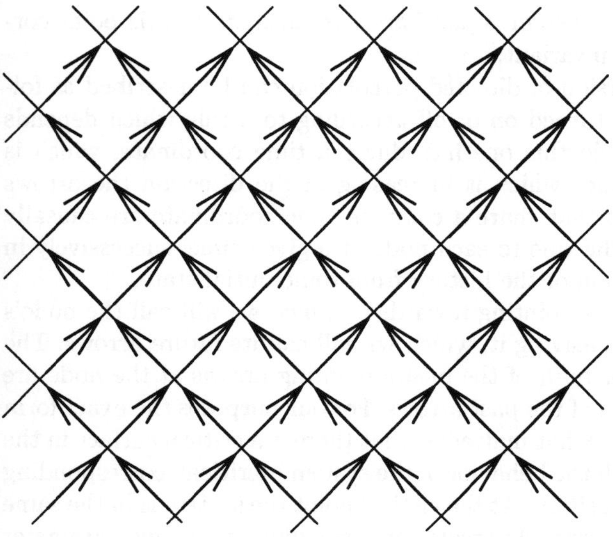

Fig. 1. A 1 + 1 dimensional directed percolation problem.

Fig. 2. One history of a directed percolation system.

connected cluster. At p^* the system is just barely connected. At this point correlation functions are scale invariant.

A more complicated version of directed percolation can be described as follows. Each diagonal link is turned on or off according to a rule which depends on several parameters. To do this one introduces a time coordinate, which is a label attached to the nodes which is increasing in the direction the arrows point and so that all nodes that share a common time coordinate are causally unrelated. We then apply the rule to each node at a given time, successively in time, generating the evolution of the history from some initial state.

Each node has two arrows pointing towards it, which we will call the node's past arrows and two arrows leaving it, which we will call its future arrows. The rule governs whether one or both of the future pointing arrows at the node are on, as a function of the state of the past arrows. For our purposes the exact form of the rules is not important, what matters is that there is a critical surface in the space of parameters at which the behavior of the system is critical, corresponding to the percolation phase transition. At the critical point the system is in the same universality class as simple directed percolation depending on the one parameter p. This second model will be called the dynamical model, as the histories evolve in time, by applying the rule to the nodes at later and later times. A dynamical model may be probabilistic or deterministic, depending on the nature of the rule applied at each node.

Notice that a history $\mathcal{M}$ of a directed percolation problem is a causal set. We will say that a node p is to the future of a node q (and write $p > q$) in a given history $\mathcal{M}$ if there is a chain of on arrows beginning at q and ending at p. A model of directed percolation in $d+1$ dimensions is then a model of dynamical causal structure for a discrete $d+1$ dimensional spacetime. A history $\mathcal{M}$ of a directed percolation model then has a causal structure and all its acutraments, including discrete spacelike surfaces, light cones, future causal domains, past causal domains, etc. In a percolation problem based on a fixed spacetime lattice as in Figure (1 we may define the *background causal structure* to be the one defined by the history in which all the arrows are on.

In particular, the values of the arrows (on or off) at one time t make a state $|\psi>$. If the model has n arrows in each constant time surface, the state space is 4^n dimensional. In the deterministic models an initial state $|\psi_0>$ evolves to a unique history $\mathcal{M}$. Thus a deterministic model of directed percolation is a cellular automata, called a Domany-Kunsel cellular automata model[33].

One way to understand what happens at the directed percolation critical point is to use the concept of damage[16]. In a deterministic model of directed percolation pick an initial state $|\psi_0>$. Evolve the system to a history $\mathcal{M}_0$. Then change one arrow a_0 in the initial state and evolve to the corresponding history $\mathcal{M}_1$. Label any arrow whose value is different in the two histories as *damaged*. The damaged arrows make a connected set $\mathcal{D}$, called the damaged set, which lie in the future causal domain of the arrow a_0 according to the background causal structure.

Hence, we see that damage corresponds to a perturbation of the discrete causal structure. It is interesting to ask how the morphology of the damaged region depends on the phase of the percolation system. Below the percolation phase transition the causal domains are finite and isolated, and the same is true for the damaged sets. Just at the phase transition point, damage is able to propagate arbitrarily far, for the first time. However, the damage is constrained to follow the background causal structure, which is the causal structure of the unperturbed history. Thus, if the theory has a continuum limit, the spread of the damage will correspond to the propagation of some causal effect. But if there is a continuum limit associated with the phase transition, then the correlation functions that measure the spread of damage will be power-law. In this case they should correspond in the continuum limit to the propagation of massless particles. Thus, if we think of the damage as the propagation of a perturbation in the causal structure, it must correspond in the continuum limit to the propagation of a graviton, which is how the propagation of a change in the causal structure is described in the perturbative theory. If the theory has a good continuum limit then the gravitons must travel arbitrarily far up the lightcones of the background causal structure. We see that this will only be possible at the critical point of the directed percolation model.

Thus, by identifying a directed percolation model with a dynamical theory of causal structure, we see that if that theory is to have a continuum limit corresponding to general relativity in 4 or more spacetime dimensions, the only possibility for the existence of such a limit is at the critical point of the directed percolation model. Thus we see that directed percolation critical phenomena must play the same role for discrete models of dynamical causal structure that ordinary second order critical behavior plays in Euclidean quantum field theory.

4 Is there quantum directed percolation?

There is however an important difference between what is required for a theory of quantum gravity and the directed percolation models so far studied by statistical physicists. In a discrete model of quantum gravity each history $\mathcal{M}$ is assigned an amplitude $\mathcal{A}[\mathcal{M}]$, which is generally a complex number. All directed percolation models so far studied (to the authors' knowledge) are either deterministic or probabilistic. In the latter case a probability $p[\mathcal{M}]$ is assigned to each history $\mathcal{M}$, which is of course a real number between 0 and 1. It is only in this case, in which each history has a probability, that we know anything about the critical phenomena associated to directed percolation. However, in quantum mechanics paths are weighed by amplitudes,which are complex numbers. Thus, it would thus be very interesting to know whether there are analogous critical phenomena in models which are set up as directed percolation models, (for example as in Figure (1), except that a complex amplitude $\mathcal{A}[e]$, rather than a probability, is assigned to the state at each node. We may call such a model a *quantum directed percolation model.* We believe that the study of such models could be very useful for understanding the conditions required for discrete models of quantum gravity to have good continuum limits.

One issue that must be stressed is that very little is actually known about the continuum limit for Lorentzian path integrals where the histories are weighed by complex phases rather than probabilities. In quantum mechanics and conventional quantum field theory the path integrals are normally *defined* by analytic continuation from Euclidean field theory, where the weights can be considered probabilities. In the absence of such a definition, one might try to define the sum over histories directly. However, one faces a serious question of whether the sums converge at all.

This problem cannot be avoided in a case such as the present, in which the system is discrete. Of course, the usual wisdom is that the classical limit will exist because the phases from histories which are far-from-classical paths interfere destructively, leaving only the contributions near-classical histories, which add constructively. The problem is that in a finite system, in which there are a finite number of histories in the sum, the cancellation coming from the destructive interference will not be complete. There will be a residue coming from the sum, with a random phase and an absolute value of order $\sqrt{n}$, if there are n far-from-classical histories[5]. This contribution must be much smaller than those coming from close to classical paths, which will have an absolute value of order m, where m is the number of close to classical paths. Thus, the existence of the classical limit seems to require that $m >> \sqrt{n}$, which means that there are many more near classical paths than far-from-classical paths. Of course, in any standard quantum system the actual situation is the opposite, there are many more far-from-classical than near-classical paths.

This argument suggests that the existence of the classical limit may require that a continuum limit has been taken in which the number of histories diverges. In this case it may be possible to tune parameters to define a limit in which the non-classical contribution to the amplitude cancels completely. In essence, this is what is forced by defining the theory in terms of an analytic continuation from a Euclidean field theory.

In the absence of a definition by an analytic continuation, the sums over causal histories may fail to have a good classical limit because they lack both an infinite sum over histories and a suitable definition of a corresponding Euclidean theory. This is perhaps the key question concerning the classical limit of such theories.

5 Discrete superspace and its structure

Having raised several issues concerned with the evaluation of the path integrals that arise in studies of evolving spin networks, we would now like to describe here a formalism and a language which may be useful for addressing them. It is convenient to consider a superspace Ω consisting of all 3 dimensional psuedomanifolds constructed with a finite number of tetrahedra. Associated to this is Ω_G, which is the space of all labeled pseudomanifolds based on the algebra G. These spaces have intrinsic structure generated by the evolution under the Pachner moves.

---

[5] This has been verified in a numerical computation by Sameer Gupta.

Consider an initial pseudomanifold Γ^0, with a finite number of tetrahedra. We then consider all pseudomanifolds γ^1_α that can be reached from Γ^0 by one instance of any of the 4 allowed moves $n \to 5 - n$. They are finite in number, and labeled by an arbitrary integer α. We will call this set $\mathcal{S}^1_{\gamma^0}$. Generalizing this, it is natural then to consider the set $\mathcal{S}^N_{\gamma^0}$ of all pseudomanifolds that can be reached from Γ^0 in N or less moves. Clearly we have $\mathcal{S}^{N-1}_{\gamma^0} \subset \mathcal{S}^N_{\gamma^0}$. We will also want to speak about the "boundary" of $\mathcal{S}^N_{\gamma^0}$, which is $\mathcal{B}^N_{\gamma^0}$, the set of all four valent graphs that can be reached from γ^0 in N moves, but cannot be reached from γ^0 by any path in fewer than N moves. A pseudomanifold in $\mathcal{B}^N_{\gamma^0}$ will be labeled $\gamma^N_{\alpha_1,\ldots\alpha_N}$ where, for example, $\gamma^2_{\alpha_1,\alpha_2}$ is the α_2'th labeled pseudomanifold that can be reached from $\gamma^1_{\alpha_1}$.

It is also convenient to use the following terminology, borrowed from considerations of combinatorial chemistry[34]. We will call the set $\mathcal{S}^1_{\gamma^0}$ the *adjacent possible* set of γ_0, as it consists of all the possible states that could directly follow γ_0. More generally, for any N, the set $\mathcal{B}^N_{\gamma^0}$ will be called the N'th *adjacent possible*, since it contains all the possible new states available to the universe after N steps that were not available after $N-1$ steps.

It is clear that the for states composed of a large number of labeled tetrahedra, the N'th adjacent possible sets grow quickly, as is typical for combinatorial systems.

We may make some straightforward observations about the sets $\mathcal{S}^N_{\gamma^0}$.

- Given two pseudomanifolds α and β in $\mathcal{S}^N_{\gamma^0}$, we will say that α generates β if there is a single move that takes α to β. (For example $\gamma^1_{\alpha_1}$ generates $\gamma^2_{\alpha_1,\alpha_2}$.) $\mathcal{S}^N_{\gamma^0}$ then has the structure of a supergraph $\mathcal{G}^N_{\gamma^0}$, which is a directed graph whose nodes consist of the elements of $\mathcal{S}^N_{\gamma^0}$, connected by directed edges that represent generation.
- A path p in $\mathcal{S}^N_{\gamma^0}$ is a list of pseudomanifolds$\gamma_1, \ldots \gamma_m$, each of whom generates the next. If there exists a path p that runs from α to δ, both elements of $\mathcal{S}^N_{\gamma^0}$ then we may say that $\alpha \leq \delta$, or "α precedes δ". $\mathcal{S}^N_{\gamma^0}$ thus has the structure of a partially ordered set.

There are corresponding statements for Ω_G, the space of all finite labeled pseudomanifolds. We may define the set $\mathcal{M}^N_{\gamma^0}$, an element of which is a labeled pseudomanifold Γ. This corresponds to all elements of Ω_G which may be reached in N steps from an initial labeled pseudomanifold γ_0. We may extend the relations just defined to the elements of $\mathcal{M}^N_{\gamma^0}$. Thus, given two labeled pseudomanifolds Γ and Δ, we may say Γ generates Δ if the graph γ of Γ generates the graph δ of Δ, with the obvious extensions to the notion of a path. Thus, $\mathcal{M}^N_{\gamma^0}$ has as well the structure of a partially ordered set. In addition, we have the "boundary" of $\mathcal{M}^N_{\gamma^0}$, consisting of all the labelings of the elements of $\mathcal{B}^N_{\gamma^0}$, which we may call $\mathcal{A}^N_{\gamma^0}$.

We may note that neither $\mathcal{M}^N_{\gamma^0}$ nor $\mathcal{S}^N_{\gamma^0}$ are causal sets, as for N large enough there will be closed paths that may begin and end on a graph $\gamma \in \mathcal{S}^N_{\gamma^0}$.

We may note that there is an obvious map $r : \Omega_G \rightarrow \Omega$ in which labels are erased.

We consider $\mathcal{M}^N_{\gamma 0}$ to be then the discrete analogue of Wheelers superspace. This is suggested by the fact that the labeled pseudomanifolds diagonalize observables that measure the three geometry. We may note that just as in the continuum case we may put a metric on $\mathcal{M}^N_{\gamma 0}$. If $\alpha > \beta$ or $\beta > \alpha$ then we may say that α and β are causally related. In this case, the metric $g(\alpha, \beta) = n$, the length of the shortest path that connects them. Thus, as in the continuum case, the metric gives the superspaces a poset structure.

6 Some simple models

We will now illustrate some of the issues involved in the continuum limit, using the frozen model as an example. This model is similar to dynamical triangulation models of Euclidean quantum gravity, but it differs from those because of the role of the causal structure. To write it down more explicitly, we let the index c take values over the four types of causal structure: $c \in \{1 \rightarrow 4, 2 \rightarrow 3, 3 \rightarrow 2, 4] \rightarrow 1\}$

There are then four amplitudes $\mathcal{A}[c]$ that must be specified. We may write them in terms of amplitudes and phases as,

$$\mathcal{A}[c] = a_c e^{\imath\theta_c} \tag{6}$$

The amplitude for a history is then given by

$$\mathcal{A}[\mathcal{M}] = \prod_c (\mathcal{A}[c])^{N_c} \tag{7}$$

where N_c is the number of occurances of the c'th causal structure in the history. These of course satisfy

$$N = \sum_c N_c. \tag{8}$$

The model has four parameters, which are the four complex numbers $\mathcal{A}[c]$. It can be further simplified so that it depends only on fewer parameters. One way to do this is to insist that the amplitude are pure phases, so that all four moves have equal probability, but with certain phases,

$$\mathcal{A}[c] = e^{\imath\theta_c} \tag{9}$$

We can further simplify by insisting that each of the pair of moves that are time reversals of each other have the same phase, this means that[6]

$$\mathcal{A}[1 \rightarrow 4] = \mathcal{A}[1 \rightarrow 4] = e^{\imath\alpha} \tag{10}$$

[6] The reader may wonder why we assign the time reversed amplitude to be equal to the original, rather than its complex conjugate. The answer is that we want a process followed, by its time reversal, to be distinct from the process where nothing happens. In general relativity a process and its time reversal are related by a diffeomorphism and thus have equal actions, thus in the quantum theory they are given by equal amplitudes.

$$\mathcal{A}[2 \to 3] = \mathcal{A}[3 \to 2] = e^{\imath\beta} \tag{11}$$

To write the amplitude let us then define $\lambda = \frac{1}{2}(\alpha + \beta)$ and $\mu = \frac{1}{2}(\alpha - \beta)$. The total amplitude of a history $\mathcal{M}$ is then,

$$\mathcal{A}[\mathcal{M}] = e^{\imath(\lambda N_{total} + \mu N_{diff})} \tag{12}$$

where

$$N_{diff} = N[1 \to 4] + N[4 \to 1] - N[2 \to 3] - N[3 \to 2] \tag{13}$$

We see that as N_{total} is proportional to the four volume, λ plays the role of a cosmological constant. It is interesting to compare this to the action for dynamical triangulations, which is of the form $S^{DT} = \lambda N_{total} + \kappa N_2$ where N_2 is the number of two simplices which is a measure of the averaged spacetime scalar curvature. This suggests that if there is a continuum limit N_{diff} might also be a measure of the averaged spacetime curvature scalar, suitable for spacetimes of Minkowskian signature.

7 The classical limit of the frozen models

Of course, the actual behavior of the evolution described by the theory will depend on the details of the amplitudes $\mathcal{A}[c]$. However, it is useful to ask whether any conclusions can be drawn about the evolution in the case that we have no information about the actual forms of the amplitudes Let us make the simplest possible assumption, which is that all the amplitudes are given by some random real phase, so that $\mathcal{A}[c] = e^{i\theta}$. Then the amplitude for any path p is $exp[i\theta n(p)]$.

In this case we can draw some simple conclusions as follows. Consider the amplitudes $\mathcal{A}[\Gamma_0 \to \Gamma_f]$ for all labeled pseudomanifolds $\Gamma_f \in \mathcal{M}^N_{\gamma 0}$. It is clear that for $\Gamma_f \in \mathcal{A}^N_{\gamma 0}$ the amplitudes $\mathcal{A}[\Gamma_0 \to \Gamma_f] = We^{\imath\theta N}$, where W is the number of inequivalent ways to reach Γ_f in N steps. Thus, the amplitudes evolve in such a way that the amplitudes for the states on the boundary is always a coherent phase.

On the other hand, consider a Γ_t which is in the interior of $\mathcal{M}^N_{\gamma 0}$. Let this be an element that is in $\mathcal{A}^M_{\gamma 0}$ for some $M << N$. There will typically be a number of different paths that reach Γ_t, with a variety of different path lengths. The number of such paths will grow rapidly with N, as long as $M << N$. The total amplitudes for such labeled pseudomanifolds to be reached after N steps then will by $\mathcal{A}[\Gamma_0 \to \Gamma_f] = \sum_r e^{\imath\theta r}$ with r a finite set of integers $M \geq r \geq N$. As N grows large this set grows, and there are typically no interesting correlations amongst them. In this case, as N grows large then $\mathcal{A}[\Gamma_0 \to \Gamma_f] \approx 0$.

This means that for N large, most of the amplitude predicted by the path integral (3) with these assumptions will be concentrated on $\mathcal{A}^M_{\gamma 0}$ and a narrow shell trailing it.

This may be considered to be a form of the classical limit, because as N grows, the amplitude to have evolved from Γ_0 to a state Γ_f by an N step path is concentrated on those states that can be reached in N steps, but no fewer. This means that as N increases the amplitude is evolving along geodesics of the metric G defined in the discrete superspace.

8 Dynamics including the parameters

In the class of theories we have formulated here the dynamics of the theory is given by four functions $A[c]$ which give the amplitude for each four simplex which is added to a history as the result of a Pachner move. These functions depend on the causal structure c and labels on the 4-simplices. By using the requirement that the functions are invariant under permutations of the elements of the four simplex that do not change the causal structure, we can reduce the functions $A[c,p]$ to particular forms which depend on a set of parameters, p, which live in a parameter space $\mathcal{P}$. The main dynamical problem is to find the set $\mathcal{P}^* \subset \mathcal{P}$ such that the amplitudes defined by the sum over histories (3) has a good classical limit.

However there is clearly something unsatisfactory about this formulation. No fundamental theory can be considered acceptable if it has a large number of parameters which must be finely tuned to some special values in order that the theory reproduces the gross features of our world. Instead, we would prefer a theory in which the critical behavior necessary for the existence of the classical limit was achieved automatically. Theories of this kind are called "self-organized critical systems".

One possibility is that the parameters p which determine the amplitudes for the different evolution moves are themselves dynamical variables which evolve during the course of the evolution of the system to values which define a critical system with a good continuum limit.

Here is one form of such a theory. We associate to each tetrahedron in the model, $\mathcal{T}_i$ a value of the parameters p_i. When a move is made it involves $n < 5$ tetrahedra. We will assume that the amplitude of the move is given by $A[c, < p >]$ where $< p >$ is the average of the p_i among the n tetrahedra involved in the move. The move creates $5-n$ new tetrahedra. We assign to each of them the new parameters $< p >$. This rule guarantees that those choices of parameters that spread the most widely through the population of tetrahedra govern the most amplitudes. In this way, the system itself may discover and select the parameters that lead to criticality, and hence a classical limit.

Other rules for the new parameters may be contemplated. Another choice is the following. The set of parameters p_α are divided randomly into n sets. The new p_α's in each of these sets are taken from the corresponding values in one of the n "parent" tetrahedra that were input into the move. This distribution of the parameters is made separately for each of the $5-n$ new tetrahedra.

The reader may object that the possibility for giving different rules for the choices of parameters violates our intention that the system choose its own laws. However, this is not the point. There is no way to avoid making a choice in giving rules to the system. What we want to avoid is the circumstance that the rules which result in a classical limit are so unlikely that it seems a miracle that they be chosen properly. What would be more comfortable is an evolution rule that has no sensitive dependence on a choice of parameters that results in the system naturally having a classical limit. By making the system choose the parameters itself, on the basis of a rule that selects those that lead to the most efficient propagation of information, we may make it possible for the system to tune itself to criticality.

9 A new approach to the problem of time

The idea that the parameters in the laws of physics may vary opens up a new possibility for the role of time in the laws of physics. If the laws evolve in time, then time must play a prefered role in the world, it cannot be just another dimension and it cannot dissapear in a timeless block universe or "wavefunction of the universe." Thus, such a possibility forces us to examine the problem of time in quantum cosmology. The following remarks, suggest that the kind of dynamics we have discussed in this paper may indeed lead to a profoundly different view of the role of time in quantum cosmology[7]

The problem of time in quantum cosmology is one of the key conceptual problems faced by theoretical physics at the present time. Although it was first raised during the 1950's, it has resisted solution, despite many different kinds of attempts[35–39]. Here we would like to propose a new kind of approach to the problem. Basically, we will argue that the problem is not with time, but with some of the assumptions that lead to the conclusion that there is a problem. These are assumptions that are quite satisfactory in ordinary quantum mechanics, but that are problematic in quantum gravity, because they may not be realizable with any constructive procedure. In a quantum theory of cosmology this is a serious problem, because one wants any theoretical construction that we use to describe the universe to be something that can be realized in a finite time, by beings like ourselves that live in that universe. If the quantum theory of cosmology requires a non-constructible procedure to define its formal setting, it is something that could only be of use to a mythical creature of infinite capability. One of the things we would like to demand of a quantum theory of cosmology is that it not make any reference to anything at all that might be posited or imagined to exist outside the closed system which is the universe itself.

We believe that this requirement has a number of consequences for the problem of constructing quantum a good quantum theory of cosmology. These have been discussed in detail elsewhere [38,40,41]. Here we would like to describe one more implication of the requirement, which appears to bear on the problem of time.

We begin by summarizing briefly the argument that time is not present in a quantum theory of cosmology. In section 3 we introduce a worry that one of the assumptions of the argument may not be realizable by any finite procedure. (Whether this is actually the case is not known presently.) We explain how the argument for the disappearance of time would be affected by this circumstance. Then we explain how a quantum theory of cosmology might be made which overcomes the problem, but at the cost of introducing a notion of time and causality at a fundamental level. As an example we refer to recent work on the path integral for quantum gravity[42], but the form of the theory we propose is more general, and may apply to a wide class of theories beyond quantum general relativity.

[7] The remained of this paper originally appeared as a separate preprint, and was published on the www.edge.com website. As it is closely related to the subject of this paper, we append it here as a last section.

9.1 The argument for the absence of time

The argument that time is not a fundamental aspect of the world goes like this[8]. In classical mechanics one begins with a space of configurations $\mathcal{C}$ of a system $\mathcal{S}$. Usually the system $\mathcal{S}$ is assumed to be a subsystem of the universe. In this case there is a clock outside the system, which is carried by some inertial observer. This clock is used to label the trajectory of the system in the configuration space $\mathcal{C}$. The classical trajectories are then extrema of some action principle, $\delta I = 0$.

Were it not for the external clock, one could already say that time has disappeared, as each trajectory exists all at once as a curve γ on $\mathcal{C}$. Once the trajectory is chosen, the whole history of the system is determined. In this sense there is nothing in the description that corresponds to what we are used to thinking of as a flow or progression of time. Indeed, just as the whole of any one trajectory exists when any point and velocity are specified, the whole set of trajectories may be said to exist as well, as a timeless set of possibilities.

Time is in fact represented in the description, but it is not in any sense a time that is associated with the system itself. Instead, the t in ordinary classical mechanics refers to a clock carried by an inertial observer, which is not part of the dynamical system being modeled. This external clock is represented in the configuration space description as a special parameterization of each trajectory, according to which the equations of motion are satisfied. Thus, it may be said that there is no sense in which time as something physical is represented in classical mechanics, instead the problem is postponed, as what is represented is time as marked by a clock that exists outside of the physical system which is modeled by the trajectories in the configuration space $\mathcal{C}$.

In quantum mechanics the situation is rather similar. There is a t in the quantum state and the Schroedinger equation, but it is time as measured by an external clock, which is not part of the system being modeled. Thus, when we write,

$$\imath\hbar\frac{d}{dt}\Psi(t) = \hat{H}\Psi(t) \tag{14}$$

the Hamiltonian refers to evolution, as it would be measured by an external observer, who refers to the external clock whose reading is t

The quantum state can be represented as a function Ψ over the configuration space, which is normalizable in some inner product. The inner product is another a priori structure, it refers also to the external clock, as it is the structure that allows us to represent the conservation of probability as measured by that clock.

When we turn to the problem of constructing a cosmological theory we face a key problem, which is that there is no external clock. There is by definition nothing outside of the system, which means that the interpretation of the theory must be made without reference to anything that is not part of the system which is modeled. In classical cosmological theories, such as general relativity applied to spatially compact universes, or models such as the Bianchi cosmologies or the Barbour-Bertotti model[43,44], this is expressed by the dynamics

[8] For more details and discussion see [35–39]

having a gauge invariance, which includes arbitrary reparameterizations of the classical trajectories. (In general relativity this is part of the diffeomorphism invariance of the theory.) As a result, the classical theory is expressed in a way that makes no reference to any particular parameterization of the trajectories. Any parameterization is as good as any other, none has any physical meaning. The solutions are then labeled by a trajectory, γ, period, there is no reference to a parameterization.

This is the sense in which time may be said to disappear from classical cosmological theories. There is nothing in the theory that refers to any time at all. At least without a good deal more work, the theory speaks only in terms of the whole history or trajectory, it seems to have nothing to say about what the world is like at a particular moment.

There is one apparently straightforward way out of this, which is to try to define an intrinsic notion of time, in terms of physical observables. One may construct parameter independent observables that describe what is happening at a point on the trajectory if that point can be labeled intrinsically by some physical property. For example, one might consider some particular degree of freedom to be an intrinsic, physical clock, and label the points on the trajectory by its value. This works in some model systems, but in interesting cases such as general relativity it is not known if such an intrinsic notion of time exists which is well defined over the whole of the configuration space.

In the quantum theory there is a corresponding phenomenon. As there is no external t with which to measure evolution of the quantum state one has instead of (14) the quantum constraint equation

$$\hat{H}\Psi = 0 \tag{15}$$

where Ψ is now just a function on the configuration space. Rather than describing evolution, eq. (15) generates arbitrary parameterizations of the trajectories. The wavefunction must be normalizable under an inner product, given by some density ρ on the configuration space. The space of physical states is then given by (15) subject to

$$1 = \int_C \rho\bar{\Psi}\Psi \tag{16}$$

We see that, at least naively time has completely disappeared from the formalism. This has led to what is called the "problem of time in quantum cosmology", which is how to either A) find an interpretation of the theory that restores a role for time or B) provide an interpretation according to which time is not part of a fundamental description of the world, but only reappears in an appropriate classical limit.

There have been various attempts at either direction. We will not describe them here, except to say that, in our opinion, so far none has proved completely satisfactory[9]. There are a number of attempts at A) which succeed when applied

[9] For good critical reviews that deflate most known proposals, see [35,36].

to either models or the semiclassical limit, but it is not clear whether any of them overcome technical obstacles of various kinds when applied to the full theory. The most well formulated attempt of type B), which is that of Barbour[39], may very well be logically consistent. But it forces one to swallow quite a radical point of view about the relationship between time and our experience.

Given this situation, we would like to propose that the problem may be not with time, but with the assumptions of the argument that leads to time being absent. Given the number of attempts that have been made to resolve the problem, which have not so far led to a good solution, perhaps it might be better to try to dissolve the issue by questioning one of the assumptions of the argument that leads to the statement of the problem. This is what we would like to do in the following.

9.2 A problem with the argument for the disappearance of time

Both the classical and quantum mechanical versions of the argument for the disappearance of time begin with the specification of the classical configuration space C. This seems an innocent enough assumption. For a system of N particles in d dimensional Euclidean space, it is simply R^{Nd}. One can then find the corresponding basis of the Hilbert space by simply enumerating the Fourier modes. Thus, for cases such as this, it is certainly the case that the configuration space and the Hilbert space structure can be specified *a priori*.

However, there are good reasons to suspect that for cosmological theories it may not be so easy to specify the whole of the configuration or Hilbert space. For example, it is known that the configuration spaces of theories that implement relational notions of space are quite complicated. One example is the Barbour-Bertotti model[43,44], whose configuration space consists of the relative distances between N particles in d dimensional Euclidean space. While it is presumably specifiable in closed form, this configuration space is rather complicated, as it is the quotient of R^{Nd} by the Euclidean group in d dimensions[39].

The configuration space of compact three geometries is even more complicated, as it is the quotient of the space of metrics by the diffeomorphism group. It is known not to be a manifold everywhere. Furthermore, it has a preferred end, where the volume of the universe vanishes.

These examples serve to show that the configuration spaces of cosmological theories are not simple spaces like R^{Nd}, but may be considerably complicated. This raises a question: could there be a theory so complicated that its space of configurations is not constructible through any finite procedure? For example, is it possible that the topology of an infinite dimensional configuration space were not finitely specifiable? And were this the case, what would be the implications for how we understand dynamics[10]?

[10] There is an analogous issue in theoretical biology. The problem is that it does not appear that a pre-specifiable set of "functionalities" exists in biology, where pre-specifiable means a compact description of an effective procedure to characterize ahead of time, each member of the set[45,41]. This problem seems to limit the pos-

We do not know whether in fact the configuration space of general relativity is finitely specifiable. The problem is hard because the physical configuration space is not the space of three metrics. It is instead the space of equivalence classes of three metrics (or connections, in some formalisms) under diffeomorphisms. The problem is that it is not known if there is any effective procedure which will label the equivalence classes.

One can in fact see this issue in one approach to describing the configuration space, due to Newman and Rovelli[46]. There the physical configuration space consists of the diffeomorphism equivalence classes of a set of three flows on a three manifold. (These come from the intersections of the level surfaces of three functions.) These classes are partially characterized by the topologies of the flow lines of the vector fields. We may note that these flow lines may knot and link, thus a part of the problem of specifying the configuration space involves classifying the knotting and linking among the flow lines.

Thus, the configuration space of general relativity cannot be completely described unless the possible ways that flow lines may knot and link in three dimensions are finitely specifiable. It may be noted that there is a decision procedure, due to Hacken, for knots, although it is very cumbersome[47]. However, it is not obvious that this is sufficient to give a decision procedure for configurations in general relativity, because there we are concerned with smooth data. In the smooth category the flow lines may knot and link an infinite number of times in any bounded region. The resulting knots may not be classifiable. All that is known is that knots with a finite number of crossings are classifiable. If these is no decision procedure to classify the knotting and linking of smooth flow lines then the points of the configuration space of general relativity may not be distinguished by any decision procedure. This means that the configuration space is not constructible by any finite procedure.

When we turn from the classical to the quantum theory the same issue arises. First of all, if the configuration space is not constructible through any finite procedure, then there is no finite procedure to define normalizable wave functions on that space. One might still wonder whether there is some constructible basis for the theory. Given the progress of the last few years in quantum gravity we can investigate this question directly, as we know more about the space of quantum states of general relativity than we do about the configuration space of the theory. This is because it has been shown that the space of spatially diffeomorphism invariant states of the quantum gravitational field has a basis which is in one to one correspondence with the diffeomorphism classes of a certain set of embedded, labeled graphs Γ, in a given three manifold Σ [3,4]. These are arbitrary graphs, whose edges are labeled by spins and whose vertices are labeled by the distinct ways to combine the spins in the edges that meet there quantum mechanically. These graphs are called *spin networks*, they were

sibilities of a formal framework for biology in which there is a pre-specified space of states which describe the functionalities of elements of a biological system. Similarly, one may question whether it is in principle possible in economic theory to give in advance an *a priori* list of all the possible kinds of jobs, or goods or services[45].

invented originally by Roger Penrose[48], and then discovered to play this role in quantum gravity[11].

Thus, we cannot label all the basis elements of quantum general relativity unless the diffeomorphism classes of the embeddings of spin networks in a three manifold Σ may be classified. But it is not known whether this is the case. The same procedure that classifies the knots is not, at least as far as is known, extendible to the case of embeddings of graphs.

What if it is the case that the diffeomorphism classes of the embeddings of spin networks cannot be classified? While it may be possible to give a finite procedure that generates all the embeddings of spin networks, if they are not classifiable there will be no finite procedure to tell if a given one produced is or is not the same as a previous network in the list. In this case there will be no finite procedure to write the completeness relation or expand a given state in terms of the basis. There will consequently be no finite procedure to test whether an operator is unitary or not. Without being able to do any of these things, we cannot really say that we have a conventional quantum mechanical description. If spin networks are not classifiable, then we cannot construct the Hilbert space of quantum general relativity.

In this case then the whole set up of the problem of time fails. If the Hilbert space of spatially diffeomorphism invariant states is not constructible, then we cannot formulate a quantum theory of cosmology in these terms. There may be something that corresponds to a "wavefunction of the universe" but it cannot be a vector in a constructible Hilbert space. Similarly, if the configuration space $\mathcal{C}$ of the theory is not constructible, then we cannot describe the quantum state of the universe in terms of a normalizable function on $\mathcal{C}$.

We may note that a similar argument arises for the path integral formulations of quantum gravity. It is definitely known that four manifolds are not classifiable; this means that path integral formulations of quantum gravity that include sums over topologies are not constructible through a finite procedure[51].

Someone may object that these arguments have to do with quantum general relativity, which is in any case unlikely to exist. One might even like to use this problem as an argument against quantum general relativity. However, the argument only uses the kinematics of the theory, which is that the configuration space includes diffeomorphism and gauge invariant classes of some metric or connection. It uses nothing about the actual dynamics of the theory, nor does it assume anything about which matter fields are included. Thus, the argument applies to a large class of theories, including supergravity.

9.3 Can we do physics without a constructible state space?

What if it is the case that the Hilbert space of quantum gravity is not constructible because embedded graphs in three space are not classifiable? How do

[11] For a review of these developments see [11]. These results have also more recently been formulated as theorems in a rigorous formulation of diffeomorphism invariant quantum field theories[50,49].

we do physics? We would like to argue now that there is a straightforward answer to this question. But it is one that necessarily involves the introduction of notions of time and causality.

One model for how to do physics in the absence of a constructible Hilbert space is seen in a recent formulation of the path integral for quantum gravity in terms of spin networks by Markopoulou and Smolin[42][12]. In this case one may begin with an initial spin network Γ_0 with a finite number of edges and nodes (This corresponds to the volume of space being finite.) One then has a finite procedure that constructs a finite set of possible successor spin networks Γ_1^α, where α labels the different possibilities. To each of these the theory associates a quantum amplitude $\mathcal{A}_{\Gamma_0 \to \Gamma_1^\alpha}$.

The procedure may then be applied to each of these, producing a new set $\Gamma_2^{\alpha\beta}$. Here $\Gamma_2^{\alpha\beta}$ labels the possible successors to each of the Γ_1^α. The procedure may be iterated any finite number of times N, producing a set of spin networks $\mathcal{S}_{\Gamma_0}^N$ that grow out of the initial spin network Γ_0 after N steps. $\mathcal{S}_{\Gamma_0}^N$ is itself a directed graph, where two spin networks are joined if one is a successor of the other. There may be more than one path in $\mathcal{S}_{\Gamma_0}^N$ between Γ_0 and some spin network Γ_{final}. The amplitude for Γ_0 to evolve to Γ_{final} is then the sum over the paths that join them in $\mathcal{S}_{\Gamma_0}^N$, in the limit $N \to \infty$ of the products of the amplitudes for each step along the way.

For any finite N, $\mathcal{S}_{\Gamma_0}^N$ has a finite number of elements and the procedure is finitely specifiable. There may be issues about taking the limit $N \to \infty$, but there is no reason to think that they are worse than similar problems in quantum mechanics or quantum field theory. In any case, there is a sense in which each step takes a certain amount of time, in the limit $N \to \infty$ we will be picking up the probability amplitude for the transition to happen in infinite time.

Each step represents a finite time evolution because it corresponds to certain causal processes by which information is propagated in the spin network. The rule by which the amplitude is specified satisfies a principle of causality, by which information about an element of a successor network only depends on a small region of the its predecessor. There are then discrete analogues of light cones and causal structures in the theory. Because the geometry associated to the spin networks is discrete[3], the process by which information at two nearby nodes or edges may propagate to jointly influence the successor network is finite, not infinitesimal.

In ordinary quantum systems it is usually the case that there is a non-vanishing probability for a state to evolve to an infinite number of elements of a basis after a finite amount of time. The procedure we've just described then differs from ordinary quantum mechanics, in that there are a finite number of possible successors for each basis state after a finite evolution. The reason is again causality and discreteness: since the spin networks represent discrete

[12] This followed the development of a Euclidean path integral by Reisenberger[52] and by Reisenberger and Rovelli[53]. Very interesting related work has also been done by John Baez[54]. We may note that the theory described in [42] involves non-embedded spin networks, which probably are classifiable, but it can be extended to give a theory of the evolution of embedded spin networks.

quantum geometries, and since information must only flow to neighboring sites of the graph in a finite series of steps, at each elementary step there are only a finite number of things that can happen.

We may note that if the Hilbert space is not constructible, we cannot ask if this procedure is unitary. But we can still normalize the amplitudes so that the sum of the absolute squares of the amplitudes to evolve from any spin network to its successors is unity. This gives us something weaker than unitarity, but strong enough to guarantee that probability is conserved locally in the space of configurations.

To summarize, in such an approach, the amplitude to evolve from the initial spin network Γ_0 to any element of $\mathcal{S}^N_{\Gamma_0}$, for large finite N is computable, even if it is the case that the spin networks cannot be classified so that the basis itself is not finitely specifiable. Thus, such a procedure gives a way to do quantum physics even for cases in which the Hilbert space is not constructible.

We may make two comments about this form of resolution of the problem. First, it necessarily involves an element of time and causality. The way in which the amplitudes are constructed in the absence of a specifiable basis or Hilbert structure requires a notion of successor states. The theory never has to ask about the whole space of states, it only explores a finite set of successor states at each step. Thus, a notion of time is necessarily introduced.

Second, we might ask how we might formalize such a theory. The role of the space of all states is replaced by the notion of the successor states of a given network. The immediate successors to a graph Γ_0 may be called the *adjacent possible*[41]. They are finite in number and constructible. They replace the idealization of all possible states that is used in ordinary quantum mechanics. We may note a similar notion of an *adjacent possible* set of configurations, reachable from a given configuration in one step, plays a role in formalizations of the self-organization of biological and other complex systems[41].

In such a formulation there is no need to construct the state space *a priori*, or equip it with a structure such as an inner product. One has simply a set of rules by which a set of possible configurations and histories of the universe is constructed by a finite procedure, given any initial state. In a sense it may be said that the system is constructing the space of its possible states and histories as it evolves.

Of course, were we to do this for all initial states, we would have constructed the entire state space of the theory. But there are an infinite number of possible initial states and, as we have been arguing, they may not be classifiable. In this case it is the evolution itself that constructs the subspace of the space of states that is needed to describe the possible futures of any given state. And by doing so the construction gives us an intrinsic notion of time.

9.4 Implications

We must emphasize first of all that these comments are meant to be preliminary. Their ultimate relevance rests partly on the issue of whether there is a decision procedure for spin networks (or perhaps for some extension of them that turns out to be relevant for real quantum gravity[11].). But more importantly, it suggests an alternative type of framework for constructing quantum theories of cosmology, in which there is no *a priori* configuration space or Hilbert space structure, but in which the theory is defined entirely in terms of the sets of adjacent possible configurations, accessible from any given configuration. Whether such formulations turn out to be successful at resolving all the problems of quantum gravity and cosmology is a question that must be left for the future[13].

There are further implications for theories of cosmology, if it turns out to be the case that their configuration space or state space is not finitely constructible. One is to the problem of whether the second law of thermodynamics applies at a cosmological scale. If the configuration space or state space is not constructible, then it is not clear that the ergodic hypothesis is well defined or useful. Neither may the standard formulations of statistical mechanics be applied. What is then needed is a new approach to statistical physics based only on the evolving set of possibilities generated by the evolution from a given initial state. It is possible to speculate whether there may in such a context be a "fourth law" of thermodynamics in which the evolution extremizes the dimension of the adjacent possible, which is the set of states accessible to the system at any stage in its evolution[41].

Finally, we may note that there are other reasons to suppose that a quantum cosmological theory must incorporate some mechanisms analogous to the self-organization of complex systems[40]. For example, these may be necessary to tune the system to the critical behavior necessary for the existence of the classical limit[56,42]. This may also be necessary if the universe is to have sufficient complexity that a four manifolds worth of spacetime events are completely distinguished by purely relational observables[38,40]. The arguments given here are complementary to those, and provide yet another way in which notions of self-organization may play a role in a fundamental cosmological theory.

Acknowledgements

We first of all thank Fotini Markopoulou, who was a coauthor of an early version of this paper, for collaboration and suggestions. We are indebted to her and to Julian Barbour for conversations which were very helpful in formulating the observations concerning time presented here. We would also like to thank John Baez, Louis Crane, Lou Kauffman and Adrian Ocneanu for discussions and help concerning the mathematical questions about classifiability. Conversations with Sameer Gupta has also been very helpful to formulating our observations on the dynamics of spin networks, and we are also endebted to him for comments on a draft of this paper. We also thank Jan Ambjorn, Kostas Anagastopolous, Per Bak, Jayanth Banavar, Roumen Borissov, Louis Crane, David Meyers, Maya Paczuski and Carlo Rovelli for helpful conversations about non-equilibrium critical phenomena and evolving spin networks. This work was supported by the

[13] We may note that the notion of an evolving Hilbert space structure may be considered apart from the issues discussed here[55].

National Science Foundation through grant PHY-9514240 to The Pennsylvania State University, a NASA grant to The Santa Fe Institute and a grant from the Jesse Phillips Foundation.

References

1. R. Penrose, *Theory of quantized directions* unpublished manuscript.; in *Quantum theory and beyond* ed T Bastin, Cambridge U Press 1971; in *Advances in Twistor Theory*, ed. L. P. Hughston and R. S. Ward, (Pitman,1979) p. 301; in *Combinatorial Mathematics and its Application* (ed. D. J. A. Welsh) (Academic Press,1971).
2. D. Finkelstein, *Spacetime Code 1-6* Phys. Rev. D5:320,1972, D5:2922,1972, Phys.Rev.D9:2219,1974, Phys.Rev.D9:2231,1974
3. C. Rovelli and L. Smolin *Discreteness of area and volume in quantum gravity* Nuclear Physics B 442 (1995) 593. Erratum: Nucl. Phys. B 456 (1995) 734.
4. C. Rovelli and L. Smolin, "Spin networks and quantum gravity" gr-qc/9505006, Physical Review D 52 (1995) 5743-5759.
5. M Kato: Phys Lett B245 (1990) 43. K Konishi G Paffuti P Provero: Phys Lett B234 (1990) 276. G Veneziano: Europhys Lett 2 (1986)199. D Amati M Ciafaloni G Veneziano Phys. Lett. B216 (1989) 41; R Guida K Konishi P Provero: Mod Phys Lett A6 (1991)1487; J J Atick E Witten, Nucl Phys B310 (1988) 291 ; D Gross and PF Mende, Phys Lett B197 (1987) 129 ; Nucl Phys B303 (1988) 407
6. J. D. Bekenstein, Lett. Nuovo Cim 11 (1974).
7. G. 'tHooft, *Dimensional reduction in quantum gravity* gr-qc/9310006.
8. L. Susskind, *The world as a hologram*, hep-th/9409089, J. Math. Phys. (1995) ; *Strings, black holes and Lorentz contractions* hep-th/9308139; Phys. Rev. Lett. 71 (1993) 2367; Phys. Rev. D 49 (1994) 6606; D 50 (1994) 2700; L. Susskind and P. Griffin, *Partons and black holes* hep-th/9410306
9. C. Rovelli, gr-qc/9710008.
10. L Smolin: in *Quantum Gravity and Cosmology*, eds J Pérez-Mercader *et al*, World Scientific, Singapore 1992.
11. L. Smolin, *The future of spin networks* gr-qc/9702030.
12. L J Garay: "Quantum gravity and minimum length", Imperial College preprint/TP/93-94/20, gr-qc/9403008 (1994).
13. L. Smolin, *Cosmology as a problem in critical phenomena* in the proceedings of the Guanajuato Conference on Complex systems and binary networks, (Springer,1995), eds. R. Lopez-Pena, R. Capovilla, R. Garcia-Pelayo, H. Waalebroeck and F. Zertuche. gr-qc/9505022.
14. S. Wolfram, *Cellular Automata and Complexity*, Perseus books, (1994); *Statistical mechanics of cellular autonata*, REv, Mod. Physa. 55 (183) 601.
15. D. Weaire and N. River, Contemp. Phys. 25 (1984) 59; J. Stavans, Rep. Prog. Phys. 54 (1993) 733; T. Aste, D. Boose and N. Rive, cond-mat/9612084;
16. S. Kauffman, J. Theor. Biology, 22 (1969) 437; *The Origins of Order* New York: Oxford University Press, 1993; *At Home in the Universe*, New York: Oxford University Press, 1996, and references theirin.
17. P. Bak, C. Tang and K. Wiesenfelf, Phys. Rev. Lett. 59 (1987) 381; Phys. Rev. A 38 (1988) 364; Per Bak *How Nature Works* Oxford: Oxford University Press, 1997.
18. F. Markopoulou and L. Smolin *Causal evolution of spin networks* gr-qc/9702025. CGPG preprint (1997), Nuclear Physics B, Nucl.Phys. B508 (1997) 409-430.

19. L. Bombelli, J. Lee, D. Meyer and R. D. Sorkin, *Spacetime as a causal set* Phys. Rev. Lett. 59 (1987) 521.
20. G. 't Hooft, *Quantum gravity: a fundamental problem and some radical ideas.* Cargèse Summer School Lectures 1978. Publ. "Recent Developments in Gravitation". Cargèse 1978. Ed. by M. Lévy and S. Deser. Plenum, New York/London, 323; *Quantization of Space and Time in 3 and in 4 Space-time Dimensions*, Lectures held at the NATO Advanced Study Institute on "Quantum Fields and Quantum Space Time", Cargèse, July 22 – August 3, 1996. gr-qc/9608037; *The scattering matrix approach for the quantum black hole: an overview.* J. Mod. Phys. A11 (1996) pp. 4623-4688. gr-qc/9607022.
21. F. Markopoulou, *Dual formulation of spin network evolution* preprint, March 1997, gr-qc/9704013.
22. T. Jacobson and L. Smolin, *Nonperturbative quantum geometries* Nuclear Physics B299 (1988) 295-345.
23. C. Rovelli and L. Smolin, Phys Rev Lett 61 (1988) 1155; Nucl Phys B133, 80 (1990)
24. S. Major and L. Smolin, *Quantum deformation of quantum gravity* gr-qc/9512020, Nucl. Phys. B 473 (1996) 267; R. Borissov, S. Major and L. Smolin, *The geometry of quantum spin networks* gr-qc/9512043, Class. and Quant. Grav. 12 (1996) 3183.
25. F. Markopoulou and L. Smolin *Quantum geometry with intrinsic local causality* preprint, Dec. 1997, gr-qc/9712067,Phys. Rev. D 58 084032 (October 15, 1998), (1998).
26. F. Markopoulou and L. Smolin *Non-perturbative dynamics of algebraic (p,q) string networks* preprint, December 1997, hep-th/9712148, Phys. Rev. D 58, 084033 (October 15, 1998).
27. L. Smolin, *A candidate for a background independent formulation of M theory*, hep-th/9903166.
28. F. Markopoulou and L. Smolin, in preparation.
29. D. Armand-Ugon, R. Gambini, O. Obregon, J. Pullin, *Towards a loop representation for quantum canonical supergravity*, hep-th/9508036, Nucl.Phys. B460 (1996) 615. See also, T. Jaocobson, *New variables for canonical supergravity* Class. Quant. Grav. 5 (1988) 923;H. Kunitomo and T. Sano *The Ashtekar formulation for canonical N=2 supergravity*, Prog. Theor. Phys. suppl. (1993) 31; Takashi Sano and J. Shiraishi, *The Non-perturbative Canonical Quantization of the N=1 Supergravity*, Nucl. Phys. B410 (1993) 423, hep-th/9211104; *The Ashtekar Formalism and WKB Wave Functions of N=1,2 Supergravities*, hep-th/9211103; K. Ezawa, *ASHTEKAR'S FORMULATION FOR N=1, N=2 SUPERGRAVITIES AS CONSTRAINED BF THEORIES*, Prog.Theor.Phys.95:863-882,1996, hep-th/9511047.T. Kadoyoshi and S. Nojiri, *N=3 AND N=4 TWO FORM SUPERGRAVITIES*, Mod.Phys.Lett.A12:1165-1174,1997, hep-th/9703149; L. F. Urrutia *Towards a loop representation of connection theories defined over a super-lie algebra* hep-th/9609010.
30. L. Smolin, *Linking topological quantum field theory and nonperturbative quantum gravity* gr-qc/9505028, J. Math. Phys. 36 (1995) 6417.
31. L. Smolin, *Holographic formulation of quantum general relativity.*
32. R. Loll, Nucl. Phys. B444 (1995) 619; B460 (1996) 143; R. DePietri and C. Rovelli, *Geometry eigenvalues and scalar product from recoupling theory in loop quantum gravity*, gr-qc/9602023, Phys.Rev. D54 (1996) 2664; Simonetta Frittelli, Luis Lehner, Carlo Rovelli, *The complete spectrum of the area from recoupling theory in loop quantum gravity* gr-qc/9608043; R. Borissov, Ph.D. thesis, Temple, (1996).
33. E. Domany and W. Kinzel, Phys. Rev. Lett. 53 (1984) 447.

34. S. Kauffman, *Investigations*, Santa Fe Preprint working paper 96-08-072.
35. C. J. Isham, Canonical Quantum Gravity and the Problem of Time, Imperial College preprint, gr-qc/9210011, /TP/91-92/25.
36. K. V. Kuchar, *Time and interpretations of quantum gravity* in the *Proceedings of the 4th Canadian Conference on General Relativity and Relativistic Astrophysics*, eds. G. Kunstatter, D. Vincent and J. Williams (World Scientific, Singapore,1992).
37. C. Rovelli, Phys. Rev. D 42 (1991) 2638; 43 (1991) 442; in *Conceptual Problems of Quantum Gravity* ed. A. Ashtekar and J. Stachel, (Birkhauser,Boston,1991); Class. Quantum Grav. 8 (1991) 317-331.
38. L. Smolin *Space and time in the quantum universe* in the proceedings of the Osgood Hill conference on *Conceptual Problems in Quantum Gravity* ed. A. Ashtekar and J. Stachel, (Birkhauser,Boston,1991); *Time, structure and evolution in cosmology*, preprint, submitted to the British Journal for the Philosophy of Science; *Time, measurement and information loss in quantum cosmology* gr-qc/9301016 in the Brill Feschrift, ed. Hu and Jacobson, Cambridge University Press, 1994.
39. J. B. Barbour,to appear in the *Procedings of the NATO Meeting on the Physical Origins of Time Asymmetry* eds. J. J. Halliwell, J. Perez-Mercader and W. H. Zurek (CAmbridge University Press, Cambridge, 1992); *On the origin of structure in the universe* in *Proc. of the Third Workshop on Physical and Philosophical Aspects of our Understanding of Space and Time* ed. I. O. Stamatescu (Klett Cotta); *Time and the interpretation of quantum gravity* Syracuse University Preprint, April 1992.
40. L. Smolin, *The Life of the Cosmos* Oxford University Press, New York and Weidenfeld and Nicolson, London, (1997).
41. S. Kauffman, *Investigations on the character of autonomous agents and the worlds they mutually create* Santa Fe Institute preprint, 1996.
42. F. Markopoulou and L. Smolin, *Causal evolution of spin networks* gr-qc/9702025. CGPG 97/2-1.
43. J. B. Barbour, Nature 249 (1974) 328 Erattum: Nature 250 (1974) 606; Nuovo Cimento 26B (1975) 16; J.B. Barbour and B. Bertotti, Nouvo Cimento 38B (1977) 1.
44. J.B. Barbour and B. Bertotti, Proc. Roy. Soc. Lond. A 382 (1982) 295.
45. S. Kauffman, *Origins of Order* Oxford University Press; *At home in the universe*, Oxford University Press, 1996.
46. E. T. Newman and C. Rovelli, Physical Review Letters, (1992)
47. J. Hemion *Classification of knots in three dimensional space* Oxford University Press.
48. R Penrose: in *Quantum theory and beyond* ed T Bastin, Cambridge U Press 1971; in *Advances in Twistor Theory*, ed. L. P. Hughston and R. S. Ward, (Pitman,1979) p. 301; in *Combinatorial Mathematics and its Application* (ed. D. J. A. Welsh) (Academic Press,1971).
49. J. Baez, Spin networks in gauge theory, Adv. Math. **117** (1996), 253-272, gr-qc/941107; Spin networks in nonperturbative quantum gravity, in *The Interface of Knots and Physics*, ed. Louis Kauffman, American Mathematical Society, Providence, Rhode Island, 1996, gr-qc/9504036.
50. A. Ashtekar and C. J. Isham, Class and Quant Grav 9 (1992) 1069; A Ashtekar J Lewandowski D Marlof J Mourãu T Thiemann: "Quantization of diffeomorphism invariant theories of connections with local degrees of freedom", gr-qc/9504018, JMP 36 (1995) 519; T. Thiemann, Quantum spin dynamics I, Harvard preprint (1996), gr-qc/9606089; Quantum spin dynamics II, Harvard preprint (1996), gr-qc/9606090; A. Ashtekar and J. Lewandowski, "Quantum Geometry I: area operator" gr-qc/9602046.

51. J.B. Hartle, *Unruly Topologies in Two Dimensional Quantum Gravity*, Class. & Quant. Grav., 2, 707, 1985b; R. Geroch and J.B. Hartle, *Computability and Physical Theories* Found. Phys., 16, 533, 1986. J.B. Hartle, *Unruly Topologies in Two Dimensional Quantum Gravity*, Class. & Quant. Grav., 2, 707, 1985b.
52. M. Reisenberger, unpublished.
53. M. Reisenberger and C. Rovelli, *"Sum over Surfaces" form of Loop Quantum Gravity*, gr-qc/9612035.
54. J. Baez, seminars, Penn State, February 1997.
55. R. Doldan, R. Gambini and P. Mora, Int. J. of Theor. Phys. 35 (1996) 2057.
56. L. Smolin, *Cosmology as a problem in critical phenomena* in the proceedings of the Guanajuato Conference on Complex systems and binary networks, (Springer,1995), eds. R. Lopez-Pena, R. Capovilla, R. Garcia-Pelayo, H. Waalebroeck and F. Zertuche. gr-qc/9505022.

Non-commutative Extensions of Classical Theories in Physics

Richard Kerner[1]

Laboratoire de Gravitation et Cosmologie Relativistes, Université Pierre-et-Marie-Curie - CNRS ESA 7065, Tour 22, 4-ème étage, Boîte 142
4, Place Jussieu, 75005 Paris, France

Abstract. We propose a short introductory overview of the non-commutative extensions of several classical physical theories. After a general discussion of the reasons that suggest that the non-commutativity is a major issue that will eventually lead to the unification of gravity with other fundamental interactions, we display examples of non-commutative generalizations of known geometries.

Finally we discuss the general properties of the algebras that could become generalizations of algebras of smooth functions on Minkowskian (Riemannian) manifolds, needed for the description of Quantum Gravity.

1 Deformations of space-time and phase space geometries

The two most important branches of modern physics created in the beginning of this century, the General Relativity and Quantum Theory, possess their well-defined classical counterparts, the Newtonian gravity theory mechanics, which are obtained as limits of these theories when the parameters c^{-1} or $\hbar$ The mathematical expression of this fact is formulated in terms of the *deformations* of the respective structures. The notion of deformation plays the central rôle in modern attempts which try to generalize the geometrical description of physical realm.

To be more precise, we can cite the example of the relation existing between the Lorentz and the Galilei groups: the Lorentz group can be considered as deformation of the Galilei group, with the characteristic parameter c^{-1}; when this parameter tends to zero, the Lorentz group is said to undergo the *contraction* into the Galilei group. Similarly, the quantization procedure proposed by J.E.Moyal [1] is a deformation of the usual Poisson algebra which is contracted back to it when the characteristic parameter of deformation which is here the Planck constant h tends to zero. Finally, Special Relativity may be considered as a contraction of General Relativity when the characteristic parameter G tends to zero (although some space-times different from the Minkowskian one can appear when the Ricci tensor is put to zero).

Now, with three fundamental constants of Nature, h, G and c^{-1} serving as deformation parameters, one can imagine seven different contractions of the hypothetical unified theory that would deserve the name of "Relativistic Quantum Gravity", and which is yet to be invented. The seven contractions correspond to the vanishing of:

a) one of the three parameters, i.e h, G, or c^{-1} only;
b) two parameters at once, i.e.(h and G), (h and c^{-1}), and (G and c^{-1});
c) all the three parameters at once, i.e. (h, G and c^{-1}).

The following Table shows the relations between the corresponding theories, as well as their usual denominations (when we know them...). We did not take into account the fact that taking the double limits might be non-commutative, which cannot be excluded *a priori* and would have made our diagram even more complicated.

Two of the theories displayed here have not found their realization yet: the "Relativistic Quantum Gravity" and the "Non-Relativistic Quantum Gravity". It is not at all clear whether these hypothetical theories can be realized without introducing some new deformation parameter depending on a new physical constant, and whether this constant should be independent or related to the three fundamental constants h, c and G or not.

It is also amusing to note that our diagram is three-dimensional - is it just a coincidence that we happen to live in three space dimensions, too? In the figure, the contractions (symbolized by the arrows coinciding with the edges of the cube) relate two-by-two different space-time or phase space geometries. The best way to describe a *geometry* is, in our sense, to define the set of variables (forming an algebra) that in a natural way would generalize the algebra of local coordinates in these spaces.

P.A.M.Dirac was already aware of the possibility of a radical modification of geometrical notions, and in his fundamental papers written in 1926 [2] he evokes the possibility of describing the phase space physics in terms of a non-commutative analogue of the algebra of functions, which he referred to as the "quantum algebra", together with its derivations, which he called "quantum differentiations". Of course, this kind of geometry seemed strange and even useless from the point of view of General Relativity. Einstein thought that further problems of physics should be solved by subsequent development of geometrical ideas, and it seemed to him that to have $a \times b$ not equal to $b \times a$ was something that does not fit very well with geometry as he understood it [3]

During several decades, mostly in the sixties and the seventies, a lot of efforts have been made in order to find a unifying approach to both these great theories.In doing so, people either tried to generalize one of the two theories so that the other one would follow, or tried to merge them together via embedding into some more general unified theory. Most of the activities in this field rather belonged to the first category.

The Hamiltonian formulation of General Relativity by R.Arnowitt, S.Deser and C.Misner [4], and later the Wheeler–De Witt equation which generalizes Schrödinger's equation for quantum wave functions describing the state of a 3-dimensional geometry of the Universe [5] can be considered as a first attempt to quantize the General Relativity. The geometric quantization developed by J.M.Souriau, D.Simms, and B.Kostant ([6], [7], [8]) tried to derive the rules of quantum mechanics by interpreting the observables and state vectors as elements of algebras of operators and functions defined on classical manifolds with sufficiently rich geometry, (e.g. symplectic manifolds, fibre bundles, jet spaces).

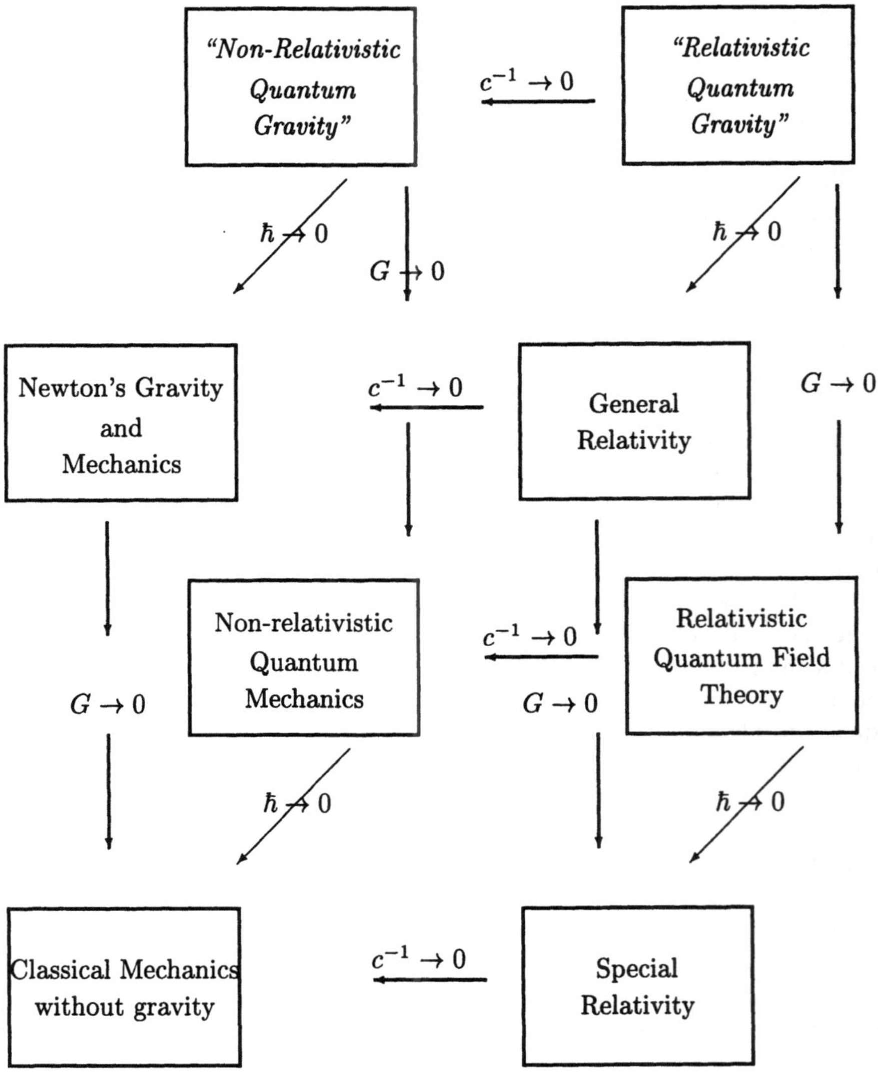

Eight limits of fundamental physical theories

Two limits (marked in italics) are still to be invented

Simultaneous consideration of the two most important new physical theories of this century, the General Relativity and Quantum Mechanics, did not bring a common tool for the description of the nature of spacetime at the microscopic level. The General Relativity develops our knowledge about global properties of space and time at very large distances, and raises the questions concerning the global topology of the Universe.

The methods of Differential Geometry which are the best adapted as the mathematical language of this theory, are very different from the methods of Quantum Physics, in which one studies the properties of the algebra of observables, considering the state vectors, as well as geometric points and trajectories, as artefacts and secondary notions. This approach has been inspired by the works of John von Neumann [9], and has much in common with the non-commutative geometry, where the very notion of a point loses its meaning.

A strong flavor of non-commutativity is also present in A. Ashtekar's approach to quantum gravity, in which the notion of coordinates becomes secondary, the only intrinsic information being encoded in the loop space (see, e.g. in A. Ashtekar [10], or C. Rovelli [11])

In the next section, we shall give a few arguments in favor of the hypothesis that the realization of a theory taking into account quantum effects in gravitation should also lead to the abandon of usual notion of coordinates and differential manifolds and to the introduction of non-commutative extensions of algebras of smooth functions on manifolds. We shall also see that such algebras can act on free modules, which becomes a natural generalization of gauge theories described mathematically as connections and curvatures on fibre bundles.

2 Why the coordinates should not commute at Planck's scale

There are several well-known arguments which suggest that the dynamical interplay between Quantum Theory and Gravitation should lead to a non-commutative version of space-time. Let us recall the few ones that are cited most frequently:

* A semi-classical argument that involves black-hole creation at very small distances: as a matter of fact, if the General Relativity remains valid at the Planck scale, then any localization of events should become impossible at the distances of the order of $\lambda_P = \sqrt{\frac{\hbar G}{c^3}}$. Indeed, according to quantum mechanical principles, lo localize an event in space-time within the radius $\Delta x^\mu \sim a$, one need to employ the energy of the order a^{-1}. When a becomes too small, the creation of a mini black hole becomes possible, thus excluding from the observation that portion of the space-time and making further localization meaningless.

Therefore, the localization is possible only if we impose the following limitation on the time interval:

$$\Delta x^0 \, (\Sigma \, \Delta x^k) \geq \lambda_P^2 \quad \text{and} \qquad \Delta x^k \, \Delta x^m \geq \lambda_P^2. \tag{1}$$

in order to avoid the black hole creation at the microscopic level.

** The topology of the space-time should be sensitive to the states of the fields which are in presence - and *vice versa*, quantum evolution of any field, including gravity, should take into account all possible field configurations, also corresponding to the fields existing in space-times with radically different topologies (a creation of a black hole is but the simplest example; one should also take into account other "exotic" configurations, such as multiple Einstein-Rosen bridges (the so-called "*wormholes*", leading in the limit of great N to the space-time *foam*.

Now, as any quantum measurement process may also lead to topological modifications, again the coordinates of an event found before and after any measurement can no more be compared, because they might refer to uncompatible coordinate patches in different local maps. As a result, quantum measures of coordinates themselves become non-commutative, and the algebra of functions on the space-time, supposed to contain also all possible local coordinates, must be replaced by its non-commutative extension, better adapted to describe the space-time foam.

*** Since the coordinates x^{μ} are endowed with a length scale, the metric must enter at certain stage in order to measure it. After quantization, the components of the tensor $g_{\mu\nu}$ become a set of dynamical fields, whose behaviour is determined by the propagators and, at least at the lowest perturbative level, by two-point correlation functions. As any other field, the components of the metric tensor will display quantum fluctuations, making impossible precise measurements of distances, and therefore, any precise definition of coordinates.

Our aim here is not to discuss all possible arguments suggesting that at the Planck scale not only the positions and momenta do not commute anymore, but also the coordinates themselves should belong to a non-commutative algebra. In what follows, we shall take it for granted that such is the case, and shall expose in a concise way, on the example of the simplest finite non-commutative algebra, which is the algebra of complex $n \times n$ matrices, how almost all the notions of usual differential geometry can be extended to the non-commutative case. We shall also show how the gauge theories and the analogs of the fibre bundle spaces and Kaluza-Klein geometries can be generalized in the non-commutative setting.

Finally, as our main subject is the hypothetical Quantum Gravity theory, and because it has to have also a limit as Relativistic Field theory when gravity is switched off, we shall analyze the conequences of the Poincaré invariance that must be imposed on any theory of this type.

3 Non-commutative differential geometry

In the examples of non-commutative generalizations of spaces of states or of algebras of observables, we have looked up to now only at the linear cases. A most general non-commutative geometry should imitate the situations encountered in the ordinary differential geometry of manifolds. Therefore, we should replace the algebra of smooth functions on a manifold, (the maximal ideals of this algebra can be identified as points of the corresponding manifold) by an more general associative algebra, which can be non-commutative. The *derivations* of this algebra will naturally generalize the notion of *vector fields*; their dual space will generalize the fields of *1-forms*, and one can continue as far as possible, trying to construct the analogues of a metric, integration, volume element, Hodge duality, Lie derivatives, connection and curvature, and so forth. It is amazing how almost all of these objects known from the classical version of differential geometry find their counterparts in the non-commutative case.

The differential algebras of this type have been studied by A.Quillen [12], A.Connes [13] and M.Dubois-Violette [14]; their application to new mathematical models of the gauge theories, including the standard model of electroweak interactions of Weinberg and Salam, has been worked out by M.Dubois-Violette *et al* [15],[16], by A.Connes and J.Lott [17], R.Coquereaux *et al* [18], and many other authors since. Here we shall give the simplest example of realization of the non-commutative geometry proposed in [15],[16], realized with the algebra of complex $n \times n$ matrices, $M_n(\mathbf{C})$. Any element of $M_n(\mathbf{C})$ can be represented as a linear combination of the unit $n \times n$ matrix $\mathbf{1}$ and (n^2-1) hermitian traceless matrices E_k, k = 1,2, ... ,(n^2-1):

$$B = \beta \ 1 + \sum \alpha^k E_k \tag{2}$$

One can choose the basis in which the following relations hold:

$$E_k E_m = (\frac{1}{n}) g_{km} 1 + S^j_{km} E_j - (\frac{i}{2}) C^j_{km} E_j \tag{3}$$

with real coefficients satisfying $S^j_{km} = S^j_{mk}$, $S^k_{km} = 0$, $C^j_{km} = -C^j_{mk}$, $C^k_{km} = 0$, and $g_{km} = C^p_{kl} C^l_{pm}$. Then C^m_{kl} are the structure constants of the Lie group $SL(n, C)$, and g_{kl} its Killing-Cartan metric tensor. All the derivations of the algebra $M_n(C)$ are *interior*, i.e. the basis of the derivations is given by the operators ∂_k such that

$$\partial_k E_m = ad(iE_k)E_m = i[E_k, E_m] = C^l_{km} E_l \tag{4}$$

By virtue of the Jacobi identity, we have

$$\partial_k \partial_m - \partial_m \partial_k = C^l_{km} \partial_l \tag{5}$$

The linear space of derivations of $M_n(C)$, denoted by $Der(M_n(C)$, *is not* a left module over the algebra $M_n(C)$ - this is the first important difference with respect to the usual differential geometry, in which a vector field can be multiplied on the left by a function, producing a new vector field. The canonical basis of 1-forms dual to the derivations ∂_k is defined formally by the relations

$$\theta^k(\partial_m) = \delta^k_m 1 \tag{6}$$

These 1-forms span a left module over $M_n(C)$, i.e. $E_l \theta^k$ is also a well-defined 1-form; indeed, $E_l \theta^k(\partial_m) = E_l \delta^k_m 1 = E_l \delta^k_m$

The exterior differential d is defined as usual, first on the 0-forms ("functions") by the identity

$$df(X) = Xf, \tag{7}$$

with f a function, X an arbitrary vector field. Here we have

$$(d1)(\partial_m) = \partial_m 1 = ad(iE_m)1 = i[E_m, 1] = 0 \tag{8}$$

so that $d1 = 0$, and

$$dE_k(\partial_m) = \partial_m(E_k) = i[E_k, E_m] = C^l_{mk} E_l \tag{9}$$

Because the Lie algebra $SL(n, C)$ is semi-simple, the matrices C^l_{km} are non-degenerate, and the above relation can be solved in θ^k's giving the explicit expression

$$dE_k = C^l_{km} E_l \theta^m \tag{10}$$

The fact that $d^2 = 0$ follows then directly from the Jacobi identity. The Grassmann algebra of p-forms is defined as usual, with the wedge product

$$\theta^k \wedge \theta^m = (\frac{1}{2})(\theta^k \otimes \theta^m - \theta^m \otimes \theta^k) \tag{11}$$

We have then

$$d\theta^k + (\frac{1}{2}) C^k_{ml} \theta^m \theta^l = 0 \tag{12}$$

If we define the *canonical 1-form* $\theta = \sum E_k \theta^k$, we can easily prove that it is coordinate-independent. Moreover, it satisfies the important relation

$$d\theta + \theta \wedge \theta = 0 \tag{13}$$

Let ω be a p -form. The anti-derivation i_X with respect to a vector field X can be defined as usual,

$$(i_X \omega)(X_1, X_2, \ldots, X_{p-1}) = \omega(X, X_1, X_2, \ldots, X_{p-1}) \tag{14}$$

The Lie derivative of a p -form ω with respect to a vector field X is defined as

$$L_X \omega = (i_X \ \ d + d \ \ i_X)\omega \tag{15}$$

It is easy to check now that the 2-form $\Omega = d\theta$ is invariant with respect to the derivations of A, i.e. that

$$L_X \Omega = 0 \tag{16}$$

for any vector field X belonging to $Der(M_n(C))$. The 2-form Ω is also non-degenerate, and it is a closed 2-form by definition, because

$$d\Omega = d^2\theta = 0 \tag{17}$$

The 2-form Ω defines a Hamiltonian structure in the algebra $M_n(C)$ in the following sense:

Let $f \in M_n(C)$ be an element of our algebra; then Ham_f is the *Hamiltonian vector field of* f defined by the equality

$$\Omega(Ham_f, X) = X\ f \tag{18}$$

for any X belonging to $Der(M_n(C))$ The *Poisson bracket* of two "functions" (observables) f and g is then defined as

$$\{f, g\} = \Omega(Ham_f, Ham_g) \tag{19}$$

A simple computation shows then that

$$\{E_k, E_m\} = \Omega(\partial_k, \partial_m) = i\ [E_k, E_m] \tag{20}$$

Therefore, in our simple version of non-commutative geometry, classical and quantum mechanics merge into one and the same structure: the Poisson bracket of any two matrix "functions" (observables) is equal, up to a factor that can be chosen as the Planck constant h , to their *commutator* .

This simple and beautiful picture is of course somewhat perturbed in the case of *infinite-dimensional algebras* for which not all the derivations are interior and might have other representations than the commutator with an observable.

The volume element induced by the canonical Cartan-Killing metric and the corresponding Hodge duality $\star$ can be also introduced in a classical manner. The volume element is given by

$$\eta = \frac{1}{(n^2-1)!} \epsilon_{i_1 i_2 \dots i_{n^2-1}} \theta^{i_1} \wedge \theta^{i_2} \wedge \dots \wedge \theta^{i_{n^2-1}} \tag{21}$$

Any $n^2 - 1$ -form is proportional to the volume element η ; the integral of such a form will be defined as the *trace* of the matrix coefficient in front of η . Then the scalar product is readily introduced for any couple of p -forms α and β as follows:

$$(\alpha, \beta) = \int (\alpha \wedge \star\beta) \tag{22}$$

With this formalism we can generalize the notions of gauge fields if we use the non-commutative matrix algebra as the analogue of the algebra of functions defined on a vertical space of a principal fibre bundle. Then we will be able to compute lagrangian densities that may be used in the variational principle producing dynamical field equations.

We shall see in the next section how this formulation of gauge theories contains besides the $SU(n)$ gauge fields also scalar multiples in the adjoint representation, which have the rôle of the Higgs fields in standard electroweak theory.

4 Non-commutative analog of Kaluza-Klein and gauge theories

At this stage we can introduce a non-commutative analogue of Kaluza-Klein type theory, which will lead to a generalization of gauge field theories. In ordinary differential geometry the fact of using a Cartesian product of two differential manifolds, or a fibre bundle locally diffeomorphic with such a product, can be translated into the language of the corresponding function algebras; as a matter of fact, in the case of the Cartesian product of two manifolds, the algebra of functions defined on it is the tensorial product of algebras of functions defined on each of the manifolds separately.

Consider the space-time manifold V_4 with its algebra of smooth functions $C^\infty(V_4)$. Let us define the tensor product

$$A = C^\infty(V_4) \otimes M_n(C) \tag{23}$$

It can be shown (cf.[13]) that

$$Der(A) = [Der(C^\infty(V_4)) \otimes 1] \oplus [C^\infty(V_4) \otimes Der(M_n(C)] \tag{24}$$

In other words, a general derivation in our tensor product algebra replacing the algebra of smooth functions on a fibre bundle space, can be written as the following vector field

$$X = X^\mu(x)\partial_\mu + \xi^k(x)\partial_k \tag{25}$$

with $\mu, \nu = 0, 1, 2, 3$; $k, l = 1, 2, \ldots, (n^2 - 1)$. A general 1-form defined on such vector fields splits naturally into *four* different components:

$$A = A^0_\mu(x)1dx^\mu + A^k_\mu(x)E_k dx^\mu + B^0_m(x)1\theta^m + B^k_m(x)E_k\theta^m \tag{26}$$

The exterior differential of a 1-form A takes into account the two kinds of differentiation; e.g. for a general matrix-valued 0-form ("function") $\Phi = \Phi^0 1(x) + \Phi^m(x)E_m$ we have

$$d(\Phi) = (\partial_\mu \Phi^0)dx^\mu + (\partial_\mu \Phi^m)E_m dx^\mu + \Phi^m C^l_{km} E_l \theta^k \tag{27}$$

The notion of *covariant derivation* can be introduced quite naturally by considering a free (right) hermitian module H over the algebra A. If we choose a *unitary element* e in H, then any element of H can be represented as $\Phi = eB$, with $B \in A$. Then the *covariant derivative* must have the following basic property:

$$\nabla(\Phi D) = (\nabla\Phi)D + \Phi \otimes dD \tag{28}$$

for arbitrary $\Phi \;\epsilon\; H$, $D \;\epsilon\; A$ Now, if $\Phi = eB$, we shall have

$$\nabla\Phi = (\nabla e)B + e \otimes dB, \tag{29}$$

and there exists a unique element $\alpha \;\epsilon\; \Lambda^1(M_n(C))$ such that

$$\nabla e = e \otimes d\alpha \tag{30}$$

satisfying the hermiticity condition $\bar{\alpha} = -\alpha$. The elements B and α are called the *components* of the field Φ and the connection ∇ in the gauge e.

Let U be a unitary matrix from the algebra A. Under a change of gauge

$$e \longrightarrow eU \tag{31}$$

the components B and α transform as follows:

$$B \longrightarrow U^{-1}B \quad , \alpha \longrightarrow U^{-1}\alpha U \; + \; U^{-1}dU \tag{32}$$

This is the analogue of the gauge theory in the non-commutative case. When applied to the connection 1-form (denoted now A instead of α), these principles lead to the following expression of the gauge field tensor $F = dA + A \wedge A$:

$$F = (F^0_{\mu\nu}1 + G^k_{\mu\nu}E_k)dx^\mu \wedge dx^\nu + [(D_\mu B^0_l)1 + (D_\mu B^m_l E_m)]dx^\mu \wedge \theta^l + G^m_{kl}E_m\theta^k \wedge \theta^l$$

where

$$F^0_{\mu\nu} = \partial_\mu A^0_\nu - \partial_\nu A^0_\mu \tag{33}$$

represents the abelian $U(1)$-gauge field;

$$G^k_{\mu\nu} = \partial_\mu A^k_\nu - \partial_\nu A^k_\mu + C^k_{lm}A^l_\mu A^m_\nu \tag{34}$$

represents the $SU(2)$-gauge field;

$$D_{mu}B^0_k = (\frac{1}{m})(\partial_\mu B^0_k) \tag{35}$$

is the derivative of the scalar triplet B^0_k ;

$$D_\mu B^m_k = (\frac{1}{m})(\partial_\mu B^m_k + C^m_{sr}A^s_\mu B^r_k) \tag{36}$$

is the covariant derivative of the scalar (Higgs type) multiplet B^m_k ; finally,

$$G^m_{kl} = (\frac{1}{m^2})(C^p_{kl}B^m_p - C^m_{sr}B^s_k B^r_l) \tag{37}$$

represents the potential contribution of the Higgs multiplet.

Here m is the dimensional parameter ($dim[m] = cm^{-1}$) introduced in order to give the proper dimension to the 1-forms θ^k. The parameter m can be later related to the characteristic mass scale of the theory. The generalized action integral is equal - in conformity with the definition of integration on the algebra of p-forms in the non-commutative case - to the *trace* of the integral over space-time V_4 of the expression $F \wedge \star F$:

$$Tr \int (F \wedge \star F)d^4x \tag{38}$$

The multiplet of scalar fields B^m_l plays here the rôle of the symmetry-breaking Higgs-Kibble field, whose quartic potential appearing in the last part of the action integrand possesses multiple local minima or maxima.

In this example, when all other fields are set equal to 0, there exist several configurations of B^m_l corresponding to vacuum states representing different gauge orbits. Indeed, it is easy to see that $G^m_{kl} = 0$ not only when $B^m_l = 0$, but also for $B^m_l = \delta^m_l$. These two vacua *can not* be transformed one into another by means of a gauge transformation, which is a novel feature when compared with the known classical versions of gauge theory coupled with Higgs fields.

Although this generalization of gauge theory including a non-commutative sector of differential geometry contains naturally the gauge group $SU(2) \times U(1)$, the Higgs multiplet arising here does not have the usually required properties, i.e. *it is not* a doublet of complex scalar fields coupled in a different way to the left- and right-handed fermions; we have instead a tensor multiplet B_l^m that admits 16 different vacuum configurations, most of them degenerate saddle points in the parameter space. Also the mass spectrum of bosons appearing in the theory is not satisfactory. Developing the bosonic fields of the model, A^0_μ, A^k_μ and B^0_k , and linearizing the equations around the vacua given by $B_l^m = 0$ or $B_l^m = \delta_l^m$ respectively, we obtain on the gauge orbit $B_l^m = 0$:

- masses of A^0_μ and A^k_μ equal zero,
- masses of B^0_l and B_l^m all equal to $\sqrt{n}\, m$;

whereas on the gauge orbit $B_l^m = \delta_l^m$:

- the $U(1)$ gauge field $A^0\mu$ remains massless while the $SU(2)$ - gauge field acquires the mass $\sqrt{2n}\, m$;
- the scalar multiplet B^0_m acquires the mass $\sqrt{2}\, m$, and the Higgs multiplet itself, B_l^m develops a mass spectrum with values 0, $\sqrt{2}\, m$ and $2\sqrt{2}\, m$.

which makes this version of unified $SU(2) \times U(1)$ theory unrealistic.

More realistic versions of non-commutative gauge models, reproducing quite well all the properties of the electroweak interactions required by the experiment, have been proposed by A.Connes and M.Lott [17], R. Coquereaux *et al.* [18], by M.Dubois-Violette *et al.*, [15], [16], and by J.Fröhlich *et al.*, [19]. In all these models the non-commutative algebra of complex matrices is tensorized with a Z_2-graded algebra, which in simplest realisation can be conceived as algebra of 2×2 matrices that splits into two linear subspaces called *"even"* (corresponding to diagonal matrices) and *"odd"* (corresponding to the off-diagonal matrices), with respective *grades* being 0 and 1, which under matrix multiplication add up *modulo* 2. The exterior derivations change the grade of an element by 1, and satisfy the *graded Leibniz rule*

$$d(AB) = (dA)B + (-1)^{grad(A)grad(B)} AdB \tag{39}$$

This enables one to represent the connection form (interpreted as the gauge-field potential) in the following form:

$$\begin{pmatrix} A & W^+ \\ W^- & Z \end{pmatrix} \tag{40}$$

where the gauge fields A and Z belong to the even part of the algebra, while the fields W^+ and W^- belong to the odd part; moreover, all these fields are themselves 2×2 matrix-valued 1-forms. Developing this theory around the appropriately chosen vacuum configuration one can quite correctly reproduce the mass spectrum, with the mass of neutral Z -boson $\frac{2}{\sqrt{3}}$ times bigger than the mass of the charged W - boson, which corresponds to the Weinberg angle of 30°. More details can be found in the papers cited above.

At this point one may try to imagine what a non-commutative extension of the General Relativity could look like ? Since a long time there exist many approaches in which the General Relativity was considered as a gauge theory, with gauge group being the infinite-dimensional group of diffeomorphisms of four-dimensional Riemannian manifolds. However, with the gauge group of this size little could be done in matter of computation and prediction, especially on the quantum level.

A more realistic direction consists in exploring the properties of linear approximation of a more complicated final version of the theory. Recently, J. Madore *et al.* in [20] have introduced the generalization of linear connections on matrix algebras defined above. With the usual definition of covariant derivation acting on the moving frame:

$$D\,\theta^{\alpha} = -\omega^{\alpha}_{\beta} \otimes \theta^{\beta} \tag{41}$$

Because the definition of covariant derivative requires that

$$D(f\xi) = df \otimes \xi + f\, D\xi, \tag{42}$$

the covariant derivative of an arbitrary 1-form $\xi_{\alpha}\theta^{\alpha}$ is

$$D\left(\xi_{\alpha}\theta^{\alpha}\right) = d\xi_{\alpha} \otimes \theta^{\alpha} - \xi_{\alpha}\,\omega^{\alpha}_{\beta}\,\theta^{\beta}$$

The covariant derivative along a vector field X is defined as

$$D_X\,\xi = i_X\,(D\xi) \tag{43}$$

and defines a mapping of $\Omega^1(V)$ on itself.

If the torsion vanishes, then one finds that

$$D^2\,\theta^{\alpha} = -\,\Omega^{\alpha}_{\beta}\,\theta^{\beta} \tag{44}$$

where $\Omega^{\alpha}_{\beta} = R^{\alpha}_{\beta\gamma\delta}\,\theta^{\gamma} \wedge \theta^{\delta}$ is the *curvature 2-form.*

The generalization of these formalism for the non-commutative case is quite obvious. We must replace the linear space of 1-forms which span the tensor and the exterior algebras by the corresponding right $\mathcal{A}$-module of 1-forms defined over our matrix algebra $\Omega^1(M_n\mathbf{C})$). In the basis introduced in the previous section, θ^k, $k = 1, 2, \ldots (n^2-1)$, we had

$$d\theta^k = -\frac{1}{2}\,C^k_{\ lm}\theta^l\theta^m, \quad \text{and} \quad df = [\theta, f].$$

It is easy to define the linear connection with vanishing torsion:

$$D\theta^r = -\omega^r_{\ s} \otimes \theta^s\,, \qquad \text{with} \qquad \omega^r_{\ s} = -\frac{1}{2}C^r_{\ st}\theta^t \tag{45}$$

Introducing the permutation operator σ as

$$\sigma(\theta^k \otimes \theta^m) = \theta^m \otimes \theta^k,$$

we can express the commutativity of the algebra $\mathcal{C}^\infty(M_4)$

$$D(\xi f) = D(f\,\xi)$$

by writing

$$D(\xi f) = \sigma(\xi \otimes df) + (D\xi)\,f.$$

The last condition can be maintained in a more general case as the requirement imposed on the connection 1-forms. It follows then that in the case of matrix algebras considered here, one has

$$D([f, \theta^k]) = [f, D\theta^k] = 0, \tag{46}$$

so that all the coefficients ω^k_{lm} must be in the center of $M_n(\mathbf{C})$, i.e. they are just complex numbers, and the torsionless connection defined above becomes unique.

The metric in the space of 1-forms over $M_n(\mathbf{C})$ has been already introduced as $g(\theta^k \otimes \theta^m) = g^{km} \in \mathbf{C}$. The fact that $\omega^k_{(lm)} = 0$ can be interpreted as the metricity of this connection. This leads to the unique definition of the corresponding curvature tensor:

$$\Omega^k_{lmn} = \frac{1}{8} C^k_{lr} C^r_{mn}$$

These constructions have been used already in [15] and [16], and can serve as the non-commutative extension of connexion and curvature on the tensor product of algebras $\mathcal{C}^\infty(M_4) \otimes M_n(\mathbf{C})$.

However, the fact that all geometrically important quantities like metric, connection and curvature coefficients, are forced to belong to the *center* of the non-commutative sector make the above generalization quite trivial and therefore unsatisfactory.

5 Minkowskian space-time as a commutative limit

In this section we shall discuss an important feature of any non-commutative geometry that contains the algebra of smooth functions on Minkowskian space-time and is supposed to be Poincaré-invariant at least in the first orders of the deformation parameter. This result has been published in 1998 (M. Dubois-Violette, J. Madore, R. Kerner, [21]). Similar ideas have been independently developed earlier by S. Doplicher, K. Fredenhagen and J.E. Roberts (cf. [22]).

The main idea is as follows. Suppose that the non-commutative geometry that is supposed to describe in an adequate way the quantum version of General Relativity contains in its center the infinite algebra of smooth functions on Minkowskian space-time. This infinite algebra serves as a representation space for the infinite-dimensional representation of the Poincaré group, in particular, the abelian group of translations, in the limit when the gravitational interaction becomes negligible, which shall correspond to the limit $\kappa \to 0$, where κ is proportional to the gravitational coupling constant G.

It seems natural to suppose that the Poincaré invariance remains still valid before the limit is attained, at least in the linear approximation with respect to the deformation parameter κ. Then an important question to be answered appears, namely, what is the dimension of the non-commutative part of the full algebra before the limit is attained? As it is shown in the reference [21], it must be infinite-dimensional. In other words, it is impossible to impose the full Poincaré invariance on a tensor product of $\mathcal{C}^\infty(M_4)$ with a *finite* non-commutative algebra, as in the example with the matrix algebras considered in previous sections. These examples can be considered only as approximations to the correct theory of non-commutative space-time and gauge field theories.

Let us consider then a one-parameter family of associative algebras, $\mathcal{A}_\kappa$, whose limit at $\kappa = O$, denoted by $\mathcal{A}_0$, admits a well-defined action of the Poincaré group on it. When $\kappa \to 0$, one should attain as a classical limit certain algebra, obviously containing $\mathcal{C}^\infty(M_4)$, the algebra of smooth functions on the Minkowskian manifold:

$$\mathcal{A}_\kappa \to \mathcal{A}_0 \supset \mathcal{C}^\infty(M_4) \tag{47}$$

The one-parameter family of associative algebras, $\mathcal{A}_\kappa$, can be analyzed with the help of the *deformation theory* developed in the well-known article by F. Bayen, M. Flato, C. Fronsdal and A. Lichnerowicz (cf. [23]). It is supposed that all $\mathcal{A}_\kappa$ coincide - as vector spaces - with a fixed vector space E. The product of any two elements f, g in $\mathcal{A}_\kappa$ can be expanded as follows:

$$(fg)_\kappa = fg + \kappa\, c(f,g) + o(\kappa^2) \tag{48}$$

where $fg = (fg)_0$ is the product in $\mathcal{A}_0$. We also assume that there is a common unit element $\mathbf{1}$ for all $\mathcal{A}_\kappa$. The commutators of any two elements f, g in $\mathcal{A}_\kappa$ and in $\mathcal{A}_O$ are related via the following equation:

$$[f,g]_\kappa = [f,g]_0 - i\,\kappa\,\{f,g\} + o(\kappa^2) \tag{49}$$

where $\{f,g\} = i\,(c(f,g) - c(g,f))$. The mapping $(f,g) \to c(f,g)$ is called a *normalized Hochschild 2-cocycle* of $\mathcal{A}_0$ with values in $\mathcal{A}_0$.

The derivation property of the commutator in $\mathcal{A}_\kappa$ should be maintained, which means that

$$[h,(fg)_\kappa]_\kappa = ([h,f]_\kappa, g)_\kappa + (f,[h,g]_\kappa)_\kappa \tag{50}$$

Then, in the first order in κ, we get

$$i\Big([h,c(f,g)] - c([h,f],g) - c(f,[h,g])\Big) = f\{h,g\} - \{h,fg\} + \{h,f\}g \tag{51}$$

This implies that if $h \in \mathcal{Z}(\mathcal{A}_0)$, the center of the algebra $\mathcal{A}_0$, then the endomorphism $\delta_h : \delta_h(f) = \{h,f\}$ is a derivation of $\mathcal{A}_0$:

$$\{h,\{f,g\}\} = \{\{h,f\},g\} + \{f,\{h,g\}\} \tag{52}$$

The *center* of the algebra $\mathcal{A}_0$, denoted by $\mathcal{Z}(\mathcal{A})$, is stable under these derivations, and therefore, it closes under the bracket $\{\,,\,\}$. This means that the Jacobi identity valid in all associative algebras $\mathcal{A}_\kappa$ remains valid, at least up to the second order in κ, in $\mathcal{A}_0$:

$$\text{from} \quad [f,[g,h]_\kappa]_\kappa + [g,[h,f]_\kappa]_\kappa + [h,[f,g]_\kappa]_\kappa = 0 \quad \text{it follows}$$

$$\{f,\{g,h\}_\kappa\}_\kappa + \{g,\{h,f\}_\kappa\}_\kappa + \{h,\{f,g\}_\kappa\}_\kappa = 0 \tag{53}$$

Summarizung up, we can make the following statement:

The center of $\mathcal{A}_0$, $\mathcal{Z}(\mathcal{A}_0)$, is a ***commutative Poisson algebra*** with the Poisson bracket given by

$$i(c(f,g) - c(g,f))$$

The linear mapping $z \to \delta_z$ maps $\mathcal{Z}(\mathcal{A}_0)$ into the Lie agebra of derivations of $\mathcal{A}_0$: $\delta_z(f) = \{z,f\}$, for $z, f \in \mathcal{A}_0$

We wish to represent the non-commutative analog of *real functions* by Hermitian elements of the extended algebra of functions. Therefore, we should impose the following *reality condition* :

- all the $\mathcal{A}_\kappa$ are complex *-algebras, whose involutive vector spaces coincide with the unique space E;

- for any $f \in E$, also $f^* \in E$; moreover, we assume that there exists a unique hermitian element which is the common unit for all these algebras, $\mathbf{1}^* = \mathbf{1}$, such that

$$(fg)^*_\kappa = (f^*g^*)_\kappa, \quad \text{and} \quad (\mathbf{1}f)_\kappa = (f\mathbf{1})_\kappa = f$$

It follows that the normalized co-cycle $c(f,g)$ satisfies natural condition

$$(c(f,g))^* = c(g^*,f^*)$$

Thus, the set $\mathcal{Z}_R(\mathcal{A}_0$ of all Hermitian elements of $\mathcal{Z}(\mathcal{A}_0$ forms naturally a real Poisson algebra, and $z \to \delta_z$ maps it into the real Lie algebra $Der(\mathcal{A}_0$ of all Hermitian derivations of $\mathcal{A}_0$.

Now comes the main point: the necessary realization of the Poincaré invariance on these algebras. The family $\mathcal{A}_\kappa$ represents non-commutative extensions of the algebra of smooth functions on space-time. Even if these algebras are not Poincaré-invariant, we wish to recover the Poincaré-invariant physics on the usual Minkowski space in the limit when $\kappa \to 0$. Therefore, we must assume that the Poincaré group $\mathcal{P}$ acts via *-automorphisms on the limit algebra $\mathcal{A}_0$:

$$(\Lambda, a) \to D_{\Lambda,a)} \in \mathcal{L}(\mathcal{A}_0, \mathcal{A}_0) \tag{54}$$

for any element $(\Lambda, a) \in \mathcal{P}$.

By hypothesis, the algebra $\mathcal{A}_0$ contains a *-subalgebra identified with the commutative algebra of smooth functions on Minkowski space, $\mathcal{C}^\infty(M_4)$. The action of $\mathcal{P}$ on $\mathcal{A}_0$ should induce the usual action of $\mathcal{P}$ on $\mathcal{C}^\infty(M_4)$ associated with the corresponding linear transformations in M_4.

We shall now argue that $\mathcal{C}^\infty(M_4)$ *can not be the whole* $\mathcal{A}_0$.

Indeed, suppose that $\mathcal{A}_0 = \mathcal{C}^\infty(M_4)$. The, in view of the our previous satement concerning the Poisson structures, there exists a Poisson bracket on M_4. This Poisson bracket must be non-trivial, since we assumed that the $\mathcal{A}_\kappa$ are all non-commutative.

On the other hand, we know that there does not exist a non-trivial Poincaré invariant bracket on M_4. Indeed, let $(f,g) \to \{f,g\}$ be such a bracket. Then, in a given coordinate patch, it can be represented analytically as

$$\{f,g\} = \Omega^{\mu\nu}\, \partial_\mu f\, \partial_\nu g \tag{55}$$

whare $\Omega^{\mu\nu} = \{x^\mu\, , \, x^\nu\}$ must be an anti-symmetric tensor field on M_4, which is constant with respect to translations and Lorentz covariant.

However, the rotational invariance already implies that the three-vectors

$$E^i = \Omega^{0i} \quad \text{and} \quad B^k = \epsilon^k_{lm}\, \Omega^{lm}, \quad (i,k,l = 1,2,3)$$

should vanish, which means that $\Omega^{\mu\nu} = 0$, and therefore, also $\{f,g\} = 0$ for all $f,g \in \mathcal{C}^\infty(M_4)$.

It seems unreasonable to suppose that the Poincaré invariance is broken at the first order in κ, because at this order we expect to recover a spin-2 Poincaré-invariant theory, coupled to other physical fields. So, if the Poincaré invariance holds at the first order in κ, it follows that the inclusion $\mathcal{C}^\infty(M_4) \subset \mathcal{A}_0$ must be a strict one, i.e. the limit $\kappa \to 0$ of $\mathcal{A}_\kappa$ must contain an extra factor besides $\mathcal{C}^\infty(M_4)$. Therefore, the normalized two-cocycle $c(\, ,\,)$ of $\mathcal{A}_0$ defined by

$$(fg)_\kappa = fg + \kappa\, c(f,g) + o(\kappa^2) \tag{56}$$

is supposed to be Poincaré-invariant, i.e. it has the property:

$$\alpha_{(\Lambda,a)}\left(c(f,g)) = c(\alpha_{(\Lambda,a)}(f), \alpha_{(\Lambda,a)}(g)\right) \tag{57}$$

which implies the invariance of the κ-bracket:

$$[f,g]_\kappa = [f,g] - i\,\{f,g\} + o(\kappa^2) \tag{58}$$

Let us consider now the elements of $\mathcal{A}_0$ that belong to $\mathcal{C}^\infty(M_4)$ and generate the commutative algebra of smooth functions on M_4 : $\quad x^\mu \in \mathcal{C}^\infty(M_4)$. By definition, we have then

$$\alpha_{(\Lambda,a)}\, x^\mu = \Lambda^{-1\mu}_{\ \ \nu}\, (x^\nu - \mathbf{1}\, a^\nu) \tag{59}$$

By choosing the origin, one can identify $\mathcal{C}^\infty(M_4)$ with the Hopf algebra of functions on the group of translations of M_4. Since $\mathcal{C}^\infty(M_4)$ is a subalgebra of $\mathcal{A}_0$, the algebra $\mathcal{A}_O$ is a bimodule over $\mathcal{C}^\infty(M_4)$. As a left $\mathcal{C}^\infty(M_4)$-module, $\mathcal{A}_0$ is isomorphic with the tensor product $\mathcal{C}^\infty(M_4) \otimes \mathcal{A}^I_0$, where $\mathcal{A}^I_0$ denotes the subalgebra of transitionally invariant elements of $\mathcal{A}_0$:

$$\mathcal{A}^I_0 = \left\{ f \in \mathcal{A}_0 \mid \alpha_{(1,a)}(f) = f \ \text{ for all } \ a \right\} \tag{60}$$

In fact, $\mathcal{A}_0$ is isomorphic with $\mathcal{C}^\infty(M_4) \otimes \mathcal{A}_0^I$ as a $(\mathcal{C}^\infty(M_4), \mathcal{A}_0^I)$-bimodule. Thus in order to recover the complete algebraic structure of $\mathcal{A}_0$, it is sufficient to describe the right multiplication by elements of $\mathcal{C}^\infty(M_4)$ of the elements of $\mathcal{A}_0^I$. The algebra $\mathcal{A}_0^I$ is stable under the derivations induced by the generators of local coordinate variables x^μ:

$$f \to ad(x^\mu)(f) = [x^\mu, f]$$

Therefore, for any $f \in \mathcal{A}_0^I$ one has

$$fx^\mu = x^\mu f - ad(x^\mu)(f)$$

or, in the tensorial representation $\mathcal{A}_0 = \mathcal{C}^\infty(M_4) \otimes \mathcal{A}_0^I$:

$$fx^\mu = x^\mu \otimes f - \mathbf{1} \otimes ad(x^\mu)(f) \quad \text{for any} \quad f \in \mathcal{A}_0^I$$

¿From this we can deduce the right multiplication of $\mathcal{C}^\infty(M_4) \otimes \mathcal{A}_0^I$ by the elements of $\mathcal{C}^\infty(M_4)$. Let us denote by X^μ the four commuting derivations of $\mathcal{A}_0^I$ induced by $ad(x^\mu)$. The algebra $\mathcal{A}_0^I$ is invariant under the action of the diffeomorphisms $\alpha_{(\Lambda,0)}$.

Let us denote by α_Λ^I the homomorphism of the Lirentz group into the group $Aut(\mathcal{A}_0^I)$ of all the *-automorphisms of $\mathcal{A}_0^I$.

Then one can summarize the above discussion of properties of our algebra by the following presentation of $\mathcal{A}_0$:

We start with a unital *-algebra $\mathcal{A}_0^I$ equipped with four commuting anti-Hermitian derivations X^μ and the action $\Lambda \to \alpha_\Lambda^I$ of the Lorentz group through the automorphisms of $\mathcal{A}_0^I$:

$$\alpha_\Lambda^I \circ X^\mu = \Lambda^{-1}\ {}^\mu_\nu\, X^\nu \circ \alpha_\Lambda^I \tag{61}$$

The entire algebra $\mathcal{A}_0$ is generated as a unital *-algebra by $\mathcal{A}_0^I$ and the four Hermitian elements x^μ which satisfy the relations:

$$x^\mu x^\nu = x^\nu x^\mu$$

$$\text{and} \qquad x^\mu f = f x^\mu + X^\mu(f) \qquad \text{if} \qquad f \in \mathcal{A}_0^I \tag{62}$$

The Poincaré group acts on $\mathcal{A}_0$ as follows:

- for $x^\mu \in \mathcal{C}^\infty(M_4)$:

$$\alpha_{(\Lambda,a)}\,(x^\mu) = \Lambda^{-1}{}^\mu_\nu(x^\nu - a^\nu\,\mathbf{1}); \tag{63}$$

- for $f \in \mathcal{A}_0^I$:

$$\alpha_{(\Lambda,a)}(f) = \alpha_\Lambda^I(f). \tag{64}$$

But we have assumed before that the bracket

$$\{f,g\} = i\left(c(f,g) - c(g,f)\right)$$

does not vanish identically on $\mathcal{C}^\infty(M_4)$. This implies that the functions $c^{\mu\nu}$ defined as

$$c^{\mu\nu} = c(x^\mu, x^\nu)$$

do not all vanish. On the other hand, these functions being Lorentz covariant must belong to $\mathcal{A}_0^I$, so that we have

$$\alpha_{\Lambda,a)}(c^{\mu\nu}) = \tilde{c}^{\mu\nu}$$

and one has

$$\alpha_\Lambda^I(c^{\mu\nu}) = \Lambda^{-1}{}_\rho^\mu \Lambda^{-1}{}_\sigma^\nu c^{\rho\sigma}, \tag{65}$$

so that the homomorphism of the Lorentz group into the group $Aut(\mathcal{A}_0^I$ of the *-automorphisms of $\mathcal{A}_0^I$ is never trivial.

This implies in turn that $\mathcal{A}_0^I$ cannot be a finite-dimensional algebra (like e.g. the complex matrix algebra discussed in our previous example), because on such an algebra all automorphisms are *inner*, and on the other hand, it is known that the Lorentz group has no non-trivial, finite dimensional unitary representations. Therefore, the extra factor that is present in $\mathcal{A}_0$ besides the usual infinite-dimensional algebra of functions (coordinates) on M_4 must be also infinite dimensional.

In view of previous analysis, the algebra $\mathcal{A}_0$ is the tensor product $\mathcal{C}^\infty(M_4) \otimes \mathcal{A}_O^I$, with the Lorentz group acting via automorphisms on $\mathcal{A}_0^I$. Since the brackets $\{x^\mu, x^\nu\} \in \mathcal{A}_0^I$, the algebra $\mathcal{A}_0^I$ must contain as a subalgebra an algebra of functions on the union of Lorentz orbits of anti-symmetric 2-tensors. The coordinates on this algebra viewed as a manifold are just the brackets $\{x^\mu, x^\nu\}$. The orbits may be labeled by the following two parameters:

$$\alpha = g_{\mu\rho}\, g_{\nu\lambda}\, \{x^\mu, x^\rho\}\{x^\nu, x^\lambda\} \qquad \text{and} \qquad \beta = \epsilon_{\mu\nu\rho\sigma}\, \{x^\mu, x^\nu\}\, \{x^\rho, x^\sigma\}. \tag{66}$$

If we want to include the definitions of time reversal and parity, we should assume that whenever a given orbit (α, β) appears in the algebra, the orbit corresponding to $(\alpha, -\beta)$ should appear as well. When one has also $\{x^\mu, \{x^\nu, x^\lambda\}\} = 0$ for all values of indeces μ, ν, λ, then $\mathcal{A}_0^I$ is equal to the above algebra.

The simplest situation occurs when $\mathcal{C}^\infty(M_4)$ belongs to the center of $\mathcal{A}_0$. In this case the cocycle c is antisymmetric (up to a co-boundary) on $\mathcal{C}^\infty(M_4)$, and also on the center $\mathcal{Z}(\mathcal{A}_0)$ itself. Then $\mathcal{A}_0$ is a commutative Poisson algebra, and the family $\mathcal{A}_\kappa$ can be obtained by its geometric quantization.

It is not difficult to give an example of such one-parameter family of algebras, containing the usual representation of the Poincaré algebra acting on smooth functions (coordinates) on M_4.

$$[x^\mu, x^\nu] = i\kappa M^{\mu\nu}$$

$$[x^\lambda, M^{\mu\nu}] = i(g^{\lambda\nu} L^\mu - g^{\lambda\mu} L^\nu)$$

$$[x^\mu, L^\nu] = i\kappa M^{\mu\nu}$$

$$[M^{\lambda\rho}, M^{\mu\nu}] = i(g^{\lambda\nu}M^{\mu\rho} - g^{\rho\nu}M^{\mu\lambda} + g^{\rho\mu}M^{\nu\lambda} - g^{\lambda\mu}M^{\nu\rho})$$

$$[L^\lambda, M^{\mu\nu}] = i(g^{\lambda\nu}L^\mu - g^{\lambda\mu}L^\nu)$$

$$[L^\mu, L^\nu] = i\kappa M^{\mu\nu} \tag{67}$$

where $g^{\mu\nu}$ denotes the Minkowskian metric $diag(-1,1,1,1)$. It follows from the above relations that for $\kappa \neq 0$ the algebras $\mathcal{A}_\kappa$ are generated by the x^μ. For any value of κ there exists an action of the Poincaré group $\mathcal{P}$ on $\mathcal{A}_\kappa$ via *-automorphisms $(\Lambda, a) \to \alpha_{(\Lambda,a)}$ defined as:

$$\alpha_{(\Lambda,a)}\, x^\mu = {\Lambda^{-1}}^\mu_\nu\,(x^\nu - a^\nu\,\mathbf{1}), \quad \alpha_{(\Lambda,a)}\, L^\mu = {\Lambda^{-1}}^\mu_\nu\, L^\nu, \quad \alpha_{(\Lambda,a)}\, I^{\mu\nu} = {\Lambda^{-1}}^\mu_\rho {\Lambda^{-1}}^\nu_\sigma\, I^{\rho\sigma}$$

The commutation relations between the $I^{\mu\nu}$ and the L^λ are the relations of the Lie algebra of $SO(4,1)$ if κ is positive, of $SO(3,2)$ if κ is negative, and of the Poincaré algebra if $\kappa = 0$. It follows that the $I^{\mu\nu}$ and the L^λ generate the corresponding enveloping algebras. The differences of the generators $x^\mu - L^\mu$ are in the center $Z(\mathcal{A}_\kappa)$ of $\mathcal{A}_\kappa$; therefore the algebra $\mathcal{A}_\kappa$ is the tensor product of the commutative algebra generated by the $(x^\mu - L^\mu)$ and the two following Casimir operators:

$$C_2 = \kappa\, g_{\mu\nu} g_{\rho\lambda}\, I^{\mu\rho} I^{\nu\lambda} + 2\, g_{\mu\nu}\, L^\mu L^\nu,$$

$$C_4 = g^{\rho\rho'}\,(\epsilon_{\rho\lambda\mu\nu} L^\lambda I^{\mu\nu})(\epsilon_{\rho'\lambda'\mu'\nu'} L^{\lambda'} I^{\mu'\nu'}) \tag{68}$$

where $\epsilon_{\mu\nu\lambda\rho}$ is the totally anti-seymmetric tensor with $\epsilon_{0123} = 1$. Therefore also $\mathcal{A}_0$ is the tensor product of the commutative algebra generated by the $(x^\mu - L^\mu)$ with the enveloping algebra of the Poincaré Lie algebra generated by the L^μ and the $I^{\mu\nu}$.

It must be stressed here that this Poincaré algebra is not the same as the Poincaré algebra acting on $\mathcal{A}_0$ (like on the space-time variables) via the automorphisms $\alpha_{(\Lambda,a)}$; this can be seen also by the fact that L^μ have the dimension of a length. This double appearance of the Poincaré algebra may be interpreted as the necessity to introduce matter besides the space-time itself as soon as we penetrate in the non-commutative sector of the great algebra containing $C^\infty(M_4)$ as a factor.

Since our Casimirs C_2 and C_4 are contained in the center of $\mathcal{A}_\kappa$, and since they are translationally invariant, we can impose some fixed values on them, thus specifying even more precisely the algebras $\mathcal{A}_\kappa$. Since the element C_2 has the dimension of a length squared, and the element C_4 that of a length to the power four, the most natural choices amount to attribute the value κ^2 to the element C_4, while the element C_2 can be given the following three particular values:

$$i)\, C_2 = \kappa; \qquad ii)\, C_2 = -\kappa, \qquad iii)\, C_2 = 0.$$

All these choices lead to $g_{\mu\nu}L^\mu L^\nu = 0$ in $\mathcal{A}_0^I$. Remembering the fact that $\mathcal{A}_0^I$ has the structure of the enveloping algebra of the Poincaré Lie algebra, the last condition is an analogue of the *zero mass* condition in the ususal case.

With the value of C_4 fixed in such a way that the representations found here are all of "zero mass" and "strictly positive spin" type, which gives the algebra $\mathcal{A}_0^I$ a characteristic two-sheet structure, corresponding to the two possible helicities, which in turn results from the fact that the Lorentz group is not simply connected.

As a concluding remark, we would like to stress the fact that in general the Poincaré covariance of $\mathcal{A}_\kappa$ is not necessary; all we need here is to ensure the Poincaré covariance of $\mathcal{A}_0$ only. Another deformation of the Poincaré algebra, called "the κ-Poincaré" has been studied in a series of papers published recently by J. Lukierski and co-authors ([24]).

Their approach is in some sense complementary to the scheme presented above: instead of considering the action of the *exact* Poincaré group on the space-time containing a non-trivial deformation because of the supposed non-commutative character of the coordinates, one chooses to consider the action of a *deformed* Poincaré group, called the κ-Poincaré, on the ordinary space-time. It seems plausible that in the linear limit both these approaches nearly coincide.

6 Quantum spaces and quantum groups

A more radical deformation of usual behaviour of functions describing the coordinates and their differentials consists in modifying the commutation relations not only between the coordinates and their differentials, but also between the coordinates themselves, and between the differentials as well, which would represent a very profound modification of the space-time structure. Moreover, if we look for the transformations that would keep these new relations invariant, we discover that such transformations can not be described by means of ordinary groups, which therefore need to be generalized. Such new generalizations have been introduced by V.Drinfeld, L.Faddeev and S.L.Woronowicz, ([25], [26], [27]). and they are known under the name of "Quantum Groups".

The litterature on this subject is very abundant; we shall cite the papers by S.L.Woronowicz [27], as well as the papers of L.C.Biedenharn [28], J.Wess and B.Zumino [29], L.A.Takhtajan [30], V.G.Kac [31]; the list is far from being exhaustive, so that we shall limit ourselves to an outline of the main idea illustrated by a simple example.

Conformally with the spirit of quantum field theories, the most important mathematical object to be studied is the algebra of observables, which are usually functions of few fundamental ones. This approach can be extended to the mathematical study of Lie groups: indeed, we can learn almost everything concerning group's structure from the algebraic structure of functions (real or complex) defined on it.

Consider a compact manifold G which is also a Lie group; let e denote its unit element. The algebra A of functions defined on G has a very particular structure, which is implemented by the following three mappings: *i*) for each $f \epsilon A$, there is an element of $A \otimes A$, denoted by Δf, such that $\Delta f(x,y) = f(xy)$; The mapping Δ :

$$A \longrightarrow A \otimes A \tag{69}$$

is called the *coproduct.* *ii*) There exists a natural mapping from A into $\mathbf{C}$ (or $\mathbf{R}$) defined by

$$\epsilon : f \longrightarrow f(e) \in C \tag{70}$$

which is called the *co-unit* *iii*) There exists a natural mapping of A into itself:

$$(Sf)(x) = f(x^{-1}) \tag{71}$$

which is called the ***antipode***

It is easy to see that in the case of the algebra of functions defined on a Lie group the co-product is non-commutative if the Lie group is non-commutative; however, the multiplication law in the algebra A itself remains commutative as long as we consider the functions taking their values in $\mathbf{C}$ or $\mathbf{R}$. This particular structure of an associative commutative algebra A with the three operations defined above, the ***co-product***, the ***co-unit*** and the ***antipode*** is called the ***Hopf algebra***. Now, the natural extension that comes to mind is to abandon the postulate of the commutativity of the product in A; in this case, the structure is named the *Quantum Group*. It should be stressed that a quantum group is not a group, but a general algebra which only in the commutative case behaves as the algebra of functions defined on a Lie group.

One of the most interesting aspects of this theory is the fact that the quantum goups arise quite naturally as the transformations of non-commutative geometries known under the name of ***quantum spaces*** introduced by Yu.Manin, J.Wess and B.Zumino, and others. We shall illustrate how a quantum group can be constructed on a simple example in two dimensions called the ***Manin plane*** ([32]).

Consider two "coordinates" x and y spanning a linear space and satisfying

$$xy \quad = q \quad yx \tag{72}$$

with a complex parameter q different from 1. Consider a transformation

$$x' = a \quad x + b \quad y, \quad y' = c \quad x + d \quad y. \tag{73}$$

which preserves the relation $xy =$ q yx, i.e. such that

$$x'y' = q \quad y'x' \tag{74}$$

We shall suppose that the quantities a, b, c, d commute with the "coordinates" x, y; the simplest realization of this requirement is achieved by assuming (disregarding the nature of the entries of the matrix) that the multiplication of x by a, b, etc. is tensorial, i.e. when we set by definition

$$x' = a \otimes x + b \otimes y. \tag{75}$$

Then the conservation of the q-commutation relations between x and y leads to the following rules for a, b, c and d:

$$ac = q \quad ca, \quad bd = q \quad db, \quad ad = da + q \quad cb - (\frac{1}{q})bc \tag{76}$$

In order to fix all possible binary relations between the coefiicients a, b, c and d we need three extra relations, which would define bc, ab and cd. Such relations can be obtained if we define the "differentials"

$$\xi = dx, \quad \eta = dy, \quad \xi^2 = 0, \quad \eta^2 = 0. \tag{77}$$

satisfying twisted p -commutation relations

$$\xi\eta + (\frac{1}{p})\eta\xi = 0 \tag{78}$$

with a new complex parameter p. Assuming that the exterior differentiation commutes with the transformation matrix and requiring the same relations for ξ' and η', we get

$$bc = (\frac{q}{p})cb, \quad ab = p \;\; ba, \quad cd = p \;\; dc. \tag{79}$$

With these relations the matrix algebra defined above becomes associative and can be given the structure of a Hopf algebra as follows:

$$\Delta \begin{pmatrix} a\, b \\ c\, d \end{pmatrix} = \begin{pmatrix} a \otimes a + b \otimes c \; a \otimes b + b \otimes d \\ c \otimes a + d \otimes c \; c \otimes b + d \otimes d \end{pmatrix}$$

The antipode S of a quantum matrix should be defined as its inverse. In order to make such a definition operational, we need a non-commutative generalization of the determinant of a matrix. Such a "(q, p)-determinant" should be defined as the combination of parameters appearing in the transformation law for the "elementary area element", i.e. the exterior product of the differentials ξ and η:

$$\xi'\eta' = D_q \xi\eta \tag{80}$$

which yields immediately

$$D_q = ad - pbc \quad = da - (\frac{1}{q})bc \tag{81}$$

The determinant D_q commutes with a and d, but has non-trivial commutation relations with the off-diagonal elements a and b (in what follows, we shall omit the subscript q for the sake of simplicity) :

$$Db = (\frac{p}{q})bD, \quad Dc = (\frac{q}{p})cD. \tag{82}$$

It should also possess an inverse D^{-1}, which in fact is a new element extending the algebra, and satisfying

$$D^{-1}D = 1 = DD^{-1} \tag{83}$$

Applying these identities to the commutation relations verified by D, one finds easily that D^{-1} commutes with a and b, and satisfies

$$bD^{-1} = (\frac{p}{q})D^{-1}b, \quad cD^{-1} = (\frac{q}{p})D^{-1}c \tag{84}$$

It is easy to see that

$$\Delta(D) = D \otimes D, \quad \Delta(D)\Delta(D^{-1}) = \Delta(1) = 1 \otimes 1 \tag{85}$$

and

$$\Delta(D^{-1}) = D^{-1} \otimes D^{-1} \tag{86}$$

The antipode of any matrix can be determined now as follows:

$$S\begin{pmatrix} a & b \\ c & d \end{pmatrix} = D^{-1}\begin{pmatrix} d & (\frac{-1}{q})b \\ -qc & a \end{pmatrix} = \begin{pmatrix} d & (\frac{1}{p})b \\ -p & a \end{pmatrix} D^{-1} \tag{87}$$

Also

$$S(D) = D^{-1}, \quad S(D^{-1}) = S(D) \tag{88}$$

but $S^2 \neq 1$. The inverse of the antipode mapping can be also defined as

$$S\begin{pmatrix} a & b \\ c & d \end{pmatrix} = D\begin{pmatrix} a & pq \\ (\frac{1}{pq})c & d \end{pmatrix} \tag{89}$$

The algebra generated by the matrices defined above, whose entries a, b, c and d satisfy the (q,p)-commutation relations is a Hopf algebra; it is denoted by $GL_{p,q}(2, \mathbf{C})$.

A differential calculus on such algebras has been developed by S.L.Woronowicz; the notion of covariant differentiation, if it can be introduced properly, may lead to new and rich extensions of the ideas of connections, curvatures and gauge fields. Here we shall give an example of the realization of covariant derivation and the curvature 2-form on the quantum plane introduced above. These results belong to M. Dubois-Violette *et al.*, published in ([33]).

The algebra of forms on the quantum plane is generated by four elements, $x, y, \xi = dx$ and $\eta = dy$, with the following commutation relations:

$$xy = q\,yx,$$

$$x\xi = q^2\xi x, \qquad x\eta = q\,\eta x + (q^2-1)\,\xi y, \qquad y\xi = q\,\xi y, \qquad y\eta = q^2\,\eta y,$$

$$\xi^2 = 0, \quad \eta^2 = 0, \quad \eta\xi + q\xi\eta = 0$$

where q is supposed not to be a root of unity. In (still hypothetical !) future physcical applications the value of the parameter q is supposed to be very close to 1, and in the linear approximation can be written as $1+\kappa$. The above conditions are of course compatible with the definitions $\xi = dx$, $\eta = dy$ and with the Leibniz rule, i.e. if we apply the operation d to the first constitutive identity $xy = q\,yx$, we obtain a relation which is a direct consequence of the four constitutive relations between x, y and their differentials ξ, η , and so forth.

All the relations between the variables x, y and their differentials ξ, η can be written in a more uniform way using a matrix notation which introduces the tensorial product of linear spaces spanned by both x, y and ξ, η variables. Denoting x and y by x^i and ξ and η by ξ^k, with $i, k = 1, 2$, we can write

$$x^i x^j - q^{-1} \hat{R}^{ij}_{\ kl}\, x^k x^l,$$

$$x^i \xi^j - q\, \hat{R}^{ij}_{\ kl}\, \xi^k x^l,$$

$$\xi^i \xi^j + q\, \hat{R}^{ij}_{\ kl}\, \xi^k \xi^l. \tag{90}$$

The tensor product of two 2-dimensional spaces is 4-dimensional, but the indices that are grouped two by two can be re-labeled with their values ranging from 1 to 4, and the R-matrix can be written as an ordinary 4×4 matrix:

$$\hat{R} = \begin{pmatrix} q & 0 & 0 & 0 \\ 0 & (q - q^{-1}) & 1 & 0 \\ 0 & 1 & 0 & 0 \\ 0 & 0 & 0 & q \end{pmatrix} \tag{91}$$

If the $SL_q(2, \mathbf{C})$ matrix (corresponding to the case $p = q^{-1}$ in the more general notation $GL_q(p, q)(2, \mathbf{C})$ introduced above) is written, with the same indices $k, l = 1, 2$ as

$$a^i_{\ k} = \begin{pmatrix} a & b \\ c & d \end{pmatrix}$$

then the invariance of the q-commutation relations with respect to the simultaneous transformation of the linear spaces x, y and ξ, η by a matrix belonging to the quantum group $SL_q(2, \mathbf{C})$ amounts to the following relation:

$$\hat{R}^{ij}_{\ kl}\, a^k_{\ m}\, a^l_{\ n} = a^i_{\ k} a^j_{\ l}\, \hat{R}^{kl}_{\ mn} \tag{92}$$

If we extend trivially the action of the differential d onto the quantum group $SL_q(2, \mathbf{C})$ itself by requiring all the coefficients $a^i_{\ k}$ to be constant,

$$d\, a^i_{\ k} = 0,$$

The coaction of $SL_q(2, \mathbf{C})$ on the x^i and the ξ^k can be defined then as follows:

$$\tilde{x}^i = a^i_{\ k} \otimes x^k, \qquad \tilde{\xi}^j = a^j_{\ m} \otimes \xi^m. \tag{93}$$

It can be found without much pain that the new variables $\tilde{x}^i$ and $\tilde{\xi}^k$ satisfy the same twisted commutation relations as formerly x^i and ξ^k.

As in the case of the matrix model of non-commutative geometry, one can introduce a *canonical* 1-form by defining

$$\theta = x\,\eta - q\,y\,\xi, \quad \text{satisfying} \quad \theta^2 = 0.$$

and is invariant under the coaction of $SL_q(2,\mathbf{C})$ with $\tilde{\theta} = \mathbf{1} \otimes \theta$ and has the following commutation relations with the variables x^k, ξ^m :

$$x^k\,\theta = q\,\theta x^k; \qquad \xi^m\,\theta = -q^3\,\theta\,\xi^m \tag{94}$$

Up to a complex multiplicative constant this is the unique element of Ω^1 (the space of q-one forms) verifying the above properties.

To define covariant derivation, we must introduce first the permutation operator σ mapping the tensor product $\Omega^1 \otimes_{\mathcal{A}} \Omega^1$ into itself. As a matter of fact, the operator σ turns out to be just the inverse of the matrix $q\,\tilde{R}^{ij}_{\ kl}$. We can write it down using the explicit indices $i, j, ..$ as follows:

$$\sigma\,(\xi \otimes \xi) = q^{-2}\,\xi \otimes \xi, \qquad \sigma\,(\xi \otimes \eta) = q^{-1}\,\eta \otimes \xi,$$

$$\sigma\,(\eta \otimes \xi) = q^{-1}\,\xi \otimes \eta - (1 - q^{-2})\,\eta \otimes \xi, \qquad \sigma\,(\eta \otimes \eta) = q^{-2}\,\eta \otimes \eta \tag{95}$$

as well as

$$\sigma\,(\xi \otimes \theta) = q^{-3}\,\theta \otimes \xi, \qquad \sigma\,(\theta \otimes \xi) = q\,\xi \otimes \theta - (1 - q^{-1})\,\theta \otimes \xi,$$

$$\sigma\,(\eta \otimes \theta) = q^{-3}\,\theta\eta, \qquad \sigma\,(\theta \otimes \eta) = q\,\eta \otimes \theta - (1 - q^{-2})\,\theta \otimes \eta. \tag{96}$$

and also

$$\sigma\,(\theta \otimes \theta) = q^{-2}\theta \otimes \theta$$

If we suppose that $q^2 \neq -1$, then the exterior algebra is obtained by dividing the tensor algebra over Ω^1 by the ideal generated by the three eigenvectors :

$$\xi \otimes \xi, \quad \eta \otimes \eta \quad \text{and} \quad \eta \otimes \xi + q\,\xi \otimes \eta,$$

corresponding to the eigenvalue q^{-2}.

The *symmetric* algebra of forms is obtained by dividing the tensor algebra over Ω^1 by the ideal generated by the eigenvector $\xi \otimes \eta - q\,\eta \otimes \xi$ corresponding to the eigenvalue -1.

There is a unique one-parameter family of covariant derivatives compatible with the algebraic structure of the algebra of forms defined above. It is given by

$$D\,\xi^k = l^{-4}\,x^k\,\theta \otimes \theta \tag{97}$$

where the parameter l must have the dimension of a length. From the invariance of θ it follows that D is invariant under the coaction of $SL_q(2,\mathbf{C})$. The analog of torsion vanishes identically.

Finally, the analog of the curvature tensor can be defined here as

$$D^2\,\xi^k = \Omega^k \otimes \theta = -\Omega^k_j \otimes \xi^l \tag{98}$$

with the curvature 2-forms given by the following matrix:

$$\Omega^i_j = l^{-4}\,(1 + q^{-2})(1 + q^{-4}) \begin{pmatrix} q^2\,xy & -q\,x^2 \\ q^2\,y^2 & -xy \end{pmatrix} \xi\,\eta \tag{99}$$

It vanishes for the particular values of q, namely, when $q = \pm i$ or $q^2 = \pm i$, but is different from zero when $q = 1$. The Bianchi identity is trivially satisfied.

No metric structure compatible with this structure can be introduced except for the trivial case when $q = 1$.

7 Conclusion

We tried to present here a few versions of non-commutative generalizations of differential geometry which are believed to serve - hopefully in some foreseable future - as new mathematical tools that will help us to describe the effects of quantum gravity. Frankly speaking, in spite of beauty and sophistication of certain models, it is hard to share this belief.

It does not mean that our efforts should be reduced or stopped at once. "*Ars longa, vita brevis*" , and there is still a lot of time ahead, especially as compared to the cosmological scale. The overall impression might be pessimistic, but there is always plenty of things to do.

For example, if we look at the diagram of Sect.1, we can note that besides the *"Relativistic Quantum Field Theory"* there is another unexplored corner, the *"Non-Relativistic Quantum Gravity"*. Maybe we should pay some more attention to this direction, too ? Or at least, if such a theory can not be formulated, try to give valuable reasons why this is the unique combination of limits of fundamental constants that can not be realized as a coherent theory ?

References

1. J.E. Moyal, *Quantum mechanics as a statistical theory*, Proc. Camb.Phil.Soc., **45**, 99 (1949)
2. P.A.M. Dirac, *The Fundamental Equations of Quantum Mechanics*, Proc.Roy.Soc., **A 109**, p.642 (1926); also *On Quantum Algebras*, Proc. Cambridge Phil.Soc., **23** p.412 (1926)
3. P.A.M. Dirac, as cited in *The Mathematical Intelligencer* **11** p.58 (1989)
4. R. Arnowitt, S. Deser, C. Misner, in *Recent Developments in General Relativity*, ed. L. Witten, Wiley, London- New York, (1962)
5. B. DeWitt, Physical Review, **160**, p.1113 (1963)
6. J.M. Souriau, *Structure des Systèmes Dynamiques*, Dunod, Paris, (1969)
7. D. Simms, in *Differential Geometrical Methors in Math. Physics*, Springer Lecture Notes in Mathematics, **170**, ed. K. Bleuler and A.Reetz, Springer-Verlag, (1970).
8. B. Kostant, *Quantization and Unitary Representations*, Lecture Notes in Mathematics, **170**, Springer Verlag p.237 (1970).
9. J. von Neumann, *Mathematical Foundations of Quantum Mechanics*, Princeton University Press, (1955)

10. A. Ashtekar, in *Mathematical Physics towards the 21-st century*, R. Sen and A. Gersten eds., Ben Gurion University of the Neguev Press, p 230 (1994).

11. C. Rovelli, Nucl. Physics B **405**, p.797 (1993), also the lectures in this volume.

12. D. Quillen, *Topology* **24**, p. 89 (1985)

13. A. Connes, *Non-Commutative Geometry*, Acad. Press., New York (1994)

14. M. Dubois-Violette, C.R.Acad.Sci. Paris, **307**, Sér.I, p.403 (1989)

15. M. Dubois-Violette, R. Kerner and J. Madore, Journ. of Math. Physics, **31** (2), p. 316 (1990)

16. M. Dubois-Violette, R. Kerner and J. Madore, *ibid*, p. 323 (1990)

17. A. Connes and J. Lott, Nucl. Phys. Proc. Suppl. B **18**, p.29 (1990)

18. R. Coquereaux, G. Esposito-Farèse and J. Vaillant, Nucl. Phys B, **353**, p.689 (1991)

19. A. Chamseddine, G. Felder and J. Fröhlich, Commun. Math. Physics, **155**, p. 205 (1993)

20. J. Madore, T. Masson and J. Mourad, Class. and Quantum Gravity, **12**, p. 1429 (1995)

21. M. Dubois-Violette, J. Madore and R. Kerner, Journal of Math. Physics, **39**, No 2, p. 730 (1998)

22. S. Doplicher, K. Friedenhagen and J.E. Roberts, Comm. Math. Phys., **172**, p. 187 (1995).

23. F. Bayen, M. Flato, C. Fronsdal, A. Lichnerowicz and D. Sternheimer, Ann. Phys. (N.Y.) **111** p. 61, p. 111 (1978)

24. J. Lukierski, A. Nowicki and H. Ruegg, Phys. Lett. B **293**, p.419 (1992)

25. V.G. Drinfeld, *Quantum Groups* , in Proceedings of Intern. Congress of Mathematics, Berkeley, p.798

26. L.D. Faddeev, Proc. of Schladming Winter School, p.89 (1989)

27. S.L. Woronowicz, Communications in Math. Phys. **111** p. 613 (1987) also: P. Podleś and S.L. Woronowicz, *ibid*, **178**, p.61 (1996)

28. L. Biedenharn, *Quantum Groups*, Lecture Notes in Physics, **370**, H.D. Doebner and J.D. Hennig eds., Springer-Verlag, p.67 (1990)

29. J. Wess and B. Zumino, Nucl. Phys. B (Proc. Suppl.) **18**, p.32 (1990)

30. L.A. Takhtajan, Adv. Studies in Pure and Applied Mathematics, **19**, p.435 (1989)

31. V.G. Kac, *Colloque Dixmier*, p. 471, Progrss in Mathematics, Vol. 92 Birkhäuser, Boston (1990); see also : C. de Concini and V.G. Kac, J. Am. Math. Soc. **5** (1), p. 151 (1992).

32. Yu. Manin, Commun. Math. Physics, **123**, p.163 (1989)

33. M. Dubois-Violette, J. Madore, T. Masson and J. Mourad, Lett. in Math.Physics, **35**, p. 351 (1995).

Conceptual Issues in Quantum Cosmology

Claus Kiefer[1,2]

[1] Fakultät für Physik, Universität Freiburg, Hermann-Herder-Str. 3, D-79104 Freiburg, Germany
[2] Institute for Advanced Study Berlin, Wallotstr. 19, D-14193 Berlin, Germany

Abstract. I give a review of the conceptual issues that arise in theories of quantum cosmology. I start by emphasising some features of ordinary quantum theory that also play a crucial role in understanding quantum cosmology. I then give motivations why spacetime cannot be treated classically at the most fundamental level. Two important issues in quantum cosmology – the problem of time and the role of boundary conditions – are discussed at some length. Finally, I discuss how classical spacetime can be recovered as an approximate notion. This involves the application of a semiclassical approximation and the process of decoherence. The latter is applied to both global degrees of freedom and primordial fluctuations in an inflationary Universe.

1 Introduction

As the title of this school indicates, a consistent quantum theory of gravity is eventually needed to solve the fundamental cosmological questions. These concern in particular the role of initial conditions and a deeper understanding of processes such as inflation. The presence of the singularity theorems in general relativity prevents the formulation of viable initial conditions in the classical theory. Moreover, the inflationary scenario can be successfully implemented only if the cosmological no-hair conjecture is imposed – a conjecture which heavily relies on assumptions about the physics at sub-Planckian scales.

It is generally assumed that a quantum theory of gravity can cure these problems. This is not a logical necessity, though, since there might exist classical theories which could achieve the same. As will be discussed in my contribution, however, one can put forward many arguments in favour of the quantisation of gravity, which is why classical alternatives will not be considered here.

Although a final quantum theory of gravity is still elusive, there exist concrete approaches which are mature enough to discuss their impact on cosmology. Here I shall focus on conceptual, rather than technical, issues that one might expect to play a role in any quantum theory of the gravitational field. In fact, most of the existing approaches leave the basic structures of quantum theory, such as its linearity, untouched.

Two aspects of quantum cosmology must be distinguished. The first is concerned with the application of quantum theory to the Universe as a whole and is independent of any particular interaction. This raises such issues as the interpretation of quantum theory for closed systems, where no external measuring agency can be assumed to exist. In particular, it must be clarified how and to what extent classical properties emerge. The second aspect deals with the peculiarities that enter through quantum aspects of the gravitational interaction. Since gravity is the dominant interaction on the largest scales, this is an important issue in cosmology. Both aspects will be discussed in my contribution.

Since many features in quantum cosmology arise from the application of standard quantum theory to the Universe as a whole, I shall start in the next section with a dicussion of the lessons that can be learnt from ordinary quantum theory. In particular, the central issue of the quantum-to-classical transition will be discussed at some length. Section 3 is then devoted to full quantum cosmology: I start with giving precise arguments why one must expect that the gravitational field is of a quantum nature at the most fundamental level. I then discuss the problem of time and related issues such as the Hilbert-space problem. I also devote some space to the central question of how to impose boundary conditions properly in quantum cosmology. The last section will then be concerned with the emergence of a classical Universe from quantum cosmology. I demonstrate how an approximate notion of a time parameter can be recovered from "timeless" quantum cosmology through some semiclassical approximation. I then discuss at length the emergence of a classical spacetime by decoherence. This is important for both the consistency of the inflationary scenario as well as for the classicality of primordial fluctuations which can serve as seeds for galaxy formation and which can be observed in the anisotropy spectrum of the cosmic microwave background.

2 Lessons from quantum theory

2.1 Superposition principle and "measurements"

The superposition principle lies at the heart of quantum theory. From a conceptual point of view, it is appropriate to separate it into a kinematical and a dynamical version (Giulini et al. 1996):

- *Kinematical version*: If Ψ_1 and Ψ_2 physical states, then $\alpha\Psi_1 + \beta\Psi_2$, where α and β are complex numbers, is again a physical state.
- *Dynamical version*: If $\Psi_1(t)$ and $\Psi_2(t)$ are solutions of the Schrödinger equation, then $\alpha\Psi_1(t) + \beta\Psi_2(t)$ is again a solution of the Schrödinger equation.

These features give rise to the ***nonseparability*** of quantum theory. If interactions between systems are present, the emergence of entangled states is unavoidable. As Schrödinger (1935) put it:

> I would not call that *one* but rather *the* characteristic trait of quantum mechanics, the one that enforces its entire departure from classical lines of thought. By the interaction the two representatives (or ψ-functions) have become entangled. ... Another way of expressing the peculiar situation is: the best possible knowledge of a *whole* does not necessarily include the best possible knowledge of all its *parts*, even though they may be entirely separated ...

Because of the superposition principle, quantum states which mimic classical states (for example, by being localised), form only a tiny subset of all possible states. Up to now, no violation of the superposition principle has been observed in quantum-mechanical experiments, and the only question is why we observe classical states at all. After all, one would expect the superposition principle to have unrestricted validity, since also macroscopic objects are composed of atoms.

The power of the superposition principle was already noted by von Neumann in 1932 when he tried to describe the measurement process consistently in quantum terms. He considers an interaction between a system and a (macroscopic) apparatus (cf. Giulini et al. 1996). Let the states of the measured system which are discriminated by the apparatus be denoted by $|n\rangle$, then an appropriate interaction Hamiltonian has the form

$$H_{int} = \sum_n |n\rangle\langle n| \otimes \hat{A}_n \ . \tag{1}$$

The operators $\hat{A}_n$, acting on the states of the apparatus, are rather arbitrary, but must of course depend on the "quantum number" n. Note that the measured "observable" is dynamically defined by the system-apparatus interaction and there is no reason to introduce it axiomatically (or as an additional concept). If the measured system is initially in the state $|n\rangle$ and the device in some initial state $|\Phi_0\rangle$, the evolution according to the Schrödinger equation with Hamiltonian (1) reads

$$\begin{aligned} |n\rangle|\Phi_0\rangle \stackrel{t}{\longrightarrow} \exp\left(-\mathrm{i}H_{int}t\right)|n\rangle|\Phi_0\rangle &= |n\rangle \exp\left(-\mathrm{i}\hat{A}_n t\right)|\Phi_0\rangle \\ &=: |n\rangle|\Phi_n(t)\rangle \ . \end{aligned} \tag{2}$$

The resulting apparatus states $|\Phi_n(t)\rangle$ are usually called "pointer positions". An analogy to (2) can also be written down in classical physics. The essential new quantum features come into play when we consider a *superposition* of different eigenstates (of the measured "observable") as initial state. The linearity of time evolution immediately leads to

$$\left(\sum_n c_n|n\rangle\right)|\Phi_0\rangle \stackrel{t}{\longrightarrow} \sum_n c_n|n\rangle|\Phi_n(t)\rangle \ . \tag{3}$$

This state does not, however, correspond to a definite measurement result – it contains a "weird" superposition of macroscopic pointer positions! This motivated von Neumann to introduce a "collapse" of the wave function, because he saw no other possibility to adapt the formalism to experience. There have been only rather recently attempts to give a concrete dynamical formulation of this collapse (see, e.g., Chap. 8 in Giulini et al. (1996)). However, none of these collapse models has yet been experimentally confirmed. In the following I shall review a concept that enables one to reconcile quantum theory with experience without introducing an explicit collapse; strangely enough, it is the superposition principle itself that leads to classical properties.

2.2 Decoherence: Concepts, examples, experiments

The crucial observation is that macroscopic objects cannot be considered as being isolated – they are unavoidably coupled to ubiquitous degrees of freedom of their einvironment, leading to quantum entanglement. As will be briefly discussed in the course of this subsection, this gives rise to classical properties for such objects – a process known as *decoherence*. This was first discussed by Zeh in the seventies and later elaborated by many authors; a comprehensive treatment is given by Giulini et al. (1996), other reviews include Zurek (1991), Kiefer and Joos (1999), see also the contributions to the volume Blanchard et al. (1999).

Denoting the environmental states with $|\mathcal{E}_n\rangle$, the interaction with system and apparatus yields instead of (3) a superposition of the type

$$\left(\sum_n c_n|n\rangle\right)|\Phi_0\rangle|\mathcal{E}_0\rangle \xrightarrow{t} \sum_n c_n|n\rangle|\Phi_n\rangle|\mathcal{E}_n\rangle . \tag{4}$$

This is again a macroscopic superposition, involving a tremendous number of degrees of freedom. The crucial point now is, however, that most of the environmental degrees of freedom are not amenable to observation. If we ask what can be seen when observing only system and apparatus, we need – according to the quantum rules – to calculate the reduced density matrix ρ that is obtained from (4) upon tracing out the environmental degrees of freedom.

If the environmental states are approximately orthogonal (which is the generic case),

$$\langle\mathcal{E}_m|\mathcal{E}_n\rangle \approx \delta_{mn} , \tag{5}$$

the density matrix becomes approximately diagonal in the "pointer basis",

$$\rho_S \approx \sum_n |c_n|^2|n\rangle\langle n| \otimes |\Phi_n\rangle\langle\Phi_n| . \tag{6}$$

Thus, the result of this interaction is a density matrix which seems to describe an ensemble of different outcomes n with the respective probabilities. One must be careful in analysing its interpretation, however: This density matrix only corresponds to an *apparent* ensemble, not a genuine ensemble of quantum states. What can safely be stated is the fact, that interference terms (non-diagonal elements) are absent locally, although they are still present in the total system, see (4). The coherence present in the initial system state in (3) can no longer be observed; it is *delocalised* into the larger system. As is well known, any interpretation of a superposition as an ensemble of components can be disproved experimentally by creating interference effects. The same is true for the situation described in (3). For example, the evolution could *in principle* be reversed. Needless to say that such a reversal is experimentally extremely difficult, but the interpretation and consistency of a physical theory must not depend on our present technical abilities. Nevertheless, one often finds explicit or implicit statements to the effect that the above processes are equivalent to the collapse of the wave function (or even solve the measurement problem). Such statements are certainly unfounded. What can safely be said, is that coherence between the subspaces of the Hilbert space spanned by $|n\rangle$ can no longer be observed in the system considered, *if* the process described by (3) is practically irreversible.

The essential implications are twofold: First, processes of the kind (3) do happen frequently and unavoidably for all macroscopic objects. Second, these processes are irreversible in practically all realistic situtations. In a normal measurement process, the interaction and the state of the apparatus are controllable to some extent (for example, the initial state of the apparatus is known to the experimenter). In the case of decoherence, typically the initial state is not known in detail (a standard example is interaction with thermal radiation), but the consequences for the local density matrix are the same: If the environment is described by an ensemble, each member of this ensemble can act in the way described above.

A complete treatment of realistic cases has to include the Hamiltonian governing the evolution of the system itself (as well as that of the environment). The exact dynamics of a subsystem is hardly manageable (formally it is given by a complicated integro-differential equation, see Chap. 7 of Giulini et al. 1996). Nevertheless, we can find important approximate solutions in some simplifying cases. One example is concerned with localisation through scattering processes and will be briefly discussed in the following. My treatment will closely follow Kiefer and Joos (1999).

Why do macroscopic objects always appear localised in space? Coherence between macroscopically different positions is destroyed *very* rapidly because of the strong influence of scattering processes. The formal description may proceed as follows. Let $|x\rangle$ be the position eigenstate of a macroscopic object, and $|\chi\rangle$ the state of the incoming particle. Following the von Neumann scheme, the scattering of such particles off an object located at position x may be written as

$$|x\rangle|\chi\rangle \stackrel{t}{\longrightarrow} |x\rangle|\chi_x\rangle = |x\rangle S_x|\chi\rangle \ , \tag{7}$$

where the scattered state may conveniently be calculated by means of an appropriate S-matrix. For the more general initial state of a wave packet we have then

$$\int \mathrm{d}^3x \ \varphi(x)|x\rangle|\chi\rangle \stackrel{t}{\longrightarrow} \int \mathrm{d}^3x \ \varphi(x)|x\rangle S_x|\chi\rangle \ , \tag{8}$$

and the reduced density matrix describing our object changes into

$$\rho(x,x') = \varphi(x)\varphi^*(x') \left\langle \chi|S_{x'}^{\dagger}S_x|\chi\right\rangle \ . \tag{9}$$

These steps correspond to the general steps discussed above. Of course, a single scattering process will usually not resolve a small distance, so in most cases the matrix element on the right-hand side of (9) will be close to unity. But if we add the contributions of many scattering processes, an exponential damping of spatial coherence results:

$$\rho(x,x',t) = \rho(x,x',0)\exp\left\{-\Lambda t(x-x')^2\right\} \ . \tag{10}$$

The strength of this effect is described by a single parameter Λ which may be called the "localisation rate" and is given by

$$\Lambda = \frac{k^2 N v \sigma_{eff}}{V} . \tag{11}$$

Here, k is the wave number of the incoming particles, Nv/V the flux, and σ_{eff} is of the order of the total cross section (for details see Joos and Zeh 1985 or Sect. 3.2.1 and Appendix 1 in Giulini et al. 1996). Some values of Λ are given in the Table.

Table 1. Localisation rate Λ in $\mathrm{cm}^{-2}\mathrm{s}^{-1}$ for three sizes of "dust particles" and various types of scattering processes (from Joos and Zeh 1985). This quantity measures how fast interference between different positions disappears as a function of distance in the course of time, see (10).

	$a = 10^{-3}$cm dust particle	$a = 10^{-5}$cm dust particle	$a = 10^{-6}$cm large molecule
Cosmic background radiation	10^{6}	10^{-6}	10^{-12}
300 K photons	10^{19}	10^{12}	10^{6}
Sunlight (on earth)	10^{21}	10^{17}	10^{13}
Air molecules	10^{36}	10^{32}	10^{30}
Laboratory vacuum (10^3 particles/cm^3)	10^{23}	10^{19}	10^{17}

Most of the numbers in the table are quite large, showing the extremely strong coupling of macroscopic objects, such as dust particles, to their natural environment. Even in intergalactic space, the 3K background radiation cannot be neglected.

In a general treatment one must combine the decohering influence of scattering processes with the internal dynamics of the system. This leads to master equations for the reduced density matrix, which can be solved explicitly in simple cases. Let me mention the example where the internal dynamics is given by the free Hamiltonian and consider the coherence length, i.e. the non-diagonal part of the density matrix. According to the Schrödinger equation, a free wave packet would spread, thereby increasing its size and extending its coherence properties over a larger region of space. Decoherence is expected to counteract this behaviour and reduce the coherence length. This can be seen in the solution shown in Fig. 1, where the time dependence of the coherence length (the width of the density matrix in the off-diagonal direction) is plotted for a truly free particle (obeying a Schrödinger equation) and also for increasing strength of decoherence. For large times the spreading of the wave packet no longer occurs and the coherence length always decreases proportional to $1/\sqrt{\Lambda t}$. More details and more complicated examples can be found in Giulini et al. (1996).

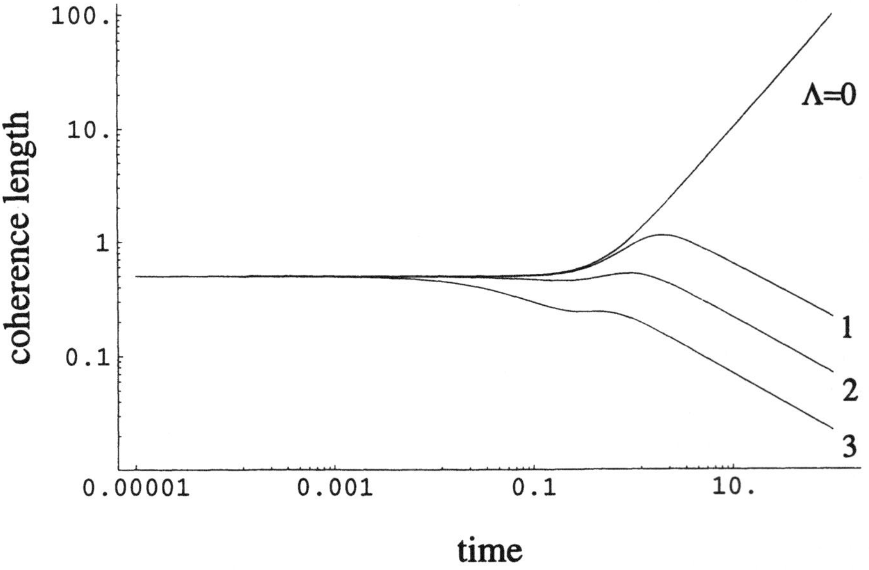

Fig. 1. Time dependence of coherence length. It is a measure of the spatial extension over which the object can show interference effects. Except for zero coupling ($\Lambda = 0$), the coherence length always decreases for large times. From Giulini et al. (1996).

Not only the centre-of-mass position of dust particles becomes "classical" via decoherence. The spatial structure of molecules represents another most important example. Consider a simple model of a chiral molecule (Fig. 2).

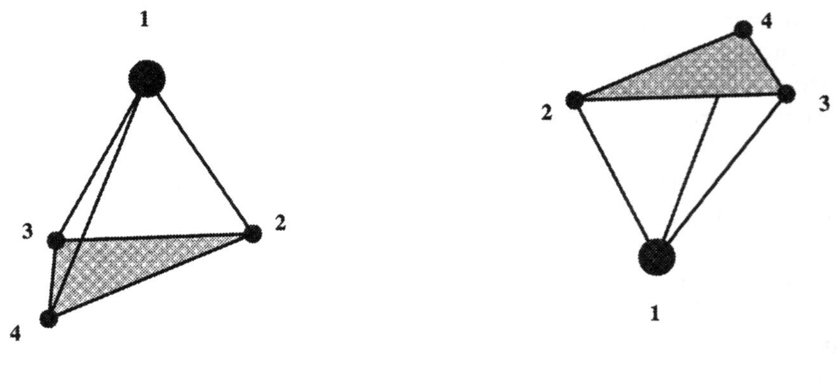

Fig. 2. Typical structure of an optically active, chiral molecule. Both versions are mirror-images of each other and are not connected by a proper rotation, if the four elements are different.

Right- and left-handed versions both have a rather well-defined spatial structure, whereas the ground state is - for symmetry reasons - a superposition of both chiral states. These chiral configurations are usually separated by a tunneling barrier (compare Fig. 3) which is so high that under normal circumstances tunneling is very improbable, as was already shown by Hund in 1929. But this alone does not explain why chiral molecules are never found in energy eigenstates! Only the interaction with the environment can lead to the localisation and the emergence of a spatial structure. We shall encounter a similar case of "symmetry breaking" in the case of quantum cosmology, see Sect. 4.2 below.

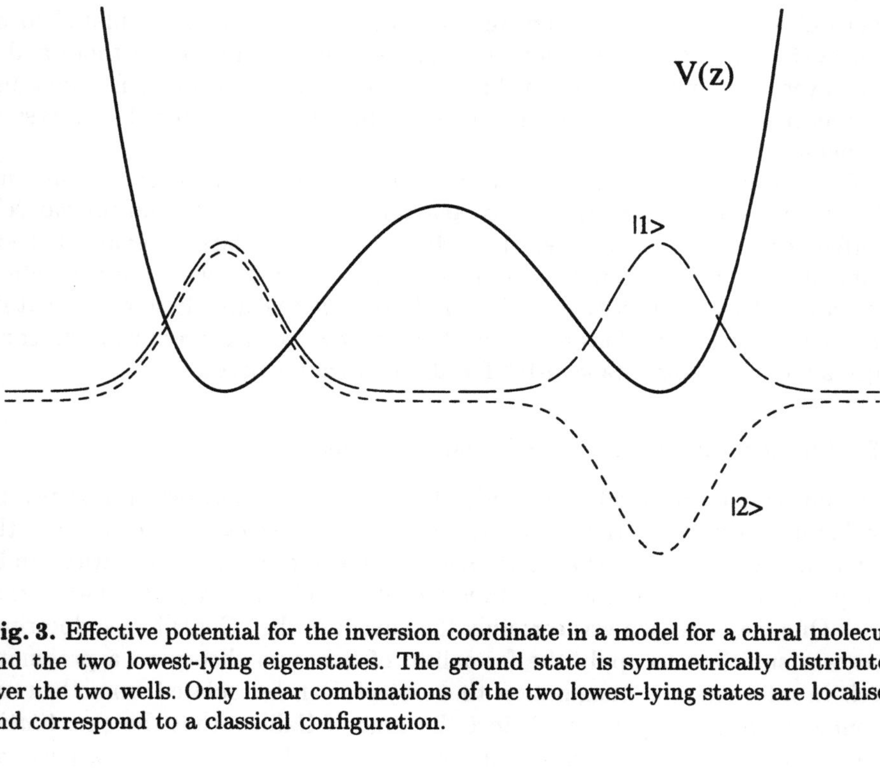

Fig. 3. Effective potential for the inversion coordinate in a model for a chiral molecule and the two lowest-lying eigenstates. The ground state is symmetrically distributed over the two wells. Only linear combinations of the two lowest-lying states are localised and correspond to a classical configuration.

I want to emphasise that decoherence should not be confused with thermalisation, although they sometimes occur together. In general, decoherence and relaxation have drastically different timescales – for a typical macroscopic situation decoherence is faster by forty orders of magnitude. This short decoherence timescale leads to the impression of discontinuities, e.g. "quantum jumps", although the underlying dynamics, the Schrödinger equation, is continuous. Therefore, to come up with a precise experimental test of decoherence, one must spend considerable effort to bring the decoherence timescale into a regime where it is comparable with other timescales of the system. This was achieved by a quantum-optical experiment that was performed in Paris in 1996, see Haroche (1998) for a review.

What is done in this experiment? The role of the system is played by a rubidium atom and its states $|n\rangle$ are two Rydberg states $|+\rangle$ and $|-\rangle$. This atom is sent into a high-Q cavity and brought into interaction with an electromagnetic field. This field plays the role of the "apparatus" and its pointer states $|\Phi_n\rangle$ are coherent states $|\alpha_+\rangle$ and $|\alpha_-\rangle$ which are correlated with the system states $|+\rangle$ and $|-\rangle$, respectively. The atom is brought into a superposition of $|+\rangle$ and $|-\rangle$ which it imparts on the coherent states of the electromagnetic field; the latter is then in a superposition of $|\alpha_+\rangle$ and $|\alpha_-\rangle$, which resembles a Schrödinger-cat state. The role of the environment is played by mirror defects and the corresponding environmental states are correlated with the respective components of the field superposition. One would thus expect that decoherence turns this superposition locally into a mixture. The decoherence time is calculated to be $t_D \approx t_R/\bar{n}$, where t_R is the relaxation time (the field-energy decay time) and $\bar{n}$ is the average photon number in the cavity. In the experiment t_R is about 160 microseconds, and $\bar{n} \approx 3.3$. These values enable one to monitor the process of decoherence as a process in time.

The decay of field coherence is measured by sending a second atom with different delay times into the cavity, playing the role of a "quantum mouse"; interference fringes are observed through two-atom correlation signals. The experimental results are found to be in complete agreement with the theoretical prediction. If a value of $\bar{n} \approx 10$ is chosen, decoherence is already so rapid that no coherence can be seen. This makes it obvious why decoherence for macroscopic objects happens "instantaneously" for all practical purposes.

2.3 On the interpretation of quantum theory[1]

It would have been possible to study the emergence of classical properties by decoherence already in the early days of quantum mechanics and, in fact, the contributions of Landau, Mott, and Heisenberg at the end of the twenties can be interpreted as a first step in this direction. Why did one not go further at that time? One major reason was certainly the advent of the "Copenhagen doctrine" that was sufficient to apply the formalism of quantum theory on a pragmatic level. In addition, the imagination that objects can be isolated from their environment was so deeply rooted since the time of Galileo, that the *quantitative* aspect of decoherence was largely underestimated. This quantitative aspect was only borne out by detailed calculations, some of which I have reviewed above. Moreover, direct experimental verification was only possible quite recently.

What are the achievements of the decoherence mechanism? Decoherence can certainly explain why and how *within* quantum theory certain objects (including fields) *appear* classical to "local" observers. It can, of course, not explain why there are such local observers at all. The classical properties are defined by the *pointer basis* for the object, which is distinguished by the interaction with the environment and which is sufficiently stable in time. It is important to emphasise that classical properties are *not* an a priori attribute of objects, but only come into being through the interaction with the environment.

[1] This is adapted from Sect. 4 of Kiefer and Joos (1999).

Because decoherence acts, for macroscopic systems, on an extremely short time scale, it appears to act discontinuously, although in reality decoherence is a smooth process. This is why "events", "particles", or "quantum jumps" are observed. Only in the special arrangement of experiments, where systems are used that lie at the border between microscopic and macroscopic, can this smooth nature of decoherence be observed.

Since decoherence studies only employ the standard formalism of quantum theory, all components characterising macroscopically different situations are still present in the total quantum state which includes system *and* environment, although they cannot be observed locally. Whether there is a real dynamical "collapse" of the total state into one definite component or not (which would lead to an Everett interpretation) is at present an undecided question. Since this may not experimentally be decided in the near future, it has been declared a "matter of taste" (Zeh 1997).

The most important feature of decoherence besides its ubiquity is its *irreversible* nature. Due to the interaction with the environment, the quantum mechanical entanglement *increases* with time. Therefore, the local entropy for subsystems increases, too, since information residing in correlations is locally unobservable. A natural prerequisite for any such irreversible behaviour, most pronounced in the Second Law of thermodynamics, is a special initial condition of very low entropy. Penrose has demonstrated convincingly that this is due to the extremely special nature of the big bang. Can this peculiarity be explained in any satisfactory way? Convincing arguments have been put forward that this can only be achieved within a quantum theory of gravity (Zeh 1999). This leads directly into the realm of quantum cosmology which is the topic of the following sections.

3 Quantum cosmology

3.1 Why spacetime cannot be classical

Quantum cosmology is the application of quantum theory to the Universe as a whole. Is such a theory possible or even – as I want to argue here – needed for consistency? In the first section I have stressed the importance of the superposition principle and the ensuing quantum entanglement with environmental degrees of freedom. Since the environment is in general also coupled to another environment, this leads ultimately to the whole Universe as the only closed quantum system in the strict sense. Therefore one must take quantum cosmology seriously. Since gravity is the dominant interaction on the largest scales, one faces the problem of quantising the gravitational field. In the following I shall list some arguments that can be put forward in support of such a quantisation, cf. Kiefer (1999):

- *Singularity theorems of general relativity*: Under very general conditions, the occurrence of a singularity, and therefore the breakdown of the theory, is unavoidable. A more fundamental theory is therefore needed to overcome these shortcomings, and the general expectation is that this fundamental theory is a quantum theory of gravity.

- *Initial conditions in cosmology*: This is related to the singularity theorems, since they predict the existence of a "big bang" where the known laws of physics break down. To fully understand the evolution of our Universe, its initial state must be amenable to a physical description.
- *Unification*: Apart from general relativity, all known fundamental theories are *quantum* theories. It would thus seem awkward if gravity, which couples to all other fields, should remain the only classical entity in a fundamental description. Moreover, it seems that classical fields cannot be coupled to quantum fields without leading to inconsistencies (Bohr-Rosenfeld type of analysis).
- *Gravity as a regulator*: Many models indicate that the consistent inclusion of gravity in a quantum framework automatically eliminates the divergences that plague ordinary quantum field theory.
- *Problem of time*: In ordinary quantum theory, the presence of an external time parameter t is crucial for the interpretation of the theory: "Measurements" take place at a certain time, matrix elements are evaluated at fixed times, and the norm of the wave function is conserved *in* time. In general relativity, on the other hand, time as part of spacetime is a dynamical quantity. Both concepts of time must therefore be modified at a fundamental level. This will be discussed in some detail in the next subsection.

The task of quantising gravity has not yet been accomplished, but approaches exist within which sensible questions can be asked. Two approaches are at the centre of current research: Superstring theory (or M-theory) and canonical quantum gravity. Superstring theory is much more ambitious and aims at a unification of all interactions within a single quantum framework (a recent overview is Sen 1998). Canonical quantum gravity, on the other hand, attempts to construct a consistent, non-perturbative, quantum theory of the gravitational field on its own. This is done through the application of standard quantisation rules to the general theory of relativity.

The fundamental length scales that are connected with these theories are the Planck length, $l_p = \sqrt{G\hbar/c^3}$, or the string length, l_s. It is generally assumed that the string length is somewhat larger than the Planck length. Although not fully established in quantitative detail, canonical quantum gravity should follow from superstring theory for scales $l \gg l_s > l_p$. One argument for this derives directly from the kinematical nonlocality of quantum theory: Quantum effects are not a priori restricted to certain scales. For example, the rather large mass of a dust grain cannot by itself be used as an argument for classicality. Rather, the process of decoherence through the environment can explain *why* quantum effects are negligible for this object, see the discussion in Sect. 2.2, in particular the quantitative aspects as they manifest themselves in the Table. Analogously, the smallness of l_p or l_s cannot by itself be used to argue that quantum-gravitational effects are small. Rather, this should be an emergent fact to be justified by decoherence (see Sect. 4). Since for scales larger than l_p or l_s general relativity is an excellent approximation, it must be clear that the canonical quantum theory must be an excellent approximation, too. The canonical theory might or might not exist on a full, non-perturbative level, but it should definitely exist as an effective theory on large scales. It seems therefore sufficient to base the following discussion on canonical quantum gravity, although I want to emphasise that the same conceptual issues arise in superstring theory.

Depending on the choice of the canonical variables, the canonical theory can be subdivided into the following approaches:

- *Quantum geometrodynamics*: This is the traditional approach that uses the three-dimensional metric as its configuration variable.
- *Quantum connection dynamics*: The configuration variable is a non-abelian connection that has many similarities to gauge theories.
- *Quantum loop dynamics*: The configuration variable is the trace of a holonomy with respect to a loop, analogous to a Wilson loop.

There exists a connection between the last two approaches, whereas their connection to the first approach is less clear. For the above reason one should, however, expect that a relation between all approaches exists at least on a semiclassical level. Here, I shall restrict myself to quantum geometrodynamics, since this seems to be the most appropriate language for a discussion of the conceptual issues. However, most of this discussion should find its pendant in the other approaches, too. A thorough discussion of these other approaches can be found in many contributions to this volume, see also Ashtekar (1999).

3.2 Problem of time

"Quantisation" is a set of heuristic recipes which allows one to guess the structure of the quantum theory from the underlying classical theory. In the canonical approach, the first step is to identify the canonical variables, the configuration and momentum variables of the classical theory. Their Poisson brackets are then translated into quantum operators. As a well-known theorem by Groenewald and van Hove states, such a translation is not possible for most of the other variables.

Details of the canonical formalism for general relativity can be found in Isham (1992), Kuchař (1992), and the references therein, and I shall give here only a brief introduction. For the definition of the canonical momenta, a time coordinate has to be distinguished. This spoils the explicit four-dimensional covariance of general relativity – the theory is reformulated to give a formulation for the dynamics of *three*-dimensional hypersurfaces. It is then not surprising that the configuration variable is the *three*-dimensional metric, $h_{ab}(\mathbf{x})$, on such hypersurfaces. The three-metric has six independent degrees of freedom. The remaining four components of the spacetime metric play the role of non-dynamical Lagrange multipliers called lapse function, $N^{\perp}(\mathbf{x})$, and shift vector, $N^{a}(\mathbf{x})$ – they parametrise, respectively, the way in which consecutive hypersurfaces are chosen and how the coordinates are selected *on* a hypersurface. The momenta canonically conjugated to the three-metric, $p^{ab}(\mathbf{x})$, form a tensor which is linearly related to the second fundamental form associated with a hypersurface – specifying the way in which the hypersurface is embedded into the fourth dimension.

In the quantum theory, the canonical variables are formally turned into operators obeying the commutation relations

$$[\hat{h}_{ab}(\boldsymbol{x}), \hat{p}^{ab}(\boldsymbol{y})] = \mathrm{i}\hbar\delta^c_{(a}\delta^d_{b)}\delta(\boldsymbol{x},\boldsymbol{y}) \ . \tag{12}$$

In a (formal) functional Schrödinger representation, the canonical operators act on wave functionals Ψ depending on the three-metric,

$$\hat{h}_{ab}(\boldsymbol{x})\Psi[h_{ab}(\boldsymbol{x})] = h_{ab}(\boldsymbol{x})\Psi[h_{ab}(\boldsymbol{x})] \tag{13}$$

$$\hat{p}^{cd}(\boldsymbol{x})\Psi[h_{ab}(\boldsymbol{x})] = \frac{\hbar}{\mathrm{i}}\frac{\delta}{\delta h_{cd}(\boldsymbol{x})}\Psi[h_{ab}(\boldsymbol{x})] \ . \tag{14}$$

A central feature of canonical gravity is the existence of constraints. Because of the four-dimensional diffeomorphism invariance of general relativity, these are four constraints per space point, one Hamiltonian constraint,

$$\hat{\mathcal{H}}_\perp\Psi = 0 \ , \tag{15}$$

and three diffeomorphism constraints,

$$\hat{\mathcal{H}}_a\Psi = 0 \ . \tag{16}$$

The total Hamiltonian is obtained by integration[2],

$$\hat{H} = \int \mathrm{d}^3x \ (N^\perp\hat{\mathcal{H}}_\perp + N^a\hat{\mathcal{H}}_a), \tag{17}$$

where $N^\perp$ and N^a denote again lapse function and shift vector, respectively. The constraints then enforce that the wave functional be annihilated by the total Hamiltonian,

$$\hat{H}\Psi = 0 \ . \tag{18}$$

The *Wheeler-DeWitt* equation (18) is the central equation of canonical quantum gravity. This also holds for quantum connection dynamics and quantum loop dynamics, although the configuration variables are different.

The Wheeler-DeWitt equation (18) possesses the remarkable property that it does not depend on any external time parameter – the t of the time-dependent Schrödinger equation has totally disappeared, and (18) looks like a stationary zero-energy Schrödinger equation. How can this be understood? In classical canonical gravity, a spacetime can be represented as a "trajectory" in configuration space – the space of all three-metrics. Although time coordinates have no intrinsic meaning in classical general relativity either, they can nevertheless be used to parametrise this trajectory in an essentially arbitrary way. Since no trajectories exist anymore in quantum theory, no spacetime exists at the most

[2] In the following I shall restrict myself to closed compact spaces; otherwise, the Hamiltonian has to be augmented by surface terms such as the ADM energy.

fundamental, and therefore also no time coordinates to parametrise any trajectory. A simple analogy is provided by the relativistic particle: In the classical theory there is a trajectory which can be parametrised by some essentially arbitrary parameter, e.g. the proper time. Reparametrisation invariance leads to one constraint, $p^2 + m^2 = 0$. In the quantum theory, no trajectory exists anymore, the wave function obeys the Klein-Gordon equation as an analogue of (18), and any trace of a classical time parameter is lost (although, of course, for the relativistic particle the background Minkowski spacetime is present, which is not the case for gravity).

Since the presence of an external time parameter is very important in quantum mechanics – giving rise to such important notions as the unitarity of states –, it is a priori not clear how to interpret a "timeless" equation of the form (18), cf. Barbour (1997) and Kiefer (1997). This is called the ***problem of time.*** A related issue is the ***Hilbert-space problem:*** What is the appropriate inner product that encodes the probability interpretation and that is conserved in time? Before discussing some of the options, it is very useful to first have a look at the explicit structure of (15) and (16). Introducing the Planck mass $m_p = (16\pi G)^{-1/2}$ and setting $\hbar = 1$, the constraint equations read

$$"\left\{ -\frac{1}{2m_p^2} G_{ab,cd} \frac{\delta^2}{\delta h_{ab} \delta h_{cd}} - m_p^2 \sqrt{h}\, {}^3R + \hat{H}_\perp^{\mathrm{mat}} \right\}" \, |\Psi[h_{ab}]\rangle = 0, \qquad (19)$$

$$"\left\{ -\frac{2}{\mathrm{i}} h_{ab} \nabla_c \frac{\delta}{\delta h_{bc}} + \hat{H}_a^{\mathrm{mat}} \right\}" \, |\Psi[h_{ab}]\rangle = 0. \qquad (20)$$

The inverted commas indicate that these are formal equations and that the factor ordering and regularisation problem have not been addressed. In these equations, ${}^3R$ and $\sqrt{h}$ denote the three-dimensional Ricci scalar and the square root of the determinant of the three-metric, respectively, and a cosmological term has not been considered here. The quantity $G_{ab,cd} = h^{-1/2}(h_{ac}h_{bd} + h_{ad}h_{bc} - h_{ab}h_{cd})$ plays the role of a metric in configuration space ("DeWitt metric"), and ∇_c denotes the covariant spatial derivative. The matter parts of the constraints, $\hat{H}_\perp^{\mathrm{mat}}$ and $\hat{H}_a^{\mathrm{mat}}$, depend on the concrete choice of matter action which we shall not specify here. Its form can be strongly constrained from general principles such as ultralocality (Teitelboim 1980). A tilde denotes a quantum operator in the standard Hilbert space of matter fields, while the bra and ket notation refers to the corresponding states.

The second equation (20) expresses the fact that the wave functional is invariant with respect to three-dimensional diffeomorphisms ("coordinate transformations"). It is for this reason why one often writes $\Psi[{}^3\mathcal{G}]$, where the argument denotes the coordinate-invariant three-*geometry*. Since there is, however, no explicit operator available which acts directly on $\Psi[{}^3\mathcal{G}]$, this is only a formal representation, and in concrete discussions one has to work with (19) and (20). It must also be remarked that this invariance holds only for diffeomorphisms that are connected with the identity; for "large" diffeomeorphism, a so-called θ-structure may arise, similarly to the θ-angle in QCD, see e.g. Kiefer (1993).

The kinetic term in (19) exhibits an interesting structure: The DeWitt metric $G_{ab,cd}$ has locally the signature diag$(-,+,+,+,+,+)$, rendering the kinetic term *indefinite*. Moreover, the one *minus sign* in the signature suggests that the corresponding degree of freedom plays the role of an "intrinsic time" (Zeh 1999). In general this does not, however, render (19) a hyperbolic equation, since even after dividing out the diffeomorphisms – going to the *superspace* of all three-geometries – there remains in general an infinite number of minus signs. In the special, but interesting, case of perturbations around closed Friedmann cosmologies, however, one global minus sign remains, and one is left with a truly hyperbolic equation (Giulini 1995). A Cauchy problem with respect to intrinsic time may then be posed. The minus sign in the DeWitt metric can be associated with the local scale part, $\sqrt{h}$, of the three-metric.

The presence of the minus sign in the DeWitt metric has an interesting interpretation: It reflects the fact that gravity is *attractive* (Giulini and Kiefer 1994). This can be investigated by considering the most general class of ultralocal DeWitt metrics which are characterised by the occurrence of some additional parameter α:

$$G^{\alpha}_{ab,cd} = h^{-1/2}(h_{ac}h_{bd} + h_{ad}h_{bc} - 2\alpha h_{ab}h_{cd}) \ , \tag{21}$$

where $\alpha = 0.5$ is the value corresponding to general relativity. One finds that there exists a critical value, $\alpha_c = 1/3$, such that for $\alpha < \alpha_c$ the DeWitt metric would become positive definite. One also finds that for $\alpha < \alpha_c$ gravity would become repulsive in the following sense: First, the second time derivative of the total volume $V = \int \mathrm{d}^3x\sqrt{h}$ (for lapse equal to one) would become, for positive three-curvature, positive instead of negative, therefore leading to an acceleration. Second, in the coupling to matter the sign of the gravitational constant would change. From the observed amount of helium one can infer that α must lie between 0.4 and 0.55.

Standard quantum theory employs the mathematical structure of a Hilbert space which is needed for the probability interpretation. Does such a structure also exist in quantum gravity? On a kinematical level, for wave functionals which are not yet necessarily solutions of the constraint equations, one can try to start with the standard Schrödinger-type inner product

$$\int \mathcal{D}h_{ab}\Psi^*[h_{ab}(x)]\Psi[h_{ab}(x)] \equiv (\Psi,\Psi)_S \ . \tag{22}$$

For wave functionals which satisfy the diffeomorphism constraints (20), this would yield divergencies since the integration runs over all "gauge orbits". In the connection representation, a preferred measure exists with respect to which the wave functionals are square integrable functions on the space of connections, see the contributions by Ashtekar, Lewandowski, and Rovelli to this volume. The construction is possible because the Hilbert space can be viewed as a limit of Hilbert spaces with finitely many degrees of freedom. It leads to interesting results for the spectra of geometric operators such as the area operator. However, no such product is known in geometrodynamics.

Since physical wave functionals have to obey (19) *and* (20), it might be sufficient if a Hilbert-space structure existed on the space of solutions, not necessarily on the space of all functionals such as in (22). Since (19) has locally the form of a Klein-Gordon equation, one might expect to use the inner product

$$\mathrm{i}\int \Pi_x \mathrm{d}\Sigma^{ab}(x)\Psi^*[h_{ab}]\left(G_{ab,cd}\frac{\overrightarrow{\delta}}{\delta h_{cd}} - \frac{\overleftarrow{\delta}}{\delta h_{cd}}G_{ab,cd}\right)\Psi[h_{ab}] \equiv (\Psi,\Psi)_{KG} \ . \tag{23}$$

The (formal) integration runs over a five-dimensional hypersurface at each space point, which is spacelike with respect to the DeWitt metric. The product (23) is invariant with respect to deformations of this hypersurface and therefore independent of "intrinsic time".

Similar to the situation with the relativistic particle, however, the inner product (23) is *not* positive definite. For the *free* relativistic particle one can perform a consistent restriction to a "positive-frequency sector" in which the analogue of (23) is manifestly positive, *provided* the spacetime background and the potential (which must be positive) are stationary, i.e., if there exists a time-like Killing vector which also preserves the potential. Otherwise, "particle production" occurs and the one-particle interpretation of the theory cannot be maintained. It has been shown that such a restriction to "positive frequencies" is *not* possible in quantum geometrodynamics (Kuchař 1992), the reason being that the Hamiltonian is not stationary. As I shall describe in Sect. 4, one can make, at least for certain states in the "one-loop level" of the semiclassical approximation, a consistent restriction to a positive-definite sector of (23).

For the relativistic particle one leaves the one-particle sector and proceeds to a field-theoretic setting, if one has to address situations where the restriction to positive frequencies is no longer possible. One then arrives at wave *functionals* for which a Schrödinger-type of inner product can be formulated. Can one apply a similar procedure for the Wheeler-DeWitt equation? Since quantum geometrodynamics is already a field theory, this would mean performing the transition to a "third-quantised" theory in which the state in (18) is itself turned into an operator. The formalism for such a theory is still in its infancy and will not be presented here (see e.g. Kuchař 1992). In a sense, superstring theory can be interpreted as providing such a framework.

All these problems could be avoided if it were possible to "solve" the constraints classically and make a transition to the physical degrees of freedom, upon which the standard Schrödinger inner product could be imposed. This would correspond to the choice of a time variable before quantisation. Formally, one would have to perform the canonical transformation

$$(h_{ab}, p^{cd}) \longrightarrow (X^A, P_A; \phi^i, p_i) \ , \tag{24}$$

where A runs from 1 to 4, and i runs from 1 to 2. X^A and P^A are the kinematical "embedding variables", while ϕ^i and p_i are the dynamical, physical, degrees of freedom. Unfortunately, such a reduction can only be performed in special situations, such as weak gravitational waves, but not in the general case, see Isham (1992) and Kuchař (1992). The best one can do is to choose the so-called "York time", but the corresponding reduction cannot be performed explicitly. Again, only on the one-loop level of the semiclassical approximation (see Sect. 4) can the equivalence of the Schrödinger product for the reduced variables and the Klein-Gordon inner product for the constrained variables be shown.

The problems of time and Hilbert space are thus not yet resolved at the most fundamental level. It is thus not clear, for example, whether (18) can sensibly be interpreted only as an eigenvalue equation for eigenvalue zero. Thus the options that will be discussed in the rest of my contribution are

- to study a semiclassical approximation and to aim at a consistent treatment of conceptual issues at that level. This is done in Sect. 4. Or
- to look for sensible boundary conditions for the Wheeler-DeWitt equation and to discuss directly solutions to this equation. This is done in the rest of this section.

3.3 Role of boundary conditions

Boundary conditions play a different role in quantum mechanics and quantum cosmology. In quantum mechanics (more generally, quantum field theory with an external background), boundary conditions can be imposed with respect to the external time parameter: Either as a condition on the wave function at a given time, or as a condition on asymptotic states in scattering situations. On the other hand, the Wheeler-DeWitt equation (18) is a "timeless" equation with a Klein-Gordon type of kinetic term.

What is the role of boundary conditions in quantum cosmology? Since the time of Newton one is accustomed to distinguish between dynamical laws and initial conditions. However, this is not a priori clear in quantum cosmology, and it might well be that boundary conditions are part of the dynamics. Sometimes quantum cosmology is even called a *theory of initial conditions* (Hartle 1997). Certainly, "initial" can here have two meanings: On the one hand, it can refer to initial condition of the classical Universe. This presupposes the validity of a semiclassical approximation (see Sect. 4) and envisages that particular solutions of (18) could *select* a subclass of classical solutions in the semiclassical limit. On the other hand, "initial" can refer to boundary conditions being imposed directly on (18). Since (18) is fundamentally timeless, this cannot refer to any classical time parameter but only to intrinsic variables such as "intrinsic time". In the following I shall briefly review some boundary conditions that have been suggested in quantum cosmology; details and additional references can be found in Halliwell (1991).

Let me start with the no-boundary proposal by Hartle and Hawking (1983). This does not yield directly boundary conditions on the Wheeler-DeWitt equation, but specifies the wave function through an integral expression – through a path integral in which only a subclass of all possible "paths" is being considered. This subclass comprises all spacetimes that have (besides the boundary where the arguments of the wave function are specified) no other boundary. Since the full quantum-gravitational path integral cannot be evaluated (probably not even be rigorously defined), one must resort to approximations. These can be semiclassical or minisuperspace approximations or a combination of both. It becomes clear already in a minisuperspace approximation that integration has to be performed over *complex* metrics to guarantee convergence. Depending on the nature of the saddle point in a semiclassical limit, the wave function can then refer to a classically allowed or forbidden situation.

Consider the example of a Friedmann Universe with a conformally coupled scalar field. After an appropriate field redefinition, the Wheeler-DeWitt equation assumes the form of an indefinite harmonic oscillator,

$$\left(\frac{\partial^2}{\partial a^2} - \frac{\partial^2}{\partial \phi^2} - a^2 + \phi^2\right)\psi(a,\phi) = 0 \; . \tag{25}$$

The implementation of the no-boundary condition in this simple minisuperspace model selects the following solutions (cf. Kiefer 1991)

$$\psi_1(a,\phi) = \frac{1}{2\pi}K_0\left(\frac{|\phi^2 - a^2|}{2}\right) \; , \tag{26}$$

$$\psi_2(a,\phi) = \frac{1}{2\pi}I_0\left(\frac{\phi^2 - a^2}{2}\right) \; , \tag{27}$$

where K_0 and I_0 denote Bessel functions. It is interesting to note that these solutions do not reflect the classical behaviour of the system (the classical solutions are Lissajous ellipses confined to a rectangle in configuration space, see Kiefer 1990) – I_0 diverges for large arguments, while K_0 diverges for vanishing argument ("light cone" in configuration space). Such features cannot always be seen in a semiclassical limit.

Another boundary condition is the so-called tunneling condition (Vilenkin 1998). It is also formulated in general terms – superspace should contain "outgoing modes" only. However, as with the no-boundary proposal, a concrete discussion can only be made within approximations. Typically, while the no-boundary proposal leads to *real* solutions of the Wheeler-DeWitt equation, the tunneling proposal predicts *complex* solutions. This is most easily seen in the semiclassical approximation (see Sect. 4), where the former predicts $\cos S$-type of solutions, while the latter predicts $\exp \mathrm{i}S$-type of solutions. (The name "tunneling proposal" comes from the analogy with situations such as α-decay in nuclear physics where an outgoing wave is present after tunneling from the nucleus.) A certain danger is connected with the word "outgoing" because it has a temporal connotation although (18) is timeless. A time parameter emerges only in a semiclassical approximation, see the next section.

A different type of boundary condition is the SIC proposal by Conradi and Zeh (1991). It demands that the wave function be simple for small scale factors, i.e. that it does not depend on other degrees of freedom. The explicit expressions exhibit many similarities to the no-boundary wave function, but since the boundary condition is directly imposed on the wave function without use of path integrals, it is much more convenient for a discussion of models which correspond to a classically recollapsing universe.

What are the physical applications that one could possibly use to distinguish between the various boundary conditions? Some issues are the following:

- *Probability for inflation*: It is often assumed that the Universe underwent a period of exponential expansion at an early stage (see also Sect. 4.3). The question therefore arises whether quantum cosmology can predict how "likely" the occurrence of inflation is. Concrete calculations address the question of the probability distribution for the initial values of certain fields that are responsible for inflation. Since such calculations necessarily involve the validity of a semiclassical approximation (otherwise the notion of inflation would not make sense), I shall give some more details in the next section.
- *Primordial black-hole production*: The production of primordial black holes during an inflationary period can in principle also be used to discriminate between boundary conditions, see e.g. Bousso and Hawking (1996).
- *Cosmological parameters*: If the wave function is peaked around definite values of fundamental fields, these values may appear as "constants of Nature" whose values can thereby be predicted. This was tentatively done for the cosmological constant (Coleman 1988). Alternatively, the anthropic principle may be invoked to select amongst the values allowed by the wave function.
- *Arrow of time*: Definite conclusions about the arrow of time in the Universe (and the interior of black holes) can be drawn from solutions to the Wheeler-DeWitt equation, see Kiefer and Zeh (1995).

Quantum cosmology is of course not restricted to quantum general relativity. It may also be discussed within effective models of string theory, see e.g. Dabrowski and Kiefer (1997), but I shall not discuss this here.

4 Emergence of a classical world

As I have reviewed in Sect. 3, there is no notion of spacetime at the full level of quantum cosmology. This was aleady anticipated by Lemaître (1931) who wrote:

> If the world has begun with a single quantum, the notions of space and time would altogether fail to have any meaning at the beginning ... If this suggestion is correct, the beginning of the world happened a little before the beginning of space and time.

It is not clear what "before" means in an atemporal situation, but it is obvious that the emergence of the usual notion of spacetime within quantum cosmology needs an explanation. This is done in two steps: Firstly, a semiclassical approximation to quantum gravity must be performed (Sect. 4.1). This leads to the recovery of an *approximate* Schrödinger equation of non-gravitational fields with respect to the semiclassical background. Secondly, the emergence of classical properties must be explained (Sect. 4.2). This is achieved through the application of the ideas presented in Sect. 2.2. A more technical review is Kiefer (1994), see also Brout and Parentani (1999). A final subsection is devoted to the emergence of classical fluctuations which can serve as seeds for the origin of structure in the Universe.

4.1 Semiclassical approximation to quantum gravity

The starting point is the observation that there occur different scales in the fundamental equations (19) and (20): The Planck mass m_p associated with the gravitational part, and other scales contained implicitly in $\hat{H}_{\perp}^{\mathrm{mat}}$. Even for "grand-unified theories" the relevant particle scales are at least three orders of magnitude smaller than m_p. For this reason one can apply Born-Oppenheimer type of techniques that are suited to the presence of different scales. In molecular physics, the large difference between nuclear mass and electron mass leads to a slow motion for the nuclei and the applicability of an adiabatic approximation. A similar method is also applied in the nonrelativistic approximation to the Klein-Gordon equation, see Kiefer and Singh (1991).

In the lowest order of the semiclassical approximation, the wave functional appearing in (19) and (20) can be written in the form

$$|\Psi[h_{ab}]\rangle = \mathrm{e}^{im_p^2 S[h_{ab}]} |\Phi[h_{ab}]\rangle \,, \tag{28}$$

where $S[h_{ab}]$ is a purely gravitational Hamilton-Jacobi function. This is a solution of the vacuum Einstein-Hamilton-Jacobi equations – the gravitational constraints with the Hamilton-Jacobi values of momenta (gradients of $S[h_{ab}]$).

Substitution of (28) into (19) and (20) leads to new equations for the state vector of matter fields $|\Phi[h_{ab}]\rangle$ depending parametrically on the spatial metric

$$\left\{\frac{1}{\mathrm{i}} G_{ab,cd} \frac{\delta S}{\delta h_{ab}} \frac{\delta}{\delta h_{cd}} + \hat{H}_{\perp}^{\mathrm{mat}}(h_{ab}) + \frac{1}{2\mathrm{i}}\, {}'' G_{ab,cd} \frac{\delta^2 S}{\delta h_{ab} \delta h_{cd}} {}'' - \frac{1}{2m_p^2}\, {}'' G_{ab,cd} \frac{\delta^2}{\delta h_{ab} \delta h_{cd}} {}'' \right\} |\Phi[h_{ab}]\rangle = 0, \tag{29}$$

$$\left\{ {}'' - \frac{2}{\mathrm{i}} h_{ab} \nabla_c \frac{\delta}{\delta h_{bc}} {}'' + \hat{H}_a^{\mathrm{mat}}(h_{ab}) \right\} |\Phi[h_{ab}]\rangle = 0. \tag{30}$$

It should be emphasised that on a formal level the factor ordering can be *fixed* by demanding the equivalence of various quantisation schemes, see Al'tshuler and Barvinsky (1996) and the references therein.

The conventional derivation of the Schrödinger equation from the Wheeler-DeWitt equation consists in the assumption of *small back reaction of quantum matter on the metric background* which at least heuristically allows one to discard the third and the fourth terms in (29). Then one considers $|\Phi[h_{ab}]\rangle$ on the solution of classical vacuum Einstein equations $h_{ab}(\mathbf{x}, t)$ corresponding to the

Hamilton-Jacobi function $S[h_{ab}]$, $|\Phi(t)\rangle = |\Phi[h_{ab}(\mathbf{x},t)]\rangle$. After a certain *choice* of lapse and shift functions $(N^{\perp}, N^a)$, this solution satisfies the canonical equations with the momentum $p^{ab} = \delta S/\delta h_{ab}$, so that the quantum state $|\Phi(t)\rangle$ satisfies the evolutionary equation obtained by using

$$\frac{\partial}{\partial t}|\Phi(t)\rangle = \int \mathrm{d}^3x\, \dot{h}_{ab}(\mathbf{x}) \frac{\delta}{\delta h_{ab}(\mathbf{x})}|\Phi[h_{ab}]\rangle \tag{31}$$

together with the truncated version of equations (29) – (30). The result is the Schrödinger equation of quantised matter fields in the external classical gravitational field,

$$\mathrm{i}\frac{\partial}{\partial t}|\Phi(t)\rangle = \hat{H}^{\mathrm{mat}}|\Phi(t)\rangle, \tag{32}$$

$$\hat{H}^{\mathrm{mat}} = \int \mathrm{d}^3x \left\{N^{\perp}(\mathbf{x})\hat{H}_{\perp}^{\mathrm{mat}}(\mathbf{x}) + N^a(\mathbf{x})\hat{H}_a^{\mathrm{mat}}(\mathbf{x})\right\}. \tag{33}$$

Here, $\hat{H}^{\mathrm{mat}}$ is a matter field Hamiltonian in the Schrödinger picture, parametrically depending on (generally nonstatic) metric coefficients of the curved spacetime background. In this way, the Schrödinger equation for non-gravitational fields has been recovered from quantum gravity as an approximation.

A derivation similar to the above can already be performed within ordinary quantum mechanics if one assumes that the total system is in a "timeless" energy eigenstate, see Briggs and Rost (1999). In fact, Mott (1931) had already considered a time-independent Schrödinger equation for a total system consisting of an α-particle and an atom. If the state of the α-particle can be described by a plane wave (corresponding in this case to high velocities), one can make an ansatz similar to (28) and derive a time-dependent Schrödinger equation for the atom alone, in which time is *defined* by the α-particle.

In the context of quantum gravity, it is most interesting to continue the semiclassical approximation to higher orders and to *derive* quantum-gravitational correction terms to (32). This was done in Kiefer and Singh (1991) and, giving a detailed interpretation in terms of a Feynman diagrammatic language, in Barvinsky and Kiefer (1998). I shall give a brief description of these terms and refer the reader to Barvinsky and Kiefer (1998) for all details.

At the next order of the semiclassical expansion, one obtains corrections to (32) which are proportional to m_p^{-2}. These terms can be added to the matter Hamiltonian, leading to an *effective* matter Hamiltonian at this order. It describes the back-reaction effects of quantum matter on the dynamical gravitational background as well as proper quantum effects of the gravitational field itself. Most of these terms are nonlocal in character: they contain the gravitational potential generated by the back reaction of quantum matter as well as the gravitational potential generated by the one-loop stress tensor of vacuum gravitons. In cases where the matter energy density is much bigger than the energy density of graviton vacuum polarisation, the dominant correction term is given by the kinetic energy of the gravitational radiation produced by the back reaction of quantum matter sources.

A possible observational test of these correction terms could be provided by the anisotropies in the cosmic microwave background (Rosales 1997). The temperature fluctuations are of the order 10^{-5} reflecting within inflationary models the ratio $m_I/m_p \approx 10^{-5}$, where m_I denotes the mass of the scalar field responsible for inflation (the "inflaton"). The correction terms would then be $(m_I/m_p)^2 \approx 10^{-10}$ times a numerical constant, which could in principle be large enough to be measurable with future satellite experiments such as MAP or PLANCK.

Returning to the "one-loop order" (28) of the semiclassical approximation, it is possible to address the issue of probability for inflation that was mentioned in Sect. 3.3, see Barvinsky and Kamenshchik (1994). In this approximation, the inner products (22) and (23) are equivalent and positive definite, see Al'tshuler and Barvinsky (1996). They can therefore be used to calculate quantum-mechanical probabilities in the usual sense.

To discuss this probability, the reduced density matrix for the inflaton, φ, should be investigated. This density matrix is calculated from the full quantum state upon integrating out all other degrees of freedom (here called f),

$$\rho_t(\varphi, \varphi') = \int \mathcal{D}f \; \psi_t^*(\varphi', f)\psi_t(\varphi, f) \; , \tag{34}$$

where ψ_t denotes the quantum state (28) after the parameter t from (32) has been used.

To calculate the probability one has to set $\varphi' = \varphi$. In earlier work, the saddle-point approximation was only performed up to the highest, tree-level, approximation. This yields

$$\rho(\varphi, \varphi) = \exp[\pm I(\varphi)] \; , \tag{35}$$

where $I(\varphi) = -3m_p^4/8V(\varphi)$ and $V(\varphi)$ is the inflationary poential. The lower sign corresponds to the no-boundary condition, while the upper sign corresponds to the tunneling condition. The problem with (3) is that ρ is not normalisable: mass scales bigger than m_p contribute significantly and results based on tree-level approximations can thus not be trusted.

The situation is improved considerably if loop effects are taken into account (Barvinsky and Kamenshchik 1994). They are incorporated by the loop effective action Γ_{loop} which is calculated on De-Sitter space. In the limit of large φ (that is relevant for investigating normalisability) this yields in the one-loop approximation

$$\Gamma_{loop}(\varphi)|_{H\to\infty} \approx Z \ln \frac{H}{\mu} \; , \tag{36}$$

where μ is a renormalisation mass parameter, and Z is the anomalous scaling. Instead of (35) one has now

$$\begin{aligned} \rho(\varphi, \varphi) &\approx \; H^{-2}(\varphi) \exp\left(\pm I(\varphi) - \Gamma_{loop}(\varphi)\right) \\ &\approx \; \exp\left(\pm \frac{3m_p^4}{8V(\varphi)}\right) \varphi^{-Z-2} \; . \end{aligned} \tag{37}$$

This density matrix is normalisable provided $Z > -1$. This in turn leads to reasonable constraints on the particle content of the theory, see Barvinsky and Kamenshchik (1994). It turns out that the tunneling wave function (with an appropriate particle content) can predict the occurrence of a sufficient amount of inflation. In earlier tree-level calculations the use of an anthropic principle was needed to get a sensible result from a non-normalisable wave function through *conditional* probabilities, see e.g. Hawking and Turok (1998). This is no longer the case here.

4.2 Decoherence in quantum cosmology[3]

As in ordinary quantum mechanics, the semiclassical limit is not yet sufficient to understand classical behaviour. Since the superposition principle is also valid in quantum gravity, quantum entanglement will easily occur, leading to superpositions of "different spacetimes". It is for this reason that the process of decoherence must be invoked to justify the emergence of a classical spacetime.

Joos (1986) gave a heuristic example within Newtonian (quantum) gravity, in which the superposition of different metrics is suppressed by the interaction with ordinary particles. How does decoherence work in quantum cosmology? In particular, what constitutes system and environment in a case where nothing is external to the Universe? The question is how to divide the degrees of freedom in the configuration space in a sensible way. It was suggested by Zeh (1986) to treat global degrees of freedom such as the scale factor (radius) of the Universe or an inflaton field as "relevant" variables that are decohered by "irrelevant" variables such as density fluctuations, gravitational waves, or other fields. Quantitative calculations can be found, e.g., in Kiefer (1987,1992).

Denoting the "environmental" variables collectively again by f, the reduced density matrix for e.g. the scale factor a is found in the usual way by integrating out the f-variables,

$$\rho(a,a') = \int \mathcal{D}f\; \Psi^*(a',f)\Psi(a,f)\;. \tag{38}$$

In contrast to the discussion following (34), the *non-diagonal* elements of the density matrix must be calculated. The resulting terms are ultraviolet-divergent and must therefore be regularised. This was investigated in detail for the case of bosons (Barvinsky et al. 1999c) and fermions (Barvinsky et al. 1999a). A crucial point is that standard regularisation schemes, such as dimensional regularisation or ζ-regularisation, do not work – they lead to $\mathrm{Tr}\rho^2 = \infty$, since the sign in the exponent of the Gaussian density matrix is changed from minus to plus by regularisation. These schemes therefore spoil one of the important properties that a density matrix must obey. This kind of problem has not been noticed before, since these regularisation schemes had not been applied to the calculation of reduced density matrices.

[3] This and the next subsection are adapted from Kiefer (1999).

How, then, can (38) be regularised? In Barvinsky et al. (1999a,c) we put forward the principle that there should be no decoherence if there is no particle creation – decoherence is an irreversible process. In particular, there should be no decoherence for static spacetimes. This has led to the use of a certain conformal reparametrisation for bosonic fields and a certain Bogoliubov transformation for fermionic fields.

As a concrete example, we have calculated the reduced density matrix for a situation where the semiclassical background is a De Sitter spacetime, $a(t) = H^{-1}\cosh(Ht)$, where H denotes the Hubble parameter. This is the most interesting example for the early Universe, since it is generally assumed that there happened such an exponential, "inflationary", phase of the Universe, caused by an effective cosmological constant. Taking various "environments", the following results are found for the main contribution to (the absolute value of) the decoherence factor, $|D|$, that multiplies the reduced density matrix for the "isolated" case:

- *Massless conformally-invariant field*: Here,
$$|D| = 1 ,$$
since no particle creation and therefore no decoherence effect takes place.
- *Massive scalar field*: Here,
$$|D| \approx \exp\left(-\frac{\pi m^3 a}{128}(a-a')^2\right) ,$$
and one notices increasing decoherence for increasing a.
- *Gravitons*: This is similar to the previous case, but the mass m is replaced by the Hubble parameter H,
$$|D| \approx \exp\left(-CH^3 a(a-a')^2\right) , \; C > 0 .$$
- *Fermions*:
$$|D| \approx \exp\left(-C' m^2 a^2 H^2 (a-a')^2\right) , \; C' > 0 .$$
For high-enough mass, the decoherence effect by fermions is thus smaller than the corresponding influence of bosons.

It becomes clear from these examples that the Universe acquires classical properties after the onset of the inflationary phase. "Before" this phase, the Universe was in a timeless quantum state which does not possess any classical properties. Viewed backwards, different semiclassical branches would meet and interfere to form this timeless quantum state (Barvinsky et al. 1999b).

For these considerations it is of importance that there is a discrimination between the various degrees of freedom. On the fundamental level of full superstring theory, for example, such a discrimination is not possible and one would therefore not expect any decoherence effect to occur at that level.

In general one would expect not only one semiclassical component of the form (28), but also many superpositions of such terms. Since (18) is a real equation, one would in particular expect to have a superposition of (28) with its complex conjugate. The no-boundary state in quantum cosmology has, for example, such a form. Decoherence also acts between such semiclassical branches, although somewhat less effective than within one branch (Barvinsky et al. 1999c). For a macroscopic Universe, this effect is big enough to warrant the consideration of only one semiclassical component of the form (28). This constitutes a symmetry-breaking effect similar to the symmetry breaking for chiral molecules: While in the former case the symmetry with respect to complex conjugation is broken, in the latter case one has a breaking of parity invariance (compare Figures 2 and 3 above).

It is clear that decoherence can only act if there is a peculiar, low-entropy, state for the very early Universe. This lies at the heart of the arrow of time in the Universe. A simple initial condition like the one in Conradi and Zeh (1991) can in principle lead to a quantum state describing the arrow of time, see also Zeh (1999).

4.3 Classicality of primordial fluctuations

According to the inflationary scenario of the early Universe, all structure in the Universe (galaxies, clusters of galaxies) arises from *quantum* fluctuations of scalar fields and scalar fluctuations of the metric. Because also fluctuations of the metric are involved, this constitutes an effect of (linear) quantum gravity.

These early fluctuations manifest themselves as anisotropies in the cosmic microwave background radiation and have been observed both by the COBE satellite and earth-based telescopes. Certainly, these observed fluctuations are classical stochastic quantities. How do the quantum fluctuations become classical?

It is clear that for the purpose of this discussion the global gravitational degrees of freedom can already by considered as classical, i.e. the decoherence process of Sect. 4.2 has already been effective. The role of the gravitational field is then twofold: firstly, the expanding Universe influences the dynamics of the quantum fluctuations. Secondly, linear fluctuations of the gravitational field are themselves part of the quantum system.

The physical wavelength of a mode with wavenumber k is given by

$$\lambda_{phys} = \frac{2\pi a}{k} \ . \tag{39}$$

Since during the inflationary expansion the Hubble parameter H remains constant, the physical wavelength of the modes leaves the particle horizon, given by H^{-1}, at a certain stage of inflation, provided that inflation does not end before this happens. Modes that are outside the horizon thus obey

$$\frac{k}{aH} \ll 1 \ . \tag{40}$$

It turns out that the dynamical behaviour of these modes lies at the heart of structure formation. These modes re-enter the horizon in the radiation-and matter-dominated phases which take place after inflation.

For a quantitative treatment, the Schrödinger equation (32) has to be solved for the fluctuations in the inflationary Universe. The easiest example, which nevertheless exhibits the same features as a realistic model, is a massless scalar field. It is, moreover, most convenient to go to Fourier space and to multiply the corresponding variable with a. The resulting fluctuation variable is called y_k, see Kiefer and Polarski (1998) for details. Taking as a natural initial state the "vacuum state", the solution of the Schrödinger equation (32) for the (complex) variables y_k reads[4]

$$\chi(y,t) = \left(\frac{1}{\pi|f|^2}\right)^{1/2} \exp\left(-\frac{1-2\mathrm{i}F}{2|f|^2}|y|^2\right) , \tag{41}$$

where

$$|f|^2 = (2k)^{-1}(\cosh 2r + \cos 2\varphi \sinh 2r), \tag{42}$$
$$F = \frac{1}{2}\sin 2\varphi \sinh 2r , \tag{43}$$

and explicit expressions can be given for the time-dependent functions r and φ. The Gaussian state (41) is nothing but a *squeezed state*, a state that is well known from quantum optics. The parameters r and φ have the usual interpretation as squeezing parameter and squeezing angle, respectively. It turns out that during the inflationary expansion $r \to \infty$, $|F| \gg 1$, and $\varphi \to 0$ (meaning here a squeezing in momentum). In this limit, the state (41) becomes also a WKB state par excellence. As a result of this extreme squeezing, this state cannot be distinguished within the given observational capabilities from a classical stochastic process, as thought experiments demonstrate (Kiefer and Polarski 1998, Kiefer et al. 1998a). In the Heisenberg picture, the special properties of the state (41) are reflected in the fact that the field operators commute at different times, i.e.

$$[\hat{y}(t_1), \hat{y}(t_2)] \approx 0 . \tag{44}$$

(Kiefer et al. 1998b). In the language of quantum optics, this is the condition for a quantum-nondemolition measurement: An observable obeying (44) can repeatedly be measured with great accuracy. It is important to note that these properties remain valid after the modes have reentered the horizon in the radiation-dominated phase that follows inflation (Kiefer et al. 1998a).

As is well known, squeezed states are very sensitive to interactions with other degrees of freedom (Giulini et al. 1996). Since such interactions are unavoidably present in the early Universe, the question arises whether they would not spoil

[4] Since there is no self-interaction of the field, different modes y_k decouple, which is why I shall suppress the index k in the following.

the above picture. However, most interactions invoke couplings in field amplitude space (as opposed to field momentum space) and therefore,

$$[\hat{y}, \hat{H}_{int}] \approx 0 , \tag{45}$$

where $\hat{H}_{int}$ denotes the interaction Hamiltonian. The field amplitudes therefore become an excellent pointer basis: This basis defines the classical property, and due to (44) this property is conserved in time. The decoherence time caused by $\hat{H}_{int}$ is very small in most cases. Employing for the sake of simplicity a linear interaction with a coupling constant g, one finds for the decoherence time scale (Kiefer and Polarski 1998)

$$t_D \approx \frac{\lambda_{phys}}{g e^r} . \tag{46}$$

For modes that presently re-enter the horizon, one has $\lambda_{phys} \approx 10^{28}$cm, $e^r \approx 10^{50}$ and therefore

$$t_D \approx 10^{-31} g^{-1} \mathrm{sec} . \tag{47}$$

Unless g is very small, decoherence acts on a very short timescale. This conclusion is enforced if higher-order interactions are taken into account. It must be noted that the interaction of the field modes with its "environment" is an *ideal* measurement – the probabilities are unchanged and the main predictions of the inflationary scenario remain the same (which manifest themselves, for example, in the form of the anisotropy spectrum of the cosmic microwave background). This would not be the case, for example, if one concluded that *particle number* instead of *field amplitude* would define the robust classical property. Realistic models of the early Universe must of course take into account complicated nonlinear interactions, see e.g. Calzetta and Hu (1995) and Matacz (1997). Although these models will affect the values of the decoherence timescales, the conceptual conclusions drawn above will remain unchanged.

The results of the last two subsections give rise to the *hierarchy of classicality* (Kiefer and Joos 1999): The global gravitational background degrees of freedom are the first variables that assume classical properties. They then provide the necessary condition for other variables to exhibit classical behaviour, such as the primordial fluctuations discussed here. These then serve as the seeds for the classical structure of galaxies and clusters of galaxies that are part of the observed Universe.

5 Acknowledgements

I thank the organisers of this school for inviting me to this interesting and stimulating meeting. I am grateful to the Institute of Advanced Study Berlin for its kind hospitality during the writing of this contribution. I also thank John Briggs, Erich Joos, Alexander Kamenshchik, Jorma Louko for their comments on this manuscript.

References

1. Al'tshuler, B.L., Barvinsky, A.O. (1996): Quantum cosmology and physics of transitions with a change of the spacetime signature. Physics-Uspekhi **39**, 429–459
2. Ashtekar, A. (1999): Quantum mechanics of geometry. Electronic report gr-qc/9901023
3. Barbour, J.B. (1997): Nows are all we need. In: *Time, Temporality, Now*, edited by H. Atmanspacher and E. Ruhnau (Springer, Berlin)
4. Barvinsky, A.O., Kamenshchik, A.Yu. (1994): Quantum scale of inflation and particle physics of the early Universe. Phys. Lett. B **332**, 270–276
5. Barvinsky, A.O., Kamenshchik, A.Yu., Kiefer, C. (1999a): Effective action and decoherence by fermions in quantum cosmology. Nucl. Phys. B **552**, 420–444
6. Barvinsky, A.O., Kamenshchik, A.Yu., Kiefer, C. (1999b): Origin of the inflationary Universe. Mod. Phys. Lett. A **14**, 1083–1088
7. Barvinsky, A.O., Kamenshchik, A.Yu., Kiefer, C., Mishakov, I.V. (1999c): Decoherence in quantum cosmology at the onset of inflation. Nucl. Phys. B **551**, 374–396
8. Barvinsky, A.O., Kiefer, C. (1998): Wheeler-DeWitt equation and Feynman diagrams. Nucl. Phys. B **526**, 509–539
9. Blanchard, P., Giulini, D., Joos, E., Kiefer, C., Stamatescu, I.-O., eds. (1999): *Decoherence: Theoretical, experimental and conceptual problems* (Springer, Berlin)
10. Bousso, R., Hawking, S.W. (1996): Pair creation of black holes during inflation. Phys. Rev. D **54**, 6312–6322
11. Briggs, J.S., Rost, J.M. (1999): Time dependence in quantum mechanics. Electronic report quant-ph/9902035
12. Brout, R., Parentani, R. (1999): Time in cosmology. Int. J. Mod. Phys. D **8**, 1–22
13. Calzetta, E., Hu, B.L. (1995): Quantum fluctuations, decoherence of the mean field, and structure formation in the early Universe. Phys. Rev. D **52**, 6770–6788
14. Coleman, S. (1988): Why there is nothing rather than something: A theory of the cosmological constant. Nucl. Phys. B **310**, 643–668
15. Conradi, H.D., Zeh, H.D. (1991): Quantum cosmology as an initial value problem. Phys. Lett. A **154**, 321–326
16. Dabrowski, M.P., Kiefer, C. (1997): Boundary conditions in quantum string cosmology. Phys. Lett. B **397**, 185–192
17. Giulini, D. (1995): What is the geometry of superspace? Phys. Rev. D **51**, 5630–5635
18. Giulini, D., Joos, E., Kiefer, C., Kupsch, J., Stamatescu, I.-O., Zeh, H.D. (1996): *Decoherence and the Appearance of a Classical World in Quantum Theory* (Springer, Berlin)
19. Giulini, D., Kiefer, C. (1994): Wheeler-DeWitt metric and the attractivity of gravity. Phy. Lett. A **193**, 21–24
20. Halliwell, J.J. (1991): Introductory lectures on quantum cosmology. In *Quantum cosmology and baby universes*, edited by S. Coleman, J.B. Hartle, T. Piran, and S. Weinberg (World Scientific, Singapore)
21. Haroche, S. (1998): Entanglement, decoherence and the quantum/classical boundary. Phys. Today **51** (July), 36–42
22. Hartle, J.B. (1997): Quantum cosmology: problems for the 21st century. In: *Physics in the 21st century*, edited by K. Kikkawa, H. Kunimoto, H. Ohtsubo (World Scientific, Singapore)
23. Hartle, J.B., Hawking, S.W. (1983): Wave function of the Universe. Phys. Rev. D **28**, 2960–2975

24. Hawking, S.W., Turok, N.G. (1998): Open inflation without false vacua. Phys. Lett. B **425**, 25–32
25. Isham, C.J. (1992): Canonical quantum gravity and the problem of time. In: *Integrable Systems, Quantum Groups, and Quantum Field Theories*, edited by L. A. Ibart and M. A. Rodrigues (Kluwer, Dordrecht)
26. Joos, E. (1986): Why do we observe a classical spacetime? Phys. Lett. A **116**, 6–8
27. Joos, E., Zeh, H.D. (1985): The emergence of classical properties through interaction with the environment. Z. Phys. B **59**, 223–243
28. Kiefer, C. (1987): Continuous measurement of mini-superspace variables by higher multipoles. Class. Quantum Grav. **4**, 1369–1382
29. Kiefer, C. (1990): Wave packets in quantum cosmology and the cosmological constant. Nucl. Phys. B **341**, 273–293
30. Kiefer, C. (1991): On the meaning of path integrals in quantum cosmology. Ann. Phys. (N.Y.) **207**, 53–70
31. Kiefer, C. (1992): Decoherence in quantum electrodynamics and quantum cosmology. Phys. Rev. D **46**, 1658–1670
32. Kiefer, C. (1993): Topology, decoherence, and semiclassical gravity. Phys. Rev. D **47**, 5414–5421
33. Kiefer, C. (1994): The semiclassical approximation to quantum gravity. In: *Canonical gravity: From Classical to Quantum*, edited by J. Ehlers and H. Friedrich (Springer, Berlin)
34. Kiefer, C. (1997): Does time exist at the most fundamental level? In: *Time, Temporality, Now*, edited by H. Atmanspacher and E. Ruhnau (Springer, Berlin)
35. Kiefer, C. (1999): Decoherence in situations involving the gravitational field. In: Blanchard et al. (1999).
36. Kiefer, C., Joos, E. (1999): Decoherence: Concepts and examples. In: *Quantum Future*, edited by P. Blanchard and A. Jadczyk (Springer, Berlin)
37. Kiefer, C., Lesgourgues, J., Polarski, D., Starobinsky, A.A. (1998a): The coherence of primordial fluctuations produced during inflation. Class. Quantum Grav. **15**, L67–L72
38. Kiefer, C., Polarski, D. (1998): Emergence of classicality for primordial fluctuations: concepts and analogies. Ann. Phys. (Leipzig) **7**, 137–158
39. Kiefer, C., Polarski, D., Starobinsky, A.A. (1998b): Quantum-to-classical transition for fluctuations in the early universe. Int. J. Mod. Phys. D **7**, 455–462
40. Kiefer, C., Singh, T.P. (1991): Quantum gravitational corrections to the functional Schrödinger equation. Phys. Rev. D **44**, 1067–1076
41. Kiefer, C., Zeh, H.D. (1995): Arrow of time in a recollapsing quantum universe. Phys. Rev. D **51**, 4145–4153
42. Kuchař, K.V. (1992): Time and interpretations of quantum gravity. In: *Proceedings of the 4th Canadian Conference on General Relativity and Relativistic Astrophysics*, edited by G. Kunstatter, D. Vincent and J. Williams (World Scientific, Singapore)
43. Lemaître, G. (1931): The beginning of the world from the point of view of quantum theory. Nature **127**, 706
44. Matacz, A. (1997): A new theory of stochastic inflation. Phys. Rev. D **55**, 1860–1874
45. Mott, N.F. (1931): On the theory of excitation by collision with heavy particles. Proc. Cambridge Phil. Soc. **27**, 553–560
46. Rosales, J.-L. (1997): Quantum state correction of relic gravitons from quantum gravity. Phys. Rev. D **55**, 4791–4794

47. Schrödinger, E. (1935): Discussion of probability relations between separated systems. Proc. Cambridge Phil. Soc. **31**, 555–563
48. Sen, A. (1998): Developments in superstring theory. To be published in the Proceedings of the 29th International Conference on High-Energy Physics, Vancouver, Canada, 23-29 July 1998, Electronic report hep-ph/9810356
49. Teitelboim, C. (1980): The Hamiltonian structure of space-time. In: *General relativity and gravitation*, edited by A. Held (Plenum Press, New York)
50. Vilenkin, A. (1998): The quantum cosmology debate. Contribution to the conference on particle physics and the early universe (COSMO 98), Monterey, CA, 15.-20.11.1998, Electronic report gr-qc/9812027
51. Zeh, H.D. (1986): Emergence of classical time from a universal wave function. Phys. Lett. A **116**, 9–12
52. Zeh, H.D. (1997): What is achieved by decoherence? In *New Developments on Fundamental Problems in Quantum Physics*, edited by M. Ferrer and A. van der Merwe (Kluwer Academic, Dordrecht)
53. Zeh, H.D. (1999): *The physical basis of the direction of time* (Springer, Berlin)
54. Zurek, W.H. (1991): Decoherence and the Transition from Quantum to Classical. Physics Today **44** (Oct.), 36–44; see also the discussion in Physics Today (letters) **46** (April), 13

Single-Exterior Black Holes

Jorma Louko

Max-Planck-Institut für Gravitationsphysik, Am Mühlenberg 5,
D-14476 Golm, Germany

Abstract. We discuss quantum properties of the single-exterior, "geon"-type black (and white) holes that are obtained from the Kruskal spacetime and the spinless Bañados-Teitelboim-Zanelli hole via a quotient construction that identifies the two exterior regions. For the four-dimensional geon, the Hartle-Hawking type state of a massless scalar field is thermal in a limited sense, but there is a discrepancy between Lorentzian and Riemannian derivations of the geon entropy. For the three-dimensional geon, the state induced for a free conformal scalar field on the conformal boundary is similarly thermal in a limited sense, and the correlations in this state provide support for the holographic hypothesis in the context of asymptotically Anti-de Sitter black holes in string theory.

1 Introduction

In quantum field theory on the Kruskal spacetime, one way to arrive at the thermal effects is through the observation that the spacetime has two exterior regions separated by a bifurcate Killing horizon. A free scalar field on the Kruskal spacetime has a vacuum state, known as the Hartle-Hawking vacuum [1,2], that is invariant under all the continuous isometries of the spacetime [3,4]. This state is pure, but the expectation values of operators with support in *one* exterior region are thermal in the Hawking temperature [1–5]. Similar observations hold for field theory on the nonextremal (2+1)-dimensional Bañados-Teitelboim-Zanelli (BTZ) black hole, both with and without spin [6], and also for conformal field theory on the conformal boundary of the BTZ hole [7–9].

In all these cases one has a vacuum state that knows about the global geometry of the spacetime, in particular about the fact that the spacetime has two exterior regions. Suppose now that we modify the spacetime in some 'reasonable' fashion so that one exterior region remains as it is, and all the modification takes place behind the Killing horizons of this exterior region. Suppose further that the modified spacetime admits a vacuum state that is, in some reasonable sense, a Hartle-Hawking type vacuum. Can we then, by probing the new vacuum in the unmodified exterior region, discover that something has happened to the spacetime behind the horizons? In particular, as the new spacetime is still a black (and white) hole, does the new vacuum exhibit thermality, and if so, at what temperature? In the (2+1)-dimensional case, the analogous questions can also be raised for conformal field theory on the conformal boundary.

These lectures address the above questions for a particular modification of the Kruskal manifold and the spinless BTZ hole: we modify the spacetimes by a quotient construction that identifies the two exterior regions with each other. For Kruskal, the resulting spacetime is referred to as the $\mathbb{RP}^3$ geon [10,11], and for BTZ, as the $\mathbb{RP}^2$ geon [9]. These spacetimes are black (and white) holes, and their only singularities are those inherited from the singularities of the two-exterior holes.

On the $\mathbb{RP}^3$ geon, a free scalar field has a vacuum induced from the Hartle-Hawking vacuum on Kruskal. The vacuum is not fully thermal for static exterior observers, but it appears thermal when probed with operators that do not see certain types of correlations, such as in particular operators with support at asymptotically late times, and the apparent temperature is then the usual Hawking temperature. However, a naive application of Euclidean-signature path-integral methods via saddle-point methods yields for the geon only half of the Bekenstein-Hawking entropy of the Schwarzschild hole with the same mass.

The situation on the conformal boundary of the $\mathbb{RP}^2$ geon is analogous. The quotient construction from the conformal boundary of Anti-de Sitter space induces on the boundary of the geon a Hartle-Hawking type vacuum that is not fully thermal, but it appears thermal when probed with operators that do not see certain types of correlations, and the apparent temperature is then the usual Hawking temperature of the BTZ hole. The properties of the boundary vacuum turn out to reflect in a surprisingly close fashion the geometry of the geon spacetime. This can be interpreted as support for the holographic hypothesis [12,13], according to which physics in the bulk of a spacetime should be retrievable from physics on the boundary of the spacetime. It further suggests that single-exterior black holes can serve as a test bed for the versions of the holographic hypothesis that arise in string theory for asymptotically Anti-de Sitter spacetimes via the Maldacena duality conjectures [7,14–16].

The material is based on joint work [9,17] with Don Marolf, whom I would like to thank for a truly delightful collaboration. I would also like to thank the organizers of the Polanica Winter School for the opportunity to present the work in a most pleasant and inspiring atmosphere.

2 Kruskal Manifold and the $\mathbb{RP}^3$ Geon

Recall that the metric on the Kruskal manifold $\mathcal{M}^L$ reads

$$ds^2 = \frac{32M^3}{r} \exp\left(-\frac{r}{2M}\right) \left(-dT^2 + dX^2\right) + r^2 d\Omega^2 , \tag{1}$$

where $d\Omega^2 = d\theta^2 + \sin^2\theta \, d\varphi^2$ is the metric on the unit two-sphere, $M > 0$, $X^2 - T^2 > -1$, and r is determined as a function of T and X by

$$\left(\frac{r}{2M} - 1\right) \exp\left(\frac{r}{2M}\right) = X^2 - T^2 . \tag{2}$$

The coordinates are global, apart from the elementary singularities of the spherical coordinates. $\mathcal{M}^L$ is manifestly spherically symmetric, and it has in addition the Killing vector

$$V^L := \frac{1}{4M} \left(X\partial_T + T\partial_X\right) , \tag{3}$$

which is timelike for $|X| > |T|$ and spacelike for $|X| < |T|$. A conformal diagram of $\mathcal{M}^L$, with the two-spheres suppressed, is shown in Fig. 1.

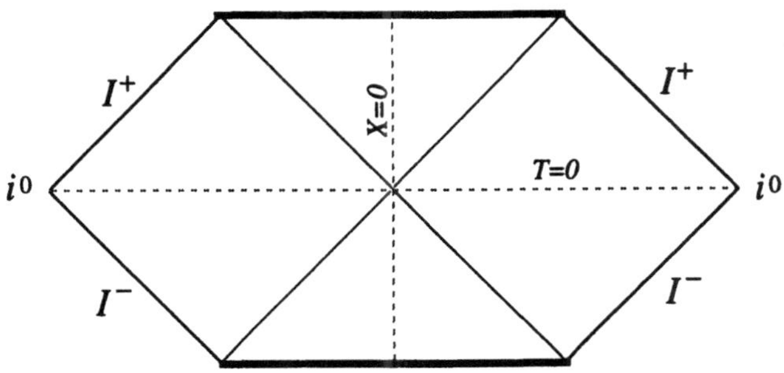

Fig. 1. Conformal diagram of the Kruskal spacetime. Each point represents a suppressed S^2 orbit of the O(3) isometry group

In each of the four quadrants of $\mathcal{M}^L$ one can introduce Schwarzschild coordinates (t, r, θ, φ) that are adapted to the isometry generated by V^L. In the "right-hand-side" exterior region, $X > |T|$, the coordinate transformation reads

$$\begin{aligned} T &= \left(\frac{r}{2M} - 1\right)^{1/2} \exp\left(\frac{r}{4M}\right) \sinh\left(\frac{t}{4M}\right) , \\ X &= \left(\frac{r}{2M} - 1\right)^{1/2} \exp\left(\frac{r}{4M}\right) \cosh\left(\frac{t}{4M}\right) , \end{aligned} \tag{4}$$

with $r > 2M$ and $-\infty < t < \infty$. The exterior metric takes then the Schwarzschild form

$$ds^2 = -\left(1 - \frac{r}{2M}\right) dt^2 + \frac{dr^2}{\left(1 - \frac{r}{2M}\right)} + r^2 d\Omega^2 , \tag{5}$$

and $V^L = \partial_t$.

Consider now on $\mathcal{M}^L$ the isometry

$$J^L : (T, X, \theta, \varphi) \mapsto (T, -X, \pi - \theta, \varphi + \pi) . \tag{6}$$

J^L is clearly involutive, it acts properly discontinuously, it preserves the time orientation and spatial orientation, and it commutes with the spherical symmetry of $\mathcal{M}^L$. The quotient space $\mathcal{M}^L/J^L$ is therefore a spherically symmetric, space and time orientable manifold. A conformal diagram of $\mathcal{M}^L/J^L$ is shown in Fig. 2. $\mathcal{M}^L/J^L$ is an inextendible black (and white) hole spacetime, and its only singularities are those inherited from the singularities of $\mathcal{M}^L$. It has only one exterior region, and its spatial topology is $\mathbb{RP}^3 \setminus \{\text{point at infinity}\}$. We refer to $\mathcal{M}^L/J^L$ as the $\mathbb{RP}^3$ geon [10,11].

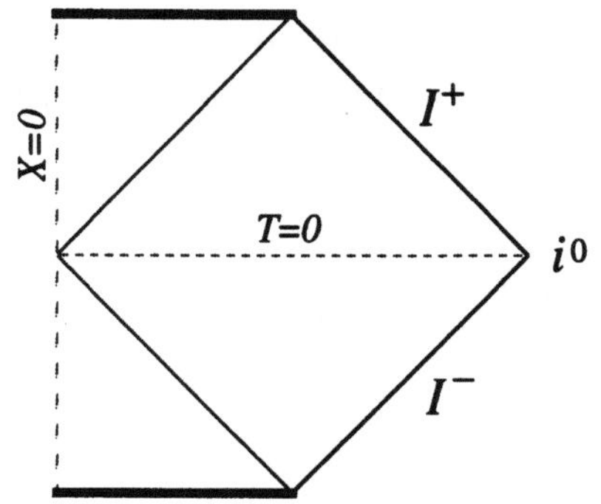

Fig. 2. Conformal diagram of the $\mathbb{RP}^3$ geon $\mathcal{M}^L/J^L$. Each point represents a suppressed orbit of the O(3) isometry group. The region $X > 0$ is isometric to the region $X > 0$ of $\mathcal{M}^L$, shown in Fig. 1, and the O(3) isometry orbits in this region are two-spheres. At $X = 0$, the O(3) orbits have topology $\mathbb{RP}^2$

The exterior region of $\mathcal{M}^L/J^L$ is clearly isometric to an exterior region of $\mathcal{M}^L$. In terms of the coordinates shown in Fig. 2, the exterior region is at $X > |T|$, and one can introduce in the exterior region standard Schwarzschild coordinates by (4). As the Killing vector V^L on $\mathcal{M}^L$ changes its sign under J^L, the timelike Killing vector ∂_t on the exterior of $\mathcal{M}^L/J^L$ can however not be continued into a globally-defined Killing vector on $\mathcal{M}^L/J^L$. This means that not all the constant t hypersurfaces in the exterior region of $\mathcal{M}^L/J^L$ are equal: among them, there is only one (in Fig. 2, the one at $T = 0$) that can be extended into a smoothly-embedded Cauchy hypersurface for $\mathcal{M}^L/J^L$.

The quotient construction from $\mathcal{M}^L$ to $\mathcal{M}^L/J^L$ can be analytically continued to the Riemannian (i.e., positive definite) sections via the formalism of (anti-)holomorphic involutions [18,19]. The Riemannian section of the Kruskal hole, denoted by $\mathcal{M}^R$, is obtained from (1) and (2) by setting $T = -i\tilde{T}$ and letting $\tilde{T}$ and X take all real values [20]. The analytic continuation of J^L, denoted by J^R, acts on $\mathcal{M}^R$ by

$$J^R : (\tilde{T}, X, \theta, \varphi) \mapsto (\tilde{T}, -X, \pi - \theta, \varphi + \pi) , \tag{7}$$

and the Riemannian section of the $\mathbb{RP}^3$ geon is $\mathcal{M}^R/J^R$.

On $\mathcal{M}^R$ we can introduce the Riemannian Schwarzschild coordinates $(\tilde{t}, r, \theta, \varphi)$, obtained from the Lorentzian Schwarzschild coordinates for $r > 2M$ by $t = -i\tilde{t}$. These Riemannian Schwarzschild coordinates are global, with the exception of a coordinate singularity at the Riemannian horizon $r = 2M$, provided they are understood with the identification $(\tilde{t}, r, \theta, \varphi) \sim (\tilde{t} + 8\pi M, r, \theta, \varphi)$ [20]. On $\mathcal{M}^R/J^R$, the Riemannian Schwarzschild coordinates need to be understood with the additional identification $(\tilde{t}, r, \theta, \varphi) \sim (\tilde{t} + 4\pi M, r, \pi - \theta, \varphi + \pi)$, which arises from the action (7) of J^R on $\mathcal{M}^R$. The Killing vector $\partial_{\tilde{t}}$ is global on $\mathcal{M}^R$, and it generates an U(1) isometry group with a fixed point at the Riemannian horizon. On $\mathcal{M}^R/J^R$, on the other hand, $\partial_{\tilde{t}}$ is global only as a line field but not as a vector field, and the analogous U(1) isometry does not exist. Embedding diagrams of $\mathcal{M}^R$ and $\mathcal{M}^R/J^R$, with the orbits of the spherical symmetry suppressed, are shown in Figs. 3 and 4.

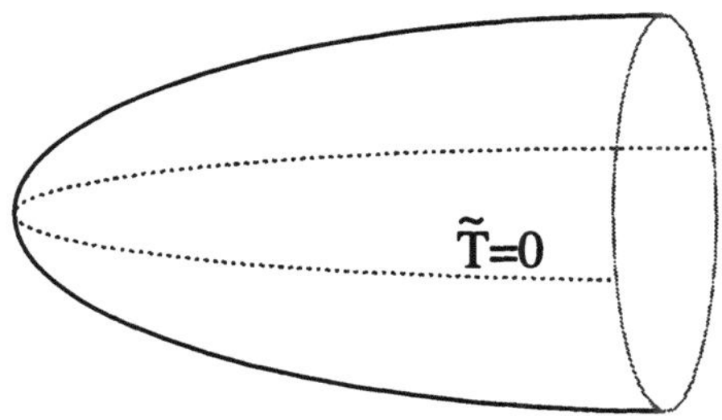

Fig. 3. A "sock" representation of the Riemannian section $\mathcal{M}^R$ of the complexified Kruskal manifold. The S^2 orbits of the O(3) isometry group are suppressed, and the remaining two dimensions $(\tilde{T}, X)$ are shown as an isometric embedding into Euclidean $\mathbb{R}^3$. The isometry generated by $\partial_{\tilde{t}}$ rotates the two shown dimensions

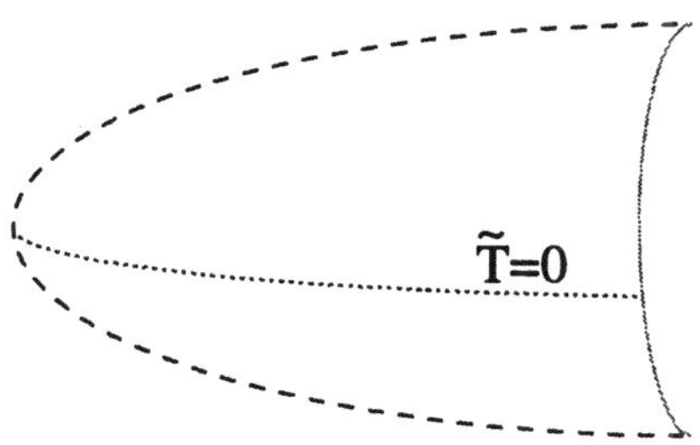

Fig. 4. A representation of the Riemannian section $\mathcal{M}^R/J^R$ of the complexified $\mathbb{RP}^3$ geon as the "front half" of the the $\mathcal{M}^R$ sock. The orbits of the O(3) isometry group are suppressed, as in Fig. 3. The generic orbits have topology S^2, but those at the "boundary" of the diagram (*dashed line*) have topology $\mathbb{RP}^2$

3 Vacua on Kruskal and on the $\mathbb{RP}^3$ Geon

We now consider a free scalar field on the Kruskal manifold and on the $\mathbb{RP}^3$ geon. For concreteness, we take here the field to be massless. The situation with a massive field is qualitatively similar [17].

Recall that the Hartle-Hawking vacuum $|0_K\rangle$ of a massless scalar field on the Kruskal manifold $\mathcal{M}^L$ can be characterized by its positive frequency properties along the affine parameters of the horizon generators [1,2,5], by the complex analytic properties of the Feynman propagator upon analytic continuation to $\mathcal{M}^R$ [1], or by the invariance under the continuous isometries of $\mathcal{M}^L$ [3,4]. $|0_K\rangle$ is regular everywhere on $\mathcal{M}^L$, but it is not annihilated by the annihilation operators associated with the future timelike Killing vectors in the exterior regions: a static observer in an exterior region sees $|0_K\rangle$ as an excited state. We have the expansion

$$|0_K\rangle = \sum_{i\cdots k} f_{i\cdots k} \left(a_i^R\right)^\dagger \left(a_i^L\right)^\dagger \cdots \left(a_k^R\right)^\dagger \left(a_k^L\right)^\dagger |0_{B,K}\rangle \ , \tag{8}$$

where the Boulware vacuum $|0_{B,K}\rangle$ is the vacuum with respect to the timelike Killing vectors in the exterior regions, $\left(a_i^R\right)^\dagger$ are the creation operators with respect to this Killing vector in the right-hand-side exterior region, and $\left(a_i^L\right)^\dagger$ are the creation operators with respect to this Killing vector in the left-hand-side exterior region.

$|0_K\rangle$ thus contains Boulware excitations in correlated pairs, such that one member of the pair has support in the right-hand-side exterior and the other member in the left-hand-side exterior. An operator with support in (say) the right-hand-side exterior does not couple to the left-hand-side excitations, and the expectation values of such operators in $|0_K\rangle$ thus look like expectation values in a mixed state. From the detailed form of the expansion coefficients $f_{i\cdots k}$ (which we do not write out here) it is seen that this mixed state is thermal, and it has at infinity the Hawking temperature $T = (8\pi M)^{-1}$.

Now, through the quotient construction from $\mathcal{M}^L$ to $\mathcal{M}^L/J^L$, $|0_K\rangle$ induces on $\mathcal{M}^L/J^L$ a Hartle-Hawking type vacuum, which we denote by $|0_G\rangle$. Again, $|0_G\rangle$ can be characterized by its positive frequency properties along the affine parameters of the horizon generators, or by the complex analytic properties of the Feynman propagator [17]. $|0_G\rangle$ has the expansion

$$|0_G\rangle = \sum_{i\cdots k} \tilde{f}_{i\cdots k} \left(\tilde{a}_i^{(1)}\right)^\dagger \left(\tilde{a}_i^{(2)}\right)^\dagger \cdots \left(\tilde{a}_k^{(1)}\right)^\dagger \left(\tilde{a}_k^{(2)}\right)^\dagger |0_{B,G}\rangle \ , \tag{9}$$

where $|0_{B,G}\rangle$ is the Boulware vacuum in the single exterior region and $\left(\tilde{a}_i^{(\alpha)}\right)^\dagger$ are the creation operators of Boulware particles in the exterior region. The indices i and α now label a complete set of positive frequency Boulware modes in the single exterior region.

We see from (9) that $|0_G\rangle$ contains Boulware excitations in correlated pairs, but the crucial point is that both members of each pair have support in the single exterior region. Consequently, the expectation values of arbitrary operators in the exterior region are not thermal. However, for operators that do not contain couplings between modes with $\alpha = 1$ and $\alpha = 2$, the expectation values turn out to be thermal, with the Hawking temperature $T = (8\pi M)^{-1}$. One class of operators for which this is the case are operators with, roughly speaking, support at asymptotically late (or early) times: the reason is that an excitation with support at asymptotically late exterior times is correlated with one with support at asymptotically early exterior times. Note that "early" and "late" here mean compared with the distinguished exterior spacelike hypersurface mentioned in Sect. 2 (in Fig. 2, the one at $T = 0$).

Thus, for a late-time observer in the exterior region of $\mathcal{M}^L/J^L$, the state $|0_{\mathrm{G}}\rangle$ is indistinguishable from the state $|0_{\mathrm{K}}\rangle$ on $\mathcal{M}^L$. This conclusion can also be reached by analyzing the response of a monopole particle detector, or from an emission-absorption analysis analogous to that performed for $|0_{\mathrm{K}}\rangle$ in [1], provided certain technical assumptions about the falloff of the two-point functions in $|0_{\mathrm{G}}\rangle$ hold [17].

4 Entropy of the $\mathbb{RP}^3$ Geon?

As explained above, for a late-time exterior observer in $\mathcal{M}^L/J^L$ the state $|0_{\mathrm{G}}\rangle$ is indistinguishable from the state $|0_{\mathrm{K}}\rangle$ on $\mathcal{M}^L$. The late-time observer can therefore promote the classical first law of black hole mechanics [21] into a first law of black hole thermodynamics exactly as for the Kruskal black hole [22–24]. The observer thus finds for the thermodynamic late time entropy of the geon the usual Kruskal value $4\pi M^2$, which is one quarter of the area of the geon black hole horizon at late times. If one views the geon as a dynamical black-hole spacetime, with the asymptotic far-future horizon area $16\pi M^2$, this is the result one might have expected on physical grounds.

On the other hand, the area-entropy relation for the geon is made subtle by the fact that the horizon area is not constant along the horizon. Away from the intersection of the past and future horizons, the horizon duly has topology S^2 and area $16\pi M^2$, just as in Kruskal. The critical surface at the intersection of the past and future horizons, however, has topology $\mathbb{RP}^2$ and area $8\pi M^2$. As it is precisely this critical surface that belongs to both the Lorentzian and Riemannian sections of the complexified manifold, and constitutes the horizon of the Riemannian section, one may expect that methods utilizing the Riemannian section of the complexified manifold [20,25] produce for the geon entropy the value $2\pi M^2$, which is one quarter of the critical surface area, and only half of the Kruskal entropy. This indeed is the case, provided the surface terms in the Riemannian geon action are handled in a way suggested by the quotient construction from $\mathcal{M}^R$ to $\mathcal{M}^R/J^R$ [17].

There are several possible physical interpretations for this disagreement between the Lorentzian and Riemannian results for the entropy. At one extreme, it could be that the path-integral framework is simply inapplicable to the geon, for reasons having to do with the absence of certain globally-defined symmetries. For instance, despite the fact that the exterior region of $\mathcal{M}^L/J^L$ is static, the restriction of $|0_{\mathrm{G}}\rangle$ to this region is not. Also, the asymptotic region of $\mathcal{M}^R/J^R$ does not have a globally-defined Killing field, and the homotopy group of any neighborhood of infinity in $\mathcal{M}^R/J^R$ is $\mathbb{Z}_2$ as opposed to the trivial group. It may well be that such an asymptotic structure does not satisfy the boundary conditions that should be imposed in the path integral for the quantum gravitational partition function.

At another extreme, it could be that the path-integral framework is applicable to the geon, and our way of applying it is correct, but the resulting entropy is physically distinct from the subjective thermodynamic entropy associated with the late-time exterior observer. If this is the case, the physical interpretation of the path-integral entropy might be in the quantum statistics in the whole exterior region, and one might anticipate this entropy to arise from tracing over degrees of freedom that are in some sense unobservable. It would thus be interesting to see see whether any state-counting calculation for the geon entropy would produce agreement with the path-integral result.

5 AdS_3, the Spinless Nonextremal BTZ Hole, and the $\mathbb{RP}^2$ Geon

We now turn to 2+1 spacetime dimensions. In this section we review how the spinless nonextremal BTZ hole and the $\mathbb{RP}^2$ geon arise as quotient spaces of the three-dimensional Anti-de Sitter space, and how this quotient construction can be extended to the conformal boundaries.

5.1 AdS_3, its Covering Space, and the Conformal Boundary

Recall that the three-dimensional Anti-de Sitter space (AdS_3) can be defined as the hyperboloid

$$-1 = -(T^1)^2 - (T^2)^2 + (X^1)^2 + (X^2)^2 \tag{10}$$

in $\mathbb{R}^{2,2}$ with the metric

$$ds^2 = -(dT^1)^2 - (dT^2)^2 + (dX^1)^2 + (dX^2)^2 \, . \tag{11}$$

We have here normalized the Gaussian curvature of AdS_3 to -1. This embedding representation makes transparent the fact that AdS_3 is a maximally symmetric space with the isometry group $O(2,2)$.

For understanding the structure of the infinity, we introduce the coordinates (t, ρ, θ) by [26]

$$\begin{aligned} T^1 &= \frac{1+\rho^2}{1-\rho^2} \cos t \, , \quad T^2 = \frac{1+\rho^2}{1-\rho^2} \sin t \, , \\ X^1 &= \frac{2\rho}{1-\rho^2} \cos\theta \, , \quad X^2 = \frac{2\rho}{1-\rho^2} \sin\theta \, . \end{aligned} \tag{12}$$

With $0 \le \rho < 1$ and the identifications $(t,\rho,\theta) \sim (t,\rho,\theta+2\pi) \sim (t+2\pi,\rho,\theta)$, these coordinates can be understood as global on AdS_3, apart from the elementary coordinate singularity at $\rho = 0$. The metric reads

$$ds^2 = \frac{4}{(1-\rho^2)^2} \left[-\frac{1}{4}(1+\rho^2)^2 dt^2 + d\rho^2 + \rho^2 d\theta^2 \right] \, . \tag{13}$$

Dropping now from (13) the conformal factor $4(1-\rho^2)^{-2}$ yields a spacetime that can be regularly extended to $\rho = 1$, and the timelike hypersurface $\rho = 1$ in this conformal spacetime is by definition the conformal boundary of AdS_3. It is a timelike two-torus coordinatized by (t, θ) with the identifications $(t,\theta) \sim (t, \theta + 2\pi) \sim (t + 2\pi, \theta)$, and it has the flat metric

$$ds^2 = -dt^2 + d\theta^2 \, . \tag{14}$$

The conformal boundary construction generalizes in an obvious way to the universal covering space of AdS_3, which we denote by $CAdS_3$. The only difference is that the coordinate t is not periodically identified. The conformal boundary of $CAdS_3$, which we denote by B_C, is thus a timelike cylinder with the metric (14) and the identification $(t,\theta) \sim (t, \theta + 2\pi)$.

5.2 The Spinless Nonextremal BTZ Hole

Let ξ_{int} be on $CAdS_3$ the Killing vector induced by the boost-like Killing vector $\xi_{\text{emb}} := -T^1\partial_{X^1} - X^1\partial_{T^1}$ of $\mathbb{R}^{2,2}$, and let D_{int} denote the largest subset of $CAdS_3$ that contains the hypersurface $t = 0$ and in which ξ_{int} is spacelike. Given a prescribed positive parameter a, the isometry $\exp(a\xi_{\text{int}})$ generates a discrete isometry group $\Gamma_{\text{int}} \simeq \mathbb{Z}$ of D_{int}. The spinless nonextremal BTZ hole is by definition the quotient space $D_{\text{int}}/\Gamma_{\text{int}}$ [6,27]. A conformal diagram, with the S^1 factor arising from the identification suppressed, is shown in Fig. 5. The horizon circumference is a, and the ADM mass is $M = a^2/(32\pi^2 G_3)$, where G_3 is the (2+1)-dimensional Newton's constant. For further discussion, including expressions for the metric in coordinates adapted to the isometries, we refer to [6,27].

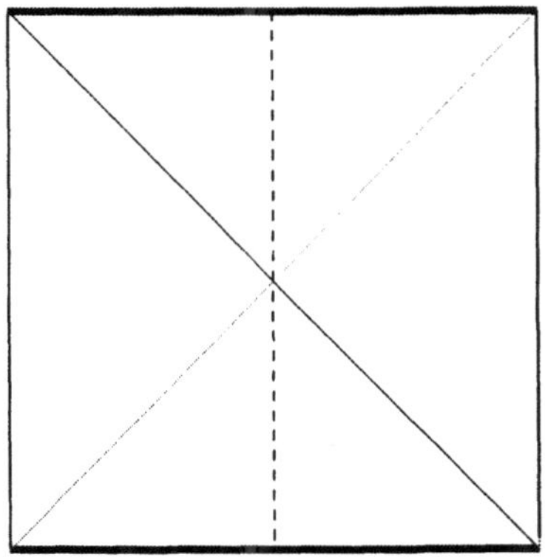

Fig. 5. A conformal diagram of the BTZ hole. Each point in the diagram represents a suppressed S^1. The involution $\tilde{J}_{\text{int}}$ introduced in Subsect. 5.3 consists of a left-right reflection about the dashed vertical line, followed by a rotation by π on the suppressed S^1

As seen in Fig. 5, the BTZ hole has two exterior regions, and the infinities are asymptotically Anti-de Sitter. The point of interest for us is that each of the infinities has a conformal boundary that is induced from B_C by the quotient construction. Technically, one observes that ξ_{int} induces on B_C the conformal Killing vector $\xi := \cos t \sin\theta\, \partial_\theta + \sin t \cos\theta\, \partial_t$, and that D_{int} reaches B_C in the two diamonds

$$\begin{aligned} D_R &:= \{(t,\theta) \mid 0 < \theta < \pi,\ |t| < \pi/2 - |\theta - \pi/2|\} \ , \\ D_L &:= \{(t,\theta) \mid -\pi < \theta < 0,\ |t| < \pi/2 - |\theta + \pi/2|\} \ . \end{aligned} \tag{15}$$

The two conformal boundaries of the BTZ hole are then the quotient spaces D_R/Γ_R and D_L/Γ_L, where Γ_R and Γ_L are the restrictions to respectively D_R and D_L of the conformal isometry group of B_C generated by $\exp(a\xi)$ [7,8]. To make this explicit, we cover D_R by the coordinates

$$\begin{aligned} \alpha &= -\ln\tan\left[(\theta - t)/2\right] \ , \\ \beta &= \ln\tan\left[(\theta + t)/2\right] \ , \end{aligned} \tag{16}$$

in which the metric induced from (14) is conformal to

$$ds^2 = -\left(\frac{2\pi}{a}\right)^2 d\alpha\, d\beta \ , \tag{17}$$

and $\xi = -\partial_\alpha + \partial_\beta$. The quotient space D_R/Γ_R, with the metric induced from (17), is thus isometric to B_C with the metric (14). In particular, it has topology $\mathbb{R}\times S^1$. It can be shown [7–9] that the conventionally-normalized Killing vector of the BTZ hole that is timelike in the exterior regions induces on D_R/Γ_R the timelike Killing vector $\eta = \partial_\alpha + \partial_\beta$. Analogous observations apply to D_L/Γ_L.

5.3 The $\mathbb{RP}^2$ Geon

The $\mathbb{RP}^2$ geon is obtained from the spinless BTZ hole in close analogy with the quotient construction used with the $\mathbb{RP}^3$ geon in Sect. 2. We denote the relevant involutive isometry of the BTZ hole by $\tilde{J}_{\text{int}}$: in the conformal diagram of Fig. 5, $\tilde{J}_{\text{int}}$ consists of a left-right reflection about the dashed vertical line, followed by a rotation by π on the suppressed S^1. A conformal diagram of the quotient space, the $\mathbb{RP}^2$ geon, is shown in Fig. 6. It is clear that the $\mathbb{RP}^2$ geon is a black (and white) hole spacetime with a single exterior region that is isometric to one exterior region of the BTZ hole. It is time orientable but not space orientable, and the spatial topology is $\mathbb{RP}^2 \setminus \{\text{point at infinity}\}$. The local and global isometries closely parallel those of the $\mathbb{RP}^3$ geon [9].

The map $\tilde{J}_{\text{int}}$ can clearly be extended to the conformal boundary of the BTZ hole, where it defines an involution $\tilde{J}$ that interchanges the two boundary components. Quotienting the conformal boundary of the BTZ hole by this involution gives the conformal boundary of the $\mathbb{RP}^2$ geon, which is thus isomorphic to one boundary component of the BTZ hole. Note that although the $\mathbb{RP}^2$ geon is not space orientable, its conformal boundary $\mathbb{R} \times S^1$ is.

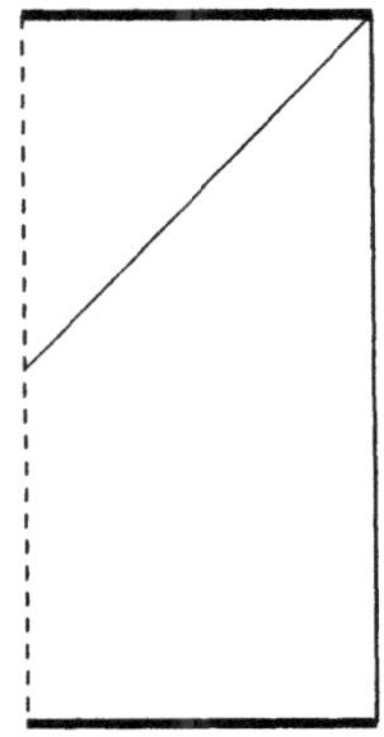

Fig. 6. A conformal diagram of the $\mathbb{RP}^2$ geon. The region not on the dashed line is identical to that in the diagram of Fig. 5, each point representing a suppressed S^1 in the spacetime. On the dashed line, each point in the diagram represents again an S^1 in the spacetime, but with only half of the circumference of the S^1's in the diagram of Fig. 5

6 Vacua on the Conformal Boundaries

We now turn to a free conformal scalar field on the boundaries of CAdS_3, the BTZ hole, and the $\mathbb{RP}^2$ geon.

Let $|0\rangle$ denote on B_C the vacuum state with respect to the timelike Killing vector ∂_t. We wish to know what kind of states $|0\rangle$ induces on the conformal boundaries of the BTZ hole and the $\mathbb{RP}^2$ geon. For concreteness, we focus the presentation on the non-zero modes of the field. The subtleties with the zero-modes are discussed in [9].

Consider the boundary of the BTZ hole. As noted above, the timelike Killing vectors on the two components do not lift to the timelike Killing vector ∂_t on B_C: the future timelike Killing vector on D_R/Γ_R lifts to $[a/(2\pi)]\eta$, and an analogous statement holds for D_L/Γ_L. To interpret the state induced by $|0\rangle$ on the BTZ hole boundary in terms of the BTZ particle modes, we must first first write the state induced by $|0\rangle$ on $D_R \cup D_L$ in terms of continuum-normalized particle states that are positive frequency with respect to η on D_R and with respect to the analogous Killing vector on D_L, and then restrict to appropriately periodic field modes in order to accommodate the identification by $\exp(a\xi)$. This calculation is quite similar to expressing the Minkowski vacuum in terms of Rindler particle modes [4,28,29]. Denoting the state induced from $|0\rangle$ by $|\mathrm{BTZ}\rangle$, we have the expansion

$$|\mathrm{BTZ}\rangle = \sum_{i\cdots k} f_{i\cdots k} \left(a_i^R\right)^\dagger \left(a_i^L\right)^\dagger \cdots \left(a_k^R\right)^\dagger \left(a_k^L\right)^\dagger |0\rangle_R \, |0\rangle_L \, , \tag{18}$$

where $|0\rangle_R$ and $|0\rangle_L$ are respectively the vacua on the two boundary components with respect to their timelike Killing vectors, $\left(a_i^R\right)^\dagger$ are the creation operators with respect to this Killing vector on D_R/Γ_R, and $\left(a_i^L\right)^\dagger$ are the creation operators with respect to this Killing vector on D_L/Γ_L. The analogy to the expansion (8) of the Hartle-Hawking vacuum on Kruskal is clear: the excitations come in correlated pairs, the two members of each pair now living on different boundary components. Restriction to one boundary component yields a thermal state, and when the normalization of the boundary timelike Killing vector is matched to that in the bulk of the spacetime, the temperature turns out to be the Hawking temperature of the BTZ hole, $a/4\pi^2$. This is the result first found in [7].

The boundary of the $\mathbb{RP}^2$ geon has a single connected component. Denoting the state induced from $|0\rangle$ by $|\mathbb{RP}^2\rangle$, we have the expansion

$$|\mathbb{RP}^2\rangle = \sum_{i\cdots k} \tilde{f}_{i\cdots k} \left(\tilde{a}_i^{(+)}\right)^\dagger \left(\tilde{a}_i^{(-)}\right)^\dagger \cdots \left(\tilde{a}_k^{(+)}\right)^\dagger \left(\tilde{a}_k^{(-)}\right)^\dagger |0\rangle_R \,, \qquad (19)$$

where $|0\rangle_R$ now denotes the geon boundary vacuum with respect to the timelike Killing vector, $\left(\tilde{a}_i^{(\alpha)}\right)^\dagger$ are the creation operators with respect to this Killing vector, and the indices i and α label the modes. The modes with $\alpha = +$ are right-movers and the modes with $\alpha = -$ are left-movers. The analogy to the expansion (9) of the Hartle-Hawking type vacuum on the $\mathbb{RP}^3$ geon is clear. For operators that do not contain couplings between modes with $\alpha = +$ and $\alpha = -$, the expectation values turn out to be thermal, with the BTZ Hawking temperature $a/4\pi^2$.

As shown in Table 1, several properties of the state $|\mathbb{RP}^2\rangle$ reflect properties of the $\mathbb{RP}^2$ geon spacetime geometry. First, $|\mathbb{RP}^2\rangle$ is a *pure* state on the boundary cylinder $\mathbb{R} \times S^1$: this follows by construction since (unlike with the BTZ hole) the single cylinder constitutes the whole conformal boundary. Second, $|\mathbb{RP}^2\rangle$ is an excited state with respect to the boundary timelike Killing field. This can be understood to reflect the fact that the spacetime attached to the boundary is not $CAdS_3$. Third, it can be shown that $|\mathbb{RP}^2\rangle$ is not invariant under translations generated by the timelike Killing vector on the boundary. This reflects the absence on the spacetime of a globally-defined Killing vector that would be timelike in the exterior region (cf. the discussion of the isometries of the $\mathbb{RP}^3$ geon in Sect. 2). Thus, $|\mathbb{RP}^2\rangle$ "knows" not just about the exterior region of the $\mathbb{RP}^2$ geon but also about the region behind the horizons.

Fourth, the correlations in $|\mathbb{RP}^2\rangle$ are between the right-movers and the left-movers. This is a direct consequence of the fact that the map $\tilde{J}_{\text{int}}$ on the BTZ hole reverses the spatial orientation, and it reflects thus the spatial nonorientability of the geon. Fifth, $|\mathbb{RP}^2\rangle$ appears thermal in the Hawking temperature for operators that do not see the correlations: this reflects the fact that the geon is a black (and white) hole spacetime.

Finally, the expectation value of the energy in $|\mathbb{RP}^2\rangle$ is, in the limit $a \gg 1$, equal to $a^2/48\pi^2$, which is quadratic in a and thus proportional to the ADM mass of the geon. The energy expectation value in the state $|\text{BTZ}\rangle$ on one boundary cylinder of the BTZ hole is also equal to $a^2/48\pi^2$, for $a \gg 1$. In this sense, the energy expectation value on a single boundary component is the same in $|0\rangle_{\mathbb{RP}^2}$

and |BTZ⟩. The analogous property in the spacetime is that the ADM mass at one infinity is not sensitive to whether a second infinity exists behind the horizons.

Table 1. Properties of the state $|\mathbb{RP}^2\rangle$, and the corresponding properties of the $\mathbb{RP}^2$ geon spacetime

$\lvert\mathbb{RP}^2\rangle$	$\mathbb{RP}^2$ geon geometry
pure state	boundary connected
excited state	not Anti-de Sitter
not static	no global KVF
correlations: left-movers with right-movers	spatially nonorientable
right-movers (left-movers) thermal, $T = a/4\pi^2$	black hole, $T_H = a/4\pi^2$
$\langle E\rangle = a^2/48\pi^2$, $a \gg 1$	$M = a^2/(32\pi^2 G_3)$

7 Holography and String Theory

We have seen that the state $|\mathbb{RP}^2\rangle$ on the boundary cylinder of the $\mathbb{RP}^2$ geon mirrors several aspects of the spacetime geometry of the $\mathbb{RP}^2$ geon. Some of this mirroring is immediate from the construction, such as the property that $|\mathbb{RP}^2\rangle$ is a pure state. Some aspects of the mirroring appear however quite nontrivial, especially the fact that the energy expectation value turned out to be proportional to the ADM mass, and with the same constant of proportionality as for the vacuum |BTZ⟩ on the boundary of the BTZ hole. One can see this as a piece of evidence in support of the holographic hypothesis [12,13], according to which physics in the bulk of a spacetime should be retrievable from physics on the boundary of the spacetime.

One would certainly not expect a free conformal scalar field on the boundary of a spacetime to carry all the information about the spacetime geometry. However, for certain spacetimes related to Anti-de Sitter space, a more precise version of the holographic hypothesis has emerged in string theory in the form of the Maldacena duality conjectures [7,14–16]. In particular, the 10-dimensional spacetime $\mathrm{CAdS}_3 \times S^3 \times T^4$, with a flat metric on the T^4 and a round metric on the S^3, is a classical solution to string theory, and the duality conjectures relate string theory on this spacetime to a certain conformal nonlinear sigma-model on the conformal boundary of the CAdS_3 component. Upon quotienting from CAdS_3 to the (in general spinning) BTZ hole, the conformal field theory on the boundary ends in a thermal state analogous to our |BTZ⟩, but with an energy expectation value that in the high temperature limit is not merely proportional to but in fact equal to the ADM mass of the hole [7]. This result can be considered a strong piece of evidence for the duality conjectures.

It would now be of obvious interest to adapt our free scalar field analysis on the boundary of the $\mathbb{RP}^2$ geon to a string theoretic context in which the duality conjectures would apply. One would expect the boundary state again to appear thermal in the Hawking temperature under some restricted set of observations. The crucial question for the the holographic hypothesis is how the correlations in the boundary state might reflect the geometry of the spacetime. As a preliminary step in this direction, a toy conformal field theory that mimics some of the anticipated features of the extra dimensions was considered in [9], and the energy expectation value in this toy theory was found to be equal to the geon ADM mass in the high temperature limit.

8 Concluding Remarks

The results presented here for the $\mathbb{RP}^3$ geon and the $\mathbb{RP}^2$ geon provide evidence that single-exterior black holes offer a nontrivial arena for scrutinizing quantum physics of black holes. It remains a subject to future work to understand to what extent the results reflect the peculiarities of these particular spacetimes, and to what extent they might have broader validity.

In some respects the $\mathbb{RP}^3$ and $\mathbb{RP}^2$ geons are certainly quite nongeneric black hole spacetimes. For example, our quotient constructions on Kruskal and the spinless BTZ hole do not immediately generalize to accommodate spin, as the putative isometry would need to invert the angular momentum. Similarly, the quotient construction on Kruskal does not immediately generalize to the Reissner-Nordström hole, as the relevant isometry would invert the electric field. Also, the spatial nonorientability of the $\mathbb{RP}^2$ geon may lead to difficulties in the string theoretic context. However, in 2+1 dimensions there exist locally Anti-de Sitter single-exterior black (and white) hole spacetimes that admit a spin, and one can choose their spatial topology to be orientable, for example $T^2\backslash\{\text{point at infinity}\}$ [26,30]. A natural next step would be to consider quantum field theory on these spinning "wormhole" spacetimes and on their conformal boundaries.

References

1. Hartle J.B., Hawking S.W. (1976) Phys. Rev. D **13**, 2188
2. Israel W. (1976) Phys. Lett. **57A**, 107
3. Kay B.S., Wald R.M. (1991) Phys. Rep. **207**, 49
4. Wald R.M. (1994) Quantum Field Theory in Curved Spacetime and Black Hole Thermodynamics. The University of Chicago Press, Chicago
5. Unruh W.G. (1976) Phys. Rev. D **14**, 870
6. Carlip S. (1995) Class. Quantum Grav. **12**, 283 [gr-qc/9506079]
7. Maldacena J., Strominger A. (1998) J. High Energy Phys. **9812**, 005 [hep-th/9804085]
8. Horowitz G.T., Marolf D. (1998) J. High Energy Phys. **9807**, 014 [gr-qc/9805207]
9. Louko J., Marolf D. (1999) Phys. Rev. D **59**, 066002 [hep-th/9808081]
10. Giulini D. (1989) PhD Thesis, University of Cambridge, Cambridge

11. Friedman J.L., Schleich K., Witt D.M. (1993) Phys. Rev. Lett. **71**, 1486; Erratum (1995) Phys. Rev. Lett. **75**, 1872 [gr-qc/9305017]
12. 't Hooft G. (1993) In: Ali A., Ellis J., Randjbar-Daemi S. (Eds.) Salamfestschrift: A Collection of Talks. World Scientific, Singapore [gr-qc/932102]
13. Susskind L. (1995) J. Math. Phys. **36** 6377 [hep-th/9409089]
14. Maldacena J. (1997) Adv. Theor. Math. Phys. **2**, 231 [hep-th/9711200]
15. Gubser S.S., Klebanov I.R., Polyakov A.M. (1998) Phys. Lett. **B428** 105 [hep-th/9802109]; Witten E. (1998) Adv. Theor. Math. Phys. **2**, 253 [hep-th/9802150]; Claus P., Kallosh R., Kumar J., Townsend P., Van Proeyen A. (1998) J. High Energy Phys. **9806**, 004 [hep-th/9801206]; Susskind L., Witten E. (1998) e-print hep-th/9805114
16. Balasubramanian V., Kraus P., Lawrence A., Trivedi S.P. (1998) e-print hep-th/9808017; Keski-Vakkuri E. (1999) Phys. Rev. D **59** 104001 [hep-th/9808037]; Danielsson U.H., Keski-Vakkuri E., Kruczenski M. (1999) J. High Energy Phys. **9901**, 002 [hep-th/9812007]; Balasubramanian V., Giddings S.B., Lawrence A. (1999) J. High Energy Phys. **9903**, 001 [hep-th/9902052]
17. Louko J., Marolf D. (1998) Phys. Rev. D **58**, 024007 [gr-qc/9802068]
18. Gibbons G.W. (1992) In: Kim J.E. (Ed.) Proceedings of the 10th Sorak School of Theoretical Physics. World Scientific, Singapore
19. Chamblin A., Gibbons G.W. (1996) Phys. Rev. D **55**, 2177 [gr-qc/9607079]
20. Gibbons G.W., Hawking S.W. (1977) Phys. Rev. D **15**, 2752
21. Bardeen J.M., Carter B., Hawking S.W. (1973) Commun. Math. Phys. **31**, 161
22. Hawking S.W. (1975) Commun. Math. Phys. **43**, 199
23. Bekenstein J.D. (1972) Nuovo Cimento Lett. **4**, 737
24. Bekenstein J.D. (1974) Phys. Rev. D **9**, 3292
25. Hawking S.W. (1979) In: Hawking S.W., Israel W. (Eds.) General Relativity: An Einstein Centenary Survey. Cambridge University Press, Cambridge
26. Åminneborg S., Bengtsson I., Brill D.R., Holst S., Peldán P. (1998) Class. Quantum Grav. **15**, 627 [gr-qc/9707036]
27. Bañados M., Teitelboim C., Zanelli J. (1992) Phys. Rev. Lett. **69**, 1849 [hep-th/9204099]; Bañados M., Henneaux M., Teitelboim C., Zanelli J. (1993) Phys. Rev. D **48**, 1506 [gr-qc/9302012]
28. Birrell N.D., Davies P.C.W. (1982) Quantum Fields in Curved Space. Cambridge University Press, Cambridge
29. Takagi S. (1986) Prog. Theor. Phys. Suppl. **88**, 1
30. Åminneborg S., Bengtsson I., Holst S. (1999) Class. Quantum Grav. **16**, 363 [gr-qc/9805028]

Dirac-Bergmann Observables for Tetrad Gravity

Luca Lusanna

Sezione INFN di Firenze,
Largo E.Fermi 2, 50125 Firenze, Italy
E-mail; lusanna@fi.infn.it

The electromagnetic, weak, strong and gravitational interactions are described by singular Lagrangians, so that their Hamiltonian formulation requires Dirac-Bergmann theory of constraints [1,2]. The requirements of gauge and/or diffeomorphism invariance, plus manifest Lorentz covariance in the case of flat spacetime, force us to work with redundant degrees of freedom. In the standard SU(3)xSU(2)xU(1) model of elementary particles in Minkowski spacetime the reduction to the physical degrees of freedom is done only at the quantum level with the BRST method. However, in this way only infinitesimal gauge transformations in the framework of local quantum field theory are considered, so that there are many open problems: the understanding of finite gauge transformations and of the associated moduli spaces, the Gribov ambiguity dependence on the choice of the function space for the fields and the gauge transformations, the confinement of quarks, the definition of relativistic bound states and how to put them among the asymptotic states, the nonlocality of charged states in quantum electrodynamics, not to speak of the foundational and practical problems posed by gravity. While behind the gauge freedom of gauge theories proper there are Lie groups acting on some internal space so that the measurable quantities must be gauge invariant, the gauge freedom of theories invariant under diffeomorphism groups of the underlying spacetime (general relativity, string theory and reparametrization invariant systems of relativistic particles) concerns the arbitrariness for the observer in the choice of the definition of "what is space and/or time" (and relative times in the case of particles), i.e. of the definitory properties either of spacetime itself or of the measuring apparatuses.

To try to clarify some of these problems, I decided to study systematically the classical Hamiltonian description of the four interactions (see Ref. [3] for a complete review of the program and Ref.[4] for previous reviews). The first stage was to understand how and when it is possible to extract a global canonical basis of physical degrees of freedom (gauge invariant Dirac's observables). It turns out that, when the configuration space is not compact, this can be achieved with special Shanmugadhasan canonical transformations both in the finite-dimensional case and in classical field theory. They replace the first and second class constraints of the theory with a set of momenta (Abelianization of first class constraints) and with pairs of conjugate canonical variables respectively. The variables conjugated to the momenta associated with the first class constraints are the gauge variables of the theory. The remaining pairs of canonical variables in the new basis are Dirac's observables. These canonical transformations are at the basis of the definition of the Faddeev-Popov measure for the path integral and trivialize the BRST construction (however, since they are generically nonlocal, we go outside local field theory). See Refs.[5,6].

By putting equal to zero the gauge variables, we get generalized global Coulomb gauges, in which the physics is described only by physical degrees of freedom. See the quoted reviews for a list of the models in Minkowski spacetime which have been treated in this way. They include: i) relativistic particle mechanics (see Refs.[7–11]); ii) the open and closed Nambu string [12]; iii) Yang-Mills theory with Grassmann-valued fermion fields in the case of a trivial principal bundle[13] with special weighted Sobolev spaces [14] in which the Gribov ambiguity is absent; iv) the Abelian and non-Abelian Higgs models [15]; v) the standard SU(3)xSU(2)xU(1) model of elementary particles [16].

However, in these generalized Coulomb gauges there is a breaking of manifest Lorentz covariance. Therefore, the next step has been to understand how we can covariantize these results in Minkowski spacetime by taking into account that the global Poincaré symmetry induces a stratification of the configurations of the ststem: they are divided in strata corresponding to the various Poincaré orbits and each stratum has a different geometry induced by the corresponding little group. To adapt the description to this geometry, for each stratum we must do a canonical transformation from the original variables to a new set consisting of center-of-mass variables x^μ, p^μ and of variables relative to the center of mass. Let us consider the stratum $p^2 > 0$. By using the standard Wigner boost $L^\mu_\nu(p,\overset{\circ}{p})$ ($p^\mu = L^\mu_\nu(p,\overset{\circ}{p})\overset{\circ}{p}^\nu$, $\overset{\circ}{p}^\mu = \eta\sqrt{p^2}(1;\mathbf{0})$, $\eta = sign\, p^o$), one boosts the relative variables at rest. The new variables are still canonical and the base is completed by p^μ and by a new center-of-mass coordinate $\tilde{x}^\mu$, differing from x^μ for spin terms. The variable $\tilde{x}^\mu$ has complicated covariance properties; instead the new relative variables are either Poincare' scalars or Wigner spin-1 vectors, transforming under the group O(3)(p) of the Wigner rotations induced by the Lorentz transformations. A final canonical transformation[17], leaving fixed the relative variables, sends the center-of-mass coordinates $\tilde{x}^\mu$, p^μ in the new set $p\cdot\tilde{x}/\eta\sqrt{p^2} = p\cdot x/\eta\sqrt{p^2}$ (the time in the rest frame), $\eta\sqrt{p^2}$ (the total mass), $\mathbf{k} = \mathbf{p}/\eta\sqrt{p^2}$ (the spatial components of the 4-velocity $k^\mu = p^\mu/\eta\sqrt{p^2}$, $k^2 = 1$), $\mathbf{z} = \eta\sqrt{p^2}(\tilde{\mathbf{x}} - \tilde{x}^o\mathbf{p}/p^o)$. $\mathbf{z}$ is a noncovariant center-of-mass canonical 3-coordinate multiplied by the total mass: it is the classical analog of the Newton-Wigner position operator (like it, $\mathbf{z}$ is covariant only under the little group O(3)(p) of the timelike Poincaré orbits).

The nature of the relative variables depends on the system. The first class constraints, once rewritten in terms of the new variables, can be manipulated to find suitable global and Lorentz scalar Abelianizations. Usually there is a combination of the constraints which determines $\eta\sqrt{p^2}$, i.e. the mass spectrum, so that the time in the rest frame $p\cdot x/\eta\sqrt{p^2}$ is the conjugated Lorentz scalar gauge variable. The other constraints eliminate some of the relative variables (in particular the relative energies for systems of interacting relativistic particles and the string): their conjugated coordinates (the relative times) are the other gauge variables: they are identified with a possible set of time parameters. The Dirac observables (apart from the center-of-mass ones $\mathbf{k}$ and $\mathbf{z}$) have to be extracted from the remaining relative variables and the construction shows that they will be either Poincare' scalars or Wigner covariant objects. In this way in each stratum preferred global Shanmugadhasan canonical transformations are identified, when no other kind of obstruction to globality is present inside the various strata.

Then, following Dirac[1] we must reformulate classical field theory on spacelike hypersurfaces foliating Minkowski spacetime M^4 [the foliation is defined by an embedding $R \times \Sigma \rightarrow M^4$, $(\tau, \vec{\sigma}) \mapsto z^{(\mu)}(\tau, \vec{\sigma}) \in \Sigma_\tau$, with Σ an abstract 3-surface diffeomorphic to R^3, with Σ_τ its copy embedded in M^4 labelled by the value τ (the Minkowski flat indices are (μ); the scalar "time" parameter τ labels the leaves of the foliation, $\vec{\sigma}$ are curvilinear coordinates on Σ_τ and $\sigma^A = (\tau, \vec{\sigma})$ are Σ_τ-adapted holonomic coordinates for M^4); this is the classical basis of Tomonaga-Schwinger quantum field theory]. In this way one gets a parametrized field theory with a covariant 3+1 splitting of Minkowski spacetime and already in a form suited to the transition to general relativity in its ADM canonical formulation (see also Ref.[18], where a theoretical study of this problem is done in curved spacetimes). The price is that one has to add as new independent configuration variables the embedding coordinates $z^{(\mu)}(\tau, \vec{\sigma})$ of the points of the spacelike hypersurface Σ_τ [the only ones carrying Lorentz indices] and then to define the fields on Σ_τ so that they know the hypersurface Σ_τ of τ-simultaneity [for a Klein-Gordon field $\phi(x)$, this new field is $\tilde{\phi}(\tau, \vec{\sigma}) = \phi(z(\tau, \vec{\sigma}))$: it contains the nonlocal information about the embedding]. Then one rewrites the Lagrangian of the given isolated system in the form required by the coupling to an external gravitational field, makes the previous 3+1 splitting of Minkowski spacetime and interpretes all the fields of the system as the new fields on Σ_τ (they are Lorentz scalars, having only surface indices). Instead of considering the 4-metric as describing a gravitational field (and therefore as an independent field as it is done in metric gravity, where one adds the Hilbert action to the action for the matter fields), here one replaces the 4-metric with the the induced metric $g_{AB}[z] = z^{(\mu)}_A \eta_{(\mu)(\nu)} z^{(\nu)}_B$ on Σ_τ [a functional of $z^{(\mu)}$; $z^{(\mu)}_A = \partial z^{(\mu)} / \partial \sigma^A$ are flat tetrad fields on Minkowski spacetime with the $z^{(\mu)}_r$'s tangent to Σ_τ] and considers the embedding coordinates $z^{(\mu)}(\tau, \vec{\sigma})$ as independent fields [this is not possible in metric gravity, because in curved spacetimes $z^\mu_A \neq \partial z^\mu / \partial \sigma^A$ are not tetrad fields so that holonomic coordinates $z^\mu(\tau, \vec{\sigma})$ do not exist]. From this Lagrangian, besides a Lorentz-scalar form of the constraints of the given system, we get four extra primary first class constraints

$$\mathcal{H}_{(\mu)}(\tau, \vec{\sigma}) = \rho_{(\mu)}(\tau, \vec{\sigma}) - l_{(\mu)}(\tau, \vec{\sigma}) T^{\tau\tau}_{sys}(\tau, \vec{\sigma}) - z_{r(\mu)}(\tau, \vec{\sigma}) T^{r\tau}_{sys}(\tau, \vec{\sigma}) \approx 0$$

[here $T^{\tau\tau}_{sys}(\tau, \vec{\sigma})$, $T^{r\tau}_{sys}(\tau, \vec{\sigma})$, are the components of the energy-momentum tensor in the holonomic coordinate system, corresponding to the energy- and momentum-density of the isolated system; one has $\{\mathcal{H}_{(\mu)}(\tau, \vec{\sigma}), \mathcal{H}_{(\nu)}(\tau, \vec{\sigma}^{\,\prime})\} = 0$] implying the independence of the description from the choice of the 3+1 splitting, i.e. from the choice of the foliation with spacelike hypersufaces. The evolution vector is given by $z^{(\mu)}_\tau = N_{[z](flat)} l^{(\mu)} + N^r_{[z](flat)} z^{(\mu)}_r$, where $l^{(\mu)}(\tau, \vec{\sigma})$ is the normal to Σ_τ in $z^{(\mu)}(\tau, \vec{\sigma})$ and $N_{[z](flat)}(\tau, \vec{\sigma})$, $N^r_{[z](flat)}(\tau, \vec{\sigma})$ are the flat lapse and shift functions defined through the metric like in general relativity: however, now they are not independent variables but functionals of $z^{(\mu)}(\tau, \vec{\sigma})$.

The Dirac Hamiltonian contains the piece $\int d^3\sigma \lambda^{(\mu)}(\tau,\vec{\sigma})\mathcal{H}_{(\mu)}(\tau,\vec{\sigma})$ with $\lambda^{(\mu)}(\tau,\vec{\sigma})$ Dirac multipliers. It is possible to rewrite the integrand in the form [${}^3g^{rs}$ is the inverse of g_{rs}]

$$\lambda_{(\mu)}(\tau,\vec{\sigma})\mathcal{H}^{(\mu)}(\tau,\vec{\sigma}) = [(\lambda_{(\mu)}l^{(\mu)})(l_{(\nu)}\mathcal{H}^{(\nu)}) - (\lambda_{(\mu)}z^{(\mu)}_r)({}^3g^{rs}z_{s(\nu)}\mathcal{H}^{(\nu)})](\tau,\vec{\sigma})$$
$$\stackrel{def}{=} N_{(flat)}(\tau,\vec{\sigma})(l_{(\mu)}\mathcal{H}^{(\mu)})(\tau,\vec{\sigma}) - N_{(flat)r}(\tau,\vec{\sigma})({}^3g^{rs}z_{s(\nu)}\mathcal{H}^{(\nu)})(\tau,\vec{\sigma})$$

with the (nonholonomic form of the) constraints

$$(l_{(\mu)}\mathcal{H}^{(\mu)})(\tau,\vec{\sigma}) \approx 0, \quad ({}^3g^{rs}z_{s(\mu)}\mathcal{H}^{(\mu)})(\tau,\vec{\sigma}) \approx 0,$$

satisfying the universal Dirac algebra of the ADM constraints. In this way we have defined new flat lapse and shift functions

$$N_{(flat)}(\tau,\vec{\sigma}) = \lambda_{(\mu)}(\tau,\vec{\sigma})l^{(\mu)}(\tau,\vec{\sigma}), \; N_{(flat)r}(\tau,\vec{\sigma}) = \lambda_{(\mu)}(\tau,\vec{\sigma})z^{(\mu)}_r(\tau,\vec{\sigma}).$$

which have the same content of the arbitrary Dirac multipliers $\lambda_{(\mu)}(\tau,\vec{\sigma})$, namely they multiply primary first class constraints satisfying the Dirac algebra. In Minkowski spacetime they are quite distinct from the previous lapse and shift functions $N_{[z](flat)}$, $N_{[z](flat)r}$, defined starting from the metric. Instead in general relativity the lapse and shift functions defined starting from the 4-metric are the coefficients (in the canonical part H_c of the Hamiltonian) of secondary first class constraints satisfying the Dirac algebra.

In special relativity, it is convenient to restrict ourselves to arbitrary spacelike hyperplanes $z^{(\mu)}(\tau,\vec{\sigma}) = x^{(\mu)}_s(\tau) + b^{(\mu)}_r(\tau)\sigma^r$. Since they are described by only 10 variables, after this restriction we remain only with 10 first class constraints determining the 10 variables conjugate to the hyperplane in terms of the variables of the system:

$$\mathcal{H}^{(\mu)}(\tau) = p^{(\mu)}_s - p^{(\mu)}_{(sys)} \approx 0, \; \mathcal{H}^{(\mu)(\nu)}(\tau) = S^{(\mu)(\nu)}_s - S^{(\mu)(\nu)}_{(sys)} \approx 0.$$

After the restriction to spacelike hyperplanes the previous piece of the Dirac Hamiltonian is reduced to $\tilde{\lambda}^{(\mu)}(\tau)\mathcal{H}_{(\mu)}(\tau) - \frac{1}{2}\tilde{\lambda}^{(\mu)(\nu)}(\tau)\mathcal{H}_{(\mu)(\nu)}(\tau)$. Since at this stage we have $z^{(\mu)}_r(\tau,\vec{\sigma}) \approx b^{(\mu)}_r(\tau)$, so that $z^{(\mu)}_\tau(\tau,\vec{\sigma}) \approx N_{[z](flat)}(\tau,\vec{\sigma})l^{(\mu)}(\tau,\vec{\sigma}) + N^r_{[z](flat)}(\tau,\vec{\sigma})\, b^{(\mu)}_r(\tau,\vec{\sigma}) \approx \dot{x}^{(\mu)}_s(\tau) + \dot{b}^{(\mu)}_r(\tau)\sigma^r = -\tilde{\lambda}^{(\mu)}(\tau) - \tilde{\lambda}^{(\mu)(\nu)}(\tau)b_{r(\nu)}(\tau)\sigma^r$, it is only now that we get the coincidence of the two definitions of flat lapse and shift functions (this point was missed in the older treatments of parametrized Minkowski theories):

$$N_{[z](flat)}(\tau,\vec{\sigma}) \approx N_{(flat)}(\tau,\vec{\sigma}) = -\tilde{\lambda}_{(\mu)}(\tau)l^{(\mu)} - l^{(\mu)}\tilde{\lambda}_{(\mu)(\nu)}(\tau)b^{(\nu)}_s(\tau)\sigma^s,$$
$$N_{[z](flat)r}(\tau,\vec{\sigma}) \approx N_{(flat)}(\tau,\vec{\sigma}) = -\tilde{\lambda}_{(\mu)}(\tau)b^{(\mu)}_r(\tau) - b^{(\mu)}_r(\tau)\tilde{\lambda}_{(\mu)(\nu)}(\tau)b^{(\nu)}_s(\tau)\sigma^s.$$

The 20 variables for the phase space description of a hyperplane are:
i) $x^{(\mu)}_s(\tau), p^{(\mu)}_s$, parametrizing the origin of the coordinates on the family of spacelike hyperplanes. The four constraints $\mathcal{H}^{(\mu)}(\tau) \approx 0$ say that $p^{(\mu)}_s$ is determined

by the 4-momentum of the isolated system.
ii) $b^{(\mu)}_A(\tau)$ (with the $b^{(\mu)}_r(\tau)$'s being three orthogonal spacelike unit vectors generating the fixed τ-independent timelike unit normal $b^{(\mu)}_\tau = l^{(\mu)}$ to the hyperplanes) and $S^{(\mu)(\nu)}_s = -S^{(\nu)(\mu)}_s$ with the orthonormality constraints $b^{(\mu)}_A\, {}^4\eta_{(\mu)(\nu)} b^{(\nu)}_B = {}^4\eta_{AB}$ [enforced by assuming the Dirac brackets $\{S^{(\mu)(\nu)}_s, b^{(\rho)}_A\} = {}^4\eta^{(\rho)(\nu)} b^{(\mu)}_A - {}^4\eta^{(\rho)(\mu)} b^{(\nu)}_A$, $\{S^{(\mu)(\nu)}_s, S^{(\alpha)(\beta)}_s\} = C^{(\mu)(\nu)(\alpha)(\beta)}_{(\gamma)(\delta)} S^{(\gamma)(\delta)}_s$ with $C^{(\mu)(\nu)(\alpha)(\beta)}_{(\gamma)(\delta)}$ the structure constants of the Lorentz algebra]. In these variables there are hidden six independent pairs of degrees of freedom. The six constraints $\mathcal{H}^{(\mu)(\nu)}(\tau) \approx 0$ say that $S^{(\mu)(\nu)}_s$ coincides the spin tensor of the isolated system. Then one has that $p^{(\mu)}_s$, $J^{(\mu)(\nu)}_s = x^{(\mu)}_s p^{(\nu)}_s - x^{(\nu)}_s p^{(\mu)}_s + S^{(\mu)(\nu)}_s$, satisfy the algebra of the Poincaré group.

Let us remark that, for each configuration of an isolated system there is a privileged family of hyperplanes (the Wigner hyperplanes orthogonal to $p^{(\mu)}_s$, existing when $p^2_s > 0$) corresponding to the intrinsic rest-frame of the isolated system. If we choose these hyperplanes with suitable gauge fixings, we remain with only the four constraints $\mathcal{H}^{(\mu)}(\tau) \approx 0$, which can be rewritten as

$$\sqrt{p^2_s} \approx [invariant\, mass\, of\, the\, isolated\, system\, under\, investigation] = M_{sys};$$
$$\mathbf{p}_{sys} = [3 - momentum\, of\, the\, isolated\, system\, inside\, the\, Wigner\, hyperplane] \approx 0.$$

There is no more a restriction on $p^{(\mu)}_s$, because $u^{(\mu)}_s(p_s) = p^{(\mu)}_s / p^2_s$ gives the orientation of the Wigner hyperplanes containing the isolated system with respect to an arbitrary given external observer.

In this special gauge we have $b^{(\mu)}_A \equiv L^{(\mu)}{}_A(p_s, \overset{\circ}{p}_s)$, $S^{(\mu)(\nu)}_s \equiv S^{(\mu)(\nu)}_{system}$, and the only remaining canonical variables are the noncovariant Newton-Wigner-like canonical "external" center-of-mass coordinate $\tilde{x}^{(\mu)}_s(\tau)$ (living on the Wigner hyperplanes) and $p^{(\mu)}_s$. Now 3 degrees of freedom of the isolated system [an "internal" center-of-mass 3-variable $\boldsymbol{\sigma}_{sys}$ defined inside the Wigner hyperplane and conjugate to $\mathbf{p}_{sys}$] become gauge variables [the natural gauge fixing is $\boldsymbol{\sigma}_{sys} \approx 0$, so that it coincides with the origin $x^{(\mu)}_s(\tau) = z^{(\mu)}(\tau, \sigma = 0)$ of the Wigner hyperplane], while the $\tilde{x}^{(\mu)}$ is playing the role of a kinematical external center of mass for the isolated system and may be interpreted as a decoupled observer with his parametrized clock (point particle clock). All the fields living on the Wigner hyperplane are now either Lorentz scalar or with their 3-indices transformaing under Wigner rotations (induced by Lorentz transformations in Minkowski spacetime) as any Wigner spin 1 index.

One obtains in this way a new kind of instant form of the dynamics (see Ref.[19]), the "Wigner-covariant 1-time rest-frame instant form"[20] with a universal breaking of Lorentz covariance. It is the special relativistic generalization of the nonrelativistic separation of the center of mass from the relative motion [$H = \frac{\mathbf{P}^2}{2M} + H_{rel}$]. The role of the center of mass is taken by the Wigner hyperplane, identified by the point $\tilde{x}^{(\mu)}(\tau)$ and by its normal $p^{(\mu)}_s$. The invariant mass M_{sys} of the system replaces the nonrelativistic Hamiltonian H_{rel} for the relative degrees of freedom, after the addition of the gauge-fixing $T_s - \tau \approx 0$ [identifying the time parameter τ, labelling the leaves of the foliation, with the Lorentz scalar time of the center of mass in the rest frame, $T_s = p_s \cdot \tilde{x}_s / M_{sys}$; M_{sys} generates the evolution in this time].

The determination of $\boldsymbol{\sigma}_{sys}$ may be done with the group theoretical methods of Ref.[21]: given a realization on the phase space of a given system of the ten Poincaré generators one can build three 3-position variables only in terms of them, which in our case of a system on the Wigner hyperplane with $\boldsymbol{p}_{sys} \approx 0$ are: i) a canonical center of mass (the "internal" center of mass $\boldsymbol{\sigma}_{sys}$); ii) a noncanonical Möller center of energy $\boldsymbol{\sigma}^{(E)}_{sys}$; iii) a noncanonical Fokker-Pryce center of inertia $\boldsymbol{\sigma}^{(FP)}_{sys}$. Due to $\boldsymbol{p}_{sys} \approx 0$, we have $\boldsymbol{\sigma}_{sys} \approx \boldsymbol{\sigma}^{(E)}_{sys} \approx \boldsymbol{\sigma}^{(FP)}_{sys}$. By adding the gauge fixings $\boldsymbol{\sigma}_{sys} \approx 0$ one can show that the origin $x^{(\mu)}_s(\tau)$ becomes simultaneously the Dixon center of mass of an extended object and both the Pirani and Tulczyjew centroids (see Ref. [22] for the application of these methods to find the center of mass of a configuration of the Klein-Gordon field after the preliminary work of Ref.[23]). With similar methods one can construct three "external" collective positions (all located on the Wigner hyperplane): i) the "external" canonical noncovariant center of mass $\tilde{x}^{(\mu)}_s$; ii) the "external" noncanonical and noncovariant Möller center of energy $R^{(\mu)}_s$; iii) the "external" covariant noncanonical Fokker-Pryce center of inertia $Y^{(\mu)}_s$ (when there are the gauge fixings $\boldsymbol{\sigma}_{sys} \approx 0$ it also coincides with the origin $x^{(\mu)}_s$). It turns out that the Wigner hyperplane is the natural setting for the study of the Dixon multipoles of extended relativistic systems[24] and for defining the canonical relative variables with respect to the center of mass. The Wigner hyperplane with its natural Euclidean metric structure offers a natural solution to the problem of boost for lattice gauge theories and realizes explicitly the machian aspect of dynamics that only relative motions are relevant.

The isolated systems till now analyzed to get their rest-frame Wigner-covariant generalized Coulomb gauges [i.e. the subset of global Shanmugadhasan canonical bases, which, for each Poincaré stratum, are also adapted to the geometry of the corresponding Poincaré orbits with their little groups (every stratum requires an independent canonical reduction); till now only the main stratum with p^2 timelike and $W^2 \neq 0$ has been investigated] are:

a) The system of N scalar particles with Grassmann electric charges plus the electromagnetic field [20]. The starting configuration variables are a 3-vector $\boldsymbol{\eta}_i(\tau)$ for each particle $[x^{(\mu)}_i(\tau) = z^{(\mu)}(\tau, \boldsymbol{\eta}_i(\tau))]$ and the electromagnetic gauge potentials $A_A(\tau, \boldsymbol{\sigma}) = \frac{\partial z^{(\mu)}(\tau,\boldsymbol{\sigma})}{\partial \sigma^A} A_{(\mu)}(z(\tau, \boldsymbol{\sigma}))$, which know the embedding of Σ_τ into M^4. One has to choose the sign of the energy of each particle, because there are not mass-shell constraints (like $p_i^2 - m_i^2 \approx 0$) among the constraints of this formulation, due to the fact that one has only three degrees of freedom for particle, determining the intersection of a timelike trajectory and of the spacelike hypersurface Σ_τ. In this way, one gets a description of relativistic particles with a given sign of the energy with consistent couplings to fields and valid independently from the quantum effect of pair production [in the manifestly covariant approach, containing all possible branches of the particle mass spectrum, the classical counterpart of pair production is the intersection of different branches deformed by the presence of interactions]. The final Dirac's observables are: i) the transverse radiation field variables $\boldsymbol{A}_{\perp}$, $\boldsymbol{E}_{\perp}$; ii) the particle canonical variables $\boldsymbol{\eta}_i(\tau)$, $\boldsymbol{\kappa}_i(\tau)$, dressed with a Coulomb cloud. The physical Hamiltonian contains the mutual instantaneous Coulomb potentials extracted from field theory and there is a regularization of the Coulomb self-energies due to the Grassmann character of the electric charges Q_i $[Q_i^2 = 0]$. In Ref.[25] there is the study of the Lienard-Wiechert potentials and of Abraham-Lorentz-Dirac equations in this rest-frame Coulomb gauge and also scalar electrodynamics is reformulated in it. Also the rest-frame 1-time relativistic statistical mechanics has been developed [20].

b) The system of N scalar particles with Grassmann-valued color charges plus the color SU(3) Yang-Mills field[26]: it gives the pseudoclassical description of the relativistic scalar-quark model, deduced from the classical QCD Lagrangian and with the color field present. With these results one can covariantize the bosonic part of the standard model given in Ref.[16].

c) The system of N spinning particles of definite energy $[(\frac{1}{2},0)$ or $(0,\frac{1}{2})$ representation of SL(2,C)] with Grassmann electric charges plus the electromagnetic field[27] and that of a Grassmann-valued Dirac field plus the electromagnetic field (the pseudoclassical basis of QED) [28]. In both cases there are geometrical complications connected with the spacetime description of the path of electric currents and not only of their spin structure: after their solution the rest-frame form of the full standard $SU(3) \times SU(2) \times U(1)$ model can be achieved.

The rest-frame description of the relativistic perfect gas is now under investigation.

All these new pieces of information will allow, after quantization of this new consistent relativistic mechanics without the classical problems connected with pair production, to find the asymptotic states of the covariant Tomonaga-Schwinger formulation of quantum field theory on spacelike hypersurfaces (to be obtained by quantizing the fields on Σ_τ): these states are needed for the theory of quantum bound states [since Fock states do not constitute a Cauchy problem for the field equations, because an in (or out) particle can be in the absolute future of another one due to the tensor product nature of these asymptotic states, bound state equations like the Bethe-Salpeter one have spurious solutions which are excitations in relative energies, the variables conjugate to relative times]. Moreover, it will be possible to include bound states among the asymptotic states.

As said in Ref.[25,26], the quantization of these rest-frame models has to overcome two problems. On the particle side, the complication is the quantization of the square roots associated with the relativistic kinetic energy terms: in the free case this has been done in Ref.[29]. On the field side (all physical Hamiltonian are nonlocal and, with the exception of the Abelian case, nonpolynomial, but quadratic in the momenta), the obstacle is the absence (notwithstanding there is no no-go theorem) of a complete regularization and renormalization procedure of electrodynamics (to start with) in the Coulomb gauge.

However, as shown in Refs.[20,13], the rest-frame instant form of dynamics automatically gives a physical ultraviolet cutoff in the spirit of Dirac and Yukawa: it is the Möller radius[30] $\rho = \sqrt{-W^2}/p^2 = |\mathbf{S}|/\sqrt{p^2}$ ($W^2 = -p^2\mathbf{S}^2$ is the Pauli-Lubanski Casimir when $p^2 > 0$), namely the classical intrinsic radius of the worldtube, around the covariant noncanonical Fokker-Pryce center of inertia $Y^{(\mu)}$, inside which the noncovariance of the canonical center of mass $\tilde{x}^\mu$ is concentrated. At the quantum level ρ becomes the Compton wavelength of the isolated system multiplied its spin eigenvalue $\sqrt{s(s+1)}$, $\rho \mapsto \hat{\rho} = \sqrt{s(s+1)}\hbar/M = \sqrt{s(s+1)}\lambda_M$ with $M = \sqrt{p^2}$ the invariant mass and $\lambda_M = \hbar/M$ its Compton wavelength. Therefore, the criticism to classical relativistic physics, based on quantum pair production, concerns the testing of distances where, due to the Lorentz signature of spacetime, one has intrinsic classical covariance problems: it is impossible to localize the canonical center of mass $\tilde{x}^\mu$ adapted to the first class constraints of the system (also named Pryce center of mass and having the same covariance of the Newton-Wigner position operator) in a frame independent way. Let us remember [20] that ρ is also a remnant in flat Minkowski spacetime of the energy conditions of general relativity: since the Möller noncanonical, noncovariant center of energy $R^{(\mu)}$has its noncovariance localized inside the same worldtube with radius ρ (it was discovered in this way) [30], it turns out that for an extended relativistic system with the material radius smaller of its intrinsic radius ρ one has: i) its peripheral rotation velocity can exceed the velocity of light; ii) its classical energy density cannot be positive definite everywhere in every frame.

Now, the real relevant point is that this ultraviolet cutoff determined by ρ exists also in Einstein's general relativity (which is not power counting renormalizable) in the case of asymptotically flat spacetimes, taking into account the Poincaré Casimirs of its asymptotic ADM Poincaré charges (when supertranslations are eliminated with suitable boundary conditions). See Ref.[3,4] for other properties of ρ.

In conclusion, the best set of canonical coordinates adapted to the constraints and to the geometry of Poincaré orbits in Minkowski spacetime and naturally predisposed to the coupling to canonical tetrad gravity is emerging for the electromagnetic, weak and strong interactions with matter described either by fermion fields or by relativistic particles with a definite sign of the energy.

Tetrad gravity is the formulation of general relativity natural for the coupling to the fermion fields of the standard model. However, we need a formulation of it, which allows to solve its constraints for doing the canonical reduction and to solve the deparametrization problem of general relativity (how to recover the rest-frame instant form when the Newton constant is put equal to zero, G=0). One also needs a formulation in which some notion of elementary particle exists so to recover Wigner's definition based on the irreducible representations of the Poincaré group in Minkowski spacetime with the further enrichment of the known good quantum numbers for their classification. Moreover, one needs some way out from the "problem of time"[31–33], since neither any consistent way to quantize time (is it a necessity?), and generically any timelike variable, nor a control on the associated problem of the relative times of a system of relativistic particles are known. Finally, one has to find a solution to the more basic problem of how to identify physically spacetime points in Einstein's formulation of general relativity, where general covariance deprives the mathematical points of the underlying 4-manifold of any physical reality [34,35], while, on the experimental side (space physics, gravitational waves detectors) , we are employing a theory of measurements of proper times and spacelike lengths which presuppones the individuation of points. This problem will appear also in the nowaday most popular program of unification of all the interactions in a supersymmetric way, i.e. in superstring theory and in its searched M-theory extension, when someone will be able to reformulate it in a background independent way.

Since neither a complete reduction of gravity with an identification of the physical canonical degrees of freedom of the gravitational field nor a detailed study of its Hamiltonian group of gauge transformations (whose infinitesimal generators are the first class constraints) has ever been pushed till the end in an explicit way, a new formulation of tetrad gravity [36–39] was developed.

To implement this program we shall restrict ourselves to the simplest class of spacetimes [time-oriented pseudo-Riemannian or Lorentzian 4-manifold $(M^4, {}^4g)$ with signature $\epsilon\,(+---)$ ($\epsilon = \pm 1$ according to either particle physics or general relativity convention) and with a choice of time orientation], assumed to be:

i) Globally hyperbolic 4-manifolds, i.e. topologically they are $M^4 = R \times \Sigma$, so to have a well posed Cauchy problem [with Σ the abstract model of Cauchy surface]: they admit regular foliations (3+1 splittings) with orientable, complete, non-intersecting spacelike 3-manifolds Σ_τ [$\tau : M^4 \rightarrow R$, $z^\mu \mapsto \tau(z^\mu)$].

ii) Asymptotically flat at spatial infinity, so to have the possibility to define asymptotic Poincaré charges [40–42]: they allow the definition of a Mωller radius also in general relativity and are a bridge towards a future soldering with the theory of elementary particles in Minkowski spacetime defined as irreducible representation of its kinematical, globally implemented Poincaré group according to Wigner. This excludes Einstein-Wheeler closed universes without boundaries (no asymptotic Poincaré charges), which were introduced to eliminate boundary conditions at spatial infinity to make the theory as machian as possible.

iii) Admitting a spinor (or spin) structure[43] for the coupling to fermion fields. Since we consider noncompact space- and time-orientable spacetimes, spinors can be defined if and only if they are "parallelizable" [44], like in our case. This implies that the orthonormal frame principal SO(3)-bundle over Σ_τ (whose connections are the spin connections determined by the cotriads) is trivial.

iv) The noncompact parallelizable simultaneity 3-manifolds (the Cauchy surfaces) Σ_τ are assumed to be topologically trivial, geodesically complete and, finally, diffeomorphic to R^3. These 3-manifolds have the same manifold structure as Euclidean spaces: a) the geodesic exponential map $Exp_p : T_p\Sigma_\tau \rightarrow \Sigma_\tau$ is a diffeomorphism ; b) the sectional curvature is less or equal zero everywhere; c) they have no "conjugate locus" [i.e. there are no pairs of conjugate Jacobi points (intersection points of distinct geodesics through them) on any geodesic] and no "cut locus" [i.e. no closed geodesics through any point].

v) Like in Yang-Mills case [13], the 3-spin-connection on the orthogonal frame SO(3)-bundle (and therefore cotriads) will have to be restricted to suited weighted Sobolev spaces to avoid Gribov ambiguities [13,45]. In turn, this implies the absence of isometries of the noncompact Riemannian 3-manifold $(\Sigma_\tau, {}^3g)$ [see for instance the review paper in Ref. [46]].

Diffeomorphisms on Σ_τ $(Diff\, \Sigma_\tau)$ are interpreted in the passive way, following Ref.[47], in accord with the Hamiltonian point of view that infinitesimal diffeomorphisms are generated by taking the Poisson bracket with the 1st class supermomentum constraints [passive diffeomorphisms are also named 'pseudodiffeomorphisms'].

By using Σ_τ-adapted holonomic coordinates for M^4, one has found a new parametrization of arbitrary tetrads and cotetrads on M^4 in terms of cotriads on Σ_τ $[{}^3e_{(a)r}(\tau,\sigma)]$, of lapse $[N(\tau,\sigma)]$ and shift $[N_{(a)}(\tau,\sigma) = \{{}^3e_{(a)r}N^r\}(\tau,\sigma)]$ functions and of 3 parameters $[\varphi_{(a)}(\tau,\sigma)]$ parametrizing point-dependent Wigner boosts for timelike Poincaré orbits. Putting these variables in the ADM action for metric gravity [40] (with the 3-metric on Σ_τ expressed in terms of cotriads: ${}^3g_{rs} = {}^3e_{(a)r}\, {}^3e_{(a)s}$ with positive signature), one gets a new action depending only on lapse, shifts and cotriads, but not on the boost parameters (therefore, there is no need to use Schwinger's time gauge). There are 10 primary and 4 secondary first class constraints and a weakly vanishing canonical Hamiltonian containing the secondary constraints like in ADM metric gravity [40]. Besides the 3 constraints associated with the vanishing Lorentz boost momenta (Abelianization of boosts), there are 4 constraints saying that the momenta associated with lapse and shifts vanish, 3 constraints describing rotations, 3 constraints generating space-diffeomorphisms on the cotriads induced by those $(Diff\, \Sigma_\tau)$ on Σ_τ (a linear combination of supermomentum constraints and of the rotation ones; a different combination of these constraints generates SO(3) Gauss law constraints for the momenta ${}^3\tilde{\pi}^r_{(a)}$ conjugated to cotriads with the covariant derivative built with the spin connection) and one superhamiltonian constraint. The six constraints connected with Lorentz boosts and rotations replace the constraints satisfying the Lorentz algebra in the older formulations. The boost parameters $\varphi_{(a)}(\tau,\sigma)$ and the three angles $\alpha_{(a)}(\tau,\sigma)$ hidden in the cotriads are the extra variables of tetrad gravity with respect to metric gravity: they allow a Hamiltonian description of the congruences of timelike accelerated observers used in the formulation of gravitomagnetism[48].

It turns out that with the technology developed for Yang-Mills theory, one can Abelianize the 3 rotation constraints and then also the space-diffeomorphism constraints so that we can arrive at a total of 13 Abelianized first class constraints. In the Abelianization of the rotation constraints one needs the Green function of the 3-dimensional covariant derivative containing the spin connection, well defined only if there is no Gribov ambiguity in the SO(3)-frame bundle and no isometry of the Riemannian 3-manifold $(\Sigma_\tau, {}^3g)$. The Green function is similar to the Yang-Mills one for a principal SO(3)-bundle [13], but, instead of the Dirac distribution for the Green function of the flat divergence, it contains the Synge-DeWitt bitensor [49] defining the tangent in one endpoint of the geodesic arc connecting two points (which reduces to the Dirac distribution only locally in normal coordinates). Moreover, the definition of the Green function now requires the geodesic exponential map.

In the resulting quasi-Shanmugadhasan canonical basis, the original cotriad can be expressed in closed form in terms of 3 rotation angles, 3 diffeomorphism-parameters and a reduced cotriad depending only on 3 independent variables (they are Dirac's observables with respect to 13 of the 14 first class constraints) and with their conjugate momenta, still subject to the reduced form of the superhamiltonian constrain: this is the phase space over the superspace of 3-geometries[50].

Till now no coordinate condition[51] has been imposed. It turns out that these conditions are hidden in the choice of how to parametrize the reduced cotriads in terms of three independent functions. The simplest parametrization (the only one studied till now) corresponds to choose a system of global 3-orthogonal coordinates on Σ_τ, in which the 3-metric is diagonal. With a further canonical transformation on the reduced cotriads and conjugate momenta, one arrives at a canonical basis containing the conformal factor $\phi(\tau,\sigma) = e^{q(\tau,\sigma)/2}$ of the 3-geometry and its conjugate momentum $\rho(\tau,\sigma)$ plus two other pairs of conjugate canonical variables $r_{\bar{a}}(\tau,\sigma)$, $\pi_{\bar{a}}(\tau,\sigma)$, $\bar{a} = 1,2$. The reduced superhamiltonian constraint, expressed in terms of these variables, turns out to be an integro-differential equation for the conformal factor (reduced Lichnerowicz equation) whose conjugate momentum is, therefore, the last gauge variable. If we replace the gauge fixing of the Lichnerowicz[52] and York[53] approach [namely the vanishing of the trace of the extrinsic curvature of Σ_τ, ${}^3K(\tau,\sigma) \approx 0$, also named the internal extrinsic York time] with the natural one $\rho(\tau,\sigma) \approx 0$ and we go to Dirac brackets, we find that $r_{\bar{a}}(\tau,\sigma)$, $\pi_{\bar{a}}(\tau,\sigma)$ are the canonical basis for the physical degrees of freedom or Dirac's observables of the gravitational field in the 3-orthogonal gauges. Let us remark that the functional form of the non-tensorial objects $r_{\bar{a}}$, $\pi_{\bar{a}}$, depends on the chosen coordinate condition.

The next step is to find the physical Hamiltonian for them and to solve the deparametrization problem. If we wish to arrive at the soldering of tetrad gravity with matter and parametrized Minkowski formulation for the same matter, we must require that the lapse and shift functions of tetrad gravity must agree asymptotically with the flat lapse and shift functions, which, however, are unambigously defined only on Minkowski spacelike hyperplanes as we have seen. In metric ADM gravity the canonical Hamiltonian is $H_{(c)ADM} = \int d^3\sigma [N\tilde{\mathcal{H}} + N_r\tilde{\mathcal{H}}^r](\tau,\sigma) \approx 0$, where $\tilde{\mathcal{H}}(\tau,\sigma) \approx 0$ and $\tilde{\mathcal{H}}^r(\tau,\sigma) \approx 0$ are the superhamiltonian and supermomentum constraints. It is differentiable and finite only for suitable $N(\tau,\sigma) = n(\tau,\sigma) \rightarrow_{|\sigma|\rightarrow\infty} 0$, $N_r(\tau,\sigma) = n_r(\tau,\sigma) \rightarrow_{|\sigma|\rightarrow\infty} 0$ defined by Beig and Ó'Murchadha[41] in suitable asymptotic coordinate systems. For more general lapse and shift functions one must add a surface term [50] to $H_{(c)ADM}$, which contains the "strong" Poincaré charges [40] P^A_{ADM}, J^{AB}_{ADM} [they are conserved and gauge invariant surface integrals]. To have well defined asymptotic Poincaré charges at spatial infinity[40,41] one needs: i) the selection of a class of coordinates systems for Σ_τ asymptotic to flat coordinates; ii) the choice of a class of Hamiltonian boundary conditions for the fields in these coordinate systems [all the fields must belong to some functional space of the type of the weighted Sobolev spaces]; iii) a definition of the Hamiltonian group $\mathcal{G}$ of gauge transformations (and in particular of proper gauge transformations) with a well defined limit at spatial infinity so to respect i) and ii). The scheme is the same needed to define the non-Abelian charges in Yang-Mills theory[13]. The delicate point is to be able to exclude supertranslations[43], because the presence of these extra asymptotic charges leads to the replacement of the asymptotic Poincaré group with the infinite-dimensional spi group[42] of asymptotic symmetries, which does not allow the definition of the Poincaré spin due to the absence of the Pauli-Lubanski Casimir. This can be done with suitable boundary conditions (in particular all the fields and gauge transformations must have direction independent limits at spatial infinity) respecting the "parity conditions" of Beig and Ó'Murchadha[41].

Let us then remark that in Ref.[54] and in the book in Ref.[1] (see also Ref.[41]), Dirac introduced asymptotic Minkowski rectangular coordinates

$$z^{(\mu)}_{(\infty)}(\tau,\boldsymbol{\sigma}) = x^{(\mu)}_{(\infty)}(\tau) + b^{(\mu)}_{(\infty)\,r}(\tau)\sigma^r$$

in M^4 at spatial infinity $S_\infty = \cup_\tau S^2_{\tau,\infty}$ For each value of τ, the coordinates $x^{(\mu)}_{(\infty)}(\tau)$ labels a point, near spatial infinity chosen as origin of Σ_τ. On it there is a flat tetrad $b^{(\mu)}_{(\infty)\,A}(\tau) = \{\, l^{(\mu)}_{(\infty)} = b^{(\mu)}_{(\infty)\,\tau} = \epsilon^{(\mu)}{}_{(\alpha)(\beta)(\gamma)} b^{(\alpha)}_{(\infty)\,1}(\tau) b^{(\beta)}_{(\infty)\,2}(\tau) b^{(\gamma)}_{(\infty)\,3}(\tau);\ b^{(\mu)}_{(\infty)\,r}(\tau)\,\}$, with $l^{(\mu)}_{(\infty)}$ τ-independent. There will be transformation coefficients $b^\mu_A(\tau,\boldsymbol{\sigma})$ from the holonomic adapted coordinates $\sigma^A = (\tau,\sigma^r)$ to coordinates $x^\mu = z^\mu(\sigma^A)$ in an atlas of M^4, such that in a chart at spatial infinity one has $z^\mu(\tau,\boldsymbol{\sigma}) = \delta^\mu_{(\mu)} z^{(\mu)}(\tau,\boldsymbol{\sigma})$ and $b^\mu_A(\tau,\boldsymbol{\sigma}) = \delta^\mu_{(\mu)} b^{(\mu)}_{(\infty)A}(\tau)$ [for $r \rightarrow \infty$ one has ${}^4g_{\mu\nu} \rightarrow \delta^{(\mu)}_\mu \delta^{(\nu)}_\nu {}^4\eta_{(\mu)(\nu)}$ and ${}^4g_{AB} = b^\mu_A\, {}^4g_{\mu\nu} b^\nu_B \rightarrow b^{(\mu)}_{(\infty)A}\, {}^4\eta_{(\mu)(\nu)} b^{(\nu)}_{(\infty)B} = {}^4\eta_{AB}$].

Dirac[54] and, then, Regge and Teitelboim[41] proposed that the asymptotic Minkowski rectangular coordinates $z^{(\mu)}_{(\infty)}(\tau,\boldsymbol{\sigma}) = x^{(\mu)}_{(\infty)}(\tau) + b^{(\mu)}_{(\infty)r}(\tau)\sigma^r$ should define 10 new independent degrees of freedom at the spatial boundary S_∞, as it happens for Minkowski parametrized theories[20] when restricted to spacelike hyperplanes [defined by $z^{(\mu)}(\tau,\boldsymbol{\sigma}) \approx x^{(\mu)}_s(\tau) + b^{(\mu)}_r(\tau)\sigma^r$]; then, 10 conjugate momenta should exist. These 20 extra variables of the Dirac proposal can be put in the form: $x^{(\mu)}_{(\infty)}(\tau)$, $p^{(\mu)}_{(\infty)}$, $b^{(\mu)}_{(\infty)A}(\tau)$ [with $b^{(\mu)}_{(\infty)\tau} = l^{(\mu)}_{(\infty)}$ τ-independent], $S^{(\mu)(\nu)}_{(\infty)}$, with Dirac brackets implying the orthonormality constraints

$$b^{(\mu)}_{(\infty)A}\, {}^4\eta_{(\mu)(\nu)} b^{(\nu)}_{(\infty)B} = {}^4\eta_{AB}$$

[so that $p^{(\mu)}_{(\infty)}$ and $J^{(\mu)(\nu)}_{(\infty)} = x^{(\mu)}_{(\infty)} p^{(\nu)}_{(\infty)} - x^{(\nu)}_{(\infty)} p^{(\mu)}_{(\infty)} + S^{(\mu)(\nu)}_{(\infty)}$ satisfy a Poincaré algebra]. In analogy with Minkowski parametrized theories restricted to spacelike hyperplanes, one expects to have 10 extra first class constraints of the type

$$p^{(\mu)}_{(\infty)} - P^{(\mu)}_{ADM} \approx 0,\ S^{(\mu)(\nu)}_{(\infty)} - S^{(\mu)(\nu)}_{ADM} \approx 0$$

with $P^{(\mu)}_{ADM}$, $S^{(\mu)(\nu)}_{ADM}$ related to the ADM Poincaré charges P^A_{ADM}, J^{AB}_{ADM}. The origin $x^{(\mu)}_{(\infty)}$ is going to play the role of a decoupled observer with his parametrized clock.

Let us remark that if we replace $p^{(\mu)}_{(\infty)}$ and $S^{(\mu)(\nu)}_{(\infty)}$, whose Poisson algebra is the direct sum of an Abelian algebra of translations and of a Lorentz algebra, with the new variables (with holonomic indices with respect to Σ_τ) $p^A_{(\infty)} = b^A_{(\infty)(\mu)} p^{(\mu)}_{(\infty)}$, $J^{AB}_{(\infty)} = b^A_{(\infty)(\mu)} b^B_{(\infty)(\nu)} S^{(\mu)(\nu)}_{(\infty)}$ $[\neq b^A_{(\infty)(\mu)} b^B_{(\infty)(\nu)} J^{(\mu)(\nu)}_{(\infty)}]$, the Poisson brackets for $p^{(\mu)}_{(\infty)}$, $b^{(\mu)}_{(\infty)A}$, $S^{(\mu)(\nu)}_{(\infty)}$ imply that $p^A_{(\infty)}$, $J^{AB}_{(\infty)}$ satisfy a Poincaré algebra. This implies that the Poincaré generators P^A_{ADM}, J^{AB}_{ADM} define in the asymptotic Dirac rectangular coordinates a momentum $P^{(\mu)}_{ADM}$ and only an ADM spin tensor $S^{(\mu)(\nu)}_{ADM}$ [to define an angular momentum tensor $J^{(\mu)(\nu)}_{ADM}$ one should find a "center of mass of the gravitational field" $X^{(\mu)}_{ADM}[{}^3g, {}^3\tilde{\Pi}]$ (see Ref.[23] for the Klein-Gordon case) conjugate to $P^{(\mu)}_{ADM}$, so that $J^{(\mu)(\nu)}_{ADM} = X^{(\mu)}_{ADM} P^{(\nu)}_{ADM} - X^{(\nu)}_{ADM} P^{(\mu)}_{ADM} + S^{(\mu)(\nu)}_{ADM}$].

The following splitting of the lapse and shift functions and the following set of boundary conditions fulfill all the previous requirements [soldering with the lapse and shift functions on Minkowski hyperplanes; absence of supertranslations [strictly speaking one gets $P^r_{ADM} = 0$ due to the parity conditions; $r = |\sigma|$]

$$
\begin{aligned}
&{}^3g_{rs}(\tau,\sigma) \rightarrow_{r\rightarrow\infty} (1+\tfrac{M}{r})\delta_{rs} + {}^3h_{rs}(\tau,\sigma) = (1+\tfrac{M}{r})\delta_{rs} + o_4(r^{-3/2}),\\
&{}^3\tilde{\Pi}^{rs}(\tau,\sigma) \rightarrow_{r\rightarrow\infty} {}^3k^{rs}(\tau,\sigma) = o_3(r^{-5/2}),
\end{aligned}
$$

$$
\begin{aligned}
&N(\tau,\sigma) = N_{(as)}(\tau,\sigma) + n(\tau,\sigma), \qquad n(\tau,\sigma) = O(r^{-(3+\epsilon)}),\\
&N_r(\tau,\sigma) = N_{(as)r}(\tau,\sigma) + n_r(\tau,\sigma), \qquad n_r(\tau,\sigma) = O(r^{-\epsilon}),\\
&N_{(as)A}(\tau,\sigma) \stackrel{def}{=} (N_{(as)}\,;\,N_{(as)r})(\tau,\sigma) = -\tilde{\lambda}_A(\tau) - \tfrac{1}{2}\tilde{\lambda}_{As}(\tau)\sigma^s,
\end{aligned}
$$

$$
\Rightarrow \; {}^3e_{(a)r}(\tau,\sigma) = (1+\tfrac{M}{2r})\delta_{(a)r} + o_4(r^{-3/2}),
$$

with ${}^3h_{rs}(\tau,-\sigma) = {}^3h_{rs}(\tau,\sigma)$, ${}^3k^{rs}(\tau,-\sigma) = -{}^3k^{rs}(\tau,\sigma)$; here ${}^3\tilde{\Pi}^{rs}(\tau,\sigma)$ is the momentum conjugate to the 3-metric ${}^3g_{rs}(\tau,\sigma)$ in ADM metric gravity.

These boundary conditions identify the class of spacetimes of Christodoulou and Klainermann[55] (they are near to Minkowski spacetime in a norm sense, contain gravitational radiation but evade the singularity theorems, because they do not satisfy the hypothesis of conformal completion to get the possibility to put control on the large time development of the solutions of Einstein's equations). These spacetimes also satisfy the rest-frame condition $P^r_{ADM} = 0$ (this requires $\tilde{\lambda}_{Ar}(\tau) = 0$ like for Wigner hyperplanes in parametrized Minkowski theories) and have vanishing shift functions (but non trivial lapse function).

After the addition of the surface term, the resulting canonical and Dirac Hamiltonians of ADM metric gravity are

$$
\begin{aligned}
&H_{(c)ADM} = \int d^3\sigma[(N_{(as)}+n)\tilde{\mathcal{H}} + (N_{(as)r}+n_r)\,{}^3\tilde{\mathcal{H}}^r](\tau,\sigma) \mapsto\\
&\mapsto H'_{(c)ADM} = \int d^3\sigma[(N_{(as)}+n)\tilde{\mathcal{H}} + (N_{(as)r}+n_r)\,{}^3\tilde{\mathcal{H}}^r](\tau,\sigma)+\\
&+\tilde{\lambda}_A(\tau)P^A_{ADM} + \tilde{\lambda}_{AB}(\tau)J^{AB}_{ADM} =\\
&= \int d^3\sigma[n\tilde{\mathcal{H}} + n_r\,{}^3\tilde{\mathcal{H}}^r](\tau,\sigma) + \tilde{\lambda}_A(\tau)\hat{P}^A_{ADM} + \tilde{\lambda}_{AB}(\tau)\hat{J}^{AB}_{ADM} \approx\\
&\approx \tilde{\lambda}_A(\tau)\hat{P}^A_{ADM} + \tilde{\lambda}_{AB}(\tau)\hat{J}^{AB}_{ADM},
\end{aligned}
$$

with the "weak conserved improper charges" $\hat{P}^A_{ADM}$, $\hat{J}^{AB}_{ADM}$ [they are volume integrals differing from the weak charges by terms proportional to integrals of the constraints]. The previous splitting implies to replace the variables $N(\tau,\vec{\sigma})$, $N_r(\tau,\vec{\sigma})$ with the ones $\tilde{\lambda}_A(\tau)$, $\tilde{\lambda}_{AB}(\tau)=-\tilde{\lambda}_{BA}(\tau)$, $n(\tau,\vec{\sigma})$, $n_r(\tau,\vec{\sigma})$ [with conjugate momenta $\tilde{\pi}^A(\tau)$, $\tilde{\pi}^{AB}(\tau)=-\tilde{\pi}^{BA}(\tau)$, $\tilde{\pi}^n(\tau,\vec{\sigma})$, $\tilde{\pi}^r_{\vec{n}}(\tau,\vec{\sigma})$] in the ADM theory.

With these assumptions one has the following form of the line element (also its form in tetrad gravity is given)

$$ds^2=\epsilon([N_{(as)}+n]^2-[N_{(as)r}+n_r]^3e^r_{(a)}\,{}^3e^s_{(a)}[N_{(as)s}+n_s])(d\tau)^2-$$
$$-2\epsilon[N_{(as)r}+n_r]d\tau d\sigma^r-\epsilon\,{}^3e_{(a)r}\,{}^3e_{(a)s}d\sigma^r d\sigma^s.$$

The final suggestion of Dirac is to modify ADM metric gravity in the following way:
i) add the 10 new primary constraints $p^A_{(\infty)}-\hat{P}^A_{ADM}\approx 0$, $J^{AB}_{(\infty)}-\hat{J}^{AB}_{ADM}\approx 0$, where $p^A_{(\infty)}=b^A_{(\infty)(\mu)}p^{(\mu)}_{(\infty)}$, $J^{AB}_{(\infty)}=b^A_{(\infty)(\mu)}b^B_{(\infty)(\nu)}S^{(\mu)(\nu)}_{(\infty)}$;
ii) consider $\tilde{\lambda}_A(\tau)$, $\tilde{\lambda}_{AB}(\tau)$, as Dirac multipliers for these 10 new primary constraints, and not as configurational (arbitrary gauge) variables coming from the lapse and shift functions [so that there are no conjugate (vanishing) momenta $\tilde{\pi}^A(\tau)$, $\tilde{\pi}^{AB}(\tau)$], in the assumed Dirac Hamiltonian [it is finite and differentiable]

$$H_{(D)ADM}=\int d^3\sigma[n\tilde{\mathcal{H}}+n_r\tilde{\mathcal{H}}^r+\lambda_n\tilde{\pi}^n+\lambda^{\vec{n}}_r\tilde{\pi}^r_{\vec{n}}](\tau,\vec{\sigma})-$$
$$-\tilde{\lambda}_A(\tau)[p^A_{(\infty)}-\hat{P}^A_{ADM}]-\tilde{\lambda}_{AB}(\tau)[J^{AB}_{(\infty)}-\hat{J}^{AB}_{ADM}]\approx 0,$$

The reduced phase space is still the ADM one: on the ADM variables there are only the secondary first class constraints $\tilde{\mathcal{H}}(\tau,\vec{\sigma})\approx 0$, $\tilde{\mathcal{H}}^r(\tau,\vec{\sigma})\approx 0$ [generators of proper gauge transformations], because the other first class constraints $p^A_{(\infty)}-\hat{P}^A_{ADM}\approx 0$, $J^{AB}_{(\infty)}-\hat{J}^{AB}_{ADM}\approx 0$ do not generate improper gauge transformations but eliminate 10 of the extra 20 variables.

In this modified ADM metric gravity, one has restricted the 3+1 splittings of M^4 to foliations whose leaves Σ_τ tend to Minkowski spacelike hyperplanes asymptotically at spatial infinity in a direction independent way. Therefore, these $\Sigma^{'}_\tau$ should be determined by the 10 degrees of freedom $x^{(\mu)}_{(\infty)}(\tau)$, $b^{(\mu)}_{(\infty)A}(\tau)$, like it happens for flat spacelike hyperplanes: this means that it must be possible to define a "parallel transport" of the asymptotic tetrads $b^{(\mu)}_{(\infty)A}(\tau)$ to get well defined tetrads in each point of $\Sigma^{'}_\tau$. While it is not yet clear whether this can be done for $\tilde{\lambda}_{AB}(\tau)\neq 0$, there is a solution for $\tilde{\lambda}_{AB}(\tau)=0$. This case corresponds to go to the Wigner-like hypersurfaces [the analogue of the Minkowski Wigner hyperplanes with the asymptotic normal $l^{(\mu)}_{(\infty)}=l^{(\mu)}_{(\infty)\Sigma}$ parallel to $\hat{P}^{(\mu)}_{ADM}$]. Following the same procedure defined for Minkowski spacetime, one gets $\bar{S}^{rs}_{(\infty)}\equiv\hat{J}^{rs}_{ADM}$ [see Ref.[20] for the definition of $\bar{S}^{AB}_{(\infty)}$], $\tilde{\lambda}_{AB}(\tau)=0$ and $-\tilde{\lambda}_A(\tau)[p^A_{(\infty)}-\hat{P}^A_{ADM}]=-\tilde{\lambda}_\tau(\tau)[\epsilon_{(\infty)}-\hat{P}^\tau_{ADM}]+\tilde{\lambda}_r(\tau)\hat{P}^r_{ADM}$ [$\epsilon_{(\infty)}=\sqrt{p^2_{(\infty)}}$], so that the final form of these four surviving constraints is ($P^r_{ADM}=0$ implies $\hat{P}^r_{ADM}\approx 0$; $M_{ADM}=\sqrt{\hat{P}^2_{ADM}}\approx \hat{P}^\tau_{ADM}$ is the ADM mass of the universe)

$$\epsilon_{(\infty)}-\hat{P}^\tau_{ADM}\approx 0,\ \hat{P}^r_{ADM}\approx 0.$$

On this subclass of foliations [whose leaves $\Sigma^{(WSW)}_\tau$ will be called Wigner-Sen-Witten hypersurfaces; they define the intrinsic asymptotic rest frame of the gravitational field] one can introduce a parallel transport by using the interpretation of Ref.[56] of the Witten spinorial method of demonstrating the positivity of the ADM energy [57]. Let us consider the Sen-Witten connection [58,57] restricted to $\Sigma^{(WSW)}_\tau$ (it depends on the trace of the extrinsic curvature of $\Sigma^{(WSW)}_\tau$) and the spinorial Sen-Witten equation associated with it. As shown in Ref.[59], this spinorial equation can be rephrased as an equation whose solution determines (in a surface dependent dynamical way) a tetrad in each point of $\Sigma^{(WSW)}_\tau$ once it is given at spatial infinity (again this requires a direction independent limit). Therefore, at spatial infinity there is a privileged congruence of timelike observers, which replaces the concept of "fixed stars" in the study of the precessional effects of gravitomagnetism on gyroscopes and whose connection with the definition of post-Newtonian coordinates has still to be explored.

On the Wigner-Sen-Witten hypersurfaces the spatial indices have become spin-1 Wigner indices [they transform with Wigner rotations under asymptotic Lorentz transformations]. As said for parametrized theories in Minkowski spacetime, in this special gauge 3 degrees of freedom of the gravitational field [an internal 3-center-of-mass variable $\boldsymbol{\sigma}_{ADM}[{}^3g,{}^3\tilde{\Pi}]$ inside the Wigner-Sen-Witten hypersurface] become gauge variables, while $\tilde{x}^{(\mu)}_{(\infty)}$ [the canonical non covariant variable replacing $x^{(\mu)}_{(\infty)}$] becomes a decoupled observer with his "point particle clock" [31,32] near spatial infinity. Since the positivity theorems for the ADM energy imply that one has only timelike or lightlike orbits of the asymptotic Poincaré group, the restriction to universes with timelike ADM 4-momentum allows to define the Möller radius $\rho_{AMD}=\sqrt{-\hat{W}^2_{ADM}/\hat{P}^2_{ADM}}$ from the asymptotic Poincaré Casimirs $\hat{P}^2_{ADM}$, $\hat{W}^2_{ADM}$.

By going from $\tilde{x}^{(\mu)}_{(\infty)}$, $p^{(\mu)}_{(\infty)}$, to the canonical basis $T_{(\infty)}=p_{(\infty)(\mu)}\tilde{x}^{(\mu)}_{(\infty)}/\epsilon_{(\infty)}=p_{(\infty)(\mu)}x^{(\mu)}_{(\infty)}/\epsilon_{(\infty)}$, $\epsilon_{(\infty)}$, $z^{(i)}_{(\infty)}=\epsilon_{(\infty)}(\tilde{x}^{(i)}_{(\infty)}-p^{(i)}_{(\infty)}\tilde{x}^{(o)}_{(\infty)}/p^{(o)}_{(\infty)})$, $k^{(i)}_{(\infty)}=p^{(i)}_{(\infty)}/\epsilon_{(\infty)}=u^{(i)}(p^{(\rho)}_{(\infty)})$, like in the flat case one finds that the final reduction requires the gauge-fixings $T_{(\infty)}-\tau\approx 0$ and $\sigma^r_{ADM}\approx 0$, where $\sigma^r=\sigma^r_{ADM}$ is a variable representing the "internal center of mass" of the 3-metric of the slice Σ_τ of the asymptotically flat spacetime M^4. Since $\{T_{(\infty)},\epsilon_{(\infty)}\}=-\epsilon$, with the gauge fixing $T_{(\infty)}-\tau\approx 0$ one gets $\tilde{\lambda}_\tau(\tau)\approx\epsilon$, and the final Dirac Hamiltonian is $H_D=M_{ADM}+\tilde{\lambda}_r(\tau)\hat{P}^r_{ADM}$ with M_{ADM} the natural physical Hamiltonian to reintroduce an evolution in the "mathematical" $T_{(\infty)}\equiv\tau$: namely in the rest-frame time identified with the parameter τ labelling the leaves $\Sigma^{(WSW)}_\tau$ of the foliation of M^4. Physical times (atomic clocks, ephemeridis time...) must be put in a local 1-1 correspondence with this "mathematical" time. This point of view excludes any Wheeler-DeWitt interpretation of an internal time (like the extrinsic York one or the WKB times), which is used in closed universes of the Einstein-Wheeler type.

All this construction holds also in our formulation of tetrad gravity (since it uses the ADM action) and in its canonically reduced form in the 3-orthogonal gauges. The final physical Hamiltonian of tetrad gravity for the physical gravitational field is the reduced volume form of the ADM energy $\hat{P}^{\tau}_{ADM}[r_{\bar{a}}.\pi_{\bar{a}}, \phi(r_{\bar{a}}, \pi_{\bar{a}})]$ with the conformal factor ϕ solution of the reduced Lichnerowicz equation in the 3-orthogonal gauge with $\rho(\tau, \sigma) \approx 0$. The Hamilton-Dirac equations generated by this Hamiltonian for $r_{\bar{a}}$, $\pi_{\bar{a}}$ generate the pair of second order equations in normal form for $r_{\bar{a}}$ hidden in the Einstein equations in this particular gauge.

Let us compare the standard generally covariant formulation of gravity based on the Hilbert action with its invariance under $Diff\, M^4$ with the ADM Hamiltonian formulation. Regarding the 10 Einstein equations of the standard approach, the Bianchi identities imply that four equations are linearly dependent on the other six ones and their gradients. Moreover, the four combinations of Einstein's equations projectable to phase space (where they become the secondary first class superhamitonian and supermomentum constraints of canonical metric gravity) are independent from the accelerations being restrictions on the Cauchy data. As a consequence the Einstein equations have solutions, in which the ten components ${}^4g_{\mu\nu}$ of the 4-metric depend on only two truly dynamical degrees of freedom (defining the physical gravitational field) and on eight undetermined degrees of freedom. This transition from the ten components ${}^4g_{\mu\nu}$ of the tensor ${}^4g$ in some atlas of M^4 to the 2 (deterministic)+8 (undetermined) degrees of freedom breaks general covariance, because these quantities are neither tensors nor invariants under diffeomorphisms (their functional form is atlas dependent).

Since the Hilbert action is invariant under $Diff\, M^4$, one usually says that a "dynamical gravitational field" is a 4-geometry over M^4, namely an equivalence class of spacetimes $(M^4, {}^4g)$, solution of Einstein's equations, modulo $Diff\, M^4$. See, however, the interpretational problems about what is observable in general relativity for instance in Refs.[34,35], in particular the facts that at least before the restriction to the solutions of Einstein's equations i) scalars under $Diff\, M^4$, like ${}^4R$, are not Dirac's observables but gauge dependent quantities; ii) the functional form of ${}^4g_{\mu\nu}$ in terms of the physical gravitational field and, therefore, the angle and distance properties of material bodies and the standard procedures of defining measures of length and time based on the line element ds^2, are gauge dependent.

Instead in the ADM formalism with the extra notion of 3+1 splittings of M^4, the (tetrad) metric ADM action (differing from the Hilbert one by a surface term) is quasi-invariant under the (14) 8 types of gauge transformations which are the pull-back of the Hamiltonian group $\mathcal{G}$ of gauge transformations, whose generators are the first class constraints of the theory . The Hamiltonian group $\mathcal{G}$ has a subgroup (whose generators are the supermomentum and superhamiltonian constraints) formed by the diffeomorphisms of M^4 adapted to its 3+1 splittings, $Diff\, M^{3+1}$ [it is different from $Diff\, M^4$]. Moreover, the Poisson algebra of the supermomentum and superhamiltonian constraints reflects the embeddability in M^4 of the foliation associated with the 3+1 splitting [60].

Now in tetrad gravity the interpretation of the 14 gauge transformations and of their gauge fixings (it is independent from the presence of matter) is the following [a tetrad in a point of Σ_τ is a local observer]: i) the gauge fixings of the gauge boost parameters associated with the 3 boost constraints and of the gauge angles associated with the 3 rotation constraints are equivalent to choose the congruence of timelike observers to be used as a standard of non rotation; ii) the gauge fixings of the 3 gauge parameters associated with the passive space diffeomorphisms [$Diff\,\Sigma_\tau$; change of coordinates charts] are equivalent to a fixation of 3 standards of length by means of a choice of a coordinate system on Σ_τ [the measuring apparatus (the "rods") should be defined in terms of Dirac's observables for some kind of matter, after its introduction into the theory]; iii) according to constraint theory the choice of 3-coordinates on Σ_τ induces the gauge fixings of the 3 shift functions [i.e. of ${}^4g_{oi}$], whose gauge nature is connected with the "conventionality of simultaneity" [61] [therefore, the gauge fixings are equivalent to a choice of synchrinization of clocks and, as a consequence, to a statement about the isotropy or anisotropy of the velocity of light in that gauge]; iv) the gauge fixing on the the momentum $\rho(\tau,\sigma)$ conjugate to the conformal factor of the 3-metric [this gauge variable is the source of the gauge dependence of 4-tensors and of the scalars under $Diff\,M^4$, together with the gradients of the lapse and shift functions] is a nonlocal statement about the extrinsic curvature of the leaves Σ_τ of the given 3+1 splitting of M^4; since the superhamiltonian constraint produces normal deformations of Σ_τ [60] and, therefore, transforms a 3+1 splitting of M^4 into another one (the ADM formulation is independent from the choice of the 3+1 splitting), this gauge fixing is equivalent to the choice of a particular 3+1 splitting; v) the previous gauge fixing induces the gauge fixing of the lapse function (which determines the packing of the leaves Σ_τ in the chosen 3+1 splitting) and, therefore, is equivalent to the fixation of a standard of proper time [again "clocks" should be built with the Dirac's observables of some kind of matter].

In the Hamiltonian formalism it is natural to define a "Hamiltonian kinematical gravitational field" as the equivalence class of spacetimes modulo the Hamiltonian group $\mathcal{G}$, and different members of the equivalence class have in general different 4-Riemann tensors [these equivalence classes are connected with the conformal 3-geometries of the Lichnerowicz-York approach and contain different gauge-related 4-geometries]. Then, a "Hamiltonian dynamical gravitational field" is defined as a Hamiltonian kinematical gravitational fields which is solution of the Hamilton-Dirac equations generated by the weak ADM energy $\hat{P}^\tau_{ADM}$. Since the Hilbert and ADM actions, even if they have different local symmetries and invariances, both generate the same Einstein equations, the equivalence classes of the "Hamiltonian dynamical gravitational fields" and of the standard "dynamical gravitational fields" (a 4-geometry solution of Einstein's equations) coincide. Indeed, on the solutions of Einstein's equations the gauge transformations generated by the superhamiltonian constraint (normal deformations of Σ_τ) and those generated by the canonical momenta of the lapse and shift functions together with the Σ_τ diffeomorphisms generated by the supermomentum constraints are restricted by the Jacobi equations associated to Einstein's equations to be those Noether symmetries of the ADM action which are also dynamical symmetries of the Hamilton equations and therefore they are a subset of the spacetime diffeomorphisms $Diff\, M^4$ (all of which are dynamical symmetries of Einstein's equations).

The 3-orthogonal gauges of tetrad gravity are the equivalent of the Coulomb gauge in classical electrodynamics (like the harmonic gauge is the equivalent of the Lorentz gauge). Only after a complete gauge fixing the 4-tensors and the scalars under $Diff\, M^4$ become measurable quantities (like the electromagnetic vector potential in the Coulomb gauge): an experimental laboratory does correspond by definition to a completely fixed gauge. At this stage it becomes acceptable the proposal of Komar[62] and Bergmann[47] of identifying the points of a spacetime $(M^4, {}^4g)$, solution of the Einstein's equations in absence of matter, in a way invariant under spacetime diffeomorphisms, by using four bilinears and trilinears in the Weyl tensors, scalar under $Diff\, M^4$ and called "individuating fields"(see also Refs.[34,35]), which do not depend on the lapse and shift functions (but only on the gauge variables corresponding to the 3-coordinates on Σ_τ and to the momentum conjugate to the conformal factor of the 3-metric, so that these fields carry the information on the choice of the 3-coordinates and of a generalized extrinsic time), to build "physical 4-coordinates" (in each completely fixed gauge they depend only on the two canonical pairs of Dirac's observables of the gravitational field), justifying a posteriori the standard measurement theory presented in all textbooks on general relativity, which presuppones the individuation of spacetime points.

Let us remember that Bergmann[47] made the following critique of general covariance: it would be desirable to restrict the group of coordinate transformations (spacetime diffeomorphisms) in such a way that it could contain an invariant subgroup describing the coordinate transformations that change the frame of reference of an outside observer (these transformations could be called Lorentz transformations; see also the comments in Ref.[63] on the asymptotic behaviour of coordinate transformations); the remaining coordinate transformations would be like the gauge transformations of electromagnetism. This is what we have done. In this way "preferred' coordinate systems will emerge (the WSW hypersurfaces with their preferred congruences of timelike observers whose 4-velocity becomes asymptotically normal to $\Sigma_\tau^{(WSW)}$ at spatial infinity), which, as said by Bergmann, are not "flat": while the inertial coordinates are determined experimentally by the observation of trajectories of force-free bodies, these intrinsic coordinates can be determined only by much more elaborate experiments (probably with gyroscopes), since they depend, at least, on the inhomogeneities of the ambient gravitational fields.

Since in the 3-orthogonal gauges we have the physical canonical basis $r_{\bar a}$, $\pi_{\bar a}$, it is possible, but only in absence of matter, to define "void spacetimes" as the equivalence class of spacetimes "without gravitational field", whose members in the 3-orthogonal gauges are obtained by adding by hand the second class constraints $r_{\bar a}(\tau,\boldsymbol{\sigma}) \approx 0$, $\pi_{\bar a}(\tau,\boldsymbol{\sigma}) \approx 0$ [one gets $\phi(\tau,\boldsymbol{\sigma}) = 1$ as the relevant solution of the reduced Lichnerowicz equation] and, in particular, their Poincaré charges vanish (this corresponds to the exceptional $p^{(\mu)} = 0$ orbit of the Poincaré group and shows the peculiarity of these solutions with zero ADM mass). It is expected that the void spacetimes can be defined in a gauge-independent way by adding to the ADM action the requirement that the leaves Σ_τ of the 3+1 splitting be 3-conformally flat, namely that the Cotton-York 3-conformal tensor vanishes. The members of this equivalence class (the extension to general relativity of the Galilean non inertial coordinate systems with their Newtonian inertial forces) are gauge equivalent to Minkowski spacetime with Cartesian coordinates and it is expected that they describe pure acceleration effects without physical gravitational field (no tidal effects).

If we add [39] to the tetrad ADM action the action for N scalar particles with positive energy in the form of Ref.[20] [where it was given on arbitrary Minkowski spacelike hypersurfaces], the only constraints which are modified are the superhamiltonian one, which gets a dependence on the matter energy density $\mathcal{M}(\tau,\sigma)$, and the 3 space diffeomorphism ones, which get a dependence on the matter momentum density $\mathcal{M}_r(\tau,\sigma)$. The canonical reduction and the determination of the Dirac observables can be done like in absence of matter. However, the reduced Lichnerowicz equation for the conformal factor of the 3-metric in the 3-orthogonal gauge and with $\rho(\tau,\sigma) \approx 0$ acquires now an extra dependence on $\mathcal{M}(\tau,\sigma)$ and $\mathcal{M}_r(\tau,\sigma)$.

Since, as a preliminary result, we are interested in identifying explicitly the instantaneous action-at-a-distance (Newton-like and gravitomagnetic) potentials among particles hidden in tetrad gravity (like the Coulomb potential is hidden in the electromagnetic gauge potential), we shall make the strong approximation of neglecting the (tidal) effects of the physical gravitational field by putting $r_{\bar a}(\tau,\sigma) \approx 0$, $\pi_{\bar a}(\tau,\sigma) \approx 0$, even if it is not strictly consistent with the Hamilton-Dirac equation (extremely weak gravitational fields). If, furthermore, we develop the conformal factor $\phi(\tau,\sigma)$ in a formal series in the Newton constant G $[\phi = 1 + \sum_{n=1}^{\infty} G^n \phi_n]$, one can find a solution $\phi = 1 + G\phi_1$ at order G (post-Minkowskian approximation) of the reduced Lichnerowicz equation where we put $r_{\bar a} = \pi_{\bar a} = 0$. However, due to a self-energy divergence in ϕ evaluated at the positions $\eta_i(\tau)$ of the particles, one needs to rescale the bare masses to physical ones, $m_i \mapsto \phi^{-2}(\tau,\eta_i(\tau)) m_i^{(phys)}$, and to make a regularization of the type defined in Refs. [64]. Then, the regularized solution for ϕ can be put in the reduced form of the ADM energy, which becomes [$\kappa_i(\tau)$ are the particle momenta conjugate to $\eta_i(\tau)$; $n_{ij} = [\eta_i - \eta_j]/|\eta_i - \eta_j|$]

$$
\begin{aligned}
\hat P^\tau_{ADM} &= \sum_{i=1}^N c\sqrt{m_i^{(phys)2}c^2 + \kappa_i^2(\tau)} - \\
&- \frac{G}{c^2}\sum_{i\neq j} \frac{\sqrt{m_i^{(phys)2}c^2+\kappa_i^2(\tau)}\,\sqrt{m_j^{(phys)2}c^2+\kappa_j^2(\tau)}}{|\eta_i(\tau)-\eta_j(\tau)|} - \\
&- \frac{G}{8c^2}\sum_{i\neq j} \frac{3\kappa_i(\tau)\cdot\kappa_j(\tau) - 5\kappa_i(\tau)\cdot n_{ij}(\tau)\kappa_j(\tau)\cdot n_{ij}(\tau)}{|\eta_i(\tau)-\eta_j(\tau)|} + O(G^2, r_{\bar a}, \pi_{\bar a}).
\end{aligned}
$$

One sees the Newton-like and the gravitomagnetic (in the sense of York) potentials (both of them need regularization) at the post-Minkowskian level (order

G but exact in c) emerging from the tetrad ADM version of Einstein general relativity when we ignore the tidal effects. For G=0 we recover N free scalar particles on the Wigner hyperplane in Minkowski spacetime, as required by de-parametrization. For $c \to \infty$, we get the post-Newtonian Hamiltonian

$$H_{PN} = \sum_{i=1}^{N} \frac{\boldsymbol{\kappa}_i^2(\tau)}{2m_i^{(phys)}} \Big(1 - \frac{2G}{c^2} \sum_{j\neq i} \frac{m_j^{(phys)}}{|\boldsymbol{\eta}_i(\tau)-\boldsymbol{\eta}_j(\tau)|}\Big) - \frac{G}{2} \sum_{i\neq j} \frac{m_i^{(phys)}\, m_j^{(phys)}}{|\boldsymbol{\eta}_i(\tau)-\boldsymbol{\eta}_j(\tau)|} -$$
$$-\frac{G}{8c^2} \sum_{i\neq j} \frac{3\boldsymbol{\kappa}_i(\tau)\cdot\boldsymbol{\kappa}_j(\tau) - 5\boldsymbol{\kappa}_i(\tau)\cdot\boldsymbol{n}_{ij}(\tau)\boldsymbol{\kappa}_j(\tau)\cdot\boldsymbol{n}_{ij}(\tau)}{|\boldsymbol{\eta}_i(\tau)-\boldsymbol{\eta}_j(\tau)|} + O(G^2, r_{\bar{a}}, \pi_{\bar{a}}),$$

which is of the type of the ones implied by the results of Refs.[64,65] [the differences are probably connected with the use of different coordinate systems and with the fact that one has essential singularities on the particle worldlines and the need of regularization].

The main open problems now under investigation are: i) the linearization of the theory in the 3-orthogonal gauges in presence of matter to find the 3-orthogonal gauge description of gravitational waves and to go beyond the previous instantaneous post-Minkowskian approximation at least in the 2-body case relevant for the motion of binaries; ii) the replacement of scalar particles with spinning ones to identify the precessional effects (like the Lense-Thirring one) of gravitomagnetism; iii) the coupling to perfect fluids for the simulation of rotating stars and for the comparison with the post-Newtonian approximations; iv) the coupling of tetrad gravity to the electromagnetic field, to fermion fields and then to the standard model, trying to make to reduction to Dirac's observables in all these cases and to study their post-Minkowskian approximations; v) the quantization of tetrad gravity in the 3-orthogonal gauge with $\rho(\tau, \sigma) \approx 0$ (namely after a complete breaking of general covariance): for each perturbative (in G) solution of the reduced Lichnerowicz equation one defines a Schroedinger equation in τ for a wave functional $\Psi[\tau; r_{\bar{a}}]$ with the associated quantized ADM energy $\hat{P}^{\tau}_{ADM}[r_{\bar{a}}, i\frac{\delta}{\delta r_{\bar{a}}}]$ as Hamiltonian; no problem of physical scalar product is present, but only ordering problems in the Hamiltonian; moreover, one has the Möller radius as a ultraviolet cutoff. Also a comparison with "loop quantum gravity" [66], which respects general covariance but only for fixed lapse and shift functions, has still to be done.

Therefore, a well defined classical stage for a unified description of the four interactions is emerging, even if many aspects have only been clarified at a heuristic level so that a big effort from both mathematical and theoretical physicists is still needed. It will be exciting to see whether in the next years some reasonable quantization picture will develop from this classical framework.

References

1. P.A.M.Dirac, Can.J.Math. **2**, 129 (1950); "Lectures on Quantum Mechanics", Belfer Graduate School of Science, Monographs Series (Yeshiva University, New York, N.Y., 1964).
2. J.L.Anderson and P.G.Bergmann, Phys.Rev. **83**, 1018 (1951). P.G.Bergmann and J.Goldberg, Phys.Rev. **98**, 531 (1955).

3. L.Lusanna, "Towards a Unified Description of the Four Interactions in Terms of Dirac-Bergmann Observables", invited contribution to the book of the Indian National Science Academy forthe International Mathematics Year 2000 AD (HEP-TH/9907081).
4. L.Lusanna, "Solving Gauss' Laws and Searching Dirac Observables for the Four Interactions", talk at the "Second Conf. on Constrained Dynamics and Quantum Gravity", S.Margherita Ligure 1996, eds. V.De Alfaro, J.E.Nelson, G.Bandelloni, A.Blasi, M.Cavaglià and A.T.Filippov, Nucl.Phys. (Proc.Suppl.) **B57**, 13 (1997) (HEP-TH/9702114). "Unified Description and Canonical Reduction to Dirac's Observables of the Four Interactions", talk at the Int.Workshop "New non Perturbative Methods and Quantization on the Light Cone', Les Houches School 1997, eds. P.Grangé, H.C.Pauli, A.Neveu, S.Pinsky and A.Werner (Springer, Berlin, 1998) (HEP-TH/9705154). "The Pseudoclassical Relativistic Quark Model in the Rest-Frame Wigner-Covariant Gauge", talk at the Euroconference QCD97, ed. S.Narison, Montpellier 1997, Nucl.Phys. (Proc. Suppl.) **B64**, 306 (1998).
5. L.Lusanna, Phys.Rep. **185**, 1 (1990). Riv. Nuovo Cimento **14**, n.3, 1 (1991). J.Math.Phys. **31**, 2126 (1990). J.Math.Phys. **31**, 428 (1990). Int.J.Mod.Phys. **A8**, 4193 (1993). M.Chaichian, D.Louis Martinez and L.Lusanna, Ann.Phys.(N.Y.)**232**, 40 (1994).
6. S.Shanmugadhasan, J.Math.Phys. **14**, 677 (1973).
7. A.Barducci, R.Casalbuoni and L.Lusanna, Nuovo Cim. **35A**, 377 (1976).A.Barducci, R.Casalbuoni, D.Dominici and L.Lusanna, Phys.Lett. **100B**, 126 (1981). A.Barducci and L.Lusanna, Nuovo Cim. **77A**, 39 (1983). A.Barducci and L.Lusanna, J.Phys. **A16**, 1993 (1983). A.Barducci, R.Casalbuoni and L.Lusanna, Nuovo Cim. **54A**, 340 (1979). L.Lusanna, Nuovo Cim. **64A**, 65 (1981).
8. Ph.Droz Vincent, Lett.Nuovo Cim. **1**, 839 (1969); Phys.Scr. **2**, 129 (1970); Rep.Math.Phys.**8**, 79 (1975). I.T.Todorov, Report Comm. JINR E2-10125,Dubna 1976 (unpublished); Ann.Inst.H.Poincaré **28A**, 207 (1978). A.Komar, Phys.Rev. **D18**, 1881 and 1887 (1978).
9. H.Sazdjian, Ann.Phys.(N.Y.) **136**, 136 (1981); Phys.Rev. **D33**, 3401 (1986); J.Math.Phys. **28**, 2618 and 1988 (1987), **29**, 1620 (1987); Ann.Phys. (N.Y.) **191**, 52 (1989); in Proc.Int. Symp. "Extended Objects and Bound Systems", eds. O.Hara, S.Ishida and S.Naka (World Scientific, Singapore, 1992). J.Bijtebier and J.Brockaert, Nuovo Cim. **A105**, 351 and 625 (1992); in Proc.Int.Symp. "Extended Objects and Bound Systems", eds. O.Hara, S.Ishida and S.Naka (World Scientific, Singapore, 1992).
10. H.W.Crater and P.Van Alstine, J.Math.Phys. **23**, 1697 (1982); Ann.Phys.(N.Y) **148**, 57 (1983); Phys.Rev.Lett. **53**, 1577 (1984); Phys.Rev. **D30**, 2585 (1984); **D34**, 1932 (1986); **D36**, 3007 (1987); **D37**, 1982 (1988); J.Math.Phys. **31**, 1998 (1990); Phys.Rev. **D46**, 766 (1992). H.W.Crater, R.L. Becker, C.Y.Wong and P.Van Alstine, Phys.Rev. **D46**, 5117 (1992); in Proc.Int.Symp. "Extended Objects and Bound Systems", eds. O.Hara, S.Ishida and S.Naka (World Scientific, Singapore, 1992). H.W.Crater and D.Yang, J.Math.Phys. **32**, 2374 (1991).
11. G.Longhi, L.Lusanna and J.M.Pons, J.Math.Phys. **30**, 1893 (1989).
12. F.Colomo, G.Longhi and L.Lusanna, Int.J.Mod.Phys. **A5**, 3347 (1990); Mod.Phys.Letters **A5**, 17 (1990). F.Colomo and L.Lusanna, Int.J.Mod.Phys. **A7**, 1705 and 4107 (1992).
13. L.Lusanna, Int.J.Mod.Phys. **A10**, 3531 and 3675 (1995).
14. V.Moncrief, J.Math.Phys. **20**, 579 (1979). M.Cantor, Bull.Am.Math.Soc. **5**, 235 (1981).

15. L.Lusanna and P.Valtancoli, Int.J.Mod.Phys. **A12**, 4769 (1997) (HEP-TH/9606078) and Int.J.Mod.Phys. **A12**, 4797 (1997). (HEP-TH/9606079).
16. L.Lusanna and P.Valtancoli, Int.J.Mod.Phys. **A13**, 4605 (1998) (HEP-TH/9707072).
17. G.Longhi and L.Lusanna, Phys.Rev. **D34**, 3707 (1986).
18. K.Kuchar, J.Math.Phys. **17**, 777, 792, 801 (1976); **18**, 1589 (1977).
19. P.A.M.Dirac, *Rev.Mod.Phys.* **21** (1949) 392.
20. L.Lusanna, Int.J.Mod.Phys. **A12**, 645 (1997).
21. M.Pauri and M.Prosperi, J.Math.Phys. **16**, 1503 (1975).
22. L.Lusanna and M.Materassi, "The Canonical Decomposition in Collective and Relative Variables of a Klein-Gordon Field in the Rest-Frame Wigner-Covariant Instant Form", Firenze Univ.preprint (HEP-TH/9904202).
23. G.Longhi and M.Materassi, J.Math.Phys. **40**, 480 (1999) (HEP-TH/9803128); "Collective and Relative Variables for a Classical Klein-Gordon Field", Firenze Univ.preprint (HEP-TH/9890024), to appear in Int.J.Mod.Phys. A.
24. W.G.Dixon, J.Math.Phys. **8**, 1591 (1967). "Extended Objects in General Relativity: their Description and Motion", in "Isolated Gravitating Systems in General Relativity", ed.J.Ehlers (North-Holland, Amsterdam, 1979).
25. D.Alba and L.Lusanna, Int.J.Mod.Phys. **A13**, 2791 (1998) (HEP-TH/9705155).
26. D.Alba and L.Lusanna, Int.J.Mod.Phys. **A13**, 3275 (1998) (HEP-TH/9705156).
27. F.Bigazzi and L.Lusanna, Int.J.Mod.Phys. **A14**, 1429 (1999) (HEP-TH/9807052).
28. F.Bigazzi and L.Lusanna, Int.J.Mod.Phys. **A14**, 1877 (1999) (HEP-TH/9807054).
29. C.Lämmerzahl, J.Math.Phys. **34**, 3918 (1993).
30. C.Möller, Ann.Inst.H.Poincaré **11**, 251 (1949); "The Theory of Relativity" (Oxford Univ.Press, Oxford, 1957).
31. C.J.Isham, "Canonical Quantum Gravity and the Problem of Time", in "Integrable Systems, Quantum Groups and Quantum Field Theories", eds.L.A.Ibort and M.A.Rodriguez, Salamanca 1993 (Kluwer, London, 1993); "Prima Facie Questions in Quantum Gravity" and "Canonical Quantum Gravity and the Question of Time", in "Canonical Gravity: From Classical to Quantum", eds. J.Ehlers and H.Friedrich (Springer, Berlin, 1994).
32. K.Kuchar, "Time and Interpretations of Quantum Gravity", in Proc.4th Canadian Conf. on "General Relativity and Relativistic Astrophysics", eds. G.Kunstatter, D.Vincent and J.Williams (World Scientific, Singapore, 1992).
33. J.Butterfield and C.J.Isham, "Space-Time and the Philosophical Challenge of Quantum Gravity", Imperial-TP-98-99-45 (gr-qc/9903072).
34. J.Stachel, in "General Relativity and Gravitation", GR11, Stockholm 1986, ed. M.A.H.Mac Callum (Cambridge Univ. Press, Cambridge, 1987); "The Meaning of General Covariance", in "Philosophical Problems of the Internal and External Worlds", Essays in the Philosophy of A.Grünbaum, eds. J.Earman, A.I.Janis, G.J.Massey and N.Rescher (Pittsburgh Univ.Press, Pittsburgh, 1993).
35. C.Rovelli, Class.Quantum Grav. **8**, 297 and 317 (1991).
36. L.Lusanna and S.Russo, "Tetrad Gravity I): A New Formulation", Firenze Univ. preprint 1998 (GR-QC/9807073).
37. L.Lusanna and S.Russo, "Tetrad Gravity II): Dirac's Observables", Firenze Univ. preprint 1998 (GR-QC/9807074).
38. R.DePietri and L.Lusanna, "Tetrad Gravity III): Asymptotic Poincaré Charges, the Physical Hamiltonian and Void Spacetimes", in preparation.
39. R.DePietri, L.Lusanna and M.Vallisneri, "Tetrad Gravity IV): The N-body Problem", in preparation.

40. R.Arnowitt, S.Deser and C.W.Misner, Phys.Rev. **117**, 1595 (1960); in "Gravitation: an Introduction to Current Research", ed.L.Witten (Wiley, New York, 1962).
41. T.Regge and C.Teitelboim, Ann.Phys.(N.Y.) **88**, 286 (1974). R.Beig and Ó Murchadha, Ann.Phys.(N.Y.) **174**, 463 (1987). L.Andersson, J.Geom.Phys. **4**, 289 (1987).
42. A.Ashtekar, "Asymptotic Structure of the Gravitational Field at Spatial Infinity", in "General Relativity and Gravitation", Vol. 2, ed.A.Held (Plenum, New York, 1980). A.Ashtekar and R.O.Hansen, J.Math.Phys. **19**, 1542 (1978). A.Ashtekar and A.Magnon, J.Math.Phys. **25**, 2682 (1984). A.Ashtekar and J.D.Romano, Class.Quantum Grav. **9**, 1069 (1992).
43. R.M.Wald, "General Relativity" (Chicago Univ.Press, Chicago, 1984).
44. R.Geroch, J.Math.Phys. **9**, 1739 (1968); **11**, 343 (1970).
45. V.Moncrief, J.Math.Phys. **16**, 1556 (1975).
46. Y.Choquet-Bruhat, A.Fischer and J.E.Marsden, "Maximal Hypersurfaces and Positivity of Mass", LXVII E.Fermi Summer School of Physics "Isolated Gravitating Systems in General Relativity", ed.J.Ehlers (North-Holland, Amsterdam, 1979).
47. P.G.Bergmann, Rev.Mod.Phys. **33**, 510 (1961).
48. R.T.Jantzen, P.Carini and D.Bini, Ann.Phys.(N.Y.) **215**, 1 (1992).I.Ciufolini and J.A.Wheeler, "Gravitation and Inertia" (Princeton Univ.Press, Princeton, 1995).
49. J.L.Synge, "Relativity: the General Theory" (North-Holland, Amsterdam, 1960). B.S.De Witt, Phys.Rev. **162**, 1195 (1967); "The Spacetime Approach to Quantum Field Theory", in "Relativity, Groups and Topology II", Les Houches 1983, eds. B.S.DeWitt and R.Stora (North-Holland, Amsterdam, 1984).
50. B.S.De Witt, Phys.Rev. **160**, 1113 (1967).
51. C.J.Isham and K.Kuchar, Ann.Phys.(N.Y.) **164**, 288 and 316 (1984). K.Kuchar, Found.Phys. **16**, 193 (1986).
52. A.Lichnerowicz, J.Math.Pure Appl. **23**, 37 (1944). Y.Choquet-Bruhat, C.R.Acad.Sci.Paris **226**, 1071 (1948); J.Rat.Mech.Anal. **5**, 951 (1956); "The Cauchy Problem" in "Gravitation: An Introduction to Current Research", ed.L.Witten (Wiley, New York, 1962).
53. J.W.York jr, Phys.Rev.Lett. **26**, 1656 (1971); **28**, 1082 (1972). J.Math.Phys. **13**, 125 (1972); **14**, 456 (1972). Ann.Ins.H.Poincaré **XXI**, 318 (1974). "Kinematics and Dynamics of General Relativity", in "Sources of Gravitational Radiation", Battelle-Seattle Workshop 1978, ed.L.L.Smarr (Cambridge Univ.Press, Cambridge, 1979). N.O'Murchadha and J.W.York jr, J.Math.Phys. **14**, 1551 (1972). Phys.Rev. **D10**, 428 (1974).
54. P.A.M.Dirac, Canad.J.Math. **3**, 1 (1951).
55. D.Christodoulou and S.Klainerman, "The Global Nonlinear Stability of the Minkowski Space" (Princeton Univ. Press, Princeton, 1993).
56. A.Ashtekar and G.T.Horowitz, J.Math.Phys. **25**, 1473 (1984).
57. E.Witten, Commun.Math.Phys. **80**, 381 (1981).
58. A.Sen, J.Math.Phys. **22**, 1781 (1981); Phys.Lett. 119B, 89 (1982).
59. J.Frauendiener,Class.Quantum Grav. **8**, 1881 (1991).
60. C.Teitelboim, "The Hamiltonian Structure of Space-Time", in "General Relativity and Gravitation", ed.A.Held, Vol.I (Plenum, New York, 1980). A.S.Hojman, K.Kuchar and C.Teitelboim, Ann.Phys. (N.Y.) **96**, 88 (1971).
61. P.Havas, Gen.Rel.Grav. **19**, 435 (1987). R.Anderson, I.Vetharaniam and G.E.Stedman, Phys.Rep. **295**, 93 (1998).
62. A.Komar, Phys.Rev. **111**, 1182 (1958). P.G.Bergmann and A.Komar, Phys.Rev.Lett. **4**, 432 (1960).

63. L.Landau and E.Lifschitz, "The Classical Theory of Fields" (Addison-Wesley, Cambridge, 1951).
64. A.Einstein, B.Hoffman and L.Infeld, Ann.Math. **39**, 66 (1938). A.Einstein and L.Infeld, Ann.Math. **41**, 797 (1940); Canad.J.Math. **1**, 209 (1949). L.Infeld, Rev.Mod.Phys. **29**, 398 (1957).
65. H.A.Lorentz and L.Droste, Amst.Akad.Versl. **26**, 392 (1917). A.Eddington and G.L.Clarke, Proc.Roy.Soc.London **A166**, 465 (1938). V.Fock, J.Phys. (U.S.S.R.) **1**, 81 (1939). A.Papapetrou, Proc.Phys.Soc.(London) **64**, 57 (1951).
66. A.Ashtekar, Phys.Rev.Lett. **57**, 2244 (1986); "New Perspectives in Canonical Gravity" (Bibliopolis, Naples, 1988); "Lectures on Non-Perturbative Canonical Gravity" (World Scientific, Singapore, 1991); "Quantum Mechanics of Riemannian Geometry", http://vishnu.nirvana.phys.psu.edu/riem_qm/riem_qm.html. C.Rovelli and L.Smolin, Nucl.Phys. **B331**, 80 (1990); **B442**, 593 (1995). C.Rovelli, "Loop Quantum Gravity", Living Reviews in Relativity http://www.livingreviews.org/Articles/Volume1/1998-1rovelli.

Meaning of Noncommutative Geometry and the Planck-Scale Quantum Group

Shahn Majid[1]

Department of Applied Mathematics and Theoretical Physics
Silver Street, Cambridge CB3 9EW

Abstract. This is an introduction for nonspecialists to the noncommutative geometric approach to Planck scale physics coming out of quantum groups. The canonical role of the 'Planck scale quantum group' $\mathbb{C}[x]\blacktriangleright\!\!\triangleleft\mathbb{C}[p]$ and its observable-state T-duality-like properties are explained. The general meaning of noncommutativity of position space as potentially a new force in Nature is explained as equivalent under quantum group Fourier transform to curvature in momentum space. More general quantum groups $\mathbb{C}(G^\star)\blacktriangleright\!\!\triangleleft U(\mathfrak{g})$ and $U_q(\mathfrak{g})$ are also discussed. Finally, the generalisation from quantum groups to general quantum Riemannian geometry is outlined. The semiclassical limit of the latter is a theory with generalised non-symmetric metric $g_{\mu\nu}$ obeying $\nabla_\mu g_{\nu\rho} - \nabla_\nu g_{\mu\rho} = 0$.

1 Introduction

There are currently several approaches to Planck-scale physics and of them 'Noncommutative geometry' is probably the most radical but also the least well-tested. As Lee Smolin in his lectures at the conference was kind enough to put it, it is 'promising but too early to tell'. Actually my point of view, which I will explain in these lectures, is that some kind of noncommutative geometry is *inevitable* whatever route we take to the Planck scale. Whether we evolve our understanding of string theory, compute quasiclassical states in loop-variable quantum gravity, or just investigate the intrinsic mathematical structure of geometry and quantum theory themselves (my own line), all roads will in my opinion lead to *some kind* of noncommutative geometry as the next more general geometry beyond nonEuclidean that is needed at the Planck scale where both quantum and gravitational effects are strong. I think the need for this and its general features can be demonstrated from simple nontechnical arguments and will try to do this here. These philosophical and conceptual issues are in Section 2.

Beyond this, and definitely a matter of opinion, it seems to me that there is are certain philosophical principles [1] which can serve as a guide to what Planck scale physics should be, in particular what I have called the *principle of representation-theoretic self-duality* (of which T-duality is one manifestation). I believe that to proceed one has to ask in fact what is the *nature of physical reality itself.* In fact I do not think that theoretical physicists can any longer afford to shy away from such questions and, indeed, with proposals for Planck-scale physics beginning to emerge it is already clear that some new philosophical basis is going to be needed which will likely be every bit as radical as those that came with quantum mechanics and general relativity. My own radical philosophy in [2][3][1] basically takes the reciprocity ideas of Mach to a modern setting. But it also suggests a different concept of reality, which I call *relative realism*, from the reductionist one that most theoretical physicists are still unwilling to give up (I said it would be radical). This might seem fanciful but what it boils down to in practice is an extension of ideas of Fourier theory to the quantum domain. Section 3 provides a modern introduction to this.

Next I will try to convince you that while there are still several different ideas for what *exactly* noncommutative geometry should be, there is slowly emerging what I call the 'quantum groups approach to noncommutative geometry' which is already *fairly* complete in the sense that it has the same degree of 'flabbiness' as Riemannian geometry (is not tied to specific integrable systems etc.) while at the same time it includes the 'zoo' of already known naturally occurring examples, mostly linked to quantum groups. Picture yourself for a moment in the times of Gauss and Riemann; clearly spheres, tori, etc., were evidently examples of something, but of what? In searching for this Riemann was able to formulate the notion of Riemannian manifold as a way to capture known examples like spheres and tori but broad enough to formulate general equations for the intrinsic structure of space itself (or after Einstein, space-time). Theoretical physics today is in a parallel situation with many naturally occurring examples from a variety of sources and a clear need for a general theory. Our approach[4] is based on fiber bundles with quantum group fiber[5], and we will come to it by the end of the lectures, in Section 6. It includes a working definition of 'quantum manifold'.

In between, I will try to give you a sense of some noncommutative geometries out there from which our intuition has to be drawn. We will 'see the sights' in the land of noncommutative geometry at least from the quantum groups point of view. Just as Lie groups are the simplest Riemannian manifolds, quantum groups are the simplest noncommutative spaces. Their homogeneous spaces are also covered, as well as quantum planes (which are more properly braided groups). We refer to [6] for more on quantum groups themselves.

At the same time, the physics reader will no doubt also want to see testable predictions, detailed models etc. While, in my opinion, it is *still* too early to rush into building models and making predictions ('one cannot run before one can walk') I will focus on at least one toy model of Planck-scale physics using these techniques. This is the Planck-scale quantum group introduced 10 years ago in [3][2] and exhibiting even then many of the features one might consider important for Planck scale physics today, including duality. This is the topic of Section 4. It is not, however, the 'theory of everything' or M-theory etc. I seriously doubt that Einstein could have formulated general relativity without the mathematical definition of a 'manifold' having been sorted out by Riemann a century before (and which I would guess had filtered down to Einstein's mathematical mentors such as Minkowski). In the same way, one really needs to sort out the correct or 'natural' definitions of noncommutative geometry some more (in particular the Ricci tensor and stress energy tensor are not yet understood) before making attempts at a full theory with testable predictions. This is on the one hand mathematics but on the other hand it has to be guided by physical intuition with or even without firm predictive models. In fact the structure of the mathematical possibilities of noncommutative geometry (which means for us results in the theory of algebra) can tell us a lot about any actual or effective theory even if it is not presently known.

The general family of *bicrossproduct quantum groups* arising in this way out of Planck scale physics contains many more examples (it is one of the two main constructions by which quantum groups originated in physics.) For example, there is a quantum group $\mathbb{C}(G^\star)\blacktriangleright\!\!\triangleleft U(\mathfrak{g})$ for every complex simple Lie algebra $\mathfrak{g}$. All these bicrossproduct quantum groups can be viewed as the actual quantum algebras of observables of actual quantum systems and can be viewed precisely as models unifying quantum and gravity-like effects [2][3]. For the record, the bicrossproduct construction $\blacktriangleright\!\!\triangleleft$ was introduced in this context at about the same time (in 1986) but independently of the more well-known quantum groups $U_q(\mathfrak{g})$ [7][8], in particular before I had even heard of V.G. Drinfeld or integrable systems. To this day the two classes of quantum groups, although constructed from the same data $\mathfrak{g}$, have never been directly related (this remains an interesting open problem). The situation is shown in Figure 1. To build a theory of noncommutative geometry we need to include naturally occurring examples such as these.

We also need to include the more traditional noncommutative algebras to which people have traditionally tried to develop geometric pictures, namely the canonical commutation relations algebra $[x,p]=\imath\hbar$ or its group version the Weyl algebra or 'noncommutative torus' $vu=e^{\imath\alpha}uv$ as in the work of A. Connes[9]. We can also consider the matrix algebras $M_n(\mathbb{C})$ as studied by Dubois-Violette, Madore and others; as we saw seen in the beautiful lecture of Richard Kerner at the conference, one can do a certain amount of noncommutative geometry for such algebras too. On the other hand, in some sense these are actually all the same example in one form or another, i.e. basically the algebra of operators on some Hilbert space (at least for generic α). These examples and the traditional ideas of vector fields as derivations and points as maximal ideals etc., come from algebraic geometry and predate quantum groups. In my opinion, however, one cannot build a valid noncommutative geometry always on the basis of essentially one example (and a lot of elegant mathematics) – one has to also include the rich vein of practical examples such as the quantum groups above. The latter have a much clearer geometric meaning but very few derivations or maximal ideals etc., i.e. we have to develop a much less obvious noncommutative differential geometry if we are to include them as well as the traditional matrix algebras and of course the commutative case corresponding to usual geometry. This is precisely what has emerged slowly in recent years and that which I will try to explain.

In Section 5 we turn for completeness, to the other and more well-known type of quantum groups, the q-deformed enveloping algebras $U_q(\mathfrak{g})$. These did not arise at all in connection with Planck scale physics or even directly as the quantisation of any physical system. Rather they are 'generalised symmetries' of quantum or lattice integrable systems. Nevertheless they are also examples

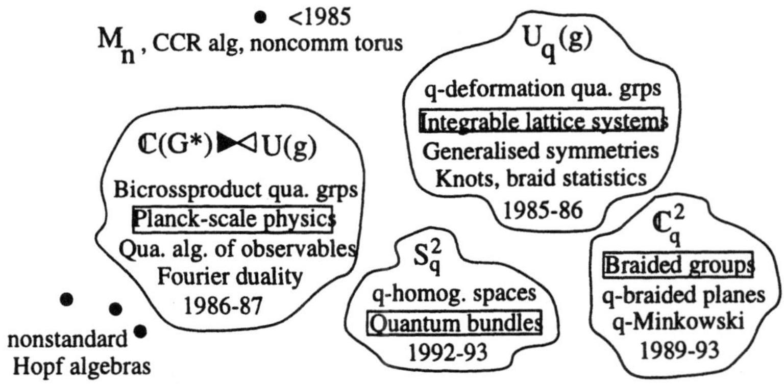

Fig. 1. The landscape of noncommutative geometry today. Some isolated 'traditional' objects such as matrix algebras and the noncomm. torus, and two classes of quantum groups

of noncommutative geometry and, if recent conjectures of Lee Smolin and collaborators prove correct, they are natural descriptions of the noncommutative geometry coming out of the loop variable approach to quantum gravity. The more established meaning of *these* quantum groups which we will focus on is that they induce braid statistics on particles transforming as their representations. In effect the dichotomy of particles into bosons (force) and fermions (matter) is broken in noncommutative geometry and in fact both are unified with each other and with quantisation. Roughly speaking the meaning of q here is a generalisation of the -1 for supersymmetry. So the natural noncommutative geometry here is 'braided geometry'[10]. Yet at the same time one may write q in terms of Planck's constant or, according to [11], the cosmological constant. It means that one physical manifestation of quantum gravity effects is as braid (e.g. fractional) statistics.

Finally, more accessible perhaps to many readers will be not so much our proposals for the full noncommutative theory but its semiclassical predictions; in order to be naturally made noncommutative one has to shift ones point of view a little and indeed move to a slightly more general notion of classical Riemannian geometry. The main prediction is that one should replace the notion of metric and its Levi-cevita connection by a notion of nondegenerate 2-tensor (not necessarily symmetric) and the notion of vanishing torsion and vanishing *cotorsion*. The cotorsion tensor associated to a 2-tensor is a new concept recently introduced in [4]. The resulting self-dual generalisation of the usual metric compatibility becomes

$$\nabla_\mu g_{\nu\rho} - \nabla_\nu g_{\mu\rho} = 0.$$

The generalisation allows a synthesis of symplectic and Riemannian geometry, which is a semiclassical analogue of the quantum-gravity unification problem. Not surprisingly, the above ideas turn out to be related at the semiclassical level to other ideas for Planck scale physics such as T-duality for sigma models on Poisson-Lie groups[12], see [13].

2 The meaning of noncommutative geometry

It stands to reason that if one seriously wants to unify quantum theory and gravity into a single theory with a single elegant point of view, one must first formulate each in the same language. On the side of gravity this is perfectly well-known and we do not need to belabour it; instead of points in a manifold one should and does speak in terms of the algebra of its coordinate functions e.g. (locally) position coordinates $\mathbf{x}$, say. Geometrical operations can then be expressed in terms of this algebra, for example a vector field might be a derivation on the algebra. 'Points' might be maximal ideals. This conventional point of view (called *algebraic geometry*) doesn't really work in practice in the noncommutative case, i.e. it needs to be modified, but it is a suitable starting point for the unification.

What about quantum mechanics? Well this too is some kind of algebra, of course noncommutative due to noncommutation relations between position and momentum. So the language we need is that of algebras. We need to modify usual algebraic geometry in such a way that it extends to algebras of observables arising in quantum systems. At the same time we should, I believe, also be guided by finding natural mathematical definitions that both include nontrivial applications in mathematics *and* encode those algebras in quantum systems which have a clear geometrical structure self-evidently in need of being encoded (perhaps even without direct physical input about Planck scale physics). For example, before the discovery of quantum groups noncommutative geometry made only minimal changes in pursuit of the above idea, e.g. to let the algebra be noncommutative but nevertheless define a vector field as a derivation. All very elegant, but not sufficient to include 'real world' examples like quantum groups.

One other general point. For classical systems we frequently make use of deep classification and other theorems about smooth manifolds; the rich structure of what is mathematically allowed e.g. by topological constraints is often a guide to building effective theories even if we do not know the details of the underlying theory. If we accept the above then the corresponding statement is that deep mathematical theorems about the classification and structure of noncommutative algebras ought to tell us about the possible effective corrections from quantum gravity even before a full theory is known (as well as be a guide to the natural structure of the full theory). We will see this in some toy examples in the next chapter. By contrast many physicists seem to believe that the only algebra in physics is the CCR algebra (or its fermionic version), or possibly Lie algebras, as if there is not in fact a much richer world of noncommutative algebras for their theories to draw upon. In fact this noncommutative world has to be at least as rich as the theory of manifolds since it must contain them in the special commutative limit. I contend that the intrinsic properties of noncommutative algebras is where we should look for new principles and ideas for the Planck scale.

2.1 Curvature in momentum space – a possible new force of nature

Before going into details of the modern approach to noncommutative geometry we want to consider some general issues about unifying quantum theory and geometry using algebra. In particular, what finally emerges as the true meaning of noncommutative geometry for Planck scale physics? In a nutshell, the answer I believe is as follows. Thus, to survive to the Planck scale we should cling to only the very deepest ideas about the nature of physics. In my opinion among the deepest is 'Born reciprocity' or the arbitrariness under position and momentum. Now, in conventional flat space quantum mechanics we take the $\mathbf{x}$ commuting among themselves and their momentum $\mathbf{p}$ likewise commuting among themselves. The commutation relation

$$[x_i, p_j] = \imath\hbar\delta_{ij} \tag{1}$$

is likewise symmetric in the roles of $\mathbf{x}, \mathbf{p}$ (up to a sign). To this symmetry may be attributed such things as wave-particle duality. A wave has localised $\mathbf{p}$ and a particle has localised $\mathbf{x}$.

Now the meaning of *curvature* in position space is, roughly speaking, to make the natural conserved $\mathbf{p}$ coordinates noncommutative. For example, when the position space is a 3-sphere the natural momentum is su_2. The enveloping algebra $U(su_2)$ should be there in the quantum algebra of observables with relations

$$[p_i, p_j] = \frac{\imath}{R}\epsilon_{ijk}p_k \tag{2}$$

where R is proportional to the radius of curvature of the S^3.

By Born-reciprocity then there should be another possibility which is *curvature in momentum space*. It corresponds under Fourier theory to noncommutativity of position space. For example if the momentum space were a sphere with m proportional to the radius of curvature, the position space coordinates would correspondingly have noncommutation relations

$$[x_i, x_j] = \frac{\imath}{m}\epsilon_{ijk}x_k. \tag{3}$$

Mathematically speaking this is surely a symmetrical and equally interesting possibility which might have observable consequences and might be observed. Note that m here is just a parameter not necessarily mass, but our use of it here does suggest the possibility of understanding the *geometry of the mass-shell* as noncommutative geometry of the position space $\mathbf{x}$. This may indeed be an interesting and as yet unexplored application of these ideas. In general terms, however, the situation is clear: for systems constrained in position space one has the usual tools of differential geometry, curvature etc., of the constrained 'surface' in position space or tools for noncommutative algebras (such as Lie algebras) in momentum space. *For systems constrained in momentum space one needs conventional tools of geometry in momentum space or, by Fourier theory, suitable tools of noncommutative geometry in position space.*

In mathematical terms, these latter two examples (2),(3) demonstrate the point of view of noncommutative geometry: we are viewing the enveloping algebra $U(su_2)$ *as if* it were the algebra of coordinates of some system, i.e. we want to answer the question

$$U(su_2) = C(?)$$

where ? will not be any usual kind of space (where the coordinates would commute). This is what we have called in [14] a 'quantum-geometry transformation' since a quantum symmetry point of view (such as the angular momentum in a quantum system) is viewed 'up-side-down' as a geometrical one. The simplest example $U(\mathfrak{b}_+)$ was studied from this point of view as a noncommutative space in [15], actually slightly more generally as $U_q(\mathfrak{b}_+)$.

For particular examples of this type we do not of course *need* any fancy noncommutative geometric point of view – Lie theory was already extensively developed just to handle such algebras. But if we wish to unify quantum and geometric effects then we should start taking this noncommutative geometric viewpoint even on such familiar algebras. What are 'vector fields' on $U(su_2)$? What is Fourier transform

$$\mathcal{F} : U(su_2) \to \mathbb{C}(SU_2)$$

from the momentum coordinates to the SU_2 position coordinates? These are nontrivial (but essentially solved) questions. Understanding them, we can proceed to construct more complex examples of noncommutative geometry which are neither $U(\mathfrak{g})$ nor $C(G)$, i.e. where noncommutative geometry is really needed and where both quantum and geometrical effects are unified. Vector fields, Fourier theory etc., extend to this domain and allow us to explore consistently new ideas for Planck scale physics. This approach to Planck scale physics based particularly on Fourier theory to extend the familiar $\mathbf{x}, \mathbf{p}$ reciprocity to the case of nonAbelian Lie algebras and beyond is due to the author in [2][16][3][1] [14] and elsewhere.

Notice also that the three effects exemplified by the three equations (1)–(3) are all independent. They are controlled by three different parameters $\hbar, R, m$ (say). Of course in a full theory of quantum gravity all three effects could exist together and be unified into a single noncommutative algebra generated by suitable $\mathbf{x}, \mathbf{p}$. Moreover, even if we do not know the details of the correct theory of quantum gravity, if we assume that something like Born reciprocity survives then all three effects indeed *should* show up in the effective theory where we consider almost-particle states with position and momenta $\mathbf{x}, \mathbf{p}$. It would require fine tuning or some special principle to eliminate any one of them. Also the same ideas could apply at the level of the quantum gravity field theory itself, but this is a different question.

2.2 Algebraic structure of quantum mechanics

In the above discussion we have assumed that quantum systems are described by algebras generated by position and momentum. Here we will examine this a little more closely. The physical question to keep in mind is the following: *what happens to the geometry of the classical system when you quantise?*

To see the problem consider what you obtain when you quantise a sphere or a torus. In usual quantum mechanics one takes the Hilbert space on position space, e.g. $\mathcal{H} = L^2(S^2)$ or $\mathcal{H} = L^2(T^2)$ and as 'algebra of observables' one takes $A = B(\mathcal{H})$ the algebra of all bounded (say) operators. It is decreed that every self-adjoint hermitian operator a (or its bounded exponential more precisely) is an observable of the system and its expectation value in state $|\psi> \in \mathcal{H}$ is

$$< a >_\psi = < \psi | a | \psi > .$$

The problem with this is that $B(\mathcal{H})$ *is the same algebra in all cases.* The quantum system does know about the underlying geometry of the configuration space or of the phase space in other ways; the choice of 'polarisation' on the phase space or the choice of Hamiltonian etc. – such things are generally defined using the underlying position or phase space geometry – but the abstract algebra $B(\mathcal{H})$ doesn't know about this. All separable Hilbert spaces are isomorphic (although not in any natural way) so their algebras of operators are also all isomorphic. In other words, whereas in classical mechanics we use extensively the detailed geometrical structure, such as the choice of phase space as a symplectic manifold, all of this is not recorded very directly in the quantum system. One more or less forgets it, although it resurfaces in relation to the more restricted kinds of questions (labeled by classical 'handles') one asks in practice about the quantum system. In other words, *the true quantum algebra of observables should not be the entire algebra $B(\mathcal{H})$ but some subalgebra $A \subset B(\mathcal{H})$*. The choice of this subalgebra is called the *kinematic structure* and it is precisely here that the (noncommutative) geometry of the classical and quantum system is encoded. This is somewhat analogous to the idea in geometry that every manifold can be visualised concretely embedded in some $\mathbb{R}^n$. Not knowing this and thinking that coordinates $\mathbf{x}$ were always globally defined would miss out on all physical effects that depend on topological sectors, such as the difference between spheres and tori.

Another way to put this is that by the Darboux theorem all symplectic manifolds are *locally* of the canonical form $\mathrm{d}x \wedge \mathrm{d}p$ for each coordinate. Similarly one should take (1) (which essentially generates all of $B(\mathcal{H})$, one way or another) only locally. The full geometry in the quantum system is visible only by considering more nontrivial algebras than this one to bring out the global structure. We should in fact consider all noncommutative algebras equipped with certain structures common to all quantum systems, i.e. inspired by $B(\mathcal{H})$ as some kind of local model or canonical example but not limited to it. The conditions on our algebras should also be enough to ensure that there *is* a Hilbert space around and that A can be viewed concretely as a subalgebra of operators on it.

Such a slight generalisation of quantum mechanics which allows this kinematic structure to be exhibited exists and is quite standard in mathematical physics circles. The required algebra is a *von Neumann* algebra or, for a slightly nicer theory, a C^*-algebra. This is an algebra over $\mathbb{C}$ with a $*$ operation and a norm $|| \ ||$ with certain completeness and other properties. The canonical example is $B(\mathcal{H})$ with the operator norm and $*$ the adjoint operation, and every other is a subalgebra.

Does this slight generalisation have observable consequences? Certainly. For example in quantum statistical mechanics one considers not only state vectors $|\psi>$ but 'density matrices' or generalised states. These are convex linear combinations of the projection matrices or expectations associated to state vectors $|\psi_i>$ with weights $s_i \geq 0$ and $\sum_i s_i = 1$. The expectation value in such a 'mixed state' is

$$<a>=\sum_i s_i <\psi_i|a|\psi_i> \tag{4}$$

In general these possibly-mixed states are equivalent to simply specifying the expectation directly as a linear map $< >: B(\mathcal{H}) \to \mathbb{C}$. This map respects the adjoint or $*$ operation on $B(\mathcal{H})$ so that $< a^*a > \geq 0$ for all operators a (i.e. a positive linear functional) and is also continuous with respect to the operator norm. Such positive linear functionals on $B(\mathcal{H})$ are precisely of the above form (4) given by a density matrix, so this is a complete characterisation of mixed states with reference only to the algebra $B(\mathcal{H})$, its $*$ operation and its norm. The expectations $< >_\psi$ associated to ordinary Hilbert space states are called the 'pure states' and are recovered as the extreme points in the topological space of positive linear functionals (i.e. those which are not the convex linear combinations of any others).

Now, if the actual algebra of observables is some subalgebra $A \subset B(\mathcal{H})$ then any positive linear functional on the latter of course restricts to one on A, i.e. defines an 'expectation state' $A \to \mathbb{C}$ which associates numbers, the expectation values, to each observable $a \in A$. But not vice-versa, i.e. the algebra A may have perfectly well-defined expectation states in this sense which are not extendable to all of $B(\mathcal{H})$ in the form (4) of a density matrix. Conversely, a pure state on $B(\mathcal{H})$ given by $|\psi> \in \mathcal{H}$ might be mixed when restricted to A. The distinction becomes crucially important for the correct analysis of quantum thermodynamic systems for example, see [17].

The analogy with classical geometry is that not every local construction may be globally defined. If one did not understand that one would miss such important things as the Bohm-Aharanov effect, for example. Although I am not an expert on the 'measurement problem' in the philosophy of quantum mechanics it does not surprise me that one would get into inconsistencies if one did not realise that the algebra of observables is a subalgebra of $B(\mathcal{H})$. And from our point of view it is precisely to understand and 'picture' the structure of the subalgebra for a given system that noncommutative geometry steps in. I would also like to add that the problem of measurement itself is a matter of matching the quantum system to macroscopic features such as the position of measuring devices. I would contend that to do this consistently one first has to know how to identify aspects of 'macrospopic structure' in the quantum system without already taking the classical limit. Only in this way can one meaningfully discuss concepts such as partial measurement or the arbitrariness of the division into measurer and measured. Such an identification is exactly the task of noncommutative geometry, which deals with extending our macroscopic intuitions and classical 'handles' over to the quantum system. Put another way, the correspondence principle in quantum mechanics typically involves choosing local coordinates like $\mathbf{x}, \mathbf{p}$ to map over. Its refinement to correspond more of the global geometry into the quantum world is the practical task of noncommutative geometry.

2.3 Principle of representation-theoretic self-duality

With the above preambles, we are in a position to consider some speculation about Planck scale physics. Personally I believe that anything we write down that is based on our past experience and not on the deepest philosophical principles is not likely to survive except as an approximation. For example, while string theory may indeed survive to models of the Planck scale as a certain approximation valid in a certain domain, it does not have enough of a radical new philosophy to provide the true conceptual leaps. I should apologise for this belief but I do not believe that Nature cares about the historically convenient route by which we might arrive at the right concepts for the Planck scale.

So as a basis we should stick only to some of the deepest principles. In my opinion one of the deepest principles concerns the nature of mathematics itself. Namely throughout mathematics one finds an intrinsic dualism between observer and observed as follows. When we think of a function f being evaluated on $x \in X$, we could equally-well think of the same numbers as x being evaluated on f a member of some dual structure $f \in \hat{X}$:

$$\text{Result} = f(x) = x(f).$$

Such a 'turning of the tables' is a mathematical fact. For any mathematical concept X one may consider maps or 'representations' from it to some self-evident class of objects (say rational numbers or for convenience real or complex numbers) wherein our results of measurements are deemed to lie. Such representations themselves form a dual structure $\hat{X}$ of which elements of X can be equally well viewed as representations. ***But is such a dual structure equally real?*** I postulated in 1987 that indeed it should be so in a complete theory. Indeed[1], *The search for a complete theory of physics is the search for a self-dual formulation in the above representation-theoretic sense (The principle of representation-theoretic self-duality).* Put another way, a complete theory of physics should admit a 'polarisation' into two halves each of which is the set of representations of the other. This division should be arbitrary – one should be able to reverse interpretations (or indeed consider canonical transformations to other choices of 'polarisation' if one takes the symplectic analogy).

Note that by completeness here I do not mean knowing in more and more detail *what* is true in the real world. That consists of greater and greater complexity but it is not *theoretical* physics. I'm considering that a theorist wants to know *why* things are the way they are. Ideally I would like on my deathbed to be able to say that I have found the right point of view or theoretical-conceptual framework from which everything else follows. Working out the details of that would be far from trivial of course. This is a more or less conventional reductionist viewpoint except that the Principle asserts that we will not have found the required point of view unless it is self-dual.

For example, there is a sense in which geometry – or 'gravity' is dual to quantum theory or matter. This is visible for some simple models such as spheres with constant curvature where it is achieved by Fourier theory. We will be saying more about this later. If we accept this then in general terms Planck scale physics has to unify these mutually dual concepts into one self-dual structure. Ideally then Einstein's equation

$$G_{\mu\nu} = T_{\mu\nu} \tag{5}$$

would appear as some kind of ***self-duality*** equation within this self-dual context. Here the stress-energy tensor $T_{\mu\nu}$ measures how matter responds to the geometry, while the Einstein tensor $G_{\mu\nu}$ measures how geometry responds to matter. This is the part of Mach's principle which apparently inspired Einstein. The question is how to make these ideas precise in a representation-theoretic sense. While this still remains a long-term goal or vision, there are some toy models[3] where some of the required features can be seen. We come to them in a later section. For the moment we note only that one needs clearly some kind of noncommutative geometry because $T_{\mu\nu}$ should really be the quantum operator stress-energy and its coupling to $G_{\mu\nu}$ through its expectation value is surely only the first approximation or semiclassical limit of an operator version of (5). But an operator version of $G_{\mu\nu}$ only makes sense in the context of noncommutative geometry. What we would hope to find, in a suitable version of these ideas, is a self-dual setting where there was a dual interpretation in which $T_{\mu\nu}$ was the Einstein tensor of some dual system and $G_{\mu\nu}$ its stress-energy. In this way the duality and self-duality of the situation would be made manifest.

This is more or less where quantum groups come in, as a simple and soluble version of the more general unification problem. The situation is shown in Figure 2. Thus, the simplest theories of physics are based on Boolean algebras (a theory consists of classification of a 'universe' set into subsets); there is a well-known duality operation interchanging a subset and its complement. The next more advanced self-dual category is that of (locally compact) Abelian groups such as $\mathbb{R}^n$. In this case the set of 1-dimensional (ir)reps is again an Abelian group, i.e. the category of such objects is self-dual. In the topological setting one has $\hat{\mathbb{R}}^n \cong \mathbb{R}^n$ so that these groups (which are at the core of linear algebra) are self-dual objects in the self-dual category of Abelian groups. Of course, Fourier theory interchanges these two. More generally, to accommodate other phenomena we step away from the self-dual axis. Thus, nonAbelian Lie groups such as SU_2 as manifolds provide the simplest examples of curved spaces. Their duals, which means constructing irreps, appear as central structures in quantum field theory (as judged by any course on particle physics in the 1960's). Wigner even defined a particle as an irrep of the Poincaré group. The unification of these two

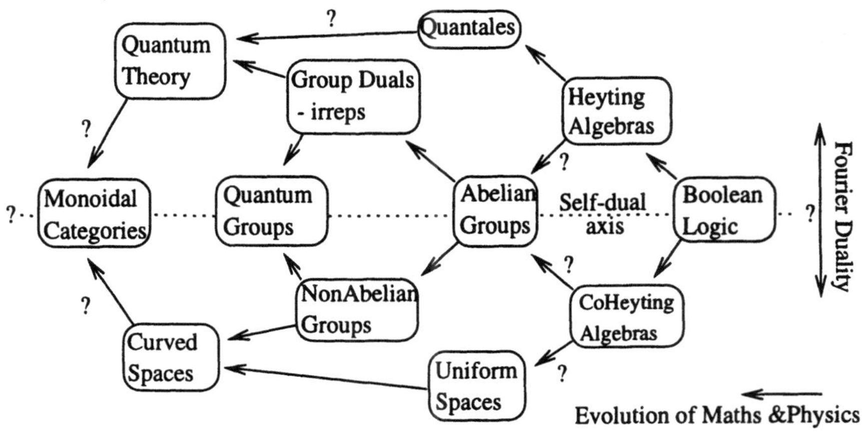

Fig. 2. Representation-theoretic approach to Planck-scale physics. The unification of quantum and geometrical effects is a drive to the self-dual axis. Arrows denote inclusion functors

concepts, groups and groups duals was for many years an open problem in mathematics. Hopf algebras or quantum groups had been invented as the next more general self-dual category containing groups and group duals (and with Hopf algebra duality reducing to Fourier duality) back in 1947 but no general classes of quantum groups going beyond groups or group duals i.e. truly unifying the two were known. In 1986 it was possible to view this open problem as a 'toy model' or microcosm of the problem of unifying quantum theory and gravity and the bicrossproduct quantum groups such as $\mathbb{C}(G^\star)\blacktriangleright\!\!\triangleleft U(\mathfrak{g})$ were introduced on this basis as toy models of Planck scale physics[3]. The construction is self-dual (the dual is of the same general form). At about the same time, independently, some other quantum groups $U_q(\mathfrak{g})$ were being introduced from a different point of view both mathematically and physically (namely as generalised symmetries). We go into details in later sections.

We end this section with some promised philosophical remarks. First of all, why the principle of self-duality? Why such a central role for Fourier theory? The answer I believe is that something very general like this (see the introductory discussion) underlies the very nature of what it means to do science. My model (no doubt a very crude one but which I think captures some of the essence of what is going one) is as follows. Suppose that some theorist puts forward a theory in which there is an actual group G say 'in reality' (this is where physics differs from math) and some experimentalists construct tests of the theory and in so doing they routinely build representations or elements of $\hat{G}$. They will end up regarding $\hat{G}$ as 'real' and G as merely an encoding of $\hat{G}$. The two points of view are in harmony because mathematically (in a topological context)

$$G\cong\hat{\hat{G}}.$$

So far so good, but through the interaction and confusion between the experimental and theoretical points of view one will eventually have to consider both, i.e. $G \times \hat{G}$ as real. But then the theorists will come along and say that they don't like direct products, everything should interact with everything else, and will seek to unify $G, \hat{G}$ into some more complicated irreducible structure G_1, say. Then the experimentalists build $\hat{G}_1$ ... and so on. This is a kind of engine for the evolution of Science.

For example, if one regarded, following Newton that space $\mathbb{R}^n$ is real, its representations $\hat{\mathbb{R}}^n$ are derived quantities $\mathbf{p} = m\dot{\mathbf{x}}$. But after making diverse such representations one eventually regards both $\mathbf{x}$ and $\mathbf{p}$ as equally valid, equivalent via Fourier theory. But then we seek to unify them and introduce the CCR algebra (1). And so on. Note that this is not intended to be a historical account but a theory for how things should have gone in an ideal case without the twists and turns of human ignorance.

One could consider this point of view as window dressing. Surely quantum mechanics was 'out there' and would have been discovered whatever route one took? Yes, but if if the mechanism is correct even as a hindsight, the same mechanism *does have predictive power* for the next more complicated theory. The structure of the theory of self-dual structures is nontrivial and not everything is possible. Knowing what is mathematically possible and combining with some postulates such as the above is not empty. For example, back in 1989 and motivated in the above manner it was shown that the category of monoidal categories (i.e. categories equipped with tensor products) was itself a monoidal category, i.e. that there was a construction $\hat{\mathcal{C}}$ for every such category $\mathcal{C}$[18]. Since then it has turned out that both conformal field theory and certain other quantum field theories can indeed be expressed in such categorical terms. Geometrical constructions can also be expressed categorically[19]. On the other hand, this categorical approach is still under-developed and its exact use and the exact nature of the required duality as a unification of quantum theory and gravity is still open. I would claim only 'something like that' (one should not expect too much from philosophy alone).

Another point to be made from Figure 2 is that if quantum theory and gravity already take us to very general structures such as categories themselves for the unifying concept then, in lay terms, what it means is that the required theory involves very general concepts indeed of a similar level to semiotics and linguistics (speaking about categories of categories etc.). It is almost impossible to conceive *within existing mathematics* (since it is itself founded in categories) what fundamentally more general structures would come after that. In other words, *the required mathematics is running out* it least in the manner that it was developed in this century (i.e. categorically) and at least in terms of the required higher levels of generality in which to look for self-dual structures. If the search for the ultimate theory of physics is to be restricted to logic and mathematics (which is surely what distinguishes science from, say, poetry), then this indeed correlates with our physical intuition that the unification of quantum theory and gravity is the last big unification for physics as we know it, or as were that theoretical physics as we know it is coming to an end. I would agree with this assertion except to say that the new theory will probably open up more questions which are currently considered metaphysics and make them physics, so I don't really think we will be out of a job even as theorists (and there will always be an infinite amount of 'what' work to be done even if the 'why' question was answered at some consensual level).

As well as seeking the 'end of physics', we can also ask more about its birth. Again there are many nontrivial and nonempty questions raised by the self-duality postulate. Certainly the key generalisation of Boolean logic to intuitionistic logic is to relax the axiom that $a \cup \tilde{a} = 1$ (that a or not a is true). Such an algebra is called a *Heyting algebra* and can be regarded as the birth of quantum mechanics. Dual to this is the notion of a *coHeyting algebra* in which we relax the law that $a \cap \tilde{a} = 0$. In such an algebra one can define the 'boundary' of a proposition as

$$\partial a = a \cap \tilde{a}$$

and show that it behaves like a derivation. This is surely the birth of geometry. How exactly this complementation duality extends to the Fourier duality for groups and on to the duality between more complex geometries and quantum theory is not completely understood, but there are conceptual 'physical' argument that this should be so, put forward in [1].

Briefly, in the simplest 'theories of physics' based only on logic one can work equally well with 'apples' or 'not-apples' as the names of subsets. *What happens to this complementation duality in more advanced theories of physics?* Apples curve space while not-apples do not, i.e. in physics one talks of apples as really existing while not-apples are merely an abstract concept. Clearly the self-duality is lost in a theory of gravity alone. But we have argued[1] that when one considers both gravity *and* quantum theory the self-duality can be restored. Thus when we say that a region is as full of apples as General Relativity allows (more matter simply forms a black hole which expands), which is the right hand limiting line in Figure 3[1], in the dual theory we might say that the region is as empty of not-apples as quantum theory allows, the limitation being the left slope in Figure 3. Here the uncertainty principle in the form of pair creation ensures that space cannot be totally empty of 'particles'. Although heuristic, these are arguments that quantum theory and gravity are dual and that this duality is an extension of complementation duality. Only a theory with both would be self-dual. Also, in view of a 'hole' moving in the opposite direction to a particle, the dual theory should also involves time reversal. The self-duality is something like CPT invariance but in a theory where gravitational and not only quantum effects are considered. We are proposing it as a key requirement for quantum-gravity.

[1] Diagrams similar to the right hand side of Figure 3 plotting the mass-energy density v size have been attributed to Brandon Carter (who was here at the conference), as a tool to plot stellar evolution

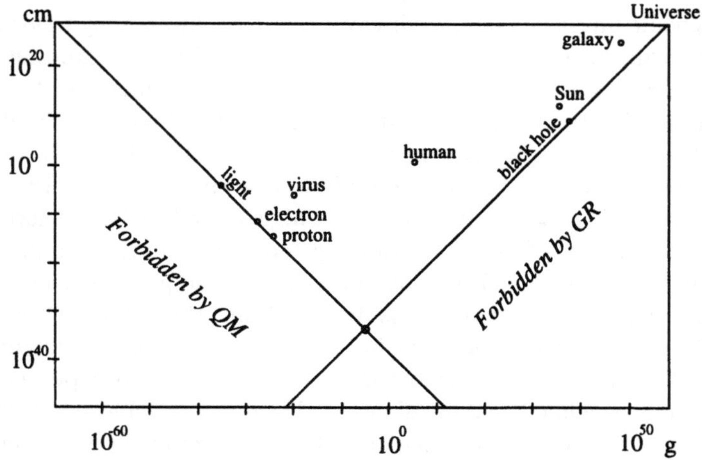

Fig. 3. Range of physical phenomena, which lie in the wedge region with us in the middle. Log plots are mass-energy v size

2.4 Relative realism

So far we have given arguments that there is at least a correlation between the mathematical structure of self-dual structures and the progressive theories of physics from their birth in 'logic' to the projected forthcoming complete theory of everything. It should at least provide a guide to the properties that should be central in unknown theories of everything, such as what have become fashionable to call 'M-theory'.

What about going further? This section will indeed be speculative but I believe it should be considered. Suppose indeed that some mathematical-structural principles (such as the principle of representation theoretic self-duality above) could exactly pin down the ultimate theory of physics along the lines discussed. This would be like giving a list of things that we expect from a complete theory – such as renormalisability, CPT-invariance, etc., except that we are considering such general versions of these 'constraints' that they are practically what it means to be a group of people following the scientific method. If this really pins down the ultimate theory then it means that *the ultimate theory of physics is no more and no less than a self-discovery of the constraints in thinking that are taken on when one decides to look at the world as a physicist.*

If this sounds cynical it is not meant to be; it is merely a Kantian or Hegelian basis of physical reality as opposed the more conventional reductionist one that most physicists take for granted. It does not mean that physics is arbitrary or random any more than the different possible manifolds 'out there' are arbitrary. The space of all possible manifolds up to equivalence has a deep and rich structure and feels every bit as real to anyone who studies it; but it is a mathematical reality 'created' when we accept the axioms of a manifold. So what we are saying is that there is not such a fundamental difference between mathematical reality and physical reality. The main difference is that mathematicians are aware of the axioms while physicists tend to discover them 'backwards' by theorising from experience. I call this point of view ***relative realism***[1]. In it, we experience reality through choices that we have forgotten about at any given moment. If we become aware of the choice the reality it creates is dissolved or 'unconstructed'. On the other hand, the reader will say that the possibility of the theory of manifolds – that the game of manifold-hunting could have been played in the first place – is itself a reality, not arbitrary. It is, but at a higher level: it is a concrete fact in a more general theory of possible axiom systems of this type. To give another example, the reality of chess is created once we chose to play the game. If we are aware that it is a game, that reality is dissolved, but the rules of chess remain a reality although not within chess but in the space of possible board games. This gives a tree-like or hierarchical structure of reality. Reality is experienced as we look down the tree while 'awareness' or enlightenment is achieved as we look up the tree. When we are born we take on millions and millions of assumptions or rules through communication, which creates our day to day perception of reality, we then spend large parts of our lives questioning and attempting to unconstruct these assumptions as we seek understanding of the world.

Ten years ago I would have had to apologise to the reader for presenting such a philosophy or 'metamodel' of physics but, as mentioned in the Introduction, now that theories of everything are beginning to be bandied about I do believe it is time to give deeper thought to these issues. As a matter of fact the paper [1] on which most of Section 2 is based was submitted in 1987 to the Canadian *Philosophy of Science Journal* where a very enthusiastic referee conditionally accepted the paper but insisted that the arguments were basically Kantian and that I had to read Kant.[2] Kant basically said that reality was a product of human thought. From this perspective the fact that life appears somewhere near the middle of Figure 3, apart from the obvious explanation that phenomena become simpler as we approach the boundaries hence most complex in the middle so this is statistically where life would develop, has a different explanation: we created our picture of physical reality around ourselves and so not surprisingly we are near the middle.

3 Fourier theory

It is now high time to turn from philosophy to more mathematical considerations. We give more details about Fourier duality and in particular how it leads to quantum groups as a concrete 'toy model' setting to explore the above ideas. At the same time it should be clear from the general nature of the discussion above that quantum groups and even noncommutative geometry itself are only relatively simple manifestations of even more general ideas that might be approached along broadly similar lines.

[2] I duly spent the entire summer of 1989 reading up Kant and revising the paper; after which the referee rejected the paper with the immortal words 'now that the basic structure of the author's case is more exposed I do not find it clarified'!

First of all, usual Fourier theory on $\mathbb{R}$ is a pairing of two groups, position x and momentum p. The momentum here labels the characters on $\mathbb{R}$, i.e the elements of the dual group $\hat{\mathbb{R}}$. The corresponding character is the plane wave

$$\chi_p(x) = e^{\imath xp}$$

The group $\hat{\mathbb{R}}$ has its group structure given by pointwise multiplication

$$\chi_p\chi_{p'}(x) = \chi_p(x)\chi_{p'}(x) = \chi_{p+p'}(x)$$

which is therefore isomorphic to $\mathbb{R}$ as the addition of momentum. Moreover, the situation is symmetrical i.e. one could regard the same plane waves as characters $\chi_x(p)$ on momentum space. The Fourier transform is a map from functions on $\mathbb{R}$ to functions on $\hat{\mathbb{R}}$,

$$\mathcal{F}(f)(p) = \int dx f(x)\chi_p(x)$$

3.1 Loop variables and Fourier duality

It is well-known that these ideas work for any locally compact Abelian group. The local-compactness is needed for the existence of a translation-invariant measure. As physicists we can also apply these ideas formally for other groups pretending that there is such a measure. For example in [20][21][22][23] we proposed a Fourier theory approach to the quantisation of photons as follows. The elements κ of the group are disjoint unions of oriented knots (i.e. links) with a product law that consists of erasing any overlapping segments of opposite orientation. The dual group is $\mathcal{A}/\mathcal{G}$ of $U(1)$ bundles and (distributional) connections A on them. Thus given any bundle and connection, the character is the holonomy

$$\chi_A(\kappa) = e^{\imath \int_\kappa A}.$$

We considered this set-up in and the inverse Fourier transform of some well-known functions on $\mathcal{A}/\mathcal{G}$ as functions on the group of knots. For example[22],

$$\mathcal{F}^{-1}(\mathrm{CS})(\kappa) = \int \mathrm{d}A\ \mathrm{CS(A)}e^{-\imath\int_\kappa A} = e^{\frac{\imath}{2\alpha}\mathrm{link}(\kappa,\kappa)} \tag{6}$$

$$\mathcal{F}^{-1}(\mathrm{Max})(\kappa) = \int \mathrm{d}A\ \mathrm{Max(A)}e^{-\imath\int_\kappa A} = e^{\frac{\imath}{2\beta}\mathrm{ind}(\kappa,\kappa)} \tag{7}$$

where

$$\mathrm{CS}(A) = e^{\frac{\alpha\imath}{2}\int A\wedge \mathrm{d}A},\quad \mathrm{Max}(A) = e^{\frac{\beta\imath}{2}\int {}^*\mathrm{d}A\wedge\mathrm{d}A}$$

are the Chern-Simmons and Maxwell actions, link denotes linking number, ind denotes mutual inductance.

The diagonal $\mathrm{ind}(\kappa,\kappa)$ is the *mutual self-inductance* i.e. you can literally cut the knot, put a capacitor and measure the resonant frequency to measure it. By the way, to make sense of this one has to use a wire of a finite thickness – the self-inductance has a log divergence. This is also the log-divergence of Maxwell theory when one tries to make sense of the functional integral, i.e. renormalisation has a clear physical meaning in this context[22].

Meanwhile, $\mathrm{link}(\kappa,\kappa)$ is the self-linking number[20][21][22] of a knot with itself, defined as follows. First of all, between two disjoint knots $\mathrm{link}(\kappa,\kappa')$ is the linking number as usual. We then introduce the following *regularised linking number*

$$\mathrm{link}_\epsilon(\kappa,\kappa')=\int_{||\boldsymbol{\epsilon}||<\epsilon}\mathrm{d}^3\boldsymbol{\epsilon}\ \mathrm{link}(\kappa,\kappa'_{\boldsymbol{\epsilon}})$$

where $\kappa'_{\boldsymbol{\epsilon}}$ is the knot displaced by the vector $\boldsymbol{\epsilon}$. The integrand is defined almost everywhere and hence integrable. Finally, we define the linking number as the limit of this as $\epsilon\to 0$, which is now defined even when knots touch or even on the same knot. At the time of [20][21][22], actually back in 1986, I made the following conjecture which is still open.

Conjecture 1. Intersections that are worse and worse (i.e. so that higher and higher derivatives coincide at the point of intersection) contribute fractions with greater and greater denominators to the regularised linking number, but the linking number remains in $\mathbb{Q}$. In the extreme limit of total overlap the self-linking number is a generic element of $\mathbb{R}$.

As evidence, if the knots intersect transversally then it is easy to see that one obtains for the regularised linking an integer $\pm\frac{1}{2}$. This is just because half the displacements will move one knot in to link more with the other, and the other half to unlink. [3] Although the conjecture remains open, it does appear that it could be interesting for loop variable quantum gravity where it would imply certain rationality properties. By the way, one might need to average over infinitesimal rotations as well as displacements to prove it.

Note also that our point of view in [20][21][22] was *distributional* because as well as considering honest smooth connections we considered 'connections' defined entirely by their holonomy. In particular, given a knot κ we defined the distribution A_κ by its character as

$$e^{\imath\int_{\kappa'}A_\kappa}=e^{\imath\mathrm{link}(\kappa,\kappa')}.$$

Such distributions are quite interesting. For example[21][23] if one formally evaluates the Maxwell action in these one has[20]

$$\mathrm{Max}(A_\kappa)=e^{\frac{1}{4\beta}\delta^2(0)\int_\kappa \mathrm{d}t\dot\kappa\cdot\dot\kappa},\tag{8}$$

[3] This result for transverse intersections, the regularised linking itself and the conjecture for higher intersections were shown to Abbay Ashtekar (and Lou Kauffman) during the ICAMP meeting in Swansea 1988 in advance of the eventual publication in [22].

the Polyakov string action. In other words, string theory can be embedded into Maxwell theory by constraining the functional integral to such 'vortex' configurations. An additional Chern-Simons term becomes similarly a 'topological mass term' link(κ,κ) that we proposed to be added to the Polyakov action.

Finally, these ideas also have analogues in the Hamiltonian formulation. Thus the CCR's for the gauge field can be equivalently formulated as

$$[\int_\kappa A, \int_\Sigma E] = 4\pi\imath\alpha \mathrm{link}(\kappa, \partial\Sigma)$$

which is a signed sum of the points of intersection of the loop with the surface. This is the point of view by which loop variables were introduced in physics in the 1970's (as an approach to QCD on lattices) by Mandelstam and others. We have observed in [20] that this has an interpretation as noncommutative geometry, generalising the noncommutative torus $v^n u^m = e^{\imath\alpha mn} u^m v^n$ to

$$v_\kappa u_{\kappa'} = e^{4\pi\imath\alpha \mathrm{link}(\kappa,\kappa')} u_{\kappa'} v_\kappa \tag{9}$$

where integers are replaced by knots or links. Here the physical picture is

$$u_\kappa = e^{\imath \int_\kappa A}, \quad v_\kappa = e^{\imath \int_\kappa \tilde{A}} \tag{10}$$

where $\tilde{A}$ is a dual connection such that $E = \mathrm{d}\tilde{A}$. So constructing the u, v is equivalent to constructing some distributional operators A, E with the usual CCR's. This point of view from [20][21] was eventually published in [23] as a noncommutative-geometric approach to the quantisation of photons.

It is also an interesting question how all of these ideas generalise from $U(1)$ to nonAbelian groups. Thus, in place of the Abelian group of knots one can first of all consider some kind of nonAbelian group of parameterized loops in the manifold, i.e. maps rather than the images of these maps. (The inequivalent classes of elements in this are the fundamental group π_1 of the manifold.) This should be paired via the Wilson loop or holonomy with nonAbelian bundles and connections. The precise groups and their duality here is a little hazy but one should think of this roughly speaking as what goes on in the construction of knot invariants from the WZW model (or from quantum group). Thus one could argue[21][22] that the relationship between the Jones polynomial J and SU_2-Chern-Simons theory should be viewed as *some kind of nonAbelian Fourier transform*

$$\mathcal{F}^{-1}(\mathrm{CS}_{SU_2})(\kappa) \sim e^{J(\kappa)} \tag{11}$$

with the Jones polynomial in the role of self-linking number[4]. We will discuss Fourier transform on nonAbelian groups in the next section using quantum group methods, though I should say that it still remains to make (11) precise along such

[4] This conjecture dates from 1986 at the time of [20] but was not published until [22], following Witten's discovery of the relation between the WZW model and the Jones polynomial at the ICAMP in Swansea in 1988

lines. The reformulation of quantum group invariants as Vassiliev invariants and the Kontsevich integrals (which generalise the linking number) could be viewed, however, as a perturbative step in this direction.

It does seems that many of these ideas have emerged in modern times in the loop variable approach to quantum gravity[24][25], with the nonAbelian group SU_2 (or another group) in place of $U(1)$. However, I want to close this section with some ideas in this area that I still did not see emerge. Indeed, what the loop variable approach tells us is that the gravitational field when recast as a spin connection is in some sense the conjugate variable to something of manifest topological and diffeomorphism-invariant meaning – knots and links in the manifold. In the same spirit it is obvious that scaler fields correspond to points in the manifold[21][23]. What about in the other direction? I would conjecture that there is another field or force in nature (possibly as yet undiscovered) corresponding to surfaces rather than loops (and so on). Then just as gauge fields tend to detect π_1, the new field would for example detect π_2. Note that in the $U(1)$ case the pairing of surfaces is of course with 2-forms (and the 2nd cohomology is the Abelianisation of π_2) – we would need a nonAbelian version of that.

Actually this conjecture was one of my main motivations back in 1986 in the slightly different context of a search for such Fourier transform or 'surface transport' methods for QCD. First of all, one can ask: if the Fourier transform of the nonAbelian Chern-Simons theory gives the quantum group link invariants as in (11), *what is the Fourier transform of the Yang-Mills action?* According to (7) it should be some kind of some kind of 'nonAbelian self-inductance'. The extra ingredient in QCD is of course confinement. Related to this is the need for some kind of 'nonAbelian Stokes theorem'. While no continuum version of the latter exists, let us suppose that is has somehow been defined, i.e. the Lie group G-valued 'parallel transport' of a nonAbelian Lie-algebra valued 2-form F over a surface such that if F is the curvature of a gauge field then

$$e^{\imath \int_{\Sigma} F} = e^{\imath \int_{\partial\Sigma} A}. \tag{12}$$

While this is not really possible (except rather artificially on a lattice by specifying paths parallel transporting back to a fixed based point) we suppose something *like* this.

Conjecture 2. With such a nonAbelian surface transport, the QCD vacuum expectation value of the flux of the quantized curvature F through a closed surface is an invariant of the surface.

The point is that one usually considers only planar spans of loops in QCD and Wilson's criterion for confinement says that these are area law. On the other hand if one considered a small planar loop spanned by a large surface 'ballooning out' from the loop one would still expect some finite result (since a large area), but on the other hand the boundary curve itself could be shrunk to zero so that its planar spanning surface also shrinks to zero and Wilson's criterion would give 1. The conjecture is that these two effects cancel out and one has in fact something that depends only on the topological class of the surface. This does require, however, making sense of (12) which might require some accompanying new fields. On the other hand, at least one standard objection to the above ideas *was* solved, namely we do not need to take traces of the holonomies etc., which means that we are considering the expectations of gauge-non-invariant operators. It was argued in [26] that one could do this in the context of a version of the background field method. This is important because one can then analyse and prove confinement locally as the statement that the expectation $< F >$ is a (nonAbelian) curvature + a non-curvature part (the latter was shown in [26] to be the skew-symmetrized gluon two-point function). The first part is 'perimeter law' and the second is 'area law' and corresponds to confinement infinitesimally. The conjecture would extend these ideas globally. At the end of the day, however, the strong force itself might emerge as related to surfaces in much the same way as gravity is to loops via the loop gravity and spin connection formalisms.

3.2 NonAbelian Fourier Transform

To generalise Fourier theory beyond Abelian groups we really have to pass to the next more general self-dual category, which is that of Hopf algebras or quantum groups. A Hopf algebra is

- A unital algebra $H, 1$ over the field $\mathbb{C}$ (say)
- A coproduct $\Delta : H \to H \otimes H$ and counit $\epsilon : H \to \mathbb{C}$ forming a *coalgebra*, with Δ, ϵ algebra homomorphisms.
- An antipode $S : H \to H$ such that $\cdot(S \otimes \mathrm{id})\Delta = 1\epsilon = \cdot(\mathrm{id} \otimes S)\Delta$.

Here a coalgebra is just like an algebra but with the axioms written as maps and arrows on the maps reversed. Thus coassociativity means

$$(\Delta \otimes \mathrm{id})\Delta = (\mathrm{id} \otimes \Delta)\Delta \tag{13}$$

etc. The axioms mean that the adjoint maps $\Delta^* : H^* \otimes H^* \to H^*$ and $\epsilon^* : \mathbb{C} \to H^*$ make H^* into an algebra. Here ϵ^* is simply ϵ regarded as an element of H^*. The meaning of the antipode S is harder to explain but it generalises the notion of inverse. It is a kind of 'linearised inversion'.

For a Hopf algebra, at least in the finite-dimensional case (i.e. with a suitable definition of dual space in general) the axioms are such that H^* is again a Hopf algebra. Its coproduct is the adjoint of the product of H and its counit is the unit of H regarded as a map on H^*. This is why the category of Hopf algebras is a self-dual one. For more details we refer to [6].

We will give examples in a moment, but basically these axioms are set up to define Fourier theory. Thinking of H as like 'functions on a group', the coproduct corresponds to the group product law by dualisation. Hence a translation-invariant integral means in general a map $\int : H \to \mathbb{C}$ such that

$$(\int \otimes \mathrm{id})\Delta = 1 \int \tag{14}$$

Meanwhile, the notion of plane wave or exponential should be replaced by the canonical element

$$\exp = \sum_a e_a \otimes f^a \in H \otimes H^* \tag{15}$$

where $\{e_a\}$ is a basis and $\{f^a\}$ is a dual basis. We can then define Fourier transform as

$$\mathcal{F} : H \to H^*, \quad \mathcal{F}(h) = \int (\exp)h = (\int \sum_a e_a h) f^a. \tag{16}$$

There is a similar formula for the inverse $H^* \to H$.

The best way to justify all this is to see how it works on our basic example for Fourier theory. Thus, we take $H = \mathbb{C}[x]$ the algebra of polynomials in one variable, as the coordinate algebra of $\mathbb{R}$. It forms a Hopf algebra with

$$\Delta x = x \otimes 1 + 1 \otimes x, \quad \epsilon x = 0 \quad Sx = -x \tag{17}$$

as an expression of the additive group structure on $\mathbb{R}$. Similarly we take $\mathbb{C}[p]$ for the coordinate algebra of another copy of $\mathbb{R}$ with generator p dual to x (the additive group $\mathbb{R}$ is self-dual).

Example 1. The Hopf algebras $H = \mathbb{C}[x]$ and $H^* = \mathbb{C}[p]$ are dual to each other with $\langle x^n, p^m \rangle = (-\imath)^n \delta_{n,m} n!$ (under which the coproduct of one is dual to the product of the other). The exp element and Fourier transform is therefore

$$\exp = \sum \imath^n \frac{x^n \otimes p^n}{n!} = e^{\imath x \otimes p}, \quad \mathcal{F}(f)(p) = \int_{-\infty}^{\infty} \mathrm{d}x f(x) e^{\imath x \otimes p}.$$

Apart from an implicit $\otimes$ symbol which one does not usually write, we recover usual Fourier theory. Both the notion of duality and the exponential series are being treated a bit formally but can be made precise.

Let is now apply this formalism to Fourier theory on classical but nonAbelian groups. We use Hopf algebra methods because Hopf algebras include both groups and group duals even in the nonAbelian case, as we have promised in Section 2. Thus, if $\mathfrak{g}$ is a Lie algebra with associated Lie group G, we have two Hopf algebras, dual to each other. One is $U(\mathfrak{g})$ the enveloping algebra with

$$\Delta \xi = \xi \otimes 1 + 1 \otimes \xi, \quad \xi \in \mathfrak{g}$$

and the other is the algebra of coordinate functions $\mathbb{C}(G)$. If G is a matrix group the functions t_{ij} which assign to a group element its ij matrix entry generate the coordinate algebra. Of course, they commute i.e. $\mathbb{C}(G)$ is the commutative polynomials in the $t^i{}_j$ modulo some other relations that characterise the group. Their coproduct is

$$\Delta t^i{}_j = t^i{}_k \otimes t^k{}_j$$

corresponding to the matrix multiplication or group law. The pairing is

$$\langle t^i{}_j, \xi\rangle = \rho(\xi)^i{}_j$$

where ρ is the corresponding matrix representation of the Lie algebra. The canonical element or exp is given by choosing a basis for $U(\mathfrak{g})$ and finding its dual basis.

Example 2. $H = \mathbb{C}(SU_2) = \mathbb{C}[a,b,c,d]$ modulo the relation $ad - bc = 1$ (and unitarity properties). It has coproduct

$$\Delta a = a\otimes a + b\otimes c, \quad \text{etc.,} \quad \Delta\begin{pmatrix} a & b\\ c & d\end{pmatrix} = \begin{pmatrix} a & b\\ c & d\end{pmatrix}\otimes\begin{pmatrix} a & b\\ c & d\end{pmatrix}.$$

It is dually paired with $H^* = U(su_2)$ in its antihermitian usual generators $\{e_i\}$ with pairing

$$\langle\begin{pmatrix} a & b\\ c & d\end{pmatrix}, e_i\rangle = \frac{\imath}{2}\sigma_i,$$

defined by the Pauli matrices. Let $\{e_1^a e_2^b e_3^c\}$ be a basis of $U(su_2)$ and $\{f^{a,b,c}\}$ the dual basis. Then

$$\exp = \sum_{a,b,c} f^{a,b,c}\otimes e_1^a e_2^b e_3^c \in \mathbb{C}(SU_2)\bar{\otimes}U(su_2)$$

$$\mathcal{F}(f) = \int_{SU_2} \mathrm{d}u f(u) f^{a,b,c}(u)\otimes e_1^a e_2^b e_3^c.$$

Here $\mathrm{d}u$ denotes the right-invariant Haar measure on SU_2. For a geometric picture one should think of e_i as noncommuting coordinates i.e. regard $U(su_2)$ as a 'noncommutative space' as in (3). An even simpler example is the Lie algebra $\mathfrak{b}_+$ with generators x, t and relations $[x,t] = \imath\lambda x$. Its enveloping algebra could be viewed as a noncommutative analogue of 1+1 dimensional space-time.

Example 3. c.f. [27] The group B_+ of matrices of the form

$$\begin{pmatrix} e^{\lambda\omega} & k\\ 0 & 1\end{pmatrix}$$

has coordinate algebra $\mathbb{C}(B_+) = \mathbb{C}[k,\omega]$ with coproduct

$$\Delta e^{\lambda\omega} = e^{\lambda\omega}\otimes e^{\lambda\omega}, \quad \Delta k = k\otimes 1 + e^{\lambda\omega}\otimes k$$

Its duality pairing with $U(\mathfrak{b}_+)$ is generated by $\langle x,k\rangle = -\imath$, $\langle t,\omega\rangle = -\imath$ and the resulting exp and Fourier transform are

$$\exp = e^{\imath k\omega}e^{\imath\omega t}, \quad \mathcal{F}(:f(x,t):) = \int_{-\infty}^{\infty}\int_{-\infty}^{\infty} \mathrm{d}x\mathrm{d}t\; e^{\imath kx}e^{\imath\omega t} f(e^{\lambda\omega}x, t)$$

where $: f(x,t) :\in U(\mathfrak{b}_+)$ by normal ordering x to the left of t.

Similarly (putting a vector $\boldsymbol{x}$ in place of x) the algebra $[\boldsymbol{x},t]=\imath\lambda\boldsymbol{x}$ is merely the enveloping algebra of the Lie algebra of the group $\mathbb{R}\rhd\!\!<\mathbb{R}^n$ introduced (for $n=2$) in [28] and could be viewed as some kind of noncommutative space-time in $1+n$ dimensions. This was justified in 1+3 dimensions in [29], where it was shown to be the correct 'kappa-deformed' Minkowski space covariant under a 'kappa-deformed' Poincaré quantum group which had been proposed earlier[30]. We see that Fourier transform then connects it to the classical coordinate algebra $\mathbb{C}(\mathbb{R}\rhd\!\!<\mathbb{R}^n)$ of the nonAbelian group $\mathbb{R}\rhd\!\!<\mathbb{R}^n$, this time with commuting coordinates $(\boldsymbol{k},\omega)$. This demonstrates in detail what we promised that noncommutavity of spacetime is related under Fourier transform to nonAbelianness (which typically means curvature) of the momentum group. Under Fourier theory it means that *all* noncommutative geometrical constructions and problems on this spacetime can be mapped over and solved as classical geometrical constructions on the nonAbelian momentum space.

This Fourier transform approach was demonstrated recently in [31], where we analyse the gamma-ray burst experiments mentioned in Giovanni Amelino-Camelia's lectures at the conference, from this point of view. In contrast to previous suggestions[32] (based on the deformed Poincaré algebra) we are able to justify the dispersion relation

$$\lambda^{-2}\left(e^{\lambda\omega}+e^{-\lambda\omega}-2\right)-\boldsymbol{k}^2e^{-\lambda\omega}=m^2 \tag{18}$$

as a well-defined mass-shell in the classical momentum group $\mathbb{R}\rhd\!\!<\mathbb{R}^3$ and give some arguments that the plane waves being of the form $e^{\imath\boldsymbol{k}\cdot\boldsymbol{x}}e^{\imath\omega t}$ above would have wave velocities given by $v_i=\frac{\partial\omega}{\partial k_i}$ (no meaningful justification for this of any kind had been given before). In particular, one has a variation in arrival time for a gamma ray emitted a displacement $\boldsymbol{L}$ away

$$\delta T=-\frac{(\boldsymbol{L}+T\boldsymbol{v})}{\omega}\cdot\delta\boldsymbol{k} \tag{19}$$

as one varies the momentum by $\delta\boldsymbol{k}$. Apparently such theoretical predictions can actually be measured for gamma ray bursts that travel cosmological distances. Of course, one needs to know the distance $\boldsymbol{L}$ and use the predicted $\boldsymbol{L}$-dependence to filter out other effects and also to filter out our lack of knowledge of the initial spectrum of the bursts. It is also conjectured in [31] that the nonAbelianness of the momentum group shows up as CPT violation and might be detected by ongoing neutral-kaon system experiments. Of course, there is nothing stopping one doing field theory in the form of Feynman rules on our classical momentum group either, except that one has to make sense of the meaning of nonAbelianess in the addition of momentum. As explained in Section 2 one can use similar techniques to those for working on curved position space, but now in momentum space, i.e. I would personally call such effects, if detected, 'cogravity'. The idea is that quantum gravity should lead to both gravitational and these more novel cogravitational effects at the macroscopic level.

Let us note finally that these nonAbelian Fourier transform ideas also work fine for finite groups and could be useful for crystallography.

4 Bicrossproduct model of Planck-scale physics

So far we have only really considered groups or their duals, albeit nonAbelian ones. The whole point of Hopf algebras, however, is that there exist examples going truly beyond these but with many of the same features, i.e. with properties of groups and group duals unified. It is high time to give some examples of Hopf algebras going beyond groups and group duals i.e. neither commutative like $\mathbb{C}(G)$ not the dual concept (cocommutative) like $U(\mathfrak{g})$, i.e. *genuine* quantum groups.

We recall from Section 2 that the unification of groups and group duals is a kind of microcosm or 'toy model' of the problem of unifying quantum theory and gravity. So our first class of quantum groups (the other to be described in a later section) come from precisely this point of view.

4.1 The Planck-scale quantum group

By 'toy model' we mean of course some kind of effective theory with stripped-down degrees of freedom but incorporating the idea that Planck scale effects would show up when we try to unify quantum mechanics and geometry through noncommutative geometry. But actually our approach can make a much stronger statement than this: we envisage that the model appears as some effective limit of an unknown theory of quantum-gravity which to lowest order would appear as spacetime and conventional mechanics on it – but even if the theory is unknown we can use the intrinsic structure of noncommutative algebras to classify *a priori* different possibilities. This is much as a phenomenologist might use knowledge of topology or cohomology to classify different *a priori* possible effective Lagrangians.

Specifically, if H_1, H_2 are two quantum groups there is a theory of the space $\mathrm{Ext}_0(H_1, H_2)$ of possible extensions

$$0 \to H_1 \to E \to H_2 \to 0$$

by some Hopf algebra E obeying certain conditions. We do not need to go into the mathematical details here but in general one can show that $E = H_1 \blacktriangleright\!\!\triangleleft H_2$ a 'bicrossproduct' Hopf algebra. Suffice it to say that the conditions are 'self-dual' i.e. the dual of the above extension gives

$$0 \to H_2^* \to E^* \to H_1^* \to 0$$

as another extension dual to the first, in keeping with our philosophy of self-duality of the category in which we work. We also note that by Ext_0 we mean quite strong extensions. There is also a weaker notion that admits the possibilities of cocycles as well, which we are excluding, i.e. this is only the topologically trivial sector in a certain nonAbelian cohomology.

Theorem 1. *[3][16]* $\mathrm{Ext}_0(\mathbb{C}[x], \mathbb{C}[p]) = \mathbb{R}\hbar \oplus \mathbb{R}G$, *i.e. the different extensions*

$$0 \to \mathbb{C}[x] \to ? \to \mathbb{C}[p] \to 0$$

of position $\mathbb{C}[x]$ by momentum $\mathbb{C}[p]$ forming a Hopf algebra are classified by two parameters which we denote $\hbar, \mathsf{G}$ and take the form

$$?\cong\mathbb{C}[x]\blacktriangleright\!\!\triangleleft_{\hbar,\mathsf{G}}\mathbb{C}[p].$$

Explicitly this 2-parameter Hopf algebra is generated by x, p with the relations and coproduct

$$[x,p]=\imath\hbar(1-e^{-\frac{x}{\mathsf{G}}}),\quad \Delta x=x\otimes 1+1\otimes x,\quad \Delta p=p\otimes e^{-\frac{x}{\mathsf{G}}}+1\otimes p.$$

This is called the *Planck scale quantum group*. It is a bit more than just some randomly chosen deformation of the coordinate algebra of the usual group $\mathbb{R}^2$ of phase space of a particle in one dimension: in physical terms what we are saying is that if we are given $\mathbb{C}[x]$ the position coordinate algebra and $\mathbb{C}[p]$ defined *a priori* as the natural momentum coordinate algebra then *all possible* quantum phases spaces built from x, p in a controlled way that preserves duality ideas (Born reciprocity) and retains the group structure of classical phase space as a quantum group are of this form labeled by two parameters $\hbar, \mathsf{G}$. We have not put these parameters in by hand – they are simply the mathematical possibilities being thrown at us. *In effect we are showing how one is forced to discover both quantum and gravitational effects from certain structural self-duality considerations.*

The only physical input here is to chose suggestive names for the two parameters by looking at limiting cases. We also should say what we mean by 'natural momentum coordinate'. What we mean is that the interpretation of p should be fixed before hand, e.g. we stipulate before hand that the Hamiltonian is $h = p^2/2m$ for a particle on our quantum phase space. Then the different commutation relations thrown up by the mathematical structure imply different dynamics. If one wants to be more conventional then one can define $\tilde{p} = p(1-e^{-\frac{x}{\mathsf{G}}})^{-1}$ with canonical commutation relations but some nonstandard Hamiltonian,

$$[x,\tilde{p}]=\imath\hbar,\quad h=\frac{\tilde{p}^2}{2m}(1-e^{-\frac{x}{\mathsf{G}}})^2.$$

Thus our approach is slightly unconventional but is motivated rather by the strong principle of equivalence that from some point of view the particle should be free. We specify x, p before-hand to be in that frame of reference and then explore their possible commutation relations. Of course the theorem can be applied in other contexts too whenever the meaning of x, p is fixed before hand, perhaps by other criteria.

The quantum flat space $\mathsf{G}\to 0$ limit Clearly in the domain where x can be treated as having values > 0, i.e. for a certain class of quantum states where the particle is confined to this region, we clearly have flat space quantum mechanics $[x,p]=\imath\hbar$ in the limit $\mathsf{G}\to 0$.

The classical $\hbar \to 0$ limit On the other hand, as $\hbar \to 0$ we just have the commutative polynomial algebra $\mathbb{C}[x,p]$ with the coalgebra shown. This is the coordinate algebra of the group B_- of matrices of the form

$$\begin{pmatrix} e^{-\frac{x}{\mathsf{G}}} & 0 \\ p & 1 \end{pmatrix}$$

which is therefore the classical phase space for general G of the system.

The dynamics The meaning of the parameter G can be identified, at least roughly, as follows. In fact the meaning of p mathematically in the construction is that it acts on the position $\mathbb{R}$ inducing a flow. For such dynamical systems the Hamiltonian is indeed naturally $h = p^2/m$ and implies that

$$\dot{p} = 0, \quad \dot{x} = \frac{p}{m}\left(1 - e^{-\frac{x}{\mathsf{G}}}\right) + O(\hbar) = v_\infty \left(1 - \frac{1}{1 + \frac{x}{\mathsf{G}} + \cdots}\right) + O(\hbar)$$

where we identify p/m to $O(\hbar)$ as the velocity $v_\infty < 0$ at $x = \infty$. We see that as the particle approaches the origin it goes more and more slowly and in fact takes an infinite amount of time to reach the origin. Compare with the formula in standard radial infalling coordinates

$$\dot{x} = v_\infty \left(1 - \frac{1}{1 + \frac{1}{2}\frac{x}{\mathsf{G}}}\right)$$

for the distance from the event horizon of a Schwarzschild black hole with

$$\mathsf{G} = \frac{G_{\mathrm{Newton}} M}{c^2},$$

where M is the background gravitational mass and c is the speed of light. Thus the heuristic meaning of G in our model is that it measures the background mass or radius of curvature of the classical geometry of which our Planck scale Hopf algebra is a quantisation.

These arguments are from [3]. Working a little harder, one finds that the quantum mechanical limit is valid (the effects of G do not show up within one Compton wavelength) if

$$mM << m_{\mathrm{Planck}}^2,$$

while the curved classical limit is valid if

$$mM >> m_{\mathrm{Planck}}^2.$$

See also [6]. *The Planck-scale quantum group therefore truly unifies quantum effects and 'gravitational' effects in the context of Figure 3.*

Of course our model is only a toy model and one cannot draw too many conclusions given that our treatment is not even relativistic. The similarity to the Schwarzschild black-hole is, however, quite striking and one could envisage more complex examples which hit that exactly on the nose. The best we can say at the moment is that the search to unify quantum theory and gravity using such methods leads to tight constraints and features such as event-horizon-like coordinate singularities. Theorem 1 says that it is not possible to make a Hopf algebra for x, p with the correct classical limit in this context without such a coordinate singularity.

The quantum-gravity $\hbar, \mathbf{G} \to \infty, \frac{\mathbf{G}}{\hbar} = \lambda$ limit Having analysed the two familiar limits we can consider other 'deep quantum-gravity' limits. For example sending both our constants to ∞ but preserving their ratio we have

$$[x,p] = \imath\lambda x, \quad \Delta x = x\otimes 1 + 1\otimes x, \quad \Delta p \otimes 1 + 1 \otimes p$$

which is once again $U(\mathfrak{b}_+)$ regarded as in Example 3 in Section 3 'up side down' as a quantum space. The higher-dimensional analogues are 'κ-deformed' Minkowski space[29] as explained in Section 3, i.e. the Planck-scale quantum group puts some flesh on the idea that this might indeed come out of quantum gravity as some kind of effective limit[27]. Time itself would have to appear as $t = p$, (or $t = \sum_i p_i$ for the higher dimensional analogues) in this limit from the momenta conjugate in the effective quantum gravity theory to the position coordinates. This speculative possibility is discussed further in [31]. At any rate this deformed Minkowski space is at least mathematically nothing but a special limit of the Planck-scale quantum group from [3]. It gives some idea how the self-duality ideas of Section 2 might ultimately connect to testable predictions for Planck scale physics e.g. testable by gamma-ray bursts of cosmological origin.

The algebraic structure and Mach's Principle The notation $\mathbb{C}[x]\blacktriangleright\!\!\triangleleft\mathbb{C}[p]$ for the Planck-scale quantum group reflects its algebraic structure. As an algebra it is a cross product $\mathbb{C}[x]\rtimes\mathbb{C}[p]$ by the action $\triangleright$ of $\mathbb{C}[p]$ on $\mathbb{C}[x]$ by

$$p\triangleright f(x) = -\imath\hbar(1 - e^{-\frac{x}{G}})\frac{\partial}{\partial x}f \tag{20}$$

which means that it is a more or less standard 'Mackey quantisation' as a dynamical system. It can also be viewed as the deformation-quantization of a certain Poisson bracket structure on $\mathbb{C}(B_-)$ if one prefers that point of view. On the other hand its coproduct is obtained in a similar but dual way as a semidirect coproduct $\mathbb{C}[x]\blacktriangleright\!\!<\mathbb{C}[p]$ by a coaction of $\mathbb{C}[x]$ on $\mathbb{C}[p]$. This coaction is induced by an action of x on functions $f(p)$ of similar form to the above but with the roles reversed. In other words, *matching the action of momentum on position is an 'equal and opposite' coaction of position back on momentum.* This is indeed inspired by the ideas of Mach[16] as was promised in Section 2.

Observable-state duality and T-duality The phrase 'equal and opposite' has a precise consequence here. Namely the algebra corresponding to the coalgebra by dualisation has a similar cross product form by an analogous action of x on p. More precisely, one can show that

$$(\mathbb{C}[x]\blacktriangleright\!\!\triangleleft_{\hbar,G}\mathbb{C}[p])^* \cong \mathbb{C}[\bar{p}]\triangleright\!\!\blacktriangleleft_{\frac{1}{\hbar},\frac{G}{\hbar}}\mathbb{C}[\bar{x}], \tag{21}$$

where $\mathbb{C}[p]^* = \mathbb{C}[\bar{x}]$ and $\mathbb{C}[x]^* = \mathbb{C}[\bar{p}]$ in the sense of an algebraic pairing as in Example 1 in Section 3. Here $\langle p, \bar{x}\rangle = \imath$ etc., which then requires a change of the parameters as shown to make the identification precise. So the Planck-scale quantum group is self-dual up to change of parameters.

This means that whereas we would look for observables $a \in \mathbb{C}[x]\blacktriangleright\!\!\triangleleft\mathbb{C}[p]$ as the algebra of observables and states $\phi \in \mathbb{C}[\bar{p}]\triangleright\!\!\blacktriangleleft\mathbb{C}[\bar{x}]$ as the dual linear space, with $\phi(a)$ the expectation of a in state ϕ (See section 2.2), there is a dual interpretation whereby

$$\text{Expectation} = \phi(a) = a(\phi)$$

for the expectation of ϕ in 'state' a with $\mathbb{C}[\bar{p}]\triangleright\!\!\blacktriangleleft\mathbb{C}[\bar{x}]$ the algebra of observables in the dual theory. More precisely, only self-adjoint elements of the algebra are observables and positive functionals states, an a state ϕ will not be exactly hermitian in the dual theory etc. But the physical hermitian elements in the dual theory will be given by combinations of such states, and vice versa. This is a concrete example of observable-state duality as promised in Section 2. It was introduced by the author in [3].

Also conjectured at the time of [3] was that this duality should be related to T-duality in string theory. As evidence is the inversion of the constant $\hbar$. In general terms coupling inversions are indicative of such dualities. Notice also that Fourier transform implements this T-duality-like transformation as

$$\mathcal{F} : \mathbb{C}[x]\blacktriangleright\!\!\triangleleft_{\hbar,G}\mathbb{C}[p] \to \mathbb{C}[\bar{p}]\triangleright\!\!\blacktriangleleft_{\frac{1}{\hbar},\frac{G}{\hbar}}\mathbb{C}[\bar{x}]$$

Explicitly, it comes out as[27]

$$\mathcal{F}(: f(x,p) :) = \int_{-\infty}^{\infty}\int_{-\infty}^{\infty} \mathrm{d}x\mathrm{d}p\; e^{-\imath(\bar{p}+\frac{\imath}{G})x} e^{-\imath\bar{x}(p+p\triangleright)} f(x,p), \tag{22}$$

where $\triangleright$ is the action (20) and $f(x,p)$ is a classical function considered as defining an element of the Planck-scale quantum group by normal ordering x to the left.

The duality here is not exactly T-duality in string theory but has some features like it. On the other hand it is done here at the quantum level and not in terms only of Lagrangians. In this sense the observable state duality can give an idea about what should be 'M-theory' in string theory. Thus, at the moment all that one knows really is that the conjectured M-theory should be some form of algebraic structure with the property that it has different semiclassical limits with different Lagrangians related to each other by S,T dualities (etc.) at the classical level. Our observable-state duality ideas [3][1][19] as well as more recent work on T-duality suggests that:

Conjecture 3. M-theory should be some kind of algebraic structure possessing one or more dualities in a representation-theoretic or observable-state sense.

Actually there is an interesting anecdote here. I once had a chance to explain the algebraic duality ideas of my PhD thesis to Edward Witten at a reception in MIT in 1988 after his colloquium talk at Harvard on the state of string theory. He asked me 'is there a Lagrangian' and when I said 'No, it is all algebraic; classical mechanics only emerges in the limits, but there are two different limits related by duality', Witten rightly (at the time) gave me a short lecture about the need for a Lagrangian. 9 years later I was visiting Harvard and Witten gave a similarly-titled colloquium talk on the state of string theory. He began by stating that there was some algebraic structure called M-theory with Lagrangians appearing only in different limits.

The noncommutative differential geometry The lack of Lagrangians and other familiar structures in the full Planck-scale theory was certainly a valid criticism back in 1988. Since then, however, noncommutative geometry has come a long way and one is able to 'follow' the geometry as we quantise the system using these modern techniques. We do not have the space to recall the whole framework but exterior algebras, partial derivatives etc., make sense for quantum groups and many other noncommutative geometries. For the Planck-scale quantum group one has[27],

$$\partial_p : f(x,p) := \frac{\mathsf{G}}{\imath\hbar} : (f(x,p) - f(x,p-\imath\frac{\hbar}{\mathsf{G}})) :, \tag{23}$$

$$\partial_x : f(x,p) :=: \frac{\partial}{\partial x} f : -\frac{p}{\mathsf{G}}\partial_p : f : \tag{24}$$

which shows the effects of $\hbar$ in modifying the geometry. Differentiation in the p direction becomes 'lattice regularised' albeit a little strangely with an imaginary displacement. In the deformed-Minkowski space setting where $p = t$ it means that the Euclidean version of the theory is related to the Minkowski one by a Wick-rotation is being lattice-regularised by the effects of $\hbar$.

Also note that for fixed $\hbar$ the geometrical picture blows up when $\mathsf{G} \to 0$. I.e the usual flat space quantum mechanics CCR algebra does not admit a deformation of conventional differential calculus on $\mathbb{R}^2$ – *one needs a small amount of 'gravity' to be present for a geometrical picture in the quantum theory.* This is also evident in the exterior algebra[27]

$$f\mathrm{d}x = (\mathrm{d}x)f, \quad f\mathrm{d}p = (\mathrm{d}p)f + \frac{\imath\hbar}{\mathsf{G}}\mathrm{d}f$$

for the relations between 'functions' f in the Planck-scale quantum group and differentials. The higher exterior algebra looks more innocent with

$$\mathrm{d}x \wedge \mathrm{d}x = 0, \quad \mathrm{d}x \wedge \mathrm{d}p = -\mathrm{d}p \wedge \mathrm{d}x, \quad \mathrm{d}p \wedge \mathrm{d}p = 0 \tag{25}$$

Starting with the differential forms and derivatives, one can proceed to gauge theory, Riemannian structures etc., in some generality. One can also write down 'quantum' Poisson brackets and Hamiltonians[27] and (in principle) Lagrangians in the full noncommutative theory. Such tools should help to bridge the gap between model building via classical Lagrangians, which I personally do not think can succeed at the Planck scale, and some of the more noncommutative-algebraic ideas in Section 2.

4.2 Higher dimensional analogue

The Planck-scale quantum group is but the simplest in a family of quantum groups with similar features and parameters. We work from now with $\mathsf{G} = \hbar = 1$ for simplicity but one can always put the parameters back.

Of course one may take the n-fold tensor product of the Planck-scale quantum group, i.e. generators x_i, p_i and different i commuting. However, in higher dimensions the Ext_0 is much bigger and I do not know of any full computation of all the possibilities for $n > 1$. More interesting perhaps are some genuinely different higher-dimensional examples along similar but nonAbelian lines, one of which we describe now. The material is covered in [6], so we will be brief.

Thus, also from 1988, there is a bicrossproduct quantum group

$$\mathbb{C}(\mathbb{R}\rtimes\mathbb{R}^2)\blacktriangleright\!\!\triangleleft U(su_2) \tag{26}$$

constructed in [28][33] (actually as a Hopf-von Neumann algebra; here we consider only the simpler algebraic structure underlying it.)

The nonAbelian group $\mathbb{R}\rtimes\mathbb{R}^2$ is the one whose enveloping algebra we have considered in Example 3 in Section 3 as noncommutative spacetime. Here, however, we take it with a Euclidean signature and a different notation. Explicitly, it consists of 3-vectors $\boldsymbol{s}$ with third component $s_3 > -1$ and with the 'curved $\mathbb{R}^3$' nonAbelian group law

$$\boldsymbol{s}\cdot\boldsymbol{t} = \boldsymbol{s} + (s_3 + 1)\boldsymbol{t}.$$

Its Lie algebra is spanned by x_0, x_i with relations $[x_i, x_0] = x_i$ for $i = 1, 2$ as discussed before (this is how this algebra appeared first, in [28], in connection this higher-dimensional version of the Planck-scale quantum group). Now, on the group $\mathbb{R}\rtimes\mathbb{R}^2$ there is an action of SU_2 by a deformed rotation. This is shown in Figure 4. The orbits are still spheres but non-concentrically nested and accumulating at $s_3 = -1$. This is a dynamical system and (26) is its Mackey quantisation as a cross product. We see that we have similar features as for the Planck-scale quantum group, including some kind of coordinate singularity as $s_3 = -1$.

At the same time there is a 'back reaction' of $\mathbb{R}\rtimes\mathbb{R}^2$ back on SU_2, which appears as a coaction of $\mathbb{C}(\mathbb{R}\rtimes\mathbb{R}^2)$ on $U(su_2)$ in the cross coalgebra structure of the quantum group. Therefore the dual system, related by Fourier theory or observable-state duality, is of the same form, namely

$$U(\mathbb{R}\rtimes\mathbb{R}^2)\triangleright\!\!\blacktriangleleft\mathbb{C}(SU_2). \tag{27}$$

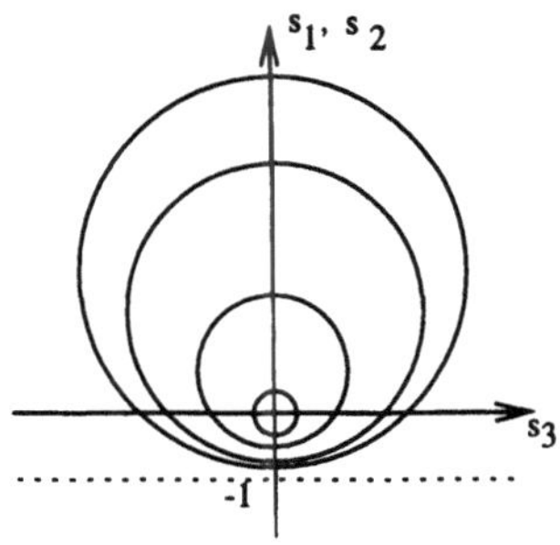

Fig. 4. Deformed action of classical SU_2 on $\mathbb{R}\bowtie\mathbb{R}^2$

It consists of a particle on SU_2 moving under the action of $\mathbb{R}\bowtie\mathbb{R}^2$. This is the dual system which, in the present case, looks quite different.

Finally, the general theory of bicrossproducts allows for a 'Schroedinger representation' of (26) on $U(\mathbb{R}\bowtie\mathbb{R}^2)$ and similarly of its dual on $U(su_2)$. Such a picture means that the 'wave functions' live in these enveloping algebras viewed as noncommutative spaces. There are also more conventional Hilbert space representations as well.

4.3 General construction

There is a general construction for bicrossproduct quantum groups of which the ones discussed so far are all examples. Thus suppose that

$$X = GM$$

is a *factorisation of Lie groups*. Then one can show that G acts on the set of M and M acts back on the set of G such that X is recovered as a double cross product (simultaneously by the two acting on each other) $X \cong G \bowtie M$. This turns out to be just the data needed for the associated cross product and cross coproduct

$$\mathbb{C}(M) \blacktriangleright\!\!\triangleleft U(\mathfrak{g}) \tag{28}$$

to be a Hopf algebra. The roles of the two Lie groups is symmetric and the dual is

$$(\mathbb{C}(M) \blacktriangleright\!\!\triangleleft U(\mathfrak{g}))^* = U(\mathfrak{m}) \triangleright\!\!\blacktriangleleft \mathbb{C}(G) \tag{29}$$

which means that there are certain families of homogeneous spaces (the orbits of one group under the other) which come in pairs, with the algebra of observables of the quantisation of one being the algebra of expectation states of the quantisation of the other. This is the more or less purest form of the ideas of Section 2 based on Mach's principle[16] and duality.

On the other hand, factorisations abound in Nature. For example every complexification of a simple Lie group factorises into its compact real form G and a certain solvable group $G^\star$, i.e. $G_{\mathbb{C}} = GG^\star$. The notation here is of a modern approach to the Iwasawa decomposition in [28]. For example, $SL_2(\mathbb{C}) = SU_2SU_2^\star$, where $SU_2^\star = \mathbb{R}{\triangleright\!\!\!<}\mathbb{R}^2$, gives the bicrossproduct quantum group (26) in the preceding section. There are similar examples

$$\mathbb{C}(G^\star){\blacktriangleright\!\!\!\triangleleft}U(\mathfrak{g}) \tag{30}$$

for all complex simple $\mathfrak{g}$. Also, slightly more general than the Iwasawa decomposition but still only a very special case of a general Lie group factorisation, let G be a Poisson-Lie group (a Lie group with a compatible Poisson-bracket). At the infinitesimal level the Poisson bracket defines a map $\mathfrak{g} \to \mathfrak{g}\otimes\mathfrak{g}$ making $\mathfrak{g}$ into a Lie bialgebra. This is an infinitesimal idea of a quantum group and is such that $\mathfrak{g}^\star$ is also a Lie algebra. In this setting there is a Drinfeld double Lie bialgebra $D(\mathfrak{g})$ and its Lie group is an example of a factorisation $GG^\star$.

By the way, this is exactly the setting for nonAbelian Poisson-Lie T-duality [12] in string theory, for classical σ-models on G and $G^\star$. The quantum groups (30) and their duals are presumably related to the quantisations of the point-particle limit of these sigma models. If so this would truly extend T-duality to the quantum case via the above observable-state duality ideas. While this is not proven exactly, something *like* this appears to be the case. Moreover, the bicrossproduct duality for (28) is much more general and is not limited to such Poisson-Lie structures on G. The group M need not be dual to G in the above sense and need not even have the same dimension. Recently it was shown that the Poisson-Lie T-duality in a Hamiltonian (but not Lagrangian) setting indeed generalises to a general factorisation like this[13].

Finally, there is one known connection between the bicrossproduct quantum groups and the more standard $U_q(\mathfrak{g})$ which we will consider next. Namely, Lukierski et al.[30] showed that a certain contraction process turned $U_q(so_{3,2})$ in a certain limit to some kind of 'κ-deformed' Poincaré algebra as mentioned below Example 3 in Section 3. It turned out later[29] that this was isomorphic to one of the bicrossproduct Hopf algebras above,

$$_\kappa\mathrm{Poincare}{\cong}\mathbb{C}(\mathbb{R}{\triangleright\!\!\!<}\mathbb{R}^3){\blacktriangleright\!\!\!\triangleleft}U(so_{3,1}).$$

The isomorphism here is nontrivial (which means in particular that κ-Poincaré certainly arose independently of the early bicrossproducts such as the 3 - dimensional case (26)). On the other hand, the bicrossproduct version of κ-Poincaré from [29] brought many benefits. First of all, the Lorentz sector is *undeformed.* Secondly, the dual is easy to compute (being an example of the general self-duality ideas above) and, finally, the Schroedinger representation means that this quantum group indeed acts covariantly on $U(\mathbb{R}{\triangleright\!\!\!<}\mathbb{R}^3)$, which should therefore be viewed as the κ-Minkowski space appropriate to this κ-Poincaré (prior to [29] one had only the noncovariant action of it on usual commutative Minkowski space, leading to a number of inconsistencies in attempting to model physics based on κ-Poincaré alone). Of course the point of view of Poincaré algebra as symmetry appears at first different from the main point of view of bicrossproducts as the quantisations of a dynamical system. However, as in Section 2 (and even for the classical Poincaré algebra) a symmetry enveloping algebra *should* also appear as part of (or all of) the quantum algebra of observables of the associated quantum theory because it should be realised among the quantum fields[14].

5 Deformed quantum enveloping algebras

No introduction to quantum groups would be complete if we did not also mention the much more well known deformations $U_q(\mathfrak{g})$ of complex simple $\mathfrak{g}$ arising from inverse scattering and the theory of solvable lattice models[7][8]. These have not, however, been very directly connected with Planck scale physics (although there are some recent proposals for this, as we saw in the lectures of Lee Smolin). They certainly did not arise that way and are not the quantum algebras of observables of physical systems. Therefore this is only going to be a lightning introduction to this topic. For more, see [6][10][34].

Rather, these quantum groups $U_q(\mathfrak{g})$ arise naturally as 'generalised' symmetries of certain spin chains and as generalised symmetries in the Wess-Zumino-Witten model conformal field theory. Just as groups can be found as symmetries of many different and unconnected systems, the same is true for the quantum groups $U_q(\mathfrak{g})$. They do, however, have a perhaps richer and more complex mathematical structure than the bicrossproducts, which is what we shall briefly outline.

As Hopf algebras one has the same duality ideas nevertheless. Thus, the quantum group $U_q(su_2)$ with generators $H, X_\pm$ and relations and coproduct

$$[H, X_\pm] = \pm X_\pm, \quad [X_+, X_-] = \frac{q^H - q^{-H}}{q - q^{-1}}$$

$$\Delta X_\pm = X_\pm \otimes q^{\frac{H}{2}} + q^{\frac{-H}{2}} \otimes X_\pm, \quad \Delta H = H \otimes 1 + 1 \otimes H$$

is dual to the quantum group $\mathbb{C}_q(SU_2)$ generated by a matrix of generators a, b, c, d. This has six relations of q-commutativity

$$ba = qab,\ ca = qac,\ bc = cb,\ dc = qcd,\ db = qbd,\ da = ad + (q - q^{-1})bc$$

and a determinant relation $ad - q^{-1}bc = 1$. The pairing is the same as in Example 2 in Section 2 at the level of generators (after a change of basis).

The main feature of these quantum groups, in contrast to the bicrossproduct ones, is that their representations form braided categories. Thus, if $V, W \in \mathrm{Rep}(U_q(\mathfrak{g}))$ then $V \otimes W$ is (as for any quantum group) also a representation. The action is

$$h{\triangleright}(v \otimes w) = (\Delta h).(v \otimes w) \tag{31}$$

for all $h \in U_q(\mathfrak{g})$, where we use the coproduct (for example the linear form of the coproduct of H means that it acts additively). The special feature of quantum groups like $U_q(\mathfrak{g})$ is that there is an element $\mathcal{R} \in U_q(\mathfrak{g})\bar{\otimes}U_q(\mathfrak{g})$ (the 'universal R-matrix or quasitriangular structure') which ensures an isomorphism of representations by

$$\Psi_{V,W} : V \otimes W \to W \otimes V, \quad \Psi_{V,W}(v \otimes w) = P \circ \mathcal{R}.(v \otimes w) \tag{32}$$

where P is the usual permutation or flip map. This ***braiding*** Ψ behaves much like the usual transposition or flip map for vector spaces but does not square to one. To reflect this one writes $\Psi = \diagup\!\!\!\diagdown$, $\Psi^{-1} = \diagdown\!\!\!\diagup$. It has properties consistent with the braid relations, i.e. when two braids coincide the compositions of Ψ, Ψ^{-1} that they represent also coincide. The fundamental braid relation of the braid group in Figure 5(a) corresponds to the famous Yang-Baxter or braid relation for the matrix corresponding to Ψ.

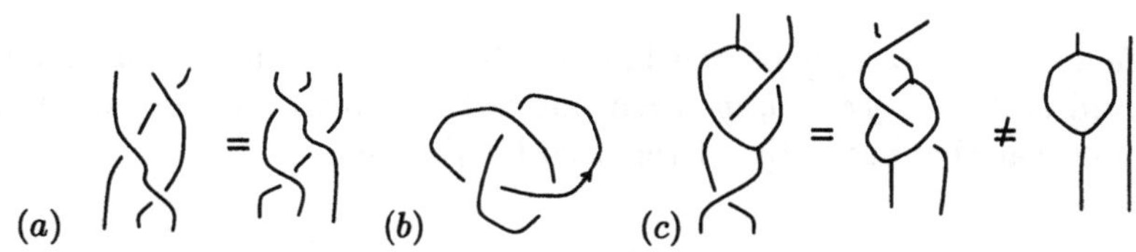

Fig. 5. (a) Braid relations (b) Trefoil knot (c) Braided algebra calculation

From this it is more or less obvious that such quantum groups lead to knot invariants. One can scan the (oriented) knot such as in Figure 5(b) from top to bottom. We choose a representation V with dual V^* and label the knot by V against a downward arc and V^* against an upward arc. As we read the knot, when we encounter an arc ${}_V\cap_{V^*}$ we let it represent the canonical element $\sum_a e_a \otimes f^a \in V \otimes V^*$. When we encounter crossings we represent them by the appropriate Ψ and finally when we encounter ${}^{V^*}\cup^{V}$ we apply the evaluation map. There is also a prescription for when we encounter ${}_{V^*}\cap_V$ and ${}^{V}\cup^{V^*}$. At the end of the day we obtain a number depending on q (which went into the braiding). This function of q is (with some fiddling that we have not discussed) an invariant of the knot regarded as a framed knot. This is not the place to give details of knot theory, but this is the rough idea. In physical terms one should think of the knot as a process in 1+1 dimensions in which a particle V and antiparticle V^* is created at an arc, some kind of scattering Ψ occurs at crossings, etc.

For standard $U_q(\mathfrak{g})$ the construction of representations is not hard, all the standard ones of $\mathfrak{g}$ just q-deform. For example, the spin $\frac{1}{2}$ representation of su_2 deforms to a 2-dimensional representation of $U_q(su_2)$. The associated knot invariant is the celebrated Jones polynomial.

5.1 Braided mathematics and braided groups

This braiding is the key property of the quantum groups $U_q(\mathfrak{g})$ and other 'quasi-triangular Hopf algebras' of similar type. It means in particular that *any algebra on which the quantum group acts covariantly becomes braided.* This is therefore indicative of a whole braided approach to noncommutative geometry or *braided geometry* via algebras or 'braided' spaces on which quantum groups $U_q(\mathfrak{g})$ act as generalised symmetries. Note that we are not so much interested in this point of view in the noncommutative geometry of the quantum groups $U_q(\mathfrak{g})$ themselves, although one can study this as a source of mathematical examples. More physical is the algebras in which these objects act.

In this approach the meaning of q is that it enters into the braiding, i.e. it generalises the -1 of supertransposition in super-geometry. This is 'orthogonal' to the usual idea of noncommutative geometry, i.e. it is not so much a property of one algebra but of composite systems, namely of the noncommutativity of tensor products. The simplest new case is where the braiding is just a factor q. To see how this works, consider the *braided line* $B = \mathbb{C}[x]$. As an algebra this is just the polynomials in one variable again.

Example 4. Let $B = \mathbb{C}[x]$ be the braided line, where independent copies x, y have braid statistics $yx = qxy$ when one is transposed past the other (c.f. a Grassmann variable but with -1 replaced by q). Then

$$\partial_q f(y) = x^{-1}\left(f(x+y) - f(x)\right)|_{x=0} = \frac{f(y) - f(qy)}{(1-q)y}$$

This is easy to see on monomials, i.e. $\partial_q y^n$ is the coefficient of the x-linear part in $(x+y)^n$ after we move all x to the left. In fact mathematicians have played with such a q-derivative since 1908[35] as having many cute properties. We see[36] that it arises very naturally from the braided point of view – one just has to realise that x is a braided variable. This point of view also leads to the correct properties of integration. Namely there is a relevant indefinite integration to go with ∂_q characterised by[37]

$$\int_0^{x+y} f(z)\mathrm{d}_q z = \int_0^y f(z)\mathrm{d}_q z + \int_0^x f(z+y)\mathrm{d}_q z \tag{33}$$

provided $yx = qxy$, $yz = qzy$ etc., during the computation. In the limit this gives the infinite Jackson integral previously known in this context. One also has braided exponentials, braided Fourier theory etc., for these braided variables.

The braided point of view is also much more powerful than simply trying to sprinkle q into formulae here and there.

Example 5. Let $B = \mathbb{C}_q^2$ be the quantum-braided plane generated by x, y with the relations $yx = qxy$, where two independent copies have the braid statistics

$$x'x = q^2xx', \quad x'y = qyx', \quad y'y = q^2yy', \quad y'x = qxy' + (q^2-1)yx'.$$

Here x', y' are the generators of the second copy of the plane. Then

$$(y + y')(x + x') = q(x + x')(y + y')$$

i.e. $x + x', y + y'$ is another copy of the quantum-braided plane. Then by similar definitions as above, one has braided partial derivatives

$$\partial_{q,x} f(x, y) = \frac{f(x, y) - f(qx, y)}{(1 - q)x}, \quad \partial_{q,y} f(x, y) = \frac{f(qx, y) - f(qx, qy)}{(1 - q)y}$$

for expressions normal ordered to x on the left. Note in the second expression an extra q as $\partial_{q,y}$ moves past the x

Thus *you can add points in the braided plane*, and then (by an infinitesimal addition) define partial derivatives etc. This is a problem (multilinear q-analysis) which had been open since 1908 and was only solved relatively recently (by the author) in [36], as a demonstration of braided mathematics. We note in passing that $yx = qxy$ is sometimes called the 'Manin plane'. Manin considered only the algebra and a quantum group action on it, without the braided point of view, without the braided addition law and without the partial derivatives.

Finally, there is a more formal way by which all such constructions are done systematically, which we now explain. It amounts to nothing less than a new kind of algebra in which algebraic symbols are replaced by braids and knots.

First of all, given two algebras B, C in a braided category (such as the representation of $U_q(\mathfrak{g})$) we have a braided tensor product $B\underline{\otimes}C$ algebra in the same category defined like a superalgebra but with -1 replaced by the braiding $\Psi_{C,B}$. Thus the tensor product becomes noncommutative (even if each algebra B, C was commutative) – the two subalgebras 'commute' up to Ψ. This is the mathematical definition of *braid statistics: the noncommutavity of the notion of 'independent' systems.* We call such noncommutativity *outer* in contrast to the inner noncommutativity of quantisation, which is a property of one algebra alone. In Example 4, the joint algebra of the independent x, y is $\mathbb{C}[x]\underline{\otimes}\mathbb{C}[y]$ with $\Psi(x \otimes y) = qy \otimes x$. In Example 5 the braided tensor product is between one copy x, y and the other x', y'. The braiding Ψ in this case is more complicated. In fact it is the same braiding from the $U_q(su_2)$ spin $\frac{1}{2}$ representation that gave the Jones polynomial. The miracle that makes knot invariants is the same miracle that allows braided multilinear algebra.

The addition law in both the above examples makes them into braided groups[38]. They are like quantum groups or super-quantum groups but with braid statistics. Thus, there is a coproduct

$$\Delta x = x \otimes 1 + 1 \otimes x, \quad \Delta y = y \otimes 1 + 1 \otimes y$$

etc., (this is a more formal way to write $x + x', y + y'$). But $\Delta : B \to B\underline{\otimes}B$ rather than mapping to the usual tensor product. We do not want to go into the whole theory of braided groups here. Suffice it to say that the theory can be developed to the same level as quantum groups: integrals, Fourier theory, etc., but using new techniques. One draws the product $B \otimes B \to B$ as a map $\curlyvee$, the coproduct as $\curlywedge$, etc. Similarly with other maps, some strands coming in for the inputs and some leaving for the outputs. We then 'wire up' an algebraic expression by wiring outputs of one operation into the inputs of others. When wires have to cross under or over we have to chose one or the other as Ψ or Ψ^{-1}. We draw such diagrams flowing down the page. An example of a braided-algebra calculation is given in Figure 5(c).

Braided groups exist in abundance. There are general arguments that every algebraic quantum field theory contains at its heart some kind of (slightly generalised) braided group[39]. Moreover, the ideas here are clearly very general: braided algebra.

5.2 Systematic *q*-Special Relativity

Clearly braided groups are the correct foundation for q-deformed geometry based on q-planes and similar q-spaces. One of their main successes in the period 1992-1994 was a more or less complete and systematic q-deformation by the team in Cambridge of the main structures of special relativity and electromagnetism, i.e. q-Minkowski space and basic structures [40][41][36][42][43][37][44][45][46]:

- q-Minkowski space as 2×2 braided Hermitian matrices
- q-addition etc., on q-Minkowski space
- q-Lorentz quantum group $\mathbb{C}_q(SU_2) \bowtie \mathbb{C}_q(SU_2)$
- q-Poincaré+scale quantum group $\mathbb{R}_q^{1,3} \rtimes \widetilde{U_q(so_{1,3})}$
- q-partial derivatives
- q-differential forms
- q-epsilon tensor
- q-metric
- q-integration with Gaussian weight
- q-Fourier theory
- q-Green functions (but no closed form)
- q-$*$ structures and q-Wick rotation

The general theory works for any braiding or 'R-matrix'. I do want to stress, however, that this project was not in a vacuum. For example, the algebra of q-Minkowski had been proposed independently of [40] in [47], but without the braided matrix or additive structures. The q-Lorentz was studied by the same authors but without its quasitriangular structure, Wess, Zumino et al.[48] studied the q-Poincaré but without its semidirect structure and action on q-Minkowski space, while Fiore[49] studied q-Gaussians in the Euclidean case, etc. More recently, we have[50][51],

- q-conformal group $\mathbb{R}_q^{1,3} \rtimes \widetilde{U_q(so_{1,3})} \ltimes \mathbb{R}_q^{1,3}$
- q-diffeomorphism group

Notably not on the list, in my opinion still open, is the correct formulation of the q-Dirac equation. Aside from this, the programme came to an end when certain deep problems emerged. In my opinion they are as follows. First of all, we ended up with formal power-series e.g. the q-Green function is the inverse Fourier transform of $(\boldsymbol{p}\cdot\boldsymbol{p}-m^2)^{-1}$ so in principle it is now defined. But not in closed form! The methods of q-analysis as in [35][52] are not yet far enough advanced to have nice names and properties for the kinds of powerseries functions encountered. This is a matter of time. Similarly, braided integration means we can in principle write down and compute braided Feynman diagrams and hence define braided quantum field theory at least operatively. Recently R. Oeckl was able[53] to apply the braided integration theory of [37] not to q-spacetime but directly to a q-coordinate algebra as the underlying vector space of fields on spacetime. Here the braided algebra B replaces the 'fields' on spacetime. Choosing a basis of such fields one can still apply braided Gaussian integration and actually compute correlation functions. So the computational problems can and are being overcome.

Secondly and more conceptual, it should be clear that when we deform classical constructions to braided ones we have to choose Ψ or Ψ^{-1} whenever wires cross. Sometimes neither will do, things get tangled up. But if we succeed it means that for every q-deformation there is another where we could have made the opposite choice in every case. *This classical geometry bifurcates into two q-deformed geometries according to Ψ or Ψ^{-1}*. Moreover, the role of the $*$ operation is that it interchanges these two[45]. Roughly speaking,

$$\text{classical geometry}\ \begin{array}{l}\nearrow\quad q-\text{geometry}\\ \qquad\qquad\ \updownarrow *\\ \searrow\ \text{conjugate } q-\text{geometry}\end{array}$$

where the conjugate is constructed by interchanging the braiding with the inverse braiding (i.e. reversing braid crossings in the diagrammatic construction). For the simplest cases like the braided line it means interchanging q, q^{-1}. This is rather interesting given that the $*$ is a central foundation of quantum mechanics and our concepts of probability. But it also means one cannot do q-quantum mechanics etc., with q-geometry alone; one needs also the conjugate geometry.

5.3 The physical meaning of q

According to what we have said above, the true meaning of q is that it generalises the -1 of fermionic statistics. That is why it is dimensionless. It is nothing other than a parameter in a mathematical structure (the braiding) in a generalisation of our usual concepts of algebra and geometry, going a step beyond supergeometry.

This also means that q is an ideal parameter for regularising quantum field theory. Since most constructions in physics q-deform, such a regularisation scheme is much less brutal than say dimensional or Pauli-Villars regularisation as it preserves symmetries as q-symmetries, the q-epsilon tensor etc. [15]. In this context it seems at first *too good* a regularisation. Something has to go wrong for anomalies to appear.

Conjecture 4. In q-regularisation the fact that only the Poincaré+scale q-deforms (the two get mixed up) typically results (when the regulator is removed after renormalisation) in a scale anomaly of some kind.

This is probably linked to a much nicer treatment of the renormalisation group that should be possible in this context. Again a lot of this must await more development of the tools of q-analysis. At any rate the result in [15] is that q-deformation does indeed regularise, turning some of the infinities from a Feynman loop integration into poles $(q-1)^{-1}$.

All of this is related to the Planck scale as follows. Thus, as well as being a good regulator one can envisage (in view of our general ideas about noncommutativity and the Planck scale) that the actual world is in reality better described by $q \neq 1$ due to Planck scale effects. In other words q-deformed geometry could indeed be the next-to-classical order approximation to the geometry coming out of some unknown theory of quantum gravity. This was the authors own personal reason[15] for spending some years q-deforming the basic structures of physics. The UV cut-off provided by a 'foam-like structure of space time' would instead be provided by q-regularisation with $q \neq 1$. Moreover, if this is so then q-deformed quantum field theory should also appear coming out of quantum gravity as an approximation one better than the usual. Such a theory would be massless according to the above remarks (because there is no q-Poincaré without the scale generator). Or at least particle masses would be small compared to the Planck mass. *How the q-scale invariance breaks would then be a mechanism for mass generation.*

There are also several other 'purely quantum' features of q-geometry not visible in classical geometry, which would likewise have consequences for Planck scale physics. One of them is:

Theorem 2. *The braided group version of the enveloping algebras $U_q(\mathfrak{g})$ and their q-coordinate algebras are isomorphic. I.e. there is essentially only one object in q-geometry with different scaling limits as $q \to 1$ to give the classical enveloping algebra of $\mathfrak{g}$ or coordinate algebra of G.*

The self-duality isomorphisms involve dividing by $q-1$ and are therefore singular when $q = 1$, i.e. this is totally alien to conventional geometric ideas. Enveloping algebras and their coordinate algebras are supposed to be dual not isomorphic. This self-duality in q-geometry is rather surprising but is fully consistent with the self-duality ideas of Section 2. In many ways q-geometry is simpler and more regular than the peculiar $q = 1$ that we are more familiar with.

Recently, it was argued[11] that since loop gravity is linked to the Wess-Zumino-Witten model, which is linked to $U_q(su_2)$ (or some other quantum group), that indeed q-geometry should appear coming out of quantum-gravity with cosomological constant Λ. There is even provided a formula

$$q = e^{\frac{2\pi\imath}{2+k}}, \quad k = \frac{6\pi}{G^2_{\text{Newton}}\Lambda}.$$

If so then the many tools of q-deformation developed in the last several years would suddenly be applicable to study the next-to-classical structure of quantum-gravity. The fact that loop variable and spin-network methods 'tap into' the revolutions that have taken place in the last decade around quantum groups, knot theory and the WZW model (this was evident for example in the black-hole entropy computation[54]) makes such a conjecture reasonable. It also indicates to me that these new quantum gravity methods are not just 'pushing some problem off to another corner' but are building on a certain genuine advance that has already revolutionised several other branches of mathematics. Usually in science when one big door is opened it has nontrivial repercussions in several fields.

One way or another the general idea is that quantum effects dominant at the Planck scale force geometry itself to be modified as we approach it such as to have a noncommutative or 'quantum' aspect expressed by $q \neq 1$. Although q is dimensionless and might be given, for example, by formulae such as the above, one can and should still think of q as behaving formally like the exponential of an *effective Planck's constant* $\hbar_0$, say. That is we can make semiclassical expansions, speak of Poisson-brackets being 'quantized' etc. This is not exactly physical quantisation except in so far as quantum effects at the Planck scale are at the root of it. The precise physical link can only be made in a full theory of quantum gravity. It is only in this sense, however, that q-geometry is 'quantum geometry' and 'quantum groups' are so called. For example, the q-coordinate algebras of $U_q(\mathfrak{g})$ are quantisations in this sense of a certain Poisson-Lie bracket on G (as mentioned in Section 4.3). Similarly for all our other q-spaces.

Example 6. [50] q-Minkowski space quantises a Poisson-bracket on $\mathbb{R}^{1,3}$ given by the action of the special conformal translations.

This again points to a remarkable interplay between q-regularisation, the renormalisation group, gravity and particle mass.

At least in this context we want to note that the braided approach of this subsection gives a new and systematic approach to the 'quantisation' problem that solves by new 'braid diagram' methods some age-old problems. Usually, one writes a Poisson bracket and tries to 'quantise' it by a noncommutative algebra. Apart from existence, *the problem often overlooked is what I call the uniformity of quantisation problem.* There is only one universe. How do we know when we have quantised this or that space separately that they are consistent with each other, i.e. that they all fit together to a single quantum universe?

Our theme in Section 2 is of course is that quantisation is not a well-defined problem. Rather one *should* have a deeper point of view which leads directly to the quantum-algebraic world – what we call geometry is then the semiclassical limit of the intrinsic structure of that, i.e. all different spaces and choices of Poisson structures on them will emerge from semiclassicalisation and not vice versa.

Braided algebra solves the uniformity problem in this way. Apart from giving the q-deformation of most structures in physics, it does it uniformly and in a generally consistent way because what what we deform is actually the category of vector spaces into a braided category. All constructions based on linear maps then deform coherently and consistently with each other as braid diagram constructions (so long as they do not get tangled). After that one inserts the formulae for specific braidings (e.g. generated by specific quantum groups) to get the q-deformation formulae. *After that* one semiclassicalises by taking commutators to lowest order, to get the Poisson-bracket that we have just quantised. Moreover, different quantum groups $U_q(\mathfrak{g})$ are all mutually consistent being related to each other by an inductive construction[55]. We have seen this with q-Minkowski space above.

In summary, the q-deformed examples demonstrate a remarkable unification of three different points of view; q as a generalisation of fermionic -1, q as a 'quantisation' (so these ideas are unified) and q as a powerful regularisation parameter in physics. By the way, these are all far from the original physical role of q, where $U_q(su_2)$ arose as a generalised symmetry of the XXY lattice model and where q measures the anisotropy due to an applied external magnetic field (rather, they are the authors' point of view developed under the heading of the braided approach to q-deformation and braided geometry).

6 Noncommutative differential geometry and Riemannian manifolds

We have promised that today there is a more or less complete theory of noncommutative differential geometry that includes most of the naturally occurring examples such as those in previous sections, but is a general theory not limited to special examples and models, i.e. has the same degree of 'flabbiness' as conventional geometry. Here I will try to convince you of this and give a working definition of a 'quantum manifold' and 'quantum Riemannian manifold'[4]. I do not want to say that this is the last word; the subject is still evolving but there is now something on the table. Among other things, our constructions are purely algebraic with operator and C^*-algebra considerations as in Connes' approach not fully worked. In any case, the reader may well want to start with the more accessible Section 6.4, where we explore the semiclassical implications at the more familiar level of the ordinary differential geometry coming out of the full noncommutative theory.

6.1 Quantum differential forms

As explained in Section 2 our task is nothing other than to give a formulation of geometry where the coordinate algebra on a manifold is replaced by a general algebra M. The first step is to *choose* the cotangent space or differential structure. Since one can multiply forms by 'functions' from the left and right, the natural definition is to define a first order calculus as a bimodule Ω^1 of the algebra M, along with a linear map $\mathrm{d}: M \to \Omega^1$ such that

$$\mathrm{d}(ab) = (\mathrm{d}a)b + a\mathrm{d}b, \quad \forall a, b \in M.$$

Differential structures are not unique even classically, and even more non-unique in the quantum case. There is, however, one universal example of which others are quotients. Here

$$\Omega^1_{\rm univ} = \ker\cdot \subset M\otimes M, \quad {\rm d}a = a\otimes 1 - 1\otimes a.$$

Classically we do not think about this much because on a group there *is* a unique translation-invariant differential calculus; since we generally work with manifolds built on or closely related to groups we tend to take the inherited differential structure without thinking. In the quantum case, i.e. when M is a quantum (or braided) group one has a similar notion [56]: a differential calculus is bicovariant if there are coactions $\Omega^1 \to \Omega^1\otimes M, \Omega^1 \to M\otimes\Omega^1$ forming a bicomodule and compatible with the bimodule structures and d.

Theorem 3. *[57] For the q-coordinate rings of the quantum groups $U_q(\mathfrak{g})$, the (co)irreducible bicovariant $(\omega^1, {\rm d})$ are essentially (for generic q) in correspondence with the irreducible representations ρ of $\mathfrak{g}$, and*

$$\Omega^1_{\rm univ} = \oplus_\rho \Omega^1_\rho.$$

The lowest spin $\frac{1}{2}$ representation of $U_q(su_2)$ defines its usual differential calculus plus a Casimir as $q\to 1$. The higher differential calculi show up in the q-geometry and correspond to higher spin. This should therefore be a step towards understanding how macroscopic differential geometry arises out of the loop gravity and spin network formalism. For example, the black-hole entropy computation[54] reported in Abbay Ashtekar's lectures at the conference was dominated by the spin $\frac{1}{2}$ states, which seems to me should be analogous to the standard differential calculus on the spin connection bundle dominating as macroscopic geometry emerges from the quantum gravity theory.

We do not have room to give more details here even of an example of Theorem 3, but see [57]. Instead we content ourselves with an even simpler and more pedagogical result.

Proposition 1. *[58] If k is a field and $M = k[x]$ the polynomials in one variable, the (co)irreducible bicovariant calculi $(\Omega^1, {\rm d})$ are in correspondence with field extensions of the form $k_\lambda = k[\lambda]$ modulo $m(\lambda) = 0$, where m is an irreducible monic polynomial. Here*

$$\Omega^1 = k_\lambda[x], \quad {\rm d}f(x) = \frac{f(x+\lambda)-f(x)}{\lambda},$$

$$f(x).g(\lambda,x) = f(x+\lambda)g(\lambda,x), \quad g(\lambda,x).f(x) = g(\lambda,x)f(x)$$

for functions f and one-forms g.

For example, over $\mathbb{C}$, $(\Omega^1, {\rm d})$ on $\mathbb{C}[x]$ are classified by $\lambda_0\in C$ and one has

$$\Omega^1 = {\rm d}x\mathbb{C}[x], \quad {\rm d}f = {\rm d}x\frac{f(x+\lambda_0)-f(x)}{\lambda_0}, \quad x{\rm d}x = ({\rm d}x)x + \lambda_0. \tag{34}$$

We see that the Newtonian case $\lambda_0 = 0$ is only one special point in the moduli space of quantum differential calculi. But if Newton had not supposed that differentials and forms commute he would have had no need to take this limit. What one finds with noncommutative geometry is that there is no need to take this limit at all. In particular, noncommutative geometry extends our usual concepts of geometry to lattice theory without taking the limit of the lattice spacing going to zero.

It is also interesting that the most important field extension in physics, $\mathbb{R} \subset \mathbb{C}$, can be viewed noncommutative-geometrically with complex functions $\mathbb{C}[x]$ the quantum 1-forms on the algebra of real functions $\mathbb{R}[x]$. As such its quantum cohomology is nontrivial, see [58].

6.2 Bundles and connections

To go further one has to have a pretty abstract view of differential geometry. For trivial bundles it is a little easier: fix a quantum group coordinate ring H. Then a gauge field is a map $H \to \Omega^1$, etc. See [59][60]. To define a manifold, however, one has to handle nontrivial bundles. In noncommutative geometry there is (as yet) no proper way to build this by patching trivial bundles. All those usual concepts involve open sets etc, not existing in the noncommutative case. Fortunately, if one thinks about it abstractly enough one can come up with a purely algebraic formulation independent of any patches or coordinate system. For simplicity we are going to limit attention to the universal calculi; the theory is know for general calculi as well.

Basically, a classical bundle has a free action of a group and a local triviality property. In our algebraic terms this translates[5][60] to an algebra P in the role of 'coordinate algebra of the total space of the bundle', a coaction $\Delta_R : P \to P \otimes H$ of the quantum group H such that the fixed subalgebra is M,

$$M = P^H = \{p \in P | \ \Delta_R p = p \otimes 1\}. \tag{35}$$

Local triviality is replaced by the requirement that

$$0 \to P(\Omega^1 M)P \to \Omega^1 P \xrightarrow{\tilde{\chi}} P \otimes \ker \epsilon \to 0 \tag{36}$$

is exact, where $\tilde{\chi} = (\cdot \otimes \mathrm{id})\Delta_R$ plays the role of generator of the vertical vector fields corresponding classically to the action of the group (for each element of H^* it maps $\Omega^1 P \to P$ like a vector field). Exactness says that the one-forms $P(\Omega^1 M)P$ lifted from the base are exactly the ones annihilated by the vertical vector fields.

An example is the quantum sphere. Classically the inclusion $U(1) \subset SU_2$ in the diagonal has coset space S^2 and defines the $U(1)$ bundle over the sphere on which the monopole lives. The same idea works here, but since we deal with coordinate algebras the arrows are reversed. The coordinate algebra of $U(1)$ is the polynomials $\mathbb{C}[g, g^{-1}]$.

Example 7. There is a projection from $\mathbb{C}_q(SU_2) \to \mathbb{C}[g, g^{-1}]$

$$\pi \begin{pmatrix} a & b \\ c & d \end{pmatrix} = \begin{pmatrix} g & 0 \\ 0 & g^{-1} \end{pmatrix}$$

Its induced coaction $\Delta_R = (\mathrm{id} \otimes \pi)\Delta$ is by the degree defined as the number of a, c minus the number of b, d in an expression. The quantum sphere S_q^2 is the fixed subalgebra i.e. the degree zero part. Explicitly, it is generated by $b_3 = ad$, $b_+ = cd$, $b_- = ab$ with q-commutativity relations

$$b_\pm b_3 = q^{\pm 2} b_3 b_\pm + (q^{\pm 2} - 1) b_3, \quad q^2 b_- b_+ = q^{-2} b_+ b_- + (q - q^{-1})(b_3 - 1)$$

and the sphere equation $b_3^3 = b_3 + q b_- b_+$, and forms a quantum bundle[5][60].

When $q \to 1$ we can write $b_\pm = \pm(x \pm \imath y)$, $b_3 = z + \frac{1}{2}$ and the sphere equation becomes $x^2 + y^2 + z^2 = \frac{1}{4}$ while the others become that x, y, z commute. The quantum sphere itself is a member of a 2-parameter family[61] of quantum spheres (the others can also be viewed as bundles in a suitable framework[62].)

One can go on and define a connection as an equivariant splitting

$$\Omega^1 P = P(\Omega^1 M)P \oplus \text{complement} \tag{37}$$

i.e. an equivariant projection Π on $\Omega^1 P$. One can show the required analogue of the usual theory, i.e. that such a projection corresponds to a connection form such that

$$\omega : \ker \epsilon \to \Omega^1 P, \quad \tilde{\chi}\omega = \mathrm{id} \tag{38}$$

where ω intertwines with the adjoint coaction of H on itself. There is such a connection on the example above – the q-monopole[5]. It is $\omega(g-1) = dda - qbdc$.

Finally, one can define associated bundles. If V is a vector space on which H coacts then we define the associated 'bundles' $E^* = (P \otimes V)^H$ and $E = \hom_H(V, P)$, the space of intertwiners. The two bundles should be viewed geometrically as 'sections' in classical geometry of bundles associated to V and V^*. Given a suitable (strong) connection one has a covariant derivative

$$D_\omega : E \to E \otimes M, \quad D_\omega = (\mathrm{id} - \Pi) \circ \mathrm{d} \tag{39}$$

This is where the noncommutative differential geometry coming out of quantum groups links up with the more traditional C^*-algebra approach of A. Connes and others. Traditionally a vector bundle over any algebra is defined as a finitely generated projective module. However, there is no notion of quantum principal bundle of course without quantum groups. The associated bundles to the q-monopole bundle are indeed finitely generated projective modules[63]. The projectors are elements of the noncommutative K-theory $K_0(S_q^2)$ and their pairing with Connes' cyclic cohomology[9] allows one to show that the bundle is non-trivial even when $q \neq 1$. Thus the quantum groups approach is compatible with Connes' approach but provides more of the (so far algebraic) infrastructure of differential geometry – principal bundles, connection forms, etc. otherwise missing.

6.3 Soldering and quantum Riemannian structure

With the above ingredients we can give a working definition of a quantum manifold. See refer to [4] for details. The idea is that the main feature of being a manifold is that, locally, one can chose a basis of the tangent space at each point (e.g. a vierbein in physics) patching up globally via GL_n gauge transformations. In abstract terms it means a frame bundle to which the tangent bundle is associated by a 'soldering form'. For a general algebra M we specify this 'frame bundle' directly as some suitable quantum group principal bundle.

Thus, we define a *frame resolution* of M as quantum principal bundle over M, (P, H, Δ_R), a comodule V and an equivariant 'soldering form' $\theta : V \to P\Omega^1 M \subset \Omega^1 P$ such that the induced map

$$E^* \to \Omega^1 M, \quad p \otimes v \mapsto p\theta(v) \tag{40}$$

is an isomorphism. Of course, all of this has to be done with suitable choices of differential calculi on M, P, H whereas we have been focusing for simplicity on the universal calculi. There are some technical problems here but the same definitions more or less work in general. Our working definition[4] of a *quantum manifold* is this data $(M, \Omega^1, P, H, \Delta_R, V, \theta)$.

The definition works in that one has many usual results. For example, a connection ω on the frame bundle induces a covariant derivative D_ω on the associated bundle E which maps over under the soldering isomorphism to a covariant derivative

$$\nabla : \Omega^1 M \to \Omega^1 M \underset{M}{\otimes} \Omega^1 M. \tag{41}$$

Its torsion is defined as corresponding similarly to $D_\omega \theta$.

Defining a Riemannian structure is harder. It turns out that it can be done in a 'self-dual' manner as follows. Given a framing, a 'generalised metric' isomorphism $\Omega^{-1}M \to \Omega^1 M$ between vector fields and one forms can be viewed as the existence of *another* framing $\theta^* : V^* \to (\Omega^1 M)P$, which we call the *coframing*, this time with V^*. Nondegeneracy of the metric corresponds to θ^* inducing an isomorphism $E \cong \Omega^1 M$.

Thus our working definition[4] of a quantum Riemannian manifold is the data $(M, \Omega^1, P, H, \Delta_R, V, \theta, \theta^*)$, where we have a framing and at the same time $(M, \Omega^1, P, H, \Delta_R, V^*, \theta^*)$ is another framing. The associated quantum metric is

$$g = \sum_a \theta^*(f^a)\theta(e_a) \in \Omega^1 M \underset{M}{\otimes} \Omega^1 M \tag{42}$$

where $\{e_a\}$ is a basis of V and $\{f^a\}$ is a dual basis (c.f. our friend the canonical element exp from Fourier theory in Section 3).

Now, this self-dual formulation of 'metric' as framing and coframing is symmetric between the two. One could regard the coframing as the framing and vice versa. From our original point of view its torsion tensor corresponding to $D_\omega \theta^*$ is some other tensor, which we call the *cotorsion tensor*[4]. We then define a generalised Levi-Civita connection on a quantum Riemannian manifold as the ∇ of a connection ω such that the torsion and cotorsion tensors both vanish.

This is about as far as this programme has reached at present. One defines curvature of course as corresponding to the curvature of ω, which is $d\omega + \omega \wedge \omega$, but before we can finish the program outlined in Section 2 we still need to understand the Ricci and Einstein tensors in this setting. For this one has to understand their classical meaning more abstractly i.e. beyond some contraction formulae even in conventional geometry. It would appear that it has a lot to do with entropy and the relation between gravity and counting (geometric) states thermodynamically.

6.4 Semiclassical limit

To get the physical meaning of the cotorsion tensor and other ideas coming out of noncommutative Riemannian geometry, let us consider the semiclassical limit. *What we find is that noncommutative geometry forces us to slightly generalise conventional Riemannian geometry itself.* If noncommutative geometry is closer to what comes out of quantum gravity then this generalisation of conventional Riemannian geometry should be needed to include Planck scale effects or at least to be consistent with them when they emerge at the next order of approximation.

The generalisation, more or less forced by the noncommutativity, is as follows:

- We have to allow any group G in the 'frame bundle', hence the more general concept of a 'frame resolution' (P, G, V, θ^a_μ) or *generalised manifold.*
- The *generalised metric* $g_{\mu\nu} = \sum_a \theta^{*a}_\mu \theta_{\nu a}$ corresponding to a coframing θ^{*a}_μ is nondegenerate but need not be symmetric.
- The *generalised Levi-Civita* connection defined as having vanishing torsion and vanishing cotorsion respects the metric only in a skew sense

$$\nabla_\mu g_{\nu\rho} - \nabla_\nu g_{\mu\rho} = 0 \tag{43}$$

- The group G is not unique (different flavours of frames are possible, e.g. an E_6-resolved manifold), not necessarily based on SO_n. This gives different flavours of covariant derivative ∇ that can be induced by a connection form ω.
- Even when G is fixed and $g_{\mu\nu}$ is fixed, the generalised Levi-Cevita condition does not fix ∇ uniquely, i.e. one should use a first order $(g_{\mu\nu}, \nabla)$ formalism.

To explain (43) we should note the general result [4] that for any generalised metric one has

$$\nabla_\mu g_{\nu\rho} - \nabla_\nu g_{\mu\rho} = \mathrm{CoTorsion}_{\mu\nu\rho} - \mathrm{Torsion}_{\mu\nu\rho}, \tag{44}$$

where we use the metric to lower all indices. Here ω gives two covariant derivatives

$$\omega \begin{array}{l} \theta \nearrow \nabla \\ \\ \theta^* \searrow {}^*\nabla \end{array}$$

depending on whether we regard θ or θ^* as the soldering form. The two are related by

$$g({}^*\nabla_X Y, Z) + g(Y, \nabla_X Z) = X(g(Y,Z)) \tag{45}$$

for vector fields X, Y, Z. The cotorsion is the torsion of ${}^*\nabla$.

Our generalisation of Riemannian geometry includes for example symplectic geometry, where the generalised metric is totally antisymmetric. So symplectic and Riemannian geometry are included as special cases and unified in our formulation. This is what we would expect if the theory is to be the semiclassicalisation of a theory unifying quantum theory and geometry. It is also remarkable that metrics with antisymmetric part are exactly what are needed in string theory to establish T-duality, which is entirely consistent with our duality ideas of Section 2.

Acknowledgements

This was a really great conference in Polanica and I'd like to thank all of the organizers and participants.

References

1. S. Majid. Principle of representation-theoretic self-duality. *Phys. Essays*, 4(3):395–405, 1991.
2. S. Majid. *Non-commutative-geometric Groups by a Bicrossproduct Construction.* PhD thesis, Harvard mathematical physics, 1988.
3. S. Majid. Hopf algebras for physics at the Planck scale. *J. Classical and Quantum Gravity*, 5:1587–1606, 1988.
4. S. Majid. Quantum and braided group Riemannian geometry. *J. Geom. Phys.*, 30:113–146, 1999.
5. T. Brzeziński and S. Majid. Quantum group gauge theory on quantum spaces. *Commun. Math. Phys.*, 157:591–638, 1993. Erratum 167:235, 1995.
6. S. Majid. *Foundations of Quantum Group Theory.* Cambridge University Press, 1995.
7. V.G. Drinfeld. Quantum groups. In A. Gleason, editor, *Proceedings of the ICM*, pages 798–820, Rhode Island, 1987. AMS.
8. M. Jimbo. A q-difference analog of $U(g)$ and the Yang-Baxter equation. *Lett. Math. Phys.*, 10:63–69, 1985.
9. A. Connes. *Noncommutative Geometry.* Academic Press, 1994.
10. S. Majid. Beyond supersymmetry and quantum symmetry (an introduction to braided groups and braided matrices). In M-L. Ge and H.J. de Vega, editors, *Quantum Groups, Integrable Statistical Models and Knot Theory*, pages 231–282. World Sci., 1993.
11. S. Major and L. Smolin. Quantum deformation of quantum gravity. *Nucl. Phys. B*, 473:267–290, 1996.
12. C. Klimcik. Poisson-Lie T-duality. *Nucl. Phys. B (Proc. Suppl.)*, 46:116–121, 1996.
13. E. Beggs and S. Majid. Poisson-lie T-duality for quasitriangular Lie bialgebras. *Preprint*, 1999.

14. S. Majid. Duality principle and braided geometry. *Springer Lec. Notes in Phys.*, 447:125–144, 1995.
15. S. Majid. On q-regularization. *Int. J. Mod. Phys. A*, 5(24):4689–4696, 1990.
16. S. Majid. Physics for algebraists: Non-commutative and non-cocommutative Hopf algebras by a bicrossproduct construction. *J. Algebra*, 130:17–64, 1990.
17. O. Bratteli and D. W. Robinson. *Operator Algebras and Quantum Statistical Mechanics*, volume II. Springer-Verlag, 1979.
18. S. Majid. Representations, duals and quantum doubles of monoidal categories. *Suppl. Rend. Circ. Mat. Palermo, Ser. II*, 26:197–206, 1991.
19. S. Majid. Some physical applications of category theory. *Springer Lec. Notes in Phys.*, 375:131–142, 1991.
20. S. Majid. Non-commutative-geometric $\mathcal{A}/\mathcal{G}$: a new approach to the quantisation of photons. *Unpublished*, 1986.
21. S. Majid. Noncommutative geometric quantisation of photons and string field theory. *Unpublished Harvard preprint*, 1988.
22. S. Majid. Fourier transforms on $\mathcal{A}/\mathcal{G}$ and knot invariants. *J. Math. Phys.*, 32:924–927, 1990.
23. S. Majid. $\mathbb{C}$-statistical quantum groups and Weyl algebras. *J. Math. Phys.*, 33:3431–3444, 1992.
24. A. Ashtekar. A 3+1 formulation of Einstein self-duality. *Contemp. Math.*, 71:39–53, 1988.
25. C. Rovelli and L. Smolin. Knot theory and quantum gravity. *Phys. Rev. Lett.*, 61:1155, 1988.
26. S. Majid. Gauge covariant vacuum expectations and glueballs. *Il Nuov. Cim.*, 104A:949–960, 1991.
27. S. Majid and R. Oeckl. Twisting of quantum differentials and the Planck scale Hopf algebra. *Commun. Math. Phys.* In Press.
28. S. Majid. Matched pairs of Lie groups associated to solutions of the Yang-Baxter equations. *Pac. J. Math.*, 141:311–332, 1990.
29. S. Majid and H. Ruegg. Bicrossproduct structure of the κ-Poincaré group and non-commutative geometry. *Phys. Lett. B*, 334:348–354, 1994.
30. J. Lukierski, A. Nowicki, H. Ruegg, and V.N. Tolstoy. q-Deformation of Poincaré algebra. *Phys. Lett. B*, 264,271:331,321, 1991.
31. G. Amelino-Camelia and S. Majid. Waves on noncommutative spacetime and gamma-ray bursts. *Preprint*, 1999.
32. G. Amelino-Camelia, J. Lukierski, and A. Nowicki. κ-deformed covariant phase space and quantum gravity uncertainty relations. *Phys. Atom. Nucl.*, 61:1811–1815, 1988.
33. S. Majid. Hopf-von Neumann algebra bicrossproducts, Kac algebra bicrossproducts, and the classical Yang-Baxter equations. *J. Funct. Analysis*, 95:291–319, 1991.
34. S. Majid. Introduction to braided geometry and q-Minkowski space. In L. Castellani and J. Wess, editors, *Proceedings of the International School 'Enrico Fermi' CXXVII*, pages 267–348. IOS Press, Amsterdam, 1996.
35. F.H. Jackson. On q-functions and a certain difference operator. *Trans. Roy. Soc. Edin.*, 46:253–281, 1908.
36. S. Majid. Braided momentum in the q-Poincaré group. *J. Math. Phys.*, 34:2045–2058, 1993.
37. A. Kempf and S. Majid. Algebraic q-integration and Fourier theory on quantum and braided spaces. *J. Math. Phys.*, 35:6802–6837, 1994.

38. S. Majid. Braided groups. *J. Pure and Applied Algebra*, 86:187–221, 1993.
39. S. Majid. Quasi-quantum groups as internal symmetries of topological quantum field theories. *Lett. Math. Phys.*, 22:83–90, 1991.
40. S. Majid. Examples of braided groups and braided matrices. *J. Math. Phys.*, 32:3246–3253, 1991.
41. S. Majid and U. Meyer. Braided matrix structure of q-Minkowski space and q-Poincaré group. *Z. Phys. C*, 63:357–362, 1994.
42. S. Majid. Free braided differential calculus, braided binomial theorem and the braided exponential map. *J. Math. Phys.*, 34:4843–4856, 1993.
43. S. Majid. q-epsilon tensor for quantum and braided spaces. *J. Math. Phys.*, 36:1991–2007, 1995.
44. S. Majid. $*$-structures on braided spaces. *J. Math. Phys.*, 36:4436–4449, 1995.
45. S. Majid. Quasi-$*$ structure on q-Poincaré algebras. *J. Geom. Phys.*, 22:14–58, 1997.
46. S. Majid. q-Euclidean space and quantum Wick rotation by twisting. *J. Math. Phys.*, 35:5025–5034, 1994.
47. U. Carow-Watamura, M. Schlieker, M. Scholl, and S. Watamura. Tensor representation of the quantum group $SL_q(2,\mathbb{C})$ and quantum Minkowski space. *Z. Phys. C*, 48:159, 1990.
48. O. Ogievetsky, W.B. Schmidke, J. Wess, and B. Zumino. q-Deformed Poincaré algebra. *Commun. Math. Phys.*, 150:495–518, 1992.
49. G. Fiore. The $SO_q(N,R)$-symmetric harmonic oscillator on the quantum Euclidean space R_q^N and its Hilbert space structure. *Int. J. Mod. Phys. A*, 26:4679–4729, 1993.
50. S. Majid. Braided geometry of the conformal algebra. *J. Math. Phys.*, 37:6495–6509, 1996.
51. S. Majid. Quantum and braided diffeomorphism groups. *J. Geom. Phys.*, 28:94–128, 1998.
52. T.H. Koornwinder. Orthogonal polynomials in connection with quantum groups. In P. Nevai, editor, *Orthogonal Polynomials: Theory and Practice*, number 294 in NATO ASI Series C, pages 257–292. Kluwer, 1990.
53. R. Oeckl. Braided quantum field theory. *Preprint*, 1999.
54. A. Ashtekar, J. Baez, A. Corichi, and K. Krasnov. Quantum geometry and black hole entropy. *Phys. Rev. Lett.*, 80:904–907, 1998.
55. S. Majid. Double bosonisation and the construction of $U_q(g)$. *Math. Proc. Camb. Phil. Soc.*, 125:151–192, 1999.
56. S.L. Woronowicz. Differential calculus on compact matrix pseudogroups (quantum groups). *Commun. Math. Phys.*, 122:125–170, 1989.
57. S. Majid. Classification of bicovariant differential calculi. *J. Geom. Phys.*, 25:119–140, 1998.
58. S. Majid. Quantum geometry of field extensions. *J. Math. Phys.*, 40:2311–2323, 1999.
59. S. Majid. Advances in quantum and braided geometry. In H.-D. Doebner and V.K. Dobrev, editors, *Quantum Group Symposium at Group XXI*, pages 11–26. Heron Press, Sofia, 1997.
60. T. Brzeziński and S. Majid. Quantum differentials and the q-monopole revisited. *Acta Appl. Math.*, 54:185–232, 1998.
61. P. Podles. Quantum spheres. *Lett. Math. Phys.*, 14:193–202, 1987.
62. T. Brzeziński and S. Majid. Quantum geometry of algebra factorisations and coalgebra bundles. *Preprint*, 1997.
63. P. Hajac and S. Majid. Projective module description of the q-monopole. *Commun. Math. Phys.* In press.

Loop Quantum Gravity and the Meaning of Diffeomorphism Invariance

Carlo Rovelli[1,2] and Marcus Gaul[3,4]

[1] Centre de Physique Théorique, CNRS Luminy, F-13288 Marseille, France
[2] Departement of Physics and Astronomy, University of Pittsburgh, Pittsburgh, PA 15260, USA
[3] Max–Planck–Institut für Physik (Werner–Heisenberg–Institut), Föhringer Ring 6, D-80805 München, Germany
[4] Sektion Physik, Ludwig–Maximilians–Universität, Theresienstr. 37, D-80333 München, Germany

Abstract. This series of lectures gives an introduction to the non-perturbative and background-independent formulation for a quantum theory of gravitation which is called *loop quantum gravity*. The Hilbert space of kinematical quantum states is constructed and a complete basis of *spin network states* is introduced. Afterwards an application of the formalism is provided by the spectral analysis of the area operator, which is the quantum analogue of the classical area function. This leads to one of the key results of loop quantum gravity obtained in the last few years: the derivation of the discreteness of the geometry and the computation of the quanta of area. Special importance is attached to the role played by the diffeomorphism group in order to clarify the notion of observability in general relativity – a concept far from being trivial. Finally an outlock onto a possible dynamical extension of the theory is given, leading to a "sum over histories" approach, namely a so-called *spin foam model*. Throughout the whole lecture great significance is attached to conceptual and interpretational issues.

1 Introduction

In the beginning of this century, physics has undergone two great conceptual changes. With the discovery of general relativity and quantum mechanics the notions of matter, causality, space and time experienced the biggest modifications since the age of Descartes, Copernicus, and Newton. However, no fully convincing synthesis of these theories exists so far. Simple dimensional analysis reveals that new predictions of a quantum theory of gravitation are expected to take place at the Planck length $l_P \equiv \left(\hbar G/c^3\right)^{1/2} \sim 10^{-35}\,\mathrm{m}$. This scale appears to be far below any current experimental technique. Nevertheless, quite recently interesting proposals and ideas to probe experimentally the physics at the Planck scale have been suggested [1,2].

From the theoretical point of view, several approaches to a theory of quantum gravity have emerged, inspired by various research fields in contemporary physics and mathematics. The most popular research direction is in the realms of string theory, followed by loop quantum gravity. Other directions range from discrete methods to non-commutative geometry. We have listed the main current approaches to a quantum theory gravity (which are moreover far from being

Table 1. Main current approaches to quantum gravity.

Traditional	**Most Popular**	**New**
Discrete methods Dynamical triangulations Regge calculus Simplicial models → 2nd *order transition?*	**String theory** → *Black hole entropy*	**Non-commutative geometry** → *Quantum theory?*
Approximate theories Euclidean quantum gravity Perturbative quantum gr. → *Woodard–Tsamis effect* QFT on curved space–times	**Loop quantum gravity** → *Black hole entropy* → *Eigenvalues of geometry:* $A_j = 16\pi\hbar G \sum_i \sqrt{j_i(j_i+1)}$	**Null surfaces** → *Observables?*
		Spin foam models → *convergence of loop, discrete, TQFT and sum-over-histories*
Unorthodox approaches Sorkin's Posets Finkelstein Twistors ...		

independent) in table 1. Despite this variety of ideas and the effort put in so far, many, many questions are still open. For an overview and a critical comparison of the different approaches, see [3]. Some of these conceptually different approaches show surprising similarities which could be a focal point of attention for the future[1].

String theory was inspired and constructed mainly by particle physicists. Its attitude towards the fundamental forces is to treat general relativity on an equal footing with the field theories describing the other interactions, the destinctive feature being the energy scale. String theory is supposed to be a theory of all interactions – electromagnetic, strong, weak and gravitational – which are treated in a unified quantum framework. Classical (super-)gravity emerges perturbatively as a low-energy limit in superstring theory.

The problem until 1995 was the lack of a non-perturbative formulation of the theory. This situation has improved with the discovery of the string dualities, "D-branes", and "M-theory" in the so-called *2nd superstring revolution*. Nevertheless despite the recent exciting discoveries in M-theory and the AdS/CFT equivalence, a complete non-perturbative or strong-coupling formulation of string/M-theory is still not in sight.

[1] For instance, string/M-theory, non-commutative geometry and loop quantum gravity seem to point to a similar *discrete* short distance space–time structure. Suggestions have been made that a complete theory must involve elements from each of the approaches. For further details we refer to [4,5].

A point that is often criticized in string theory by relativists is the lack of a background independent formulation, i.e. invariance under active diffeomorphisms, which is one of the fundamental principles of general relativity. String/M-theory is formulated on a (implicitly) fixed background geometry which is itself not dynamical. In a truely background independent formulation, no reference to any classical metric should enter neither the definition of the state space nor the dynamical variables of the theory. Rather the metric should appear as an operator allowing for quantum states which may themselves be superpositions of different backgrounds.

In fact, relativists do not view general relativity as an additional item in the list of the field theories describing fundamental forces, but rather as a major change in the manner space and time are described in physics. This point is often misunderstood, and is often a source of confusion; it might be worthwhile spending a few additional words. The key point is not that the gravitational force, by itself, must necessarilly be seen as different from the other forces: the point of view that the gravitational force is just one (and the weakest) among the interactions is certainly viable and valuable. Rather, the key point is that, with general relativity, we have understood that the world is *not* a non-dynamical metric manifold with dynamical fields living over it. Rather, it is a collection of dynamical fields living, so to say, in top of each other. The gravitational field can be seen – if one wishes so – as one among the fields. But the definiton of the theory over a given background is, from a fundamental point of view, physically incorrect.

Loop quantum gravity is a background independent approach to quantum gravity. For many details on this approach, and for complete references, see [6]. Loop quantum gravity has been developed *ab initio* as a non-perturbative and background independent canonical quantum theory of gravity. Besides ordinary general relativity and quantum mechanics no additional input is needed. The approach makes use of the reformulation of general relativity as a dynamical theory of connections. Due to this choice of variables the phase space of the theory resembles at the kinematical level closely that of conventional $SU(2)$ Yang–Mills theory. The main ingredient of the appraoch is the choice of holonomies of the connections – the loop variables – as the fundamental degrees of freedom of quantum gravity.

The philosophy behind this approch is different from string theory as one considers here standard four dimensional general relativity trying to develop a theory of quantum gravity in its proper meaning without claiming to describe a unified picture of all interactions. Loop quantum gravity is extremely successful in describing Planck-scale phenomena. The main open problem, on the other hand, is the connection with low-energy phenomena. In this respect loop quantum gravity has opposite strength and weakness than string theory.

One might wonder how one can hope to have a consistent non-perturbative formulation of quantum gravity when perturbative quantization of covariant general relativity is non-renormalizable. However, the basic assumption in proving the non-renormalizability of general relativity is the availability of a Minkowskian space–time at arbitrarily short distances, an assumption which is certainly not correct in a theory of quantized gravity, i.e. a quantum space–time regime. As will be discussed later, one of the key results obtained so far in loop quantum gravity has been the calculation of the quanta of geometry [7], i.e. the spectra of the quantum analogues to the classical area and volume functionals. Remarkably they turned out to be discrete! This result indicates the existence of a quantum space–time structure at the Planck scale which doesn't have to be continuous anymore. More specifically this implies the emergence of a natural cut-off in quantum gravity that might also account as a regulator of the ultraviolet divergencies plaguing the standard model. Thus, standard perturbative techniques in field theory cannot be taken for granted at scales where quantum effects of gravity are expected to dominate.

General relativity is a constrained theory. Classically, the constraints are equivalent to the dynamical equations of motion. The transition to the quantum theory is carried out using canonical quantization by appliying the algorithm developed by Dirac [8]. In the loop approach, the unconstrained classical theory is quantized, requiring the implementation of quantum constraint operators afterwards. Despite many results obtained in the last few years, a complete implementation of all constraints including the Hamiltonian constraint, which is the generator of "time evolution", i.e. the dynamical part of the theory, is still elusive. This is not surprising, since we do not expect to be able to obtain a complete solution of a highly non-trivial and non-linear theory.

To address this issue, covariant methods to understand the dynamics have been developed in the last few years. These can be obtained from a "sum over histories" approach, derived from the canonical formulation. This development has led to the so-called *spin foam models*, in which spin networks are loosely speaking "propagated in time", leading to a space–time formulation of loop quantum gravity. This formulation of the theory provides a starting point for approximations, offers a more intuitive understanding of quantum space–time, and is much closer to particle physics methods. A brief description of these models will be given below in sect. 5.2.

These lectures are organized as follows. We start in sect. 2 with the basic mathematical framework of loop quantum gravity. We will end up with the definition of the kinematical Hilbert space of quantum gravity. In the next section an application of these tools is provided by constructing the basic operators on this Hilbert space. We calculate in a simple manner the spectrum of what is going to be physically interpreted as the area operator. Section 4 deals with the important question of observability in classical and quantum gravity, a topic which is far from being trivial, and the meaning of diffeomorphism invariance in this context. In the end of theses notes, we will give the prospects for a dynamical description of loop quantum gravity, which is encoded in the concept of *spin foam models*. One such ansatz is briefly discussed in sect. 5. The following final section concludes with future perspectives and open problems.

For the sake of completeness, we give in table 2 a short historical survey of the main achievements in canonical quantum gravity since the reformulation of general relativity in terms of connection variables. A more detailled discussion of some of these various aspects is given in [6].

Table 2. Short historical survey of canonical quantum gravity.

'86	Classical Connection Variables	Ashtekar, Sen
'87	Lattice loop states solve $\hat{H}$	Jacobsen, Smolin
'88	Loop Quantum Gravity	Smolin, Rovelli
'92	Weave States	Ashtekar, Rovelli, Smolin
'95	Spin Network States	Rovelli, Smolin
'95	Volume and Area Operators	Rovelli, Smolin
'95	Measure and Functional Calculus	Isham, Baez, Thiemann, Marolf, Mourau, Ashtekar, Lewandowski
'98	Hamiltonian Operator	Thiemann
'98	Black Hole Entropy	Krasnov, Rovelli, Baez, Corichi, Ashtekar
'98	Spin Foam Formulation	Reisenberger, Rovelli, Baez

2 Basic Formalism of Loop Quantum Gravity

Our attention in this lecture will be focused on conceptual foundations and the development of the main ideas behind loop quantum gravity. However, because of the highly mathematical nature of the subject some technical details are unavoidable, thus this section is devoted to the essential mathematical foundations.

The reader is not assumed to be familiar with the connection variables, which constitute the basis for all the effort in canonical quantum gravity since 1986. Thus we start by considering the canonical formalism in the connection approach, which is reviewed in [9]. For a recent overview of loop quantum gravity and a comprehensive list of references we refer to [6].

2.1 A brief Outline of the Connection Formalism

In loop quantum gravity, we construct the quantum theory using canonical quantization. This is analogous to ordinary field theory is the functional Schrödinger representation. The approach may be called conservative in the sense that originally no new structures like supersymmetry[2], extended objects, or extra dimensions other than four are postulated. (It is important to emphasize, in this context, the fact, although sometimes forgotten these days, that supersymmetry, extended objects or extra dimensions are interesting theoretical hypotheses, not established properties of Nature!). The approach aims at unifying quantum

[2] Nevertheless, there exist extensions of the Ashtekar variables to supergravity [10], and quite recently $N = 1$ supersymmetry was introduced in the context of spin networks [11].

mechanics and general relativity by developing new non-perturbative techniques from the outset and by staying as close as possible to the conventional settings of quantum theory and experimentally tested general relativity.

The foundations of the formalism date back to the early 60s when the "old" Hamiltonian or canonical formulation of classical general relativity, known as ADM formalism, was constructed. The canonical scheme is based on the construction of the phase space Γ which is a covariant notion. It is the space of solution of the equations of motion, modulo gauges. However, in order to coordinatize Γ explicitly, one usually breakes explicitly covariance and splits 4-dimensional space–time $\mathcal{M}$ into 3-dimensional space plus time. We insist on the fact that this breaking of covariance is not structurally needed in order to set up the canonical formalism; rather it is only an artefact of the coordinatization we chose for the phase space. We cover $\mathcal{M}$, which is choosen to have topology $\mathbb{R}\times M$, with a foliation M_t, where M is the 3-manifold which represents "space" and $t \in \mathbb{R}$ is a (unphysical) time parameter. The basic variables on phase space are taken to be the induced 3-metric $q_{ab}(x)$ on M and the extrinsic curvature K_{ab} of M.

The easiest construction of the connection variables is given by first reformulating the ADM–formalism of canonical gravity in terms of (local) triads $e^i_a(x)$, which satisfy $q_{ab}(x) = e^i_a(x)e^i_b(x)$. This introduces an additional local $SU(2)$ gauge symmetry into the theory, which corresponds geometrically to arbitrary local frame rotations. One obtains $\left(E^a_i(x), K^i_a(x)\right)$ as the new canonical pair on phase space Γ. E^a_i is the *inverse* densitized triad[3], i.e. a vector with respect to $SU(2)$ and density weight one. The densitized triad itself is defined by $E^i_a := e\, e^i_a$, where e is the determinant of e^i_a. The indices $a, b, c, \ldots$ refer to spatial tangent space components, while $i, j, k, \ldots$ are internal $SU(2)$ indices. The inverse triad E^a_i is the square-root of the 3-metric in the sense that

$$E^a_i(x)E^b_i(x) = q\, q^{ab}(x) \,, \tag{1}$$

where q is the determinant of the 3-metric $q_{ab}(x)$. The canonically conjugate variable $K^i_a(x)$ of $E^a_i(x)$ is again closely related to the extrinsic curvature of M via $K^i_a = K_{ab}E^{bi}/\sqrt{q}$.

Finally the transition to the connection variables is made using a canonical transformation on the (real) phase space,

$$A^i_a(x) = \Gamma^i_a(x) + \beta\, K^i_a(x) \,. \tag{2}$$

Here $\Gamma^i_a(x)$ is the $SU(2)$ spin connection compatible with the triad, and β, the Immirzi parameter, is an arbitrary real constant. The original Ashtekar–Sen connection $A(x)$ was introduced in 1982 as a complex selfdual connection on the spatial 3-manifold M, corresponding to $\beta = i$ in (2). Nevertheless, we will use the real formulation.

[3] In the literature one often finds tensor densities marked with an upper tilde for each positive density weight and a lower tilde for each negative weight, such that the densitized triad is written as $\widetilde{E}^i_a(x) := \sqrt{q(x)}\, e^i_a(x)$.

Ashtekar discovered in 1986 that $(A_a^i(x), E_i^a(x)) \in \Gamma$ form a canonical pair on the phase space Γ of general relativity. Here A_a^i has to be considered as the new configuration variable, while the inverse densitized triad E_i^a corresponds to the canonically conjugate momentum. With this reformulation classical general relativity has the same kinematical phase space structure as an $SU(2)$ Yang–Mills theory.

The Poisson algebra generated by the new variables is

$$\{E_i^a(x), E_j^b(y)\} = 0\,, \quad \left\{A_a^i(x), A_b^j(y)\right\} = 0\,,$$
$$\{A_a^i(x), E_j^b(y)\} = \beta\, G\, \delta_j^i\, \delta_a^b\, \delta^3(x,y)\,, \tag{3}$$

where G is the usual gravitational constant. It arises because the conjugate momentum of the configuration variable A_a^i (obtained as the derivative of the Lagrangian with respect to the velocities) is actually given by $1/G \times E_i^a$. As a consequence of the 4-dimensional diffeomorphism invariance of general relativity, the (canonical) Hamiltonian vanishes weakly[4]. The full dynamics of the theory is encoded in so-called first-class constraints which are functions on phase space that vanish for physical configurationes. The constraints generate transformations between those classical configurations that are physically indistinguishable. The first-class constraints of canonical general relativity are the familiar Gauss law of Yang–Mills theory, which generates local $SU(2)$ gauge transformations, the diffeomorphism constraint generating 3-dimensional diffeomorphisms of the manifold M, and the Hamiltonian constraint, which is the generator of the evolution of the inital spatial slice M in coordinate time. The Gauss constraint enters the theory as a result of the choice of triads and it makes general relativity resemble a Yang–Mills gauge theory. Indeed the phase spaces of both theories are similar. The constrained surface of general relativity is embedded into that of Yang–Mills theory apart from the additional local restrictions which appear in gravity besides the Gauss law.

The use of Ashtekar's original set of canonical variables involving a *complex* connection $A(x)$ leads to a simplification of the Hamiltonian constraint. With the use of a *real* connection, the constraint loses its simple polynomial form. At first, this was considered as a serious obstacle for the quantization. However, T. Thiemann succeeded in constructing a Lorentzian quantum Hamiltonian constraint [12] in spite of the non-polynomiality of the classical expression. His work has prompted the wide use of the real connection, a use which was first advocated by F. Barbero.

We now briefly describe the quantum implementation of the above described kinematical setting. The canonical variables A and E (or functions of these), which are defined in the unconstrained (thus unphysical) phase space are replaced by quantum operators acting on some Hilbert space of states. This results

[4] This is true for *any* generally covariant theory, which means that a theory whose gauge group contains the diffeomorphism group of the underlying manifold has a weakly vanishing Hamiltonian. Here weakly vanishing refers to vanishing on physical configurations.

in the promotion of Poisson brackets to commutators. In other words, an algebra of observables should act on a Hilbert space. More precisely, one establishes an isomorphism between the Poisson algebra of classical variables and the algebra generated by the corresponding Hermitian operators by introducing a linear operator representation of this Poisson algebra. In the simplest case, the quantum states are normalizable functionals over configuration space, i.e. functionals of the connection $\Psi(A)$. The subset of physical states is obtained from the set of all wavefunctions on M by imposing the quantum analogues of the contraints, i.e. by requiring the physical states to lie in the kernel of all quantum constraint operators[5]

The space of physical states must have the structure of a Hilbert space, namely a scalar product, in order to be able to compute expectation values. This Hilbert structure is determined by the requirement that *real* physical observables correspond to *self-adjoint* operators. In order to define a Hilbert structure on the space of *physical* states, it is convenient (althought not strictly necessary) to define first a Hilbert structure on the space of *unconstrained* states. This is because we have a much better knowledge of the unconstrained observables than of the physical ones. If we choose a scalar product on the unconstrained state space which is gauge invariant, then there exist standard techniques to "bring it down" to the space of the physical states. Thus, we need a gauge and diffeomorphism invariant scalar product, with respect to which real observables are self-adjoint operators.

2.2 Basic Definitions

In this subsection we start with the actual topic of the lecture, the construction of *loop quantum gravity*. Space–time is assumed to be a 4-dimensional Lorentzian manifold $\mathcal{M}$ with topology $\mathbb{R} \times M$, where M is a real analytic and orientable 3-manifold. For simplicity we consider it as compact, e.g. S^3. Loosely speaking M represents "space" while $\mathbb{R}$ refers to "time".

On M we define a smooth, Lie-algebra valued connection one-form A, i.e. $A(x) = A^i_a(x)\,\tau_i\,dx^a$, where x are local coordinates on M, $A^i_a(x) \in C^\infty(M)$, and $\tau_i = (i/2)\,\sigma_i$ are the $SU(2)$ generators (in the fundamental representation) with σ_i being the Pauli matrices. The indices $a, b, c = 1, 2, 3$ play the role of tangent space indices while $i, j, k = 1, 2, 3$ are abstract internal $su(2) := Lie(SU(2))$ indices[6]. We call $\mathcal{A} = \{A\}$ the space of smooth connections on M and denote continuous functionals on $\mathcal{A}$ as $\Psi(A)$. These functionals build a topological vector space L under the pointwise topology.

[5] A distinct quantization method is the *reduced phase space quantization*, where the physical phase space is constructed classically by solving the constraints and factoring out gauge equivalence prior to quantization. But for a theory as complicated as general relativity it seems impossible to construct the reduced phase space. The two methods could lead to inequivalent quantum theories. Of course, it is possible, in principle, that more than one consistent quantum theory having general relativity as its classical limit might exist.

[6] Mathematically more sophisticated one would consider principal G-bundles over M, with structural group $G = SU(2)$ (i.e. compact and connected). Let $\mathcal{A}$ be the space of

2.3 The Construction of a Hilbert Space $\mathcal{H}$

In order to define a Hilbert space $\mathcal{H}$ based on the above linear space L of quantum states $\Psi(A)$, one needs to introduce an inner product, i.e. an appropriate measure on the space of quantum states for which the appearance of the *compact* gauge group $SU(2)$ turns out to be essential. Furthermore we demand the following properties of $\mathcal{H}$:

- $\mathcal{H}$ should carry a unitary representation of $SU(2)$
- $\mathcal{H}$ should carry a unitary representation of $Diff(M)$.

We consider a special class of functions of the connection in L, the ***cylindrical functions***. For their construction we need some tools, namely ***holonomies*** and ***graphs***.

Holonomy. Let a curve γ be defined as a continuous, piecewise analytic map from the intervall $[0,1]$ into the 3-manifold M,

$$\gamma : [0,1] \longrightarrow M \tag{4a}$$

$$s \longmapsto \{\gamma^a(s)\}\ ,\ a = 1,2,3\ . \tag{4b}$$

The holonomy or parallel propagator $U[A,\gamma]$, respectively, of the connection A along the curve γ is defined by

$$U[A,\gamma](s) \in SU(2)\ , \tag{5a}$$

$$U[A,\gamma](0) = \mathbf{1}\ , \tag{5b}$$

$$\frac{d}{ds}U[A,\gamma](s) + A_a\big(\gamma(s)\big)\,\dot{\gamma}^a(s)\,U[A,\gamma](s) = 0\ , \tag{5c}$$

where $\dot{\gamma}(s) := \frac{d\gamma(s)}{ds}$ is the tangent to the curve. The formal solution of (5c) is given by

$$U[A,\gamma](s) = \mathcal{P}\exp\int_\gamma A = \mathcal{P}\exp\int_\gamma ds\,\dot{\gamma}^a\,A_a^i\big(\gamma(s)\big)\,\tau_i\ , \tag{6}$$

in such a way that for any matrix-valued function $A\big(\gamma(s)\big)$, which is defined along γ, the path ordered expression (6) is given in terms of the power series expansion

$$\mathcal{P}\exp\int_0^1 ds\,A\big(\gamma(s)\big)$$

$$= \mathbf{1} + \int_0^1 ds\,A\big(\gamma(s)\big) + \int_0^1 ds\int_0^s dt\,A\big(\gamma(t)\big)\,A\big(\gamma(s)\big) + \ldots\ . \tag{7}$$

---

smooth connections on the bundle. As a result of the orientability of M the principal $SU(2)$-bundles are topologically trivial. Hence the $SU(2)$ connections on the bundle can be represented by a $su(2)$-valued 1-form on M.

The effect of path ordering $\mathcal{P}$ appears in the product of matrices which are always ordered according to the modulus of the parameter, i.e. in the third term of (7) t is always smaller (or equal) than s.

In a later section we will focus our attention to Wilson loops, which are traces of the holonomy of A along a closed curve γ, $\gamma(0) = \gamma(1)$, i.e. a loop, which in the following is denoted by α,

$$T[A, \alpha] = \mathrm{Tr}\, U[A, \alpha] \, . \tag{8}$$

They are by construction gauge invariant functionals of the connection.

The key successful idea of the loop approach to quantum gravity [13] is to choose the loop states, namely the states (9) as the basis states for quantum gravity, i.e.

$$\Psi_\alpha(A) = \mathrm{Tr}\, U[A, \alpha] \, . \tag{9}$$

These states are extended to disconnected loops, or multiloops, respectively, (collections of a finite number of closed curves $\{\alpha_1, \ldots, \alpha_n\}$), which are also denoted with α, by defining $\Psi_\alpha(A) = \prod_i \mathrm{Tr}\, U[A, \alpha_i]$. These states have a number of remarkable features: they allow us to control completely the solution of the diffeomorphism constraint, and they "largely" solve the Hamiltonian constraint, as we will see later. In QCD, states of this kind are unphysical, because they have infinite norm (they are "too concentrated", or "not sufficiently smeared"). If in QCD we artificially declare these states to have finite norm, we end up with an unphysically huge, non-separable Hilbert space. In gravity, on the other hand, these states, or, more precisely, the equivalence classes of these states under diffeomorphisms, define finite norm states. They are not too concentrated since in a sense they are – by diffeomorphism invariance – "smeared all over the manifold". Thus, they provide a natural and physical way to represent quantum excitations of the gravitational field.

For some time, however, a serious problem for loop quantum gravity was given by the fact that the states (9) form an overcomplete basis. The problem was solved in [14] by introducing the spin network states, which are combinations of loop states that form a genuine (non-overcomplete) basis. In the sequel, these spin network states are going to be constructed.

Graphs. The next important tools that we need are *graphs*. A graph $\Gamma_n = \{\gamma_1, \ldots, \gamma_n\}$ is a finite collection of n (oriented) piecewise analytic curves or edges γ_i, $i = 1, \ldots, n$, respectively, embedded in the 3-manifold M, that meet, if at all, only at their endpoints. As an example, consider the graph Γ_3 in Fig. 1 which is constructed of three curves γ_i, denoted as *links*.

Cylindrical Functions. Now pick a graph Γ_n as defined above. For each of the n links γ_i of Γ_n consider the holonomy $U_i(A) := U[A, \gamma_i]$ of the connection A along γ_i. Every (smooth) connection assigns a group element $g_i \in SU(2)$ to each link γ_i of Γ via the holonomy, $g_i \equiv U_i(A) = \mathcal{P} \exp \int_\gamma A$. Thus an element of

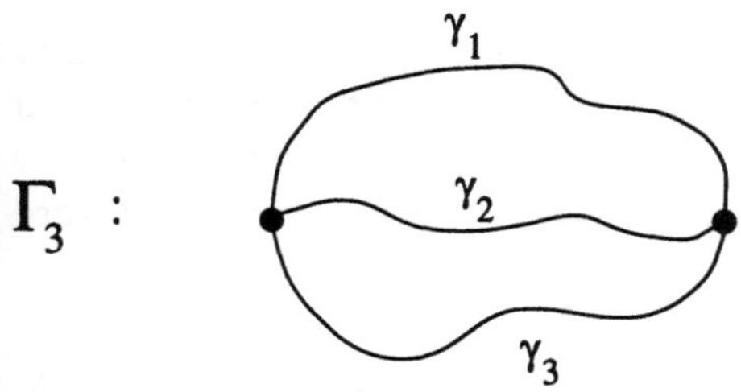

Fig. 1. A simple example of a graph.

$[SU(2)]^n$ is assigned to the graph Γ_n. The next step is to consider complex-valued functions $f_n(g_1, \dots, g_n)$ on $[SU(2)]^n$,

$$f_n : [SU(2)]^n \to \mathbb{C} . \tag{10}$$

These functions are Haar-integrable by construction, i.e. finite with respect to the Haar measure of $[SU(2)]^n$ which is induced by that of $SU(2)$ as a natural extension in terms of products of copies of it.

Hence, given any graph Γ_n and a function f_n, it is now straightforward to define the states that are required for the construction of the Hilbert space as

$$\Psi_{\Gamma_n, f_n}(A) := f_n(U_1, \dots, U_n) . \tag{11}$$

These functionals, which depend on the connection only via the holonomies, are the so-called *cylindrical functions*[7]. They form a dense subset of states in L, the above defined space of continuous smooth functions on $\mathcal{A}$. Thus the exclusive use of this special class of functions is justified for the construction of the Hilbert space.

As an example we consider the cylindrical function corresponding to the graph Γ_3 in Fig. 1. Let f_3 be defined as

$$f_3(U_1, U_2, U_3) := \mathrm{Tr}(U_1 U_2 U_3) . \tag{12}$$

Hence it follows that the cylindrical function corresponding to Γ_3 is given by

$$\Psi_{\Gamma_3, f_3}(A) = \mathrm{Tr}\left(U[A, \gamma_1] U[A, \gamma_2] U[A, \gamma_3]\right) . \tag{13}$$

An important property of cylindrical functions which turns out to be essential for the definition of an inner product is the following. A cylindrical function based on a graph Γ can always be rewritten as one which is defined according to $\tilde{\Gamma}$, where $\Gamma \subseteq \tilde{\Gamma}$, i.e. there exists a bigger graph $\tilde{\Gamma}$ that contains Γ as a subgraph. One obtaines

$$\Psi_{\Gamma, f} = \Psi_{\tilde{\Gamma}, \tilde{f}} \tag{14}$$

[7] The name *cylindrical function* stems from integration theory on infinite-dimensional manifolds, where they are introduced to define cylindrical measures. One can view the cylindrical function associated to a given graph as being constant with respect to some (in fact, most) of the dimensions of the space of connections, i.e. as a cylinder on that space.

by simply taking $\tilde{f}$ to depend only on those group elements U_i that belong to the links in Γ but not to $\tilde{\Gamma}$. In other words, any two cylindrical functions can always be viewed as being defined on the same graph which is just constructed as the union of the original ones. Given this property, it is now straightforward to define a scalar product for any two cylindrical functions f and g by

$$\langle \Psi_{\Gamma,f} | \Psi_{\Gamma,g} \rangle := \int_{[SU(2)]^n} dU_1 \ldots dU_n \, \overline{f(U_1, \ldots, U_n)} \, g(U_1, \ldots, U_n) \, . \tag{15}$$

Here $dU_1 \ldots dU_n$ is a shorthand notation for the induced Haar measure on $[SU(2)]^n$. Furthermore, the scalar product (15) is invariant under local $SU(2)$ transformations and diffeomorphisms.

The required unconstrained Hilbert space $\mathcal{H}$ of quantum states is obtained by completing the space of all finite linear combinations of cylindrical functions (for which the scalar product is also defined) in the norm which is induced by the quadratic form (15) on a cylindrical function as $||\Psi_{\Gamma,f}|| = \langle \Psi_{\Gamma,f} | \Psi_{\Gamma,f} \rangle^{1/2}$. This Hilbert space $\mathcal{H}$ of loop quantum gravity (which is non-separable) has the properties we required in the beginning of the section, namely

- $\mathcal{H}$ carries a unitary representation of local $SU(2)$
- $\mathcal{H}$ carries a unitary representation of $Diff(M)$.

The unitary representations on $\mathcal{H}$ are naturally realized on the quantum states $\psi(A) \in \mathcal{H}$ in terms of the transformation of their arguments. Remember that the states were defined in (11) as cylindrical functions, i.e. they depend on the connections only via holonomies along the links of the underlying graph.

Under local $SU(2)$ gauge transformations the connection A transforms inhomogeneously like a gauge potential. Nevertheless, the holonomy turns out to have a homogeneous transformation rule. Similarly, if one considers spatial diffeomorphisms $\phi : M \to M$, one finds that the connection transforms as a one-form. This induces the transformation of the holonomy as a distortion of the curve γ along which it is defined, and thus of the graph Γ which underlies the cylindrical functions. In other words, a representation of the diffeomorphism group is imprinted on the holomonies.

These transformation rules give rise to the above mentioned representations on the Hilbert space $\mathcal{H}$. The fact that $\mathcal{H}$ carries *unitary* representations stems from the invariance of the scalar product (15) under local $SU(2)$ transformations and spatial diffeomorphisms.

There are several mathematical developements connected with the construction given above. They involve projective families and projective limits, generalized connections, representation theory of C^*-algebras, measure theoretical techniques, and others. These further developements, however, are not needed for the following, and for understanding the basic physical results of loop quantum gravity. For details and references on these developements, see [15].

2.4 A Basis in the Hilbert Space

We now construct an orthonormal basis in the Hilbert space $\mathcal{H}$. We begin by defining a *spin network*, which is an extension of the notion of graph, namely a colored graph. Consider a graph Γ with n links γ_i, $i = 1, \dots, n$, embedded in the 3-manifold M. To each link γ_i we assign a non-trivial irreducibel representation of $SU(2)$ which is labeled by its spin j_i or equivalentely by $2j_i$, an integer which is called the *color* of the link. The Hilbert space on which this irreducible spin-j_i representation is defined is denoted by $\mathcal{H}_{j_i}$.

Next, consider a particular node p, say a k-valent one. There are k links $\gamma_1, \dots, \gamma_k$ that meet at this node. They are colored as $j_1, \dots, j_k$. Let $\mathcal{H}_{j_1}, \dots, \mathcal{H}_{j_k}$ be the Hilbert spaces of the representations associated to the k links. Consider the tensor product of these Hilbert spaces

$$\mathcal{H}_p = \mathcal{H}_{j_1} \otimes \dots \otimes \mathcal{H}_{j_k} \,. \tag{16}$$

Fix, once and for all, an orthonormal basis in $\mathcal{H}_p$. Now, the choice of an element N_p of this basis is called a *coloring of the node p*.

A (non-gauge invariant) *spin network* S is then defined as a colored embedded graph, namely it is a graph embedded in space in which links as well as nodes are colored. That is, it is an embedded graph plus the assignement of a spin j_i to each link γ_i and the assigmenet of a basis element N_p to each node p. A spin network is thus a triple $S = (\Gamma, \boldsymbol{j}, \boldsymbol{N})$. The vector notations $\boldsymbol{j}$ and $\boldsymbol{N}$ are abbreviations for $\boldsymbol{j} = \{j_i\}$, $i = 1, \dots, n$, the collection of all irreducible $SU(2)$ representations associated to the n links in Γ, and $\boldsymbol{N} = \{N_p\}$ stands for the basis elements attached to the nodes.

Now we are able to define a *spin network state* $\Psi_S(A)$ as a cylindrical function f_S associated to the spin network S whose graph is Γ, as

$$\Psi_S(A) = \Psi_{\Gamma, f_S}(A) = f_S(U[A, \gamma_1], \dots, U[A, \gamma_n]) \,. \tag{17}$$

The cylindrical function f_S is constructed by taking the holonomy along each link of the graph in that irreducible representation of $SU(2)$ which is associated to the link, and contracting the holonomy matrices with the vector N_p at each node p where the links meet, giving

$$f_S(U_1, U_2, \dots, U_n) = \prod_{\text{links } i} R^{j_i}(U_i) \otimes \prod_{\text{nodes } p} N_p \,. \tag{18}$$

Here $R^{j_i}(U_i)$ is the representation matrix of the holonomy U_i in the spin-j_i irreducible representation of $SU(2)$ associated to a link γ_i, $i = 1, \dots, n$. The contraction is possible because N_p is an element of the tensor product of the Hilbert spaces associated to the links that meet at the node. Therefore it can be seen as a tensor with one index in each of these spaces. On the other hand, the holonomies in the representation j_i can be seen as matrices between the same spaces. A moment of reflection shows that the indices match exactly.

By varying the graph, the colors of the links and the basis elements at the nodes, we can construct a family of states, which can all be normalized. At last, using the well-known Peter–Weyl theorem, it can easily be shown that any two distinct states Ψ_S are orthonormal,

$$\langle\Psi_S|\Psi_{S'}\rangle = \delta_{SS'} \tag{19a}$$
$$= \delta_{\Gamma,\Gamma'}\delta_{j,j'}\delta_{N,N'} \,, \tag{19b}$$

and that if Ψ is normal to every spin network state, then $\Psi = 0$. Therefore the spin network states form a complete orthonormal basis in the kinematical Hilbert space $\mathcal{H}$.

2.5 The $SU(2)$ Gauge Constraint

The physical quantum state space $\mathcal{H}_{phys}$ is obtained by imposing the quantum constraint equations on the Hilbert space $\mathcal{H}$. We want to impose the quantum constraints one after another as it is shown in Fig. 2.

$$\begin{array}{ccccccc}
\mathcal{H} & \xrightarrow{SU(2)} & \mathcal{H}_0 & \xrightarrow{Diff(M)} & \mathcal{H}_{diff} & \xrightarrow{\hat{H}} & \mathcal{H}_{phys} \\
\downarrow & & \downarrow & & \downarrow & & \downarrow \\
\Psi(A) & \longrightarrow & |S> & \longrightarrow & |s> & \longrightarrow & ?
\end{array}$$

Fig. 2. A step by step construction of the physical Hilbert space.

In this diagram the first line refers to the imposition of the quantum constraints yielding the appropriate invariant Hilbert spaces, while the second line gives the corresponding basis. The question mark stands for the fact that the explicit construction of the states in the physical Hilbert space is not yet understood. This is not surprising, since having the complete set of these states explicitely would amount to having *solved* the theory completely, a much stronger result than what we are looking for.

We begin here the process of solving the constraint by first considering the $SU(2)$ gauge constraint. An $SU(2)$ gauge transformation $\lambda(x)$ acts on the connection in the well known fashion $A \to A_\lambda$, and they act on the quantum wave functionals $\Psi(A)$ by transforming the argument A. Since the basis states we are considering depend on the connection A through the holonomy, which transforms homogeneously, the transformation properties of the states are easy to work out. In fact, a moment of reflection shows that an $SU(2)$ gauge transformation acts on a spin network state simply by $SU(2)$ transforming the coloring of the nodes N_p. More precisely, the spaces $\mathcal{H}_p$ carry a representation of $SU(2)$, and are transformed by the $SU(2)$ element $\lambda(x_p)$, where x_p is the point of the manifold in which the node p lies.

It is then easy to find the complete set of gauge invariant states. The Hilbert space $\mathcal{H}_p$, being a tensor product of irreducible representations, can be decomposed into its irreducible parts,

$$\mathcal{H}_{j_1} \otimes \ldots \otimes \mathcal{H}_{j_k} = \bigoplus_J (\mathcal{H}_J)^{k_J} \ . \tag{20}$$

Here k_J denotes the multiplicity of the spin-J irreducible representation. Among all subspaces of this decomposition we are interested in the SU(2) gauge invariant one (the singlet), i.e. the $J = 0$ subspace, denoted as $(\mathcal{H}_0)^{k_0}$. We pick an arbitrary basis in this subspace, and assign one basis element N_p to the node p. A spin network in which the coloring of the nodes is given by such invariant tensors N_p is called a ***gauge invariant spin network*** (often, the expression spin network is used for the gauge invariant ones). The spin network states constructed in terms of gauge invariant spin networks solve the gauge constraint and form a complete orthonormal basis in $\mathcal{H}_0$, the $SU(2)$ gauge invariant Hilbert space.

The quantities N_p are called ***intertwiners.*** They are invariant tensors with indices in different irreducible $SU(2)$ representations. They provide the possibilities to couple representations of $SU(2)$, i.e. they map the incoming irreducible representations at a node to the outgoing ones[8]. Thus they are given by standard Clebsch–Gordon theory.

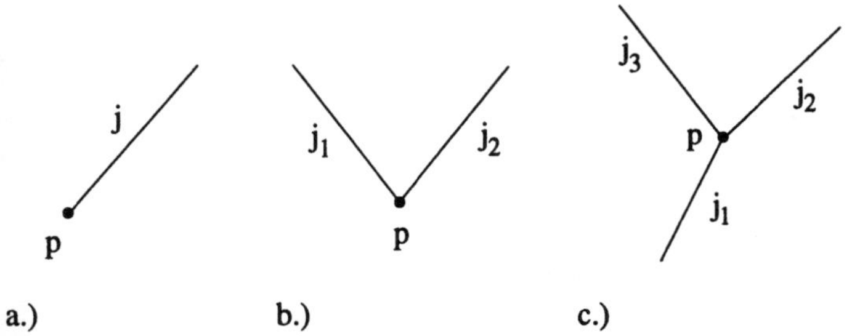

Fig. 3. Invariant tensors at $n = 1, 2, 3$-valent nodes.

To clarify the mathematics of the intertwiners, consider some examples. In the case of a 1-valent node as shown in Fig. 3a, there is no invarinat tensor, hence the dimensionality of the corresponding Hilbert space $\mathcal{H}_0$ is zero. Considering on the other hand Fig. 3b with a 2-valent node p, there exists a single intertwiner

[8] The denotation "incoming" and "outgoing" are just convenient labelings here. Actually one may wonder why we don't really care about the orientation of the links in the graph. This happens just because it can be neglected in the case where $SU(2)$ acts as gauge group, being a consequence of the following. In general, an inversion of the orientation of a link γ with associated irreducible representation j would lead to a change of this representation to its conjugate j^*. But since for $SU(2)$ j and j^* are unitary equivalent, considerations concerning the orientation are simplified, and we don't really worry about it in these lectures.

only if the colors of the links are equal, which is

$$(N_p)_{j_1 j_2} = \delta_{j_1 j_2} \ . \tag{21}$$

The last and most interesting example is Fig. 3c with a trivalent node, which corresponds to the coupling of three spins, well-known from the quantum theory of angular momentum. As long as the representations associated to the links satisfy the Clebsch–Gordan condition $|j_2 - j_3| \leq j_1 \leq j_2 + j_3$, once j_2 and j_3 are fixed (analogously for any other pair), a unique intertwiner exists because there is only one way of combining three irreducible representations in order to obtain a singlet representation. The invariant tensor is then given by nothing but the familiar Wigner $3j$-coefficient, which is (apart from normalization)

$$(N_p)_{n_1 n_2 n_3} = \begin{pmatrix} j_1 & j_2 & j_3 \\ n_1 & n_2 & n_3 \end{pmatrix} , \tag{22}$$

otherwise the dimension of $\mathcal{H}_0$ is zero again.

Let's now consider a simple example of a spin network state. We take the spin network in Fig. 4 corresponding to a graph with two trivalent nodes and three links joining them. Let two of the links carry (fundamental) spin-1/2 representations, while the third link has a spin-1 representation attached to it.

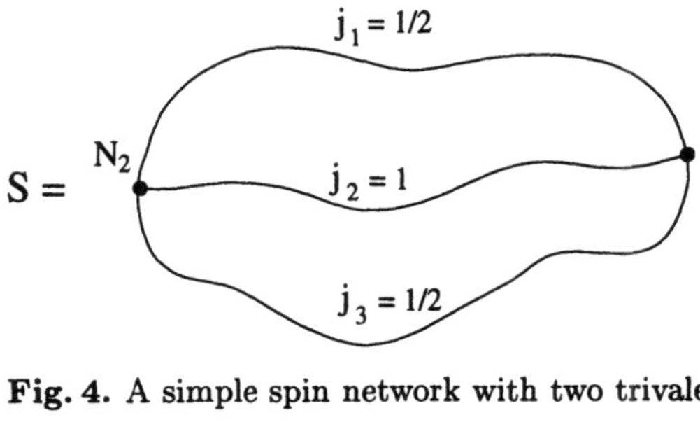

Fig. 4. A simple spin network with two trivalent nodes.

The elements N_1 and N_2 of an appropriate basis of invariant tensors are assigned to the nodes. The corresponding spin network state then reads explicitely

$$\begin{aligned} \Psi_S(A) &= R^{\frac{1}{2}}[U_1]_A{}^B \, R^1[U_2]_i{}^j \, R^{\frac{1}{2}}[U_3]_C{}^D \, (N_1)^{AiC} \, (N_2)_{BjD} \\ &= (U_1)_A{}^B \, R^1[U_2]_i{}^j \, (U_3)_C{}^D \begin{pmatrix} \frac{1}{2} & 1 & \frac{1}{2} \\ A & i & C \end{pmatrix} \begin{pmatrix} \frac{1}{2} & 1 & \frac{1}{2} \\ B & j & D \end{pmatrix} . \end{aligned} \tag{23}$$

Here $i, j = 1, 2, 3$ denote vector and $A, B, \ldots = 1, 2$ spinor indices, respectively. The holonomy is abbreviated as $U_k := U[A, \gamma_k]$.

Finally, we mention that a spin network state can be decomposed into loop states. This decomposition can be done in general by using the following rule, which follows from well-known properties of $SU(2)$ representation theory. Replace each link of the graph with associated spin j with $2j$ parallel strands. Antisymmetrize these strands along each link (obtaining a formal linear combination of drawings). The intertwiners at the nodes can be represented as collections of segments joining the strands of different links. By joining these segments with the strands one obtains a linear combination of multiloops. The spin network states can then be expanded in the corresponding loop states. For details of this construction, see [14].

Applying this rule to the above example (23), we obtain the following. Writing out the explicit expression for the spin-1 representation in terms of spin-1/2 representations (which we will not do here), and using the explicit form of the Clebsch–Gordan coefficient, it is not hard to see that

$$\Psi_S(A) = \Psi_\alpha - \Psi_\beta \ , \tag{24}$$

where β is the loop obtained by joining the four segments γ_1, γ_2, γ_3, γ_2, and α is the double loop $\{\alpha_1, \alpha_2\}$. Here α_1 is obtained by joining γ_1 and γ_2, while α_2 is obtained by joining γ_2 and γ_3. For a graphical illustration, see Fig. 5.

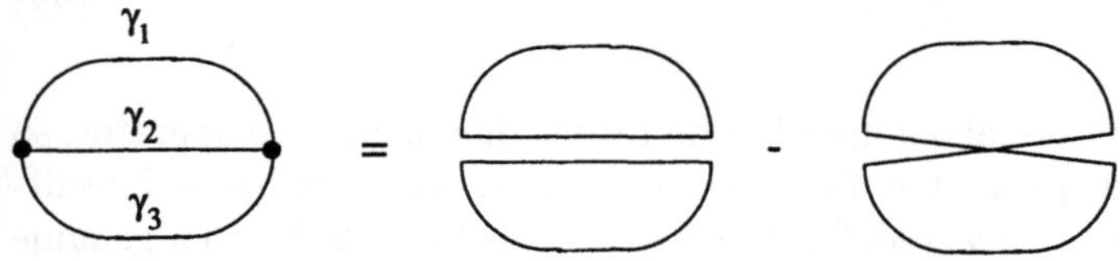

Fig. 5. The decomposition of the spin network state (23) into loop states.

2.6 Operators on $\mathcal{H}$

We now have a gauge invariant kinematical Hilbert space of quantum gravity including an orthonormal basis of spin network states at our disposal. Below, we want to construct self-adjoint $SU(2)$ gauge invariant operators, i.e. linear maps $L : \mathcal{H} \to \mathcal{H}$, that might even be of physical interest.

In this paragraph we will straightforwardly construct well-defined gauge invariant operators and think about their physical interpretation in the next section. We proceed as in usual quantum mechanics by constructing multiplicative and derivative operators, corresponding to "position" and "momentum", respectively. See also [17] and [18].

The simplest operator one can imagine is given by the holonomy itself. Given a curve γ, take the holonomy of the connection along γ to define the multiplicative operator

$$\hat{U}(\gamma) = U[A, \gamma] \ . \tag{25}$$

However, this operator is not $SU(2)$ gauge invariant. In order to obtain gauge invariance, we simply consider the trace of the holonomy along a loop γ, resulting in the operator

$$\hat{T}[\gamma] = \mathrm{Tr}\, U[\gamma] \ . \tag{26}$$

This definition provides a well-defined, gauge invariant and multiplicative operator[9] acting on a state functional as

$$\hat{T}[\gamma]\,\Psi_S(A) = \left(\mathcal{P}\exp\int_\gamma A\right)\Psi_S(A) = U[A,\gamma]\,\Psi_S(A)\,. \tag{27}$$

Hence the definition of multiplicative operators didn't seem to cause any problems.

The construction of a gauge invariant derivative operator turns out to be more subtle. The configuration variable in our approach is the connection $A(x)$, thus the conjugate momentum operator would be some functional derivative with respect to it. The same statement is obtained by considering the Poisson algebra (3) and proceeding as usual in quantum field theories, i.e. formally replacing E^a_i with the functional derivative (we neglect the Immirzi parameter β),

$$E^a_i(x) \;\longrightarrow\; -i\hbar G\,\frac{\delta}{\delta A^i_a(x)}\,. \tag{28}$$

This object is an operator-valued distribution rather than a genuine operator, so it has to be integrated against test functions or, in other words, it has to be suitably smeared in order to be well-defined. Thus, to transform (28) into a genuine operator and regularize expressions involving it, an appropriate smearing over a surface Σ has to be performed. Roughly, this can be seen as follows. The functional derivative (28) with respect to the connection one-form $A(x)$ is a vector density of weight one, or equivalentely a two-form, since it is always possible to transform vector densities into a two-forms by contraction with the Levi–Civita density. Furthermore, since two-forms are naturally integrated against surfaces, a natural geometrical, i.e. coordinate independent regularization scheme is suggested. And indeed, this turns out to be the right way of handling the problem! In fact, smearing (28) as described above, will give us a well-defined operator which is *coordinate invariant* and *finite*.

We start by considering a surface, that is a two-dimensional manifold Σ embedded in M. We use local coordinates x^a, $a = 1,2,3$, on M and let $\boldsymbol{\sigma} = (\sigma^1, \sigma^2)$ be coordinates on the surface Σ. Thus the embedding is given by

$$\Sigma:\; (\sigma^1,\sigma^2) \mapsto x^a(\sigma^1,\sigma^2)\,. \tag{29}$$

We define an operator (using $G = \hbar = 1$)

$$\hat{E}^i(\Sigma) := -i\int_\Sigma d\sigma^1 d\sigma^2\, n_a(\boldsymbol{\sigma})\,\frac{\delta}{\delta A^i_a(x(\boldsymbol{\sigma}))}\,, \tag{30}$$

[9] It is really the choice of this so-called Wilson loops which is the characteristic feature of loop quantum gravity. Indeed, the loop approach can be built on Wilson loops and appropriate momentum operators (the so-called loop variables) which form a closed algebra and thus were used as the starting point for canonical quantization.

where

$$n_a(\sigma) = \epsilon_{abc} \frac{\partial x^b(\sigma)}{\partial \sigma^1} \frac{\partial x^c(\sigma)}{\partial \sigma^2} \tag{31}$$

is the normal one-form on Σ and ϵ_{abc} is the Levi-Civita tensor of density weight (-1).

The next step is to compute the action of this operator on holonomies $U[A,\gamma]$, which are the basic building blocks of the gauge invariant state functionals, i.e. the spin network states. The coordinates of the curve γ in M, which is parametrized by s, will be denoted in the following as $x^a(s) := \gamma^a(s)$.

We begin with the functional derivative of holonomies. A detailed derivation of the relevant formulas can be obtained using the first variation of the defining differential equation (5c) of the holonomy with respect to the connection, see [19]. Consider the surface Σ and a curve γ along which the holonomy was constructed in the simplest case where they have only one individual point of intersection P, which is not supposed to lie at the endpoints of γ, as shown in Fig. 6.

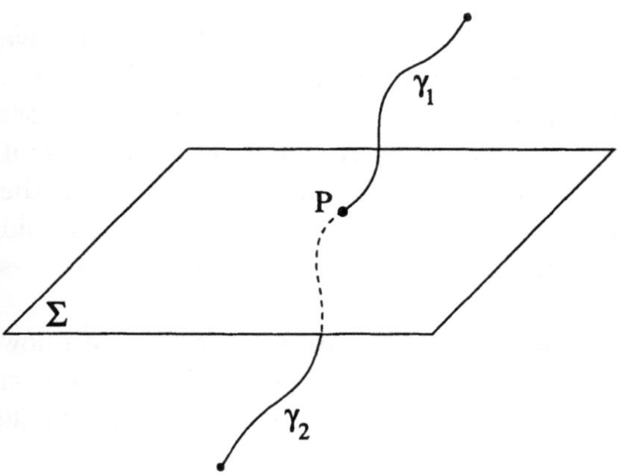

Fig. 6. A curve that intersects the surface in an individual point.

For later convenience the curve is devided into two parts, $\gamma = \gamma_1 \cup \gamma_2$, one lying "above", the other "below" the surface. We get

$$\begin{aligned} \frac{\delta}{\delta A^i_a(x(\sigma))} U[A,\gamma] &= \frac{\delta}{\delta A^i_a(x(\sigma))} \left(\mathcal{P} \exp \int_\gamma ds\, \dot{x}^a\, A^i_a(x(s))\, \tau_i \right) \\ &= \int_\gamma ds\, \frac{\partial x^a}{\partial s}\, \delta^3(x(\sigma), x(s))\, U[A,\gamma_1]\, \tau_i\, U[A,\gamma_2] \,. \end{aligned} \tag{32}$$

Here $U[A,\gamma_1]$ and $U[A,\gamma_2]$ are the parallel propagators along those segments of γ which "start" or "end", respectively, on $P = \Sigma \cap \gamma \neq \emptyset$, see Fig. 6. In order to avoid confusion, recall that $x(\sigma)$ are the coordinates of the surface Σ embedded in the 3-manifold M, while $x(s)$ are the coordinates of $\gamma = \gamma_1 \cup \gamma_2$, just as defined in the beginning of sect. 2.3.

We are now prepared to care about the action of the operator $\hat{E}^i(\Sigma)$ on $U[A,\gamma]$. Using (32), the result can immediately be evaluated,

$$\begin{aligned}
&\hat{E}^i(\Sigma)\, U[A,\gamma] \\
&\quad = -i \int_\Sigma d\sigma^1 d\sigma^2\, \epsilon_{abc} \frac{\partial x^a(\sigma)}{\partial \sigma^1} \frac{\partial x^b(\sigma)}{\partial \sigma^2} \frac{\delta}{\delta A^i_c(x(\sigma))} U[A,\gamma] \\
&\quad = -i \int_\Sigma \int_\gamma d\sigma^1 d\sigma^2 ds\, \epsilon_{abc} \frac{\partial x^a}{\partial \sigma^1} \frac{\partial x^b}{\partial \sigma^2} \frac{\partial x^c}{\partial s}\, \delta^3\big(x(\sigma), x(s)\big) \times \\
&\qquad\qquad \times U[A,\gamma_1]\, \tau^i\, U[A,\gamma_2] \,. \qquad (33)
\end{aligned}$$

A closer look at this result reveals a great simplification of the last integral since one notices the appearance of the Jacobian J for the coordinate transformation $(\sigma^1, \sigma^2, s) \to (x^1, x^2, x^3)$, namely

$$J \equiv \frac{\partial\,(\sigma^1, \sigma^2, s)}{\partial\,(x^1, x^2, x^3)} = \epsilon_{abc} \frac{\partial x^a}{\partial \sigma^1} \frac{\partial x^b}{\partial \sigma^2} \frac{\partial x^c}{\partial s} \,. \qquad (34)$$

In our case, we may assume that the Jacobian is non-vanishing, since we have required that a single, non-degenerate point of intersection of Σ and γ exists. The Jacobian (34) and the integral (33) would vanish, if the tangent vectors given by the partial derivatives in (34), would be coplanar, i.e. if a tangent $\partial x^{a,b}(\sigma)/\partial \sigma^{1,2}$ to the surface would be parallel to the tangent $\partial x^c(s)/\partial s$ of the curve. This happens, for instance, if the curve lies entirely in Σ. Then there would be of course no individual *point* of intersection. We will discuss the various cases a little more at the end of this section.

But let's come back to the actual topic – the simplification of (33). We now carry out the described coordinate transformation, which puts us in the position to integrate out the 3-dimensional δ-distribution. We get for the case of a single intersection the interesting result[10]

$$\int_\Sigma \int_\gamma d\sigma^1 d\sigma^2 ds\, \epsilon_{abc} \frac{\partial x^a(\sigma)}{\partial \sigma^1} \frac{\partial x^b(\sigma)}{\partial \sigma^2} \frac{\partial x^c(s)}{\partial s}\, \delta^3\big(x(\sigma), x(s)\big) = \pm 1 \,, \qquad (35)$$

where the sign depends on the relative orientation of the surface to the curve (this sign will soon become irrelevant). Hence we obtain the simple result

$$\hat{E}^i(\Sigma)\, U[A,\gamma] = \pm i\, U[A,\gamma_1]\, \tau^i\, U[A,\gamma_2] \,. \qquad (36)$$

So we see that the action of the operator $\hat{E}^i(\Sigma)$ on holonomies consists of inserting the matrix $(\pm i\, \tau^i)$ at the point of intersection. Taking advantage of this result, the generalization to the case of more than one single point of intersection is trivial – it is just the sum of all such insertions.

[10] The integer-valued integral is indeed an analytic (coordinate independent) expression for the intersection number of the surface Σ and the curve γ. It is zero in the case of no intersection at all.

Putting all this together, and using P to denote different separate points of intersection, we have:

$$\hat{E}^i(\Sigma)\,U[A,\gamma] = \begin{cases} 0 & \text{if } \;\Sigma \cap \gamma = \emptyset \\ \sum_P \pm i\,U[A,\gamma_1^P]\,\tau^i\,U[A,\gamma_2^P] & \text{if } \;P \in \Sigma \cap \gamma\,. \end{cases} \tag{37}$$

A further generalization of (37) is needed in view of spin networks, where arbitrary irreducible spin-j representations are associated to links and the accompanying holonomies, denoted by $R^j(U[A,\gamma])$. We obtain easily (again for just one single point of intersection, which may be extended analogously to (37))

$$\hat{E}^i(\Sigma)\,R^j(U[A,\gamma]) = \pm i\,R^j(U[A,\gamma_1])\,{}^{(j)}\tau^i\,R^j(U[A,\gamma_2])\,. \tag{38}$$

Here ${}^{(j)}\tau^i$ is the corresponding $SU(2)$ generator in the spin-j representation.

We now have a well-defined operator at our disposal. One may wonder why this smearing scheme gives a well-defined operator, since we have used only a two-dimensional smearing over a surface Σ instead of a three-dimensional one over M, as one might have expected. The answer is that the state functionals have support on one dimension, or in other words, they contain just one-dimensional excitations.

The action of $\hat{E}^i(\Sigma)$ on a spin network state $\Psi_S(A)$ follows immediately from the above considerations. We take a gauge invariant spin network S which intersects the surface Σ at a single point. The holomony along the crossing link γ being in the spin-j representation of $SU(2)$. Then we split[11] the spin network state $\Psi_S(A)$ into a part consisting of this holonomy $R^j(U[A,\gamma])$ along γ, and the "rest" of the state, which is denoted by $\Psi_{S-\gamma}(A)$. Thus we obtain

$$\Psi_S(A) = \Psi^{mn}_{S-\gamma}(A)\,R^j(U[A,\gamma])_{mn}\,. \tag{39}$$

We used the index notation with m and n being indices in the Hilbert space that is attached to γ. Obviously $\Psi_{S-\gamma}(A)$ is not gauge invariant any more. Using (38) we get immediately

$$\hat{E}^i(\Sigma)\,\Psi_S(A) = \pm i\Big[R^j(U[A,\gamma_1])\,{}^{(j)}\tau^i\,R^j(U[A,\gamma_2])\Big]_{mn}\Psi^{mn}_{S-\gamma}(A)\,. \tag{40}$$

Eventually, we see that $\hat{E}^i(\Sigma)$ spoils gauge invariance, since the resulting functional is not an element of $\mathcal{H}_0$ any more. We construct a gauge invariant derivative operator in the next paragraph.

An $SU(2)$ Gauge Invariant Operator. Gauge invariance is spoiled in (40) by the insertion of a matrix τ_i (which is gauge covariant, but not gauge invariant)

[11] To see how this is possible recall the definitions (17) and (18) of a spin network state as a cylindrical function.

at the point of intersection. We can try to construct a gauge invariant operator simply by squaring this matrix, namely by defining

$$\hat{E}^2(\Sigma) := \hat{E}^i(\Sigma)\,\hat{E}^i(\Sigma)\,, \tag{41}$$

where summation over $i = 1, \ldots, 3$ is assumed. Let us compute the action of this operator on a spin network that has only a single point of intersection with Σ. Using the same notation as above, we obtain

$$\begin{aligned}
&\hat{E}^2(\Sigma)\,\Psi_S(A) \\
&\quad = -\Big[R^j\big(U[A,\gamma_1]\big)\,{}^{(j)}\tau^i\,{}^{(j)}\tau^i\,R^j\big(U[A,\gamma_2]\big)\Big]_{mn}\Psi^{mn}_{S-\gamma}(A) \\
&\quad = \Big[R^j\big(U[A,\gamma_1]\big)\,j(j+1)\,R^j\big(U[A,\gamma_2]\big)\Big]_{mn}\Psi^{mn}_{S-\gamma}(A) \\
&\quad = j(j+1)\Big[R^j\big(U[A,\gamma_1]\big)\,R^j\big(U[A,\gamma_2]\big)\Big]_{mn}\Psi^{mn}_{S-\gamma}(A) \\
&\quad = j(j+1)\,\Psi_S(A)\,.
\end{aligned} \tag{42}$$

Here $C := {}^{(j)}\tau^i\,{}^{(j)}\tau^i = -j(j+1)\times\mathbf{1}$ is the Casimir operator of $SU(2)$.

Thus it seems we are lucky this time. We have found the important result that the spin network state is an eigenstate of this seemingly gauge invariant operator and even calculated its eigenvalues. But we have calculated this result in case of a *single* intersection between S and Σ. It is easy to convince oneself that for *several* points of intersection crossterms would appear that again spoil the gauge invariance of $\hat{E}^2(\Sigma)\,\Psi_S(A)$. However, using a simple trick, these crossterms may be eliminated in order to construct a genuinely $SU(2)$ gauge invariant operator in the following way.

Since we have shown that in the case of a *single* intersection $\hat{E}^2(\Sigma)$ turns out to be an operator of the type we are looking for, it is natural to consider a partition ρ of Σ into n small surfaces Σ_n, where $\bigcup_n \Sigma_n = \Sigma$, in such a way that for any given spin network S all different points of intersection P lie in distinct surfaces Σ_n, as it is shown in Fig. 7 for a curve γ which intersects the surface several times. Clearly $n = n(\rho)$ depends on the degree of refinement of the partition.

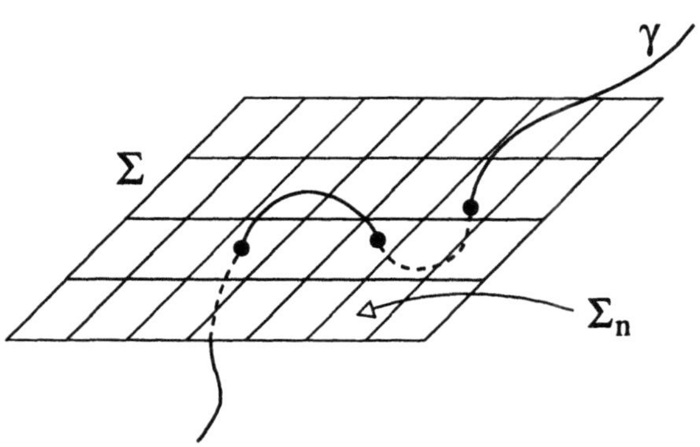

Fig. 7. A partition of Σ.

Hence, we obtain a new operator $\hat{A}(\Sigma)$ which is defined in the limit of infinitely fine triangulations or partitions of Σ, respectively, as

$$\hat{A}(\Sigma) := \lim_{\rho\to\infty} \sum_{n=n(\rho)} \sqrt{\hat{E}^i(\Sigma_n)\,\hat{E}^i(\Sigma_n)}\,. \tag{43}$$

The square root is introduced for later convenience. It can furthermore be shown, that this operator is defined independently of the partition ρ chosen. For simplicity, we disregard spin networks that have either a node lying on Σ or a continuous, i.e. infinite number of intersection points with it, cf. Fig. 8.

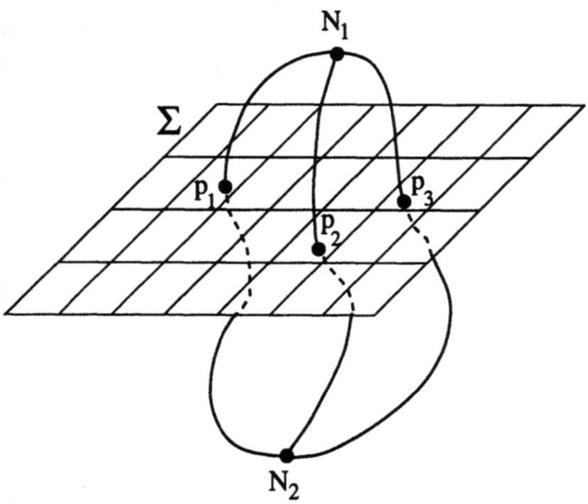

Fig. 8. A simple spin network S intersecting the surface Σ.

Then, using (42) we obtain immediately the action of $\hat{A}(\Sigma)$ on a spin network state as

$$\hat{A}(\Sigma)\,\Psi_S(A) = \sum_{P\in S\cap\Sigma} \sqrt{j_P(j_P+1)}\,\Psi_S(A)\,. \tag{44}$$

Hence, each link of the spin network S labelled by the irreducible representation j of $SU(2)$ which crosses the surface transversely in the small surface Σ_n, would contribute a factor of $\sqrt{j(j+1)}$. Other subsurfaces $\Sigma_{n'}$ that have no intersection with a link of S would give no contribution. Since the operator is diagonal on spin network states and real on this basis, it is also self-adjoint.

To summarize, we have obtained for each surface $\Sigma \in M$ a well-defined $SU(2)$ gauge invariant and self-adjoint operator $\hat{A}(\Sigma)$, which is diagonalized in the spin network basis on $\mathcal{H}_0$, the Hilbert space of gauge invariant state functionals. The corresponding spectrum (with the restrictions mentioned) is labeled by multiplets $\boldsymbol{j} = (j_1, \dots, j_n)$, $i = 1, \dots, n$, and n arbitrary, of positive half integers j_i. This is called *main sequence* of the spectrum and is given (up to constant factors) by

$$A_j(\Sigma) = \sum_i \sqrt{j_i(j_i+1)}\,. \tag{45}$$

As mentioned, (45) is not the result of the most general case, since we excluded crossings of S and Σ, in which the intersection points P may be nodes p of the spin network. To complete the picture and include all cases, we finally give the full spectrum of $\hat{A}(\Sigma)$, which was calculated in [20] directly in the loop representation and in [21] in the connection representation. In the general case we may divide the links that meet at the nodes on the surface into three classes according to their relative position with respect to the surface, see Fig. 9. First, there are the "tangential" (t) links which lie entirely in Σ. The remaining two classes are given by the "up" (u) and "down" (d) links according to the (arbitrary) side of Σ they lie on.

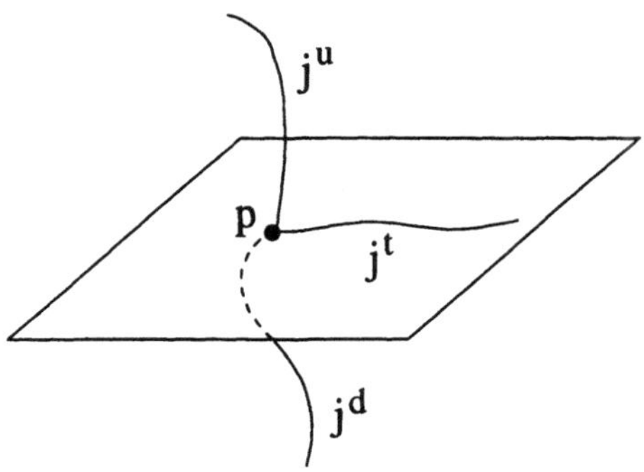

Fig. 9. The three classes of links that meet in a node on the surface.

The *full* spectrum of (43) (the so-called *second sequence*) is labeled by n-tuplets of triplets of positive half integers j_i, namely $j_i = (j_i^u, j_i^d, j_i^t)$, $i = 1, \ldots, n$, and n arbitrary. It is given by

$$A_{j_i}(\Sigma) = \frac{1}{2}\sum_i \sqrt{2j_i^u(j_i^u+1) + 2j_i^d(j_i^d+1) - j_i^t(j_i^t+1)}\,. \tag{46}$$

This is the complete spectrum. It contains of course the previous case (45) corresponding to to $j_i^u = j_i^d$ and $j_i^t = 0$.

3 Quantization of the Area

In the previous section we have described the construction and diagonalization of the $SU(2)$ gauge invariant and self-adjoint operator $\hat{A}(\Sigma)$ using a basis of spin network states in the kinematical gauge invariant Hilbert space $\mathcal{H}_0$. The physical interpretation of this operator was totally disregarded. The operator we have studied is explicitely

$$\hat{A}(\Sigma) := \lim_{\rho\to\infty} \sum_{n=n(\rho)} \sqrt{\hat{E}^i(\Sigma_n)\,\hat{E}^i(\Sigma_n)}\,. \tag{47}$$

We now search the corresponding classical quantity. Just as in usual quantum mechanics this amounts to replacing the quantum operators $\hat{E}^i(\Sigma)$ by their classical analogues.

The conjugate momentum operator, which is essentially given by $\delta/\delta A^i_a(x)$, is the quantum analogue of the (smooth) inverse densitized triad $E^a_i(x)$, i.e. we have the correspondence

$$E^a_i(x) \longleftrightarrow -i\hbar G\frac{\delta}{\delta A^i_a(x)}\,, \tag{48}$$

between classical and quantum quantities, as we already stated in sect. 2.6. We replace the operator (47) in the classical limit by its analogue (48),

$$A(\Sigma) := \lim_{\rho\to\infty} \sum_{n=n(\rho)} \sqrt{E^i(\Sigma_n)E^i(\Sigma_n)}\,, \tag{49}$$

and study its physical meaning. Here we use again $\hbar = G = 1$. Moreover,

$$E^i(\Sigma_n) = \int_{\Sigma_n} d\sigma^1 d\sigma^2\, n_a(\sigma)\, E^{ia}(x(\sigma)) \tag{50}$$

is the classical analogue of the smeared version (30) of the operator $\hat{E}^i(\Sigma_n)$ defined on one specific subsurface Σ_n of the triangulation ρ of Σ, and

$$n_a(\sigma) = \epsilon_{abc}\frac{\partial x^b(\sigma)}{\partial\sigma^1}\frac{\partial x^c(\sigma)}{\partial\sigma^2} \tag{51}$$

is the normal to Σ_n. For a sufficiently fine partition ρ, i.e. arbitrarily small surfaces Σ_n, the integral (50) can be approximated by

$$E^i(\Sigma_n) \approx \Delta\sigma^1\Delta\sigma^2\, n_a(\sigma)\, E^{ai}(x_n(\sigma)), \tag{52}$$

where x_n is an arbitrary point in Σ_n and $(\Delta\sigma^1\Delta\sigma^2)$ denotes its coordinate area. Inserting this result back into the classical expression (49) gives

$$A(\Sigma) = \lim_{\rho\to\infty} \sum_{n=n(\rho)} \Delta\sigma^1\Delta\sigma^2 \sqrt{n_a(\sigma)E^{ai}(x_n(\sigma))\, n_b(\sigma)E^{bi}(x_n(\sigma))} \tag{53a}$$

$$= \int_\Sigma d^2\sigma \sqrt{n_a(\sigma)E^{ai}(x(\sigma))\, n_b(\sigma)E^{bi}(x(\sigma))}\,. \tag{53b}$$

The second line (53b) follows immediately by noting that (53a) is nothing but the definition of the Riemann integral. For its evaluation we choose local coordinates in such a way that $x^3(\sigma) = 0$ on Σ and furthermore $x^1(\sigma) = \sigma^1$, $x^2(\sigma) = \sigma^2$, resulting in $n_a = n_b = (0,0,1)$. We obtain

$$A(\Sigma) = \int_\Sigma d^2\sigma \sqrt{E^{3i}(x)\, E^{3i}(x)} \tag{54a}$$

$$= \int_\Sigma d^2\sigma \sqrt{\det g(x)\, g^{33} x} \tag{54b}$$

$$= \int_\Sigma d^2\sigma \sqrt{g_{11}g_{22} - g_{12}g_{21}} \tag{54c}$$

$$= \int_\Sigma d^2\sigma \sqrt{\det\,(^2 g)}\ . \tag{54d}$$

For the derivation of (54b) we used relation (1) between the 3-metric and the triad variables, which is $g^{ab}(x) \det g(x) = E^{ai}(x) E^{bi}(x)$, while the transition to the next equation is made by using the definition for the inverse of a matrix. Noting that $(^2 g)$ is the two-dimensional metric induced by g_{ab} on Σ, one recognizes the result (54d) as the covariant expression for the *area* of Σ.

In fact, since the classical geometrical observable "area of a surface" is a functional of the metric, i.e. of the gravitational field, in a quantum theory of gravity, where the metric is an operator, the area turns into an operator as well. If this operator reveals a discrete spectrum, this would, according to quantum mechanics, also imply the discreteness of physical areas at the Planck length. Thus the area is quantized!

Restoring all neglected constants (and a factor of 16π which occurs in a detailed calculation) and using the notation and results we obtained in sect. 2.6, namely the discreteness of the spectrum of $\hat{A}(\Sigma)$, which from now on will be denoted as ***area operator***, gives the (main sequence of) eigenvalues of the area as

$$A(\Sigma) = 16\pi\hbar G \sum_i \sqrt{j_i(j_i+1)}\ . \tag{55}$$

This formula gives the area of a surface Σ that is intersected by a spin network S without having nodes lying in it. The quanta are labeled by multiplets j of half integers as already realized in sect. 2.6. The generalization to the case where nodes of S are allowed to lie in Σ, yielding the second sequence of eigenvalues, is given by (46).

4 The Physical Contents of Quantum Gravity and the Meaning of Diffeomorphism Invariance

Some questions arise immediately from the results we discussed in the last section.

Is $A(\Sigma)$ observable in quantum gravity?

or equivalently:

What should a quantum theory of gravitation predict?

These questions are intimately related to the issue of observability in both classical and quantum gravity – an issue which is far from trivial. Let us begin with an examination of the classical theory. For a closer look at this topic, we refer to [22–24].

4.1 Passive and Active Diffeomorphism Invariance

We consider ordinary classical general relativity formulated on a 4-dimensional manifold $\mathcal{M}$ on which we introduce local coordinates x^μ, $\mu = 0, \ldots, 3$, abbreviated by x.

The Einstein equations

$$R_{\mu\nu} - \frac{1}{2} R\, g_{\mu\nu} = 0 \tag{56}$$

are invariant under the group of diffeomorphisms $Diff(\mathcal{M})$ of $\mathcal{M}$. Recall that a diffeomorphism ϕ is a C^∞ map between manifolds that is one-to-one, onto and has a C^∞ inverse. In other words, the diffeomorphism group is formed by the set of mappings $\phi : \mathcal{M} \rightarrow \mathcal{M}$ which preserve the structure of $\mathcal{M}$. We consider diffeomorphisms which are given in local coordinates by the smooth maps $x'^\mu = x'^\mu(x^\nu)$. The inverse transformations are $x^\mu = x^\mu(x'^\nu)$.

Suppose a solution $g_{\mu\nu}(x)$ of Einstein's equations (56) is given, then due to diffeomorphism invariance, $\tilde{g}_{\mu\nu}(x)$ is also a solution, where

$$\tilde{g}_{\mu\nu}(x') = \frac{\partial x^\rho}{\partial x'^\mu} \frac{\partial x^\sigma}{\partial x'^\nu}\, g_{\rho\sigma}\big(x(x')\big) \,. \tag{57}$$

There are two geometrical interpretations of (57) known as *passive* and *active* diffeomorphisms.

Passive diffeomorphism invariance refers to invariance under change of coordinates, i.e. the same object is represented in different coordinate systems. Choose a (local) coordinate system $\{x^\mu\}$ in which the metric is $g_{\mu\nu}(x)$. In a second system $\{x'^\mu\}$ the metric is given by $\tilde{g}_{\mu\nu}(x')$. Satisfying (57), both of them represent the *same* metric on $\mathcal{M}$.

Active diffeomorphisms on the other hand relate different objects in $\mathcal{M}$ in the same coordinate system. This means that $x'^\mu(x)$ is viewed as a map associating one point of the manifold to another point of the manifold. Take for example two points $P, Q \in \mathcal{M}$ and consider two metrics $g_{\mu\nu}(x)$ and $\tilde{g}_{\mu\nu}(x)$, which are both solutions of (56). Then the distance d between P and Q computed using the two metrics is different, i.e. $d_g(P,Q) \neq d_{\tilde{g}}(P,Q)$. We have two *distinct* metrics on $\mathcal{M}$ which both solve Einstein's equations. These two metrics might still be related by equation (57). They are related by an active diffeomorphism.

The relations between active and passive diffeomorphisms, as well as the choice of coordinates, is clarified in Fig. 10.

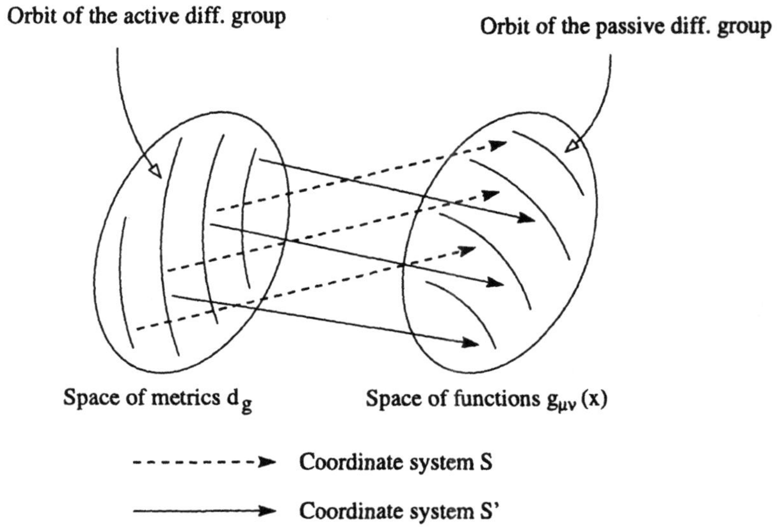

Fig. 10. The relation between active and passive diffs and the choice of coordinates.

In order to avoid confusion with regard to passive and active diffeomorphisms in coordinate-dependent considerations, we simply drop coordinates and pass over to the coordinate-free formulation. Thus we consider the mainfold $\mathcal{M}$ with metric g, defined as the map

$$g : \mathcal{M} \times \mathcal{M} \to \mathbb{R} \tag{58a}$$

$$(P,Q) \mapsto d_g\,(P,Q)\ , \tag{58b}$$

where $P, Q \in \mathcal{M}$. Suppose d_g solves Einstein's equations. A diffeomorphism $\phi : \mathcal{M} \to \mathcal{M}$ acts as a smooth displacement over the manifold, resulting in $d_{\tilde{g}}$,

$$d_{\tilde{g}}\,(P,Q) = d_g\left(\phi^{-1}(P), \phi^{-1}(Q)\right)\ . \tag{59}$$

Active diffeomorphism invariance is the fact that if d_g is a solution of the Einstein theory, so is $d_{\tilde{g}}$. This shows that Einstein's theory is invariant under (active!) diffeomorphisms even in a coordinate free formulation.

General relativity is distinguished from other dynamical field theories by its invariance under *active* diffeomorphisms. Any theory can be made invariant under *passive* diffeomorphisms. Passive diffeomorphism invariance is a property of the *formulation* of a dynamical theory, while active diffeomorphism invariance is a property of the dynamical theory *itself*. Invariance under smooth displacements of the *dynamical* fields holds only in general relativity and in any general relativistic theory. It does not hold in QED, QCD, or any other theory on a fixed (flat or curved) background.

4.2 Dirac Observables

Consider a classical dynamical system whose equations of motion do not uniquely determine its evolution, as pictorially illustrate in Fig. 11. The two solutions $\varphi(t)$ and $\tilde{\varphi}(t)$ which evolve from the same set of initial data, separate at some later time $\hat{t}$, i.e.

$$\varphi(t) = \tilde{\varphi}(t) \quad \text{if} \quad t < \hat{t} \tag{60a}$$
$$\varphi(t) \neq \tilde{\varphi}(t) \quad \text{if} \quad t \geq \hat{t} \geq 0 \,. \tag{60b}$$

Then, as first accurately argued by Dirac, $\varphi(t)$ and $\tilde{\varphi}(t)$ must be physically indistinguishable or *gauge-related*, respectively. Otherwise determinism, which is a basic principle in classical physics, would be lost. Dirac gave the definition of observables respecting determinism in the following way. A *gauge invariant* or *Dirac observable* is a function $\mathcal{O}$ of the dynamical variables that does not distinguish $\varphi(t)$ and $\tilde{\varphi}(t)$, i.e.

$$\mathcal{O}\big(\varphi(t)\big) = \mathcal{O}\big(\tilde{\varphi}(t)\big) \,. \tag{61}$$

In other words, only those observables that have the same values on the solutions $\varphi(t)$ and $\tilde{\varphi}(t)$ can be observed. Hence the theory can predict only Dirac observables.

Does this imply that any physical quantity that we measure is necessarily a Dirac observable? It turns out that one has to answer in the negative. To understand this sublety, consider the example of a simple pendulum described by the variable α which is the deflection angle out of equilibrium. The motion of the pendulum is given by the evolution of α in time t, namely by $\alpha(t)$. Since $\alpha(t)$ is predicted by the equation of motion for any time t once the initial data

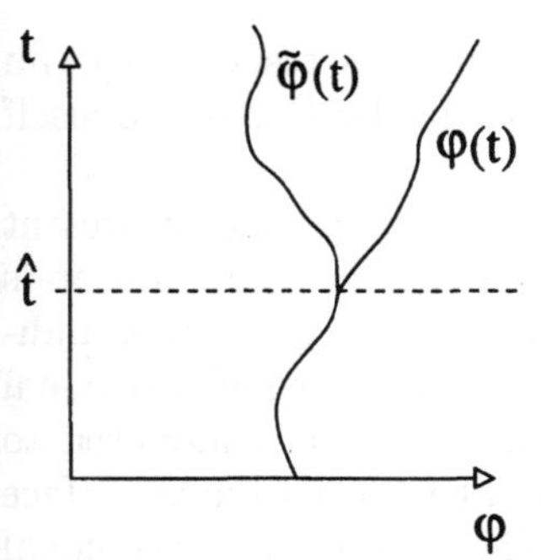

Fig. 11. An example for a not uniquely determined evolution of a dynamical system.

set is fixed, it is a Dirac observable. One should notice that we are actually describing a system in terms of two physical quantities rather than one, namely the pendulum itself, described by position α, and a clock measuring the time t. However, in contrast to position, there is no way how time could be predicted. It simply tells us "when" we are. Therefore, t is a measureable quantity but it is not a Dirac observable. To state this more precisely, we introduce the notion of *partial observables.* We call t an *independent partial observable* and α a *dependent partial observable.* The Dirac observable is given by $\alpha(t)$.

There is an important relation between Dirac observables and the Hamitonian formalism. Dirac observables are characterized by having vanishing Poisson brackets with the constraints. In fact, the entire constrained system formalism was built by Dirac with the purpose of characterizing the gauge invaraint observables (the Dirac observables). To elucidate this feature, consider a classical dynamical system with canonical Hamiltonian H_0, as well as k additional constraints

$$C_m = 0\,, \qquad m = 1, \ldots, k\,, \tag{62}$$

defined on phase space. The complete Hamiltonian, which is defined on the full phase space, is given by

$$H = H_0 + N_m(t)\, C_m\,, \tag{63}$$

with k arbitrary functions $N_m(t)$ that can be interpreted as Lagrangian multipliers. The dynamics of an observable $\mathcal{O}$ is given by the Hamiltonian equations

$$\dot{\mathcal{O}} = \{\mathcal{O}, H\} + N^m(t)\{\mathcal{O}, C_m\}\,. \tag{64}$$

From this one recognizes immediately that the evolution is deterministic, and thus $\mathcal{O}$ a Dirac observable, only if

$$\{\mathcal{O}, C_m\} = 0 \qquad \forall\, m\,, \tag{65}$$

just as claimed.

4.3 The Hole Argument

Dirac's postulate that only gauge invariant or Dirac observables, respectively, can be measurable quantities, was applied to general relativity by Einstein himself in his famous "hole argument" from 1912.

Suppose we have a space–time $\mathcal{M}$ including other structures that represent matter (e.g. scalar fields or particles). Suppose that the matter configuration is such that there is a *hole* in space–time, i.e. a region without matter, as indicated in Fig. 12. Let $g_{\mu\nu}(x)$ and $\tilde{g}_{\mu\nu}(x)$ be two distinct metrics which are equal everywhere in $\mathcal{M}$ except for the hole, but nevertheless, both are supposed to solve Einstein's equations. Now we introduce a spacelike (initial data) surface such that the hole is entirely in the future of it. Since the metrics are equal everywhere outside, they do have the same set of initial data on the surface.

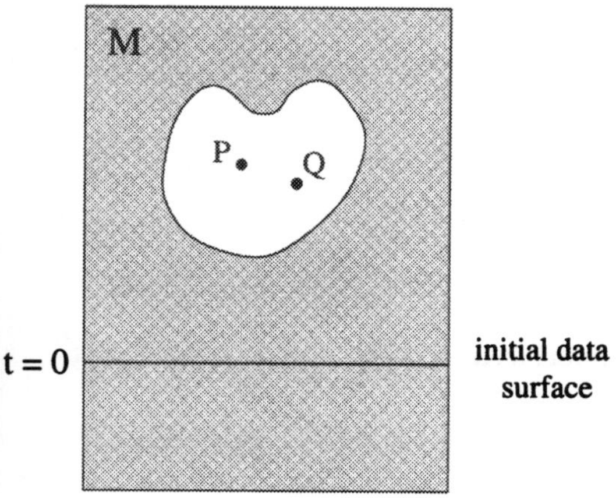

Fig. 12. The hole argument.

If we now consider the distance between two distinct points P and Q which are both inside the hole, we note immediately that $d_g(P,Q) \neq d_{\tilde{g}}(P,Q)$, although the metrics have the same inital conditions. Hence, according to the discussion in the previous section, d_g is *not* a Dirac observable. So it seems that we uncovered a mystery of the theory! The distance is not an observable predicted by the theory. Then the obvious question we have to ask is:

"*What* is predicted by general relativity at all?"

Einstein was so impressed by this conclusion, that he claimed in 1912 that general covariance could *not* be a property of the theory of gravity. It took some time – three years – until Einstein presented the solution to this puzzle, and thus got back to general covariance, in 1915. To illustrate his strategy, we consider a setting similar to the one above, which corresponds to Fig. 12. More precisely, we consider general relativity and 4 particles denoted by A, B, C and D. Their trajectories are determined by the equations of motion and they are supposed to start at the spacelike inital surface, as shown in Fig. 13. Furthermore, we suppose that A and B meet in i inside the hole, and C and D meet in j inside the hole as well. Consider now the distance d between the point i and the point j. Is d a Dirac observable? At first sight, we are in the same situation as above, but there is an essential subtle difference. Consider now the diffeomorphism that sends $g_{\mu\nu}(x)$ into $\tilde{g}_{\mu\nu}(x)$. Since the theory is invariant only under a diffeomorphism that acts on *all* its dynamical variables, $\tilde{g}_{\mu\nu}(x)$ is a solution of the Einstein equations only if the diffeomorphism displaces the trajectories of the particles as well. Thus i and j will also be displaced by the diffeomorphism. Then, after having performed the active diffeomorphism, the new distance between the intersection points is

$$\tilde{d} = d_{\tilde{g}}\left(\phi(i), \phi(j)\right) = d_g\left(\phi^{-1}\phi(i), \phi^{-1}\phi(j)\right) = d_g(i,j) = d\,. \tag{66}$$

Hence it follows that this distance is gauge invariant. The distance d between the itersection points is a Dirac observable.

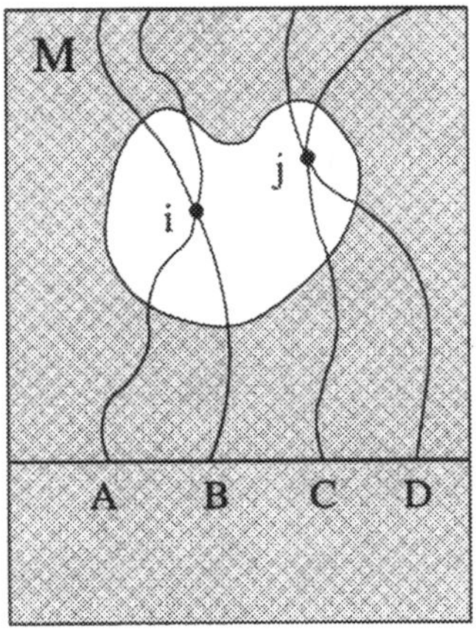

Fig. 13. A solution to the hole argument.

One can extend this setting also to cases which involve fields. As an example, consider general relativity and 2 additional fields, namely $g_{\mu\nu}(x)$, $\varphi_t(x)$, and $\varphi_z(x)$. Then the area A of the $\varphi_t = \varphi_z = 0$ surface is a Dirac observable as well, and is given by

$$A = \int_{\substack{\varphi_t=0\\ \varphi_z=0}} d^2\sigma \sqrt{\det {}^2g} \, . \tag{67}$$

For the slightly generalized case of general relativity and three fields, i.e. $g_{\mu\nu}(x)$, $\varphi_t(x)$, $\varphi_z(x)$, and $\varphi_\Sigma(x)$, the area $A(\Sigma)$ of the surface determined by

$$\varphi_t = \varphi_z = 0, \quad \text{and} \quad \varphi_\Sigma \geq 0 \tag{68}$$

is again a Dirac observable. It is in this case given by

$$A(\Sigma) = \int_{\substack{\varphi_t=0\\ \varphi_z=0}} d^2\sigma \, \delta(\varphi_\Sigma) \sqrt{\det {}^2g} \, . \tag{69}$$

In general, to define "local" Dirac observables in general relativity we have to use some of the degrees of freedom of the theory (the particles, the fields) for localizing a spacetime point or a spacetime region. It is important to notice that in principle we do not need *matter* or fields to do so. Instead, we can use part of the degrees of freedom of the graviational field itself. This strategy was followed for instance by Komar and Bergman by defining 4 curvature scalars and using them as physically defined coordinates. While formally correct, the use of gravitational degrees of freedom for defining observables in general relativity leads us far away from observables concretely used in the realistic applications of general relativity, all of which use matter degrees of freedom for localizing the observables. An example of a realistic observable used in physical applications of general relativity is the physical distance between two spacetime events, one on a Global Positioning System (GPS) satellite and one on a Earth based GPS station. In this case, matter degrees of freedom (coupled to gravity) localize two spacetime points and the distance between them is a Dirac observable.

To sum up, we have seen that the puzzle of the hole argument can be resolved. Physical quantities predicted by general relativity, i.e. Dirac observables, can be defined inside the hole. But in order to "localize" points, we have to use some dynamical quantity. The most realistic way of doing so is to use matter. In other words, Dirac observables are defined in space–time regions which are determined by dynamical objects.

In the following section, we will see that this definition of localization, which is necessary in general relativity, implies a profound change of our notions of space and time.

4.4 The Physical Interpretation

Before considering the conceptual changes in the notions of space and time brought by general relativity, it is instructive to reflect on the main modifications that these concepts have undergone in the historical development of physics. The key developments in this business are related to the names of Descartes (and Aristotle), Newton, and last but not least, Einstein.

According to Descartes, there is no "space" at all, but only physical objects which can be in touch with each other. The "position" or location, respectively, of an object is only defined by the naming of other physical objects close to it, i.e. the position of a body is the set of those objects to which the body is contiguous. Equally important is the concept of "motion", which is defined as the change of position. Thus motion is determined by the change of contiguity, i.e. only in relation to other objects. This point of view is denoted *relationalism*. Descartes' definitions of space, position and motion are by the way essentially the same that were given by Aristotle.

An important historical step was then provided by Newton's definition of physical space. According to Newton, "space" exists by itself, independently of the objects in it. Motion of a body can be defined with respect to space alone, irrespectlively whether other objects are present. Newton insists on this points, on the ground that acceleration can be defined absolutely. In fact, it is only thanks to the fact that acceleration is defined in absolute terms, that the entire structure of Newton's mechanics ($F = ma$) holds. Newton discussed the fact that acceleration is absolute in the famous example of the rotating bucket, which shows that the absolute rotation of the water, and not the rotation with respect to the bucket, has observable consequences. Thus, according to Newton, space exists independently of objects, weather they are present or not. The location of objects is the part of space that they occupy. This implies that motion can be understood without regard to surrounding objects. Similarly, Newton uses absolute time, leading to a space–time picture which provides an always present fixed background over which physics takes place. Objects can always be localized in space and time with respect to this fixed non-dynamical background.

But if there is "space" which is always present, how can it be captured, or observed? This can be done by using reference systems. The great idea was to select some physical bodies (like walls, rules or clocks) and treat them as reference systems. Physically one has to distinguish the dynamical objects that one wants to study from reference system objects. They are dynamically decoupled.

In the language introduced earlier, the dynamical objects define dependent partial observables, while the objects referred to as reference system define independent partial observables. Examples for dynamical objects may be the deflection angle α of a pendulum, or the position x of a particle. The Dirac observables would then just be $\alpha(t)$ and $x(t)$.

As an example, we take the case of a pendulum. The differential equation governing this dynamical system is (for small oscillations) just $\ddot{\alpha}(t) = -\omega^2 \alpha(t)$. The solution is

$$\alpha(t) = A \sin(\omega t + \varphi) \ . \tag{70}$$

A *state* is determined by the constants A and φ, or equivalently, by initial position and velocity at some fixed time. Once the state (A and φ) is known, the functional dependence $\alpha(t)$ between the dependent and independent observables can be computed. In fact, it is given by (70).

Thus, in the Newtonian scheme, we have a fixed space and a fixed time, revealed by the objects of the reference system. The objects forming the reference system determine localization in space and in time and define partial observables (t, above) which are not dynamical variables in the dynamical models one considers.

In general relativity things change profoundly. We have seen in the discussion of the hole argument and its solution, that the theory does not distinguish reference system objects from dynamical objects. This means that independent and dependent physical observables are not distinguished any more! The reference system can not be decoupled from the dynamics. Therefore, in the Einsteinian framework the notion of "dynamical object" has to be extended compared to the Newtonian case, since now also the reference system objects are included as dynamical variables. Localization of observables is determined by other variables of the theory. Therefore:

Position and Motion are fully relational in general relativity!

This important statement is the same as provided in the Cartesian–Aristotelian picture.

The essential consequence of the fact that localization of dynamical objects in general relativity is defined only with respect to each other, is the appearance of the diffeomorphism group. Indeed, if we displace all *dynamical* objects in the manifold at once, we generate nothing but an equivalent mathematical description of the same physical state, because localization with respect to the manifold is irrelevant. Only relative localization is relevant. This is precisely the claim of active diffeomorphism invariance of the theory. Hence, a physical state is not located somewhere.

In a quantum theory of gravity, we should not expect quantum exitations *on* space–time, as the Newtonian point of view would imply, rather we should expect quantum excitations *of* space–time.

The challenge in the construction of quantum gravity is to find a quantum field theory in which position and motion are fully relational, i.e. a quantum field theory without an *a priori* space–time localization. Here the wheel turns full circle, and we return to loop quantum gravity, which implements precisely these requirements.

5 Dynamics, True Observables and Spin Foams

The analysis of the important question of observability in general relativity led to the insight that spatiotemporal relationalism à la Descartes plays a major role in the formulation of the theory.

In this section we will return to the quantum theory and firstly focus on the implementation of relationalism into the framework of canonical quantum gravity. Secondly, we will investigate the dynamics and the true, i.e. physical observables of the theory, which formally amounts to the still open problem of solving the Hamiltonian constraint. Instead, we will construct a projection operator onto the physical states of loop quantum gravity, which will lead to a covariant space–time formulation and a relation to the so-called spin foam models. For a more detailed analysis of this topic we refer to [25].

As we mentioned at the end of the last section, loop quantum gravity is well-suited to tackle the matters discussed there. Thus, the starting point of our considerations is the implementation of the concept of ***non-localizability*** into the framework of loop quantum gravity. And of course, as one might have expected, this is achieved by solving the diffeomorphism constraint!

Recall from sect. 2.5 that the basis in the gauge invariant Hilbert space $\mathcal{H}_0$ is given by the spin network states $\Psi_S(A)$. In the following we adopt Dirac's bra-ket notation and denote an abstract basis state Ψ_S as $|S\rangle$, which would be given in the connection representation by

$$\Psi_S(A) = \langle A|S\rangle \; . \tag{71}$$

5.1 The Diffeomorphism Constraint

We described in sect. 2.3 that the Hilbert space $\mathcal{H}$ carries a natural unitary representation $U(Diff)$ of the diffeomorphism group $Diff(M)$ of the 3-manifold M. In the following we will outline the construction of the diffeomorphism invariant Hilbert space $\mathcal{H}_{diff}$ (recall Fig. 2), which can be considered as the space $\mathcal{H}/Diff(M)$ of solutions of the quantum diffeomorphism contraint.

Let's consider a finite action of a unitary representation $U(Diff)$ of $Diff(M)$ on a spin network state $|S\rangle$:

$$U(\phi)|S\rangle = |\phi \cdot S\rangle \; , \qquad \phi \in Diff(M) \; . \tag{72}$$

Thus, U sends a state of the spin network basis into another one. To obtain states which are invariant under U, one has to solve

$$U\Psi = \Psi \; . \tag{73}$$

However, there is no finite norm state invariant under the action of the diffeomorphism group. This is not surprising, as the gauge group is not compact, and leads us to a familiar situation in quantum theory. The way out is to use generalized states techniques. The simplest manner of doing so is to solve (73) in $\mathcal{H}^*$, the space dual to the space of finite linear combinatons of spin network states. We construct $\mathcal{H}_{diff}$ as the $Diff(M)$ invariant part of $\mathcal{H}^*$.

Now, let s be an equivalence class of embedded spin networks S under $Diff(M)$, i.e. $S, S' \in s$, if there exists a $\phi \in Diff(M)$, such that $S' = \phi \cdot S$. An equivalence class s or abstract spin network, respectively, is a spin network which is "smeared" over M. It is usually called s-knot. We define

$$\langle s|S\rangle = \begin{cases} 0 & \text{if} \quad S \notin s \\ 1 & \text{if} \quad S \in s \,. \end{cases} \tag{74}$$

Then the $\langle s|$ span $\mathcal{H}_{diff}$, in which the scalar product is defined as

$$\langle s|s'\rangle = \begin{cases} 0 & \text{if} \quad s \neq s' \\ c(s) & \text{if} \quad s = s \,. \end{cases} \tag{75}$$

Here $c(s)$ is the number of *discrete* symmetries on the abstract s-knot under a diffeomorphism. Accordingly, the states $\frac{1}{c(s)}\,|s\rangle$ form an orthonormal basis (notice that we freely interchange bra's and ket's).

The states $|s\rangle$ are the diffeomorphism invariant quantum states of the gravitational field. They are described by abstract, non-embedded (knotted, colored) graphs s. As we have seen above, each link of the graph can be seen as carrying a quantum of area. As shown for instance in [17], a similar results holds for the volume: in this case, they are the nodes that carry quanta of volume. Thus, and abstract graph can be seen as an elementary quantum excitation of space formed by "chunks" of space (the nodes) with quantized volume, separated by sheets of surface (corresponding to the links), with quantized area. The key point is that the graph does not live on a manifold. The quantized space does reside "somewhere". Instead, it defines the "where" by itself. This is the picture of quantm spacetime that emerges from loop quantum gravity.

Formal Manipulations. We close the discussion on the diffeomorphism constraint by reexpressing the diffeomorphism invariant states using some intriguing formal expressions that will lead us to dealing with the hamiltonian constraint.

Although we noticed that $\mathcal{H}_{diff}$ is not a subspace of $\mathcal{H}$, there exists nevertheless a "projection operator" Π,

$$\Pi \;:\; \mathcal{H} \rightarrow \mathcal{H}_{diff} \,. \tag{76}$$

It acts as

$$\Pi|S\rangle = |s\rangle \,, \tag{77}$$

i.e. a spin network state is mapped to the corresponding diffeomorphism invariant equivalence class. In the following, its construction will be given. We start by formally defining a measure on $Diff(M)$, which is required to satisfy

$$\int_{Diff} [d\phi] = 1\,, \tag{78}$$

and

$$\int_{Diff} [d\phi]\, \delta_{S,\phi\cdot S} = c(s)\,. \tag{79}$$

Loosely speaking, (79) refers to taking an embedded spin network, acting with diffeomorphisms on it (i.e. displace it smoothly in the manifold), and finally moving it back to the spin network one started with. Then $c(s)$ just counts the number of ways one can do this.

Now, a diffeomorphism invariant knot state $|s\rangle$ can be written as

$$|s\rangle = \int_{Diff} [d\phi]\, |\phi\cdot S\rangle\,, \qquad S \in s\,. \tag{80}$$

Using only the definitions (78)–(80), one immediately concludes (74) and (75). Furthermore, we can also derive a more explicit expression of the projection operator Π. Therefore, note that the generator of the diffeomorphism constraint $\boldsymbol{D}[\boldsymbol{f}]$, which is a smooth vector field $\boldsymbol{f}$ on M, is an element of the Lie-algebra of $Diff(M)$. Then, from (72) and (80) one concludes

$$|s\rangle = \int_{Diff} [d\phi]\, U(\phi)|S\rangle = \int [d\boldsymbol{f}]\, e^{i\boldsymbol{f}\boldsymbol{D}}\, |S\rangle\,, \tag{81}$$

where in the second step we have expressed $U(\phi) \in U(Diff)$ as the exponential of an algebra element, and formally integrate over the algebra rather that the group. From this and (77) we can immediately read off the projector

$$\Pi = \int [d\boldsymbol{f}]\, e^{i\boldsymbol{f}\boldsymbol{D}}\,. \tag{82}$$

Finally, we also obtain a diffeomorphism invariant quadratic form[12] on $\mathcal{H}$,

$$\langle S|S'\rangle_{diff} \equiv \langle S|\Pi|S'\rangle = \int [d\boldsymbol{f}]\, \langle S|e^{i\boldsymbol{f}\boldsymbol{D}}|S'\rangle\,, \tag{83}$$

where one should notice that the spin networks S and S' are not diffeomorphism invariant itselves. Hence it follows (roughly) that the knowledge of the matrix elements $\langle S|\Pi|S'\rangle$ of the projection operator is equivalent to the solution of the diffeomorphism constraint!

[12] The quadratic form $\langle\ |\ \rangle_{diff}$ is highly degenerate, of course.

5.2 The Hamiltonian Constraint, Spin Foam, and Physical Observables

In Fig. 2 we illustrated the plan for a step by step construction of the physical Hilbert space by solving the quantum constraint operators successively. Carrying this out, we were led from the unconstrained Hilbert space $\mathcal{H}$ firstly to the gauge invariant space $\mathcal{H}_0$, equipped with an orthonormal basis of spin network states $|S\rangle$, and secondly, as we described in sect. 5.1, to the diffeomorphism invariant Hilbert space $\mathcal{H}_{diff}$, for which it was also possible to define an orthonormal basis $|s\rangle$ of s-knot states. The final step, marked with a question mark in Fig. 2, remains to be done: the physical states of the theory should lie in the kernel of the quantum Hamiltonian constraint operator. Of course, we do not expect to find a complete solution of the Hamiltonian constraint, which would correspond to a complete solution of the theory. Rather, we need a well posed definition of the Hamiltonian constraint, and a strategy to compute with it and to unravel its physical content.

Here, we will give only a sketchy account of the definition of the Hamiltonian constraint. On the other hand, we will illustrate the way of using this constraint a bit more in detail. The idea we will illustrate is to search the solution of the constraint by constructing a projector on physical states, in the same fashion as we did in the last paragraph on the diffeomorphism constraint. This construction will lead us to the so-called spin foam models, which represent a covariant formulation of the dynamics of quantum gravity, and provide the most exciting and promising of the recent developements in his subject.

A Simple Example. We have already considered the construction of a projection operator in relation to the diffeomorphism constraint in sect. 5.1. Nevertheless, it is instructive to give here a simple toy example, that should explain the procedure in a more accurate way.

Consider a simple dynamical quantum mechanical system with an unconstrained Hilbert space of square integrable functions over $\mathbb{R}^2$, i.e. $\mathcal{H} = L^2(\mathbb{R}^2)$. The system is constrained by demanding invariance with respect to rotations around the z-axis, i.e the angular momentum operator

$$\hat{J} := \hat{J}_z = i\,(x\,\partial_y - y\,\partial_x) \tag{84a}$$

$$\mathrel{\widehat{=}} \hat{J}_\varphi = i\,\partial_\varphi \tag{84b}$$

is the quantum constraint. In (84a) and (84b) we considered two representations, namely the cartesian and the polar coordinate representation, in which the wave functions appear as $\Psi(x,y)$ or $\Psi(r,\varphi)$, respectively. We will confine ourselves to the latter one. The physical Hilbert space $\mathcal{H}_{phys}$ is given as the subspace of $\mathcal{H}$ subject to

$$\hat{J}\Psi = 0\,, \tag{85}$$

i.e. the physical state functionals are required to lie in the kernel of the quantum constraint operator $\hat{J}$. We know that $\hat{J}$ is the generator of the group $U(1)$ with parameter α, acting as

$$U(1) \times \mathbb{R}^2 \to \mathbb{R}^2 \tag{86a}$$
$$(\alpha, (r, \varphi)) \mapsto (r, \varphi + \alpha) \,. \tag{86b}$$

Due to compactness of $U(1)$, the constraint equation (85) could be solved directly. However, we will follow a different path. We try to solve the problem using a projection operator

$$\Pi \; : \quad \mathcal{H} \to \mathcal{H}_{phys} \,. \tag{87}$$

Note, that the (finite) action of the constrait on a general state functional $\Psi(r, \varphi)$ is given by

$$e^{i\alpha\hat{J}}\, \Psi(r, \varphi) = \Psi(r, \varphi + \alpha) \,. \tag{88}$$

Hence, the projection operator Π on physical states is defined as

$$\Pi = \int_0^{2\pi} d\alpha\, e^{i\alpha\hat{J}} \,. \tag{89}$$

It acts on $\Psi(r, \varphi) \in \mathcal{H}$ as follows,

$$\Pi\, \Psi(r, \varphi) = \int_0^{2\pi} d\alpha\, e^{i\alpha\hat{J}} \Psi(r, \varphi) \tag{90a}$$
$$= \int_0^{2\pi} d\alpha\, \Psi(r, \varphi + \alpha) = \Psi(r) \,, \tag{90b}$$

resulting in a function $\Psi(r)$ which is independent of φ, just as one might have expected for the physical states. However, because of the projector Π, there is no need to perform calculations in the physical subspace $\mathcal{H}_{phys}$, rather one can stay in the unconstrained Hilbert space $\mathcal{H}$ – a remarkable simplification! Using the scalar product

$$\langle \Psi | \Phi \rangle = \int_0^{2\pi} \int_0^{\infty} d\varphi\, dr\, \overline{\Psi(r, \varphi)}\, \Phi(r, \varphi) \,, \tag{91}$$

in $\mathcal{H}$, one arrives at the important result

$$\langle \Psi | \Phi \rangle_{phys} \equiv \langle \Psi | \Pi | \Phi \rangle \,. \tag{92}$$

This equation is similar to the result obtained in the discussion of the diffeomorphism constraint in sect. 5.1. The quadratic form $\langle \, | \, \rangle_{phys}$ in $\mathcal{H}_{phys}$ is indeed expressed as a scalar product over states lying in $\mathcal{H}$. Thus, knowing the matrix elements (92) of the projection operator in the unconstrained Hilbert space is equivalent to having solved the constraint!

It is worth mentioning, that a similar scheme can be applied to operators. Suppose there exists a non gauge invariant[13] operator $O = O(r, \varphi)$ on $\mathcal{H}$. Then a fully gauge invariant operator $R = R(r)$ in $\mathcal{H}_{phys}$ can be constructed by defining

$$R := \Pi \, O \, \Pi \, . \tag{93}$$

The calculation of matrix elements of the physical operator R is then reduced to a calculation in the unconstrained Hilbert space, obtaining finally

$$\langle \Psi | O | \Phi \rangle_{phys} \equiv \langle \Psi | \Pi \, O \, \Pi | \Phi \rangle \, . \tag{94}$$

The Hamiltonian Constraint. Let us now proceed with the application of the projector method to the Hamiltonian constraint in quantum gravity, the only constraint which is left for the definition of the dynamical theory. The following discussion is mainly based on plausibility considerations resulting in rough arguments. A more complete treatment would include an exponential growing expenditure of energy, which is beyond the scope of this lecture.

Here, for simplicity, we deal only with the Euclidean part of the Hamiltonian constraint. In the classical theory, this is

$$H_{cl} \simeq F_{ab} E^a E^b + \text{Lorentzian part} \, . \tag{95}$$

$E^{a,b}$ are the triads, and F_{ab} is the curvature of the connection, which, we recall, is an antisymmetric tensor.

When passing over to the quantum theory $\mathcal{H}_{cl}$ turns into an operator and has to suitably regularized. A typical regularization process consists of the following steps. First, we introduce a regularization parameter ϵ, and we replace the expression (95) with a regularized, ϵ dependent one, written in terms of quantities that we know how to promote to quantum operators, and which tends to H_{cl} as ϵ tends to zero. In the second step, replace the classical quantities with their quantum analogues, leading to the Hamiltonian operator $\hat{H}_\epsilon$. In the last step, the parameter is forced to go to zero, $\epsilon \to 0$, yielding a well-defined quantum Hamiltonian operator $\hat{H}_\epsilon \to \hat{H}$.

We will not carry out explicitely this construction here. We only mention that there exist *several* different versions of it. The first completely consistent construction, yielding a well-defined and finite operator, was obtained by Thiemann in [12], building on the results and ideas in [13] and [16].

The key point which is common to all different regularization procedures is the vanishing of the action of the Hamiltonian operator on the holonomy $U[A, \gamma]$. That is

$$\hat{H}(x) \, U[A, \gamma] = 0 \, , \tag{96}$$

if x is on an interior point of the curve γ. The reason for this can roughly be understood as follows. If we replace the triad in (95) with its quantum analogue

[13] For later convenience, we refer to the symmetry in this example as a "gauge" symmetry.

and apply the resulting operator to the holonomy (without bothering about regularization), we obtain

$$F_{ab}\,\frac{\delta}{\delta A_a}\frac{\delta}{\delta A_b}\,U[A,\gamma]\;\sim\;F_{ab}\,\dot{\gamma}^a\dot{\gamma}^b=0\,. \tag{97}$$

While F_{ab} is an antisymmetric tensor, the product of $\dot{\gamma}^a$ and $\dot{\gamma}^b$, which are tangent to γ, is symmetric, on the other hand. Thus the result is zero, since we contract an antisymmetric tensor with a symmetric quantity. This derivation is formal only, since an infinite coefficient multiplies the right hand side of (97), but a rigorous regularized calculation yields the same result.

However, if one calculates the action of $\hat{H}$ on spin network states, the result turns out to be not equal to zero,

$$\hat{H}(x)\,\Psi_S\neq 0\,, \tag{98}$$

due to the end points of the links, namely the nodes. The tangent vectors at a node in (97) can refer to different links, and thus be distinct. So there are terms with non-zero contributions. From this considerations, it follows immediately that the Hamiltonian constraint operator acts on the *nodes* only.

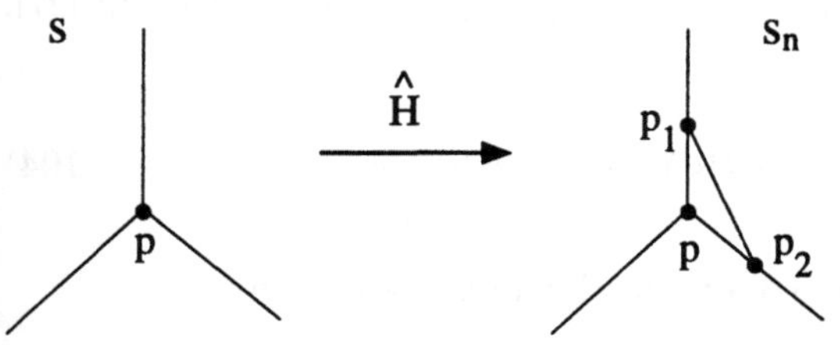

Fig. 14. The action of the Hamiltonian constraint on a trivalent node.

The action of the operator on a single node is illustrated in Fig. 14. It acts by creating an extra link which joins two points p_1 and p_2 lying on distinct links adjacent to the node p. The color of the link between p and p_1, as well as between p and p_2 is altered and the state is multiplied by a coefficient A. The explicit expressions are computed in [26]. We obtain for the action of $\hat{H}$ on a spin networks state $|s\rangle$,

$$\hat{H}[N]\,|s\rangle=\sum_{\text{nodes } n \text{ of } s} A_n\,N(x_n)\,|s_n\rangle\,, \tag{99}$$

where x_n refers to the point in which the node n is located, and s_n is the spin network in which the node n is altered as in Fig. 14. Furthermore, the smeared Hamiltonian constraint $\hat{H}[N]$, with smearing function $N(x)$, is given by

$$\hat{H}[N]=\int d^3x\,N(x)\,\hat{H}(x)\,. \tag{100}$$

Spin Foam. Now we want to define the physical Hilbert space $\mathcal{H}_{phys}$ using the projector method explained above, starting from the diffeomorphism invariant Hilbert space $\mathcal{H}_{diff}$ by considering s-knot states. Similar to (82) or (89), respectively, we construct the projection operator

$$P = \int [dN]\, e^{i\hat{H}[N]} = \int [dN]\, e^{i\int N\hat{H}} . \tag{101}$$

In the abstract spin network basis, the matrix elements of P are

$$\langle s|P|s'\rangle = \langle s| \int [dN]\, e^{i\int N\hat{H}} |s'\rangle . \tag{102}$$

It can be shown, that a diffeomorphism invariant notion of integration exists for this functional integral [25]. According to (83) or (92), respectively, the matrix elements of P are used to define the quadratic form

$$\langle s|s'\rangle_{phys} = \langle s|P|s'\rangle . \tag{103}$$

The physical Hilbert space $\mathcal{H}_{phys}$ is then defined over $\mathcal{H}_{diff}$, from which we started, by this quadratic form.

In order to calculate the matrix elements (102) of the projector, we expand the exponent. We neglect here many technicalities, which can be found in [27]. The expansion has the structure

$$\langle s|P|s'\rangle \sim \langle s|s'\rangle + \int [dN] \left(\langle s|\hat{H}|s'\rangle + \langle s|\hat{H}\hat{H}|s'\rangle + \ldots \right) . \tag{104}$$

Using now the action (99) of $\hat{H}$ on spin network states, we obtain

$$\langle s|s'\rangle_{phys} = \langle s|P|s'\rangle \sim \langle s|s'\rangle + \sum_{\text{nodes } n \text{ of } s'} A_n \langle s|s'_n\rangle + \ldots , \tag{105}$$

where we "integrated out" integrals of the type

$$\int [dN]\, (N \cdots N) . \tag{106}$$

Equation (105) admits an extremely compelling graphical interpretation as a sum over histories of evolutions of s-knot states. Thus one can regard the projector as a propagator in accordance with Feynman.

To see this, consider the 4-manifold $\mathcal{M} = \Sigma \times [0,1]$, where we denote the hypersurfaces of the boundary of $\mathcal{M}$, corresponding to 0 and 1 in the interval, as Σ_i and Σ_f, respectively. We define the "initial state" on Σ_i as $s_i := s'$, and the "final state" on Σ_f as $s_f := s$. Then the term $\langle s_f|s_i\rangle$, which is of order zero in the expansion (105), is non-vanishing only if $s_f = s_i$, i.e. the corresponding graphs have to be continuously deformable into each other, such that the colors of the links and nodes match. Graphically, this is expressed by sweeping out a surface $\sigma = \sigma_i \times [0,1]$, as shown in Fig. 15. The surface is formed by two-dimensional

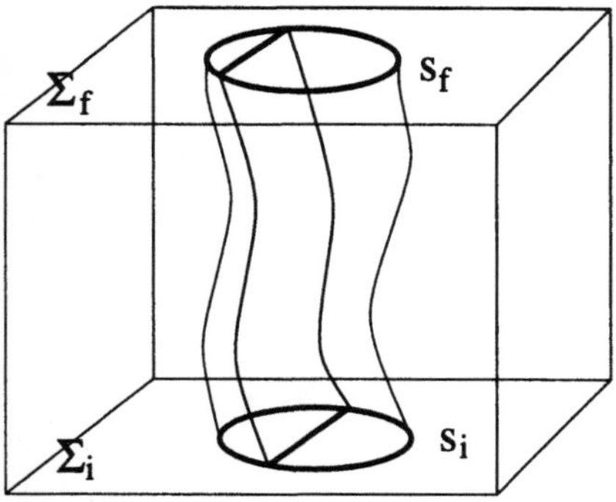

Fig. 15. The diagram corresponding to a term of order zero.

submanifolds of $\mathcal{M}$ – so-called faces – which join in edges. The faces are swept out by spin network links and the edges by the nodes. Thus, every face of σ is colored just as the underlying link, and to every edge the intertwiner of the underlying node is associated.

Next, we consider a first order term $\langle s|s'_n\rangle$ in (105). It appears because of a single action of the Hamiltonian constraint, i.e. it corresponds to adding (or removing) one link, or, equivalentely, two nodes into a spin network, cf. Fig. 14. The situation is similar to the one described for the term of order zero, but now at some point p of σ the surface branches as shown in Fig. 16. That's the reason why the graph of s_i is not equal any more to the one which is associated to s_f. The surfaces are again colored corresponding to the underlying links.

The picture one should have in mind is the following. $\mathcal{M}$ can be imagined as a spacetime, and s_i is a spin network that evolves continuously in a coordinate denoted as "time" up to a point p where the spin network branches because of the Hamiltonian constraint. This means, that the single node p_i degenerates in the sense of being transformed into two nodes which are connected by a link. The accompanying branching of the surface in p is called the *elementary vertex* of the theory, which is at the same time the simplest geometric vertex, see Fig. 17.

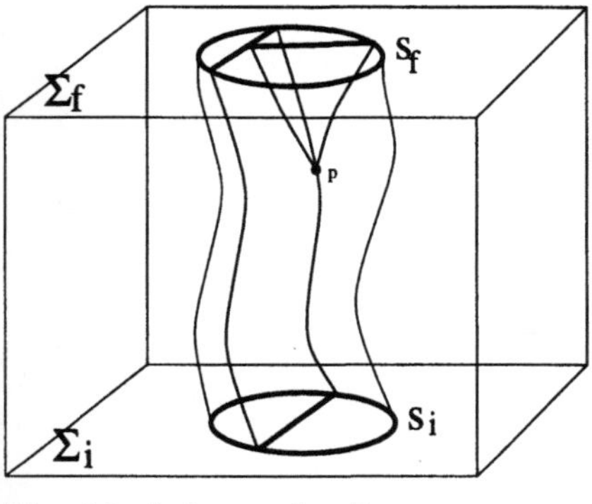

Fig. 16. A first order diagram.

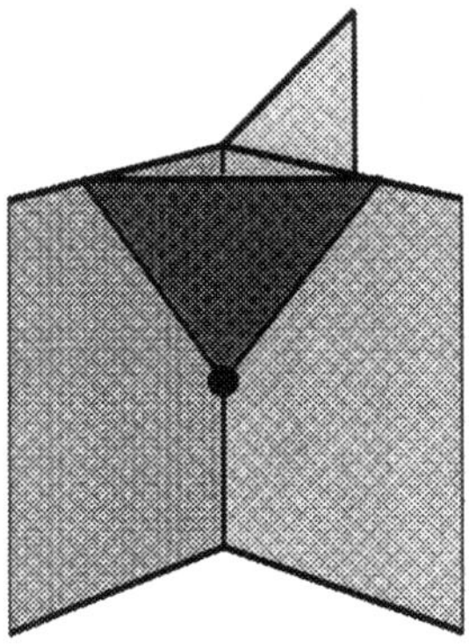

Fig. 17. The elementary vertex.

Finally, we consider a term of second order and give the coloring of the surfaces explicitely, see Fig. 18. Let us look at the transition from an (abstract) spin network s_i with two trivalent nodes connected by three links with colors $(3, 5, 7)$ to the s-knot with the same graph but coloring $(3, 6, 8)$. The intermediate step is an s-knot with four nodes, such that an elementary creation, as well as an elementary annihilation vertex occur.

Despite of the above simplified considerations, it is plausible that the expansion (105) of $\langle s|P|s'\rangle$ can be written as a sum over topologically inequivalent branched colored surfaces σ, the so-called ***spin foams*** [27,28], which are bounded by s_i and s_f. Each surface σ represents the history of the initial s-knot state, and is weighted by the product of coefficients A_ν, which are associated to the vertices of σ. Recall that these coefficients appeared in (99) from the action of the Hamiltonian constraint on s-knot states. They depend only on the coloring

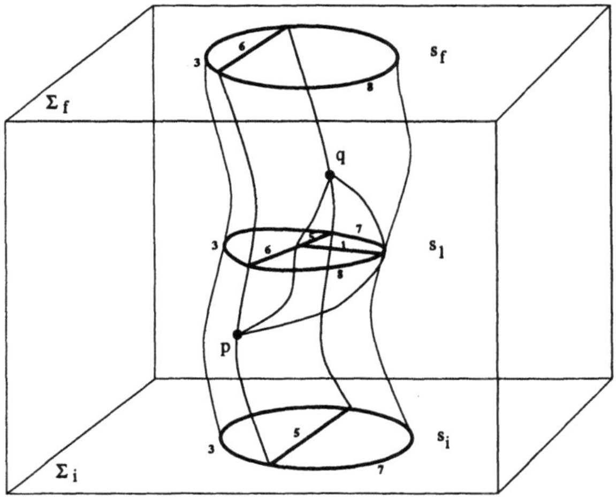

Fig. 18. A term of second order.

of the faces and edges adjacent to the relevant vertices. In the end, we obtain

$$\langle s|s'\rangle_{phys} = \langle s|P|s'\rangle = \sum_{\substack{\text{spin}\\ \text{foams}\,\sigma}} \prod_{\substack{\text{vertices}\\ \nu\in\sigma}} A_\nu \tag{107}$$

for the transition amplitude between the two (abstract) spin network states s_i and s_f. We call it a transition amplitude because of the obvious formal analogy to the expressions in standard quantum field theory, which gives rise to the interpretation of Fig. 15, 16 and 18 roughly as "Feynman diagrams" of quantum gravity.

This interpretation is reinforced by a number of independent results. For instance, certain discretized covariant approaches to quantum gravity lead precisely to a "sum over discretized 4-geometries", very similar to (107). See for instance ([29]). Inspired by the construction above, J. Baez has defined a general notion of *spin foam model* and studied the structure of these models in general. See [28] and references therein.

Physical Observables. To round off this section, we briefly comment on an application of the projector method to the calculation of physical observables. As before, we will not consider problems that arise with the normalization process or ill-defined expressions that might occur, but rather concentrate on the conceptual framework. A more detailed account can be found in [25].

In the spirit of this section, we try to construct a physical observable, i.e. a self-adjoint operator which is invariant under the full symmetry group of the theory, by starting from a 3-diffeomorphism invariant operator O which acts on $\mathcal{H}_{diff}$. We have seen in (93) that the fully gauge invariant observable is then given by

$$R = POP\,, \tag{108}$$

where the projector P onto the physical Hilbert space is defined in (101). Indeed, R is invariant under four-dimensional diffeomorphisms. Thus, the expectation value in a physical state is given by

$$\langle s|O|s\rangle_{phys} := \langle s|P\,O\,P|s\rangle\,. \tag{109}$$

Performing now a similar treatment of the matrix elements (109) as above, we obtain the expression for the expectation values in the spin foam version as

$$\langle s|O|s\rangle_{phys} \sim \sum_{\substack{\text{spin}\\ \text{foams}\,\sigma}} \left(\sum_{\tilde{s}} O(\tilde{s})\right) \prod_{\substack{\text{vertices}\\ \nu\in\sigma}} A_\nu\,. \tag{110}$$

For simplicity, we have chosen O to be diagonal. Furthermore, $\tilde{s}$ are all possible spin networks that cut a spin foam σ (i.e. a branched colored two-surface) in two parts, a future and a past one. These $\tilde{s}$ may be considered as spacelike slices, which cut a given spin foam.

On closer examination of (110) one recognizes that the first summation has to be performed over all possible spin foams σ, and on top of that, for each of these spin foams, all of its spacelike slices have to be summed up. Without going into details, we briefly mention that this has the appealing geometrical interpretation of an "integration over space–time", or a bit more precisely, of an "integration" over the location of the ADM surfaces in (the quantum version of the classical) four-dimensional space–time. Thus, expectation values of physical observables are given as averages over the spin foam in an intuitively similar manner as one is used to from standard quantum field theory. Moreover, this method provides a framework for the non-perturbative, space–time covariant formulation of a diffeomorphism invariant quantum field theory.

However, so far we didn't mention any problems that arise. Recall first, that we considered only the Lorentzian part of the Hamiltonian constraint. Furthermore, it is unclear what shape physical observables O, which are at least required to yield finite results, should take. Intuitively, we might expect that observables of the form

$$O = \tilde{O} \times \delta(\text{something}) \,, \tag{111}$$

might be finite, and might correspond to the realistic relational observables discussed above. But so far the topic of this final subsection is still very little explored territory and our considerations may at best give some rough ideas of what remains to be done.

6 Open Problems and Future Perspectives

This series of lectures was devoted to loop quantum gravity, a non-perturbative canonical formulation for a quantum theory of gravitation. We introduced the basic principles of the theory in the kinematical regime, including spin network states which provide an orthonormal basis in the gauge invariant Hilbert space. As an application, one of the most exiting results obtained in the last few years, the discreteness of geometry, was examined by considering the quantization of the area. Furthermore, by taking the basic principles of general relativity seriously, we have shown by discussing the topics of diffeomorphism invariance and observability in general relativity, that loop quantum gravity is well-adapted for a quantum theory of gravitation.

Finally, in order to examine also the non-perturbative *dynamics* of quantum gravity a little, an ansatz for the construction of the physical Hilbert space by means of a projection method was explained. We tried to clarify its interpretation in terms of a spin foam model, in which the projection operator itself plays the role of a propagator for the space–time evolution of (abstract) spin networks. Its Feynman diagram like graphic representation was presented as well. We also gave the prospects for a possible calculation of expectation values of operators representing physical observables, by using the spin foam formalism.

There are several open questions which remain to be explored. We mentioned that, because of different regularization schemes, there exist several versions of the Hamiltonian constraint. Thus, one of the most intriguing questions would certainly be to find the "right" consistent Hamiltonian constraint, i.e. the one which has the correct classical limit. Closely related is the question of how such a classical limit should be studied at all. What are the coherent states? What is the ground state of the theory? Does a notion of "ground state" make sense at all, in a general covariant theory?

The problem of constructing four-dimensional diffeomorphism invariant observables is crucial. We do know many four-dimensional diffeomorphism invariant observables in general relativity: in fact, we use them in the classical applications of general relativity, which are nowdays extremely numerous. But to express such observables in the quantum theory is still technically hard. In particular, in order to compare loop quantum gravity with particle physics approaches, and to make contact with traditional quantum field theory, it would be extremely useful to be able to compute scattering amplitudes in an asymptotically flat context. Some kind of perturbation expansion should be used for such a project. But in this context the notion of "expansion", and "perturbative" are delicate (expand around what?). For these problems, the spin foam formalism may turn out to be essential, since it provides a space–time formulation of a diffeomorphism invariant theory.

We close these letures by expressing the wish that some of the students that so enthusistically attended them will be the ones able to solve these problems, to give us a fully convincing quantum theory of spacetime, and thus push forward this extraordinary beautiful adventure, which is exploring Nature and its marvellous and disconcerting secrets.

References

1. G. Amelino–Camelia, *Gravity wave interferometers as probes of a low-energy effective quantum gravity*. `e-print: gr-qc/9903080`.
   G. Amelino–Camelia, Contribution to: "XXXV Karpacz Winter School on Theoretical Physics: From Cosmology to Quantum Gravity," Polanica, Poland, February 2–12, 1999, in this volume.
2. J. Ellis, N.E. Mavromatos and D.V. Nanopoulos, *Search for Quantum Gravity*. `e-print: hep-th/9905048`.
3. C. Rovelli, *Strings, loops and others: A Critical survey of the present approaches to quantum gravity*. To appear in: 15th International Conference on General Relativity and Gravitation (GR15) in Pune, India, Dec. 16-21, 1997, `e-print: gr-qc/9803024`.
4. L. Smolin, Contribution to: "XXXV Karpacz Winter School on Theoretical Physics: From Cosmology to Quantum Gravity," Polanica, Poland, February 2–12, 1999, in this volume.
5. L. Smolin, *The future of spin networks*. `e-print: gr-qc/9702030`.
6. C. Rovelli, *Loop Quantum Gravity*. Living Reviews in Relativity, 1998-1, `http://www.livingreviews.org/` and `e-print: gr-qc/9710008`.
7. C. Rovelli and L. Smolin, *Discreteness of Area and Volume in Quantum Gravity*, Nucl. Phys. **B442** (1995) 593, `e-print: gr-qc/9411005`; Erratum: Nucl. Phys. **B456** (1995) 734.

8. P.A.M. Dirac, *Lectures on Quantum Mechanics*. Yeshiva University, New York, 1964.
9. C. Rovelli, *Ashtekar formulation of general relativity and loop space nonperturbative quantum gravity: A Report*. Class. Quant. Grav. **8** (1991) 1613.
10. T. Jacobson, *New variables for canonical supergravity*. Class. Quant. Grav. **5** (1988) 923.
11. Y. Ling and L. Smolin, *Supersymmetric Spin Networks and Quantum Supergravity*. e-print: hep-th/9904016.
12. T. Thiemann, *Quantum Spin Dynamics (QSD)*. Class. Quant. Grav. **15** (1998) 839, e-print: gr-qc/9606089.
13. C. Rovelli and L. Smolin, *Loop representation of quantum general relativity*. Nucl. Phys. **B331** (1990) 80.
14. C. Rovelli and L. Smolin, *Spin Networks and Quantum Gravity*. Phys. Rev. **D52** (1995) 5743, e-print: gr-qc/9505006.
15. A. Ashtekar, *Mathematical Problems of Non-perturbative Quantum General Relativity*. In: J. Zinn-Justin and B. Julia (eds.), "Les Houches Summer School on Gravitation and Quantizations," Les Houches, France, Jul. 5–Aug. 1, 1992. North-Holland, 1995, e-print: gr-qc/9302024.
16. C. Rovelli and L. Smolin, *The physical hamiltonian in nonperturbative quantum gravity*. Phys. Rev. Lett. **72** (1994) 446, e-print: gr-qc/9308002.
17. R. De Pietri and C. Rovelli, *Geometry Eigenvalues and Scalar Product from Recoupling Theory in Loop Quantum Gravity*. Phys. Rev. **D54** (1996) 2664, e-print: gr-qc/9602023.
18. C. Rovelli and P. Upadhya, *Loop quantum gravity and quanta of space: a primer*. e-print: gr-qc/9806079.
19. J. Lewandowski, E.T. Newman and C. Rovelli, *Variations of the parallel propagator and holonomy operator and the Gauss law constraint*. J. Math. Phys. **34** (1993) 4646.
20. S. Fritelli, L. Lehner and C. Rovelli, *The complete spectrum of the area from recoupling theory in loop quantum gravity*. Class. Quant. Grav. **13** (1996) 2921, e-print: gr-qc/9608043.
21. A. Ashtekar and J. Lewandowski, *Quantum Theory of Geometry I: Area Operators*. Class. Quant. Grav. **14** (1997) A55, e-print: gr-qc/9602046.
22. C. Rovelli, *What is observable in classical and quantum gravity?* Class. Quant. Grav. **8** (1991) 297
23. C. Rovelli, *Halfway Through the Woods: Contemporary Research on Space and Time*. In: J. Earman and J.D. Norton (eds.), "The Cosmos of Science," University of Pittsburgh Press and Universitäts-Verlag Konstanz, Konstanz, 1997.
24. C. Rovelli, *Quantum spacetime: what do we know?* e-print: gr-qc/9903045.
25. C. Rovelli, *The projector on physical states in loop quantum gravity*. Phys. Rev. **D59** (1999) 104015, e-print: gr-qc/9806121.
26. R. Borissov, R. DePietri and C. Rovelli, *Matrix Elements of Thiemann's Hamiltonian Constraint in Loop Quantum Gravity*. Class. Quant. Grav. **14** (1997) 2793, e-print:gr-qc/9703090.
27. M. Reisenberger and C. Rovelli, *"Sum over Surfaces" form of Loop Quantum Gravity*. Phys. Rev. **D56** (1997) 3490, e-print: gr-qc/9612035.
28. J.C. Baez, *Spin Foam Models*. Class.Quant.Grav. **15** (1998) 1827, e-print: gr-qc/9709052.
29. M.P. Reisenberger, *A lattice worldsheet sum for 4-d Euclidean general relativity*. e-print: gr-qc/9711052.

Black Holes in String Theory

Kostas Skenderis

Spinoza Institute, University of Utrecht,
Leuvenlaan 4, 3584 CE Utrecht, The Netherlands

K.Skenderis@phys.uu.nl

Abstract. This is a set of introductory lecture notes on black holes in string theory. After reviewing some aspects of string theory such as dualities, brane solutions, supersymmetric and non-extremal intersection rules, we analyze in detail extremal and non-extremal $5d$ black holes. We first present the D-brane counting for extremal black holes. Then we show that $4d$ and $5d$ non-extremal black holes can be mapped to the BTZ black hole (times a compact manifold) by means of dualities. The validity of these dualities is analyzed in detail. We present an analysis of the same system in the spirit of the adS/CFT correspondence. In the "near-horizon" limit (which is actually a near inner-horizon limit for non-extremal black holes) the black hole reduces again to the BTZ black hole. A state counting is presented in terms of the BTZ black hole.

1 Introduction

The physics of 20th century is founded on two pillars: quantum theory and general theory of relativity. Quantum theory has been extremely successful in describing the physics at microscopic scales while general relativity has been equally successful with physics at cosmological scales. However, attempts to construct a quantum theory of gravity stumble upon the problem of the non-renormalizability of the theory. Is it really necessary to have a quantum theory of gravity? Why not having gravity classical and matter quantized? Is it just an aesthetic question or is there an internal inconsistency if some of the physical laws are classical and some quantum? If some of the interactions are classical then one could use only these interactions in order to arbitrarily obtain the position and the velocity of particles, thus violating Heiseberg's uncertainty principle. Therefore, at the fundamental level, if some of the physical laws are quantum, all of them have to be quantum.

It is amusing to see what happens if we insist on both classical general relativity and the uncertainty principle. Suppose we want to measure a spacetime coordinate with accuracy δx, then by the uncertainty principle there will be energy of order $1/\delta x$ localized in this region. But if δx is very small then the energy will be so large that a black hole will be formed, and the spacetime point will be hidden behind a horizon! One can estimate[1] that the scale that leads to a black hole formation (through the uncertainty principle) is of order of the Planck length l_p. Therefore, classical general relativity and quantum mechanics become incompatible at scales of order l_p.

One of the most fascinating objects that general relativity predicts is black holes. Classically, black holes are completely black. Objects inside their event horizon are eternally trapped. Even light rays are confined by the gravitational force. In addition, there is a singularity hidden behind the horizon. In the early seventies, a number of laws that govern the physics of black holes were established[2–4]. In particular, it was found that there is a very close analogy between these laws and the four laws of thermodynamics[3]. The black hole laws become that of thermodynamics if one replaces the surface gravity κ of the black hole by the temperature T of a body in thermal equilibrium, the area of the black hole A by the entropy S[4], the mass of the black hole M by the energy of the system E etc. It is natural to wonder whether this formal similarity is more than just an analogy. At the classical level one immediately runs into a problem if one tries to take this analogy seriously: classically black holes only absorb so their temperature is strictly zero. In a seminal paper[5], however, Hawking showed that quantum mechanically black holes emit particles with thermal spectrum. The temperature was found to be $T = \kappa/2\pi$! Then from the first law follows the "Bekenstein-Hawking entropy formula",

$$S = \frac{A}{4G_N} \tag{1}$$

where G_N is Newton's constant. Having established that black hole laws are thermodynamic in nature one would like to understand what is the underlying microscopic theory. What are the microscopic degrees of freedom that make up the black hole?

Since black holes radiate, they lose mass and they may eventually evaporate. Observing such a phenomenon is rather unlikely since one can estimate the lifetime of a black hole of stellar mass to have lifetime[1] longer than the age of the universe. The fact, however, that black holes Hawking radiate and may eventually evaporate leads to an important paradox. The matter that falls into black holes has structure. The outgoing radiation, however, is structure-less since it is thermal. What happens to the information stored in the black hole if the black hole completely evaporates? If it gets lost then the evolution is not unitary. Hawking argued that these considerations imply that quantum mechanics has to be modified. There is great controversy over the question of the final state of black holes, and there is no completely satisfactory scenario. We will not enter into this question in these lectures. Let us note, however, that the resolution of this problem may be related to the question of understanding the microscopic description of black holes. Radiation from stars also has a thermal spectrum. However, we do not claim that information is lost in stars. The thermal spectrum is due to averaging over microscopic states.

We have seen that semi-classical considerations yield a number of important issues. Any consistent quantum theory of gravity should provide answers to the questions raised in the previous paragraphs. The leading candidate for a quantum theory of gravity is string theory. Therefore, string theory ought to resolve these issues. Issues involving black holes are non-perturbative in nature.

[1] For black holes of mass M the Hawking temperature is of order $T \sim 10^{-6}(M_\odot/M)\ K$ and their lifetime of order $10^{71}(M_\odot/M)^{-3}\ s$.

Up until recently, however, we only had a perturbative formulation of string theory. The situation changed dramatically over the last few years. Dualities have led to a unified picture of all string theories [6, 7]. Moreover, new non-perturbative objects, the D-branes, were discovered[8]. These new ingredients made possible to tackle some of the problems mentioned above.

In these lectures we review recent progress in understanding black holes using string theory. Previous reviews for black holes in string theory include [9–11]. We start by briefly reviewing perturbative strings, D-branes and dualities in section 2. In particular, we review in some detail T-duality in backgrounds with isometries. In section 3 we present the brane solutions of type II and eleven dimensional supergravity, their connections through dualities, and a set of intersection rules that yields new solutions describing configurations of intersecting branes. We use these results in section 4 in order to study extremal and non-extremal black holes. In section 4.1 we analyze extremal $5d$ black holes. We show that one can derive the Bekenstein-Hawking entropy formula by counting D-brane states. In section 4.2 we show that $4d$ and $5d$ non-extremal black holes can be mapped to the BTZ black hole[27, 13] (times a compact manifold) by means of dualities. We show that a general U-duality transformation preserves the thermodynamic characteristics of black holes. Then we critically examine the so-called shift transformation that removes the constant part from harmonic functions. We show that this transformation also preserves the thermodynamic characteristics of the original black hole. In general, however, it is not a symmetry of the theory. Section 4.3 contains a short introduction to adS/CFT duality, and its application to black holes. The low-energy decoupling limit employed in the adS/CFT correspondence (which is a near inner-horizon limit for non-extremal black holes) also yields a connection with the BTZ black hole. We use the connection to the BTZ black hole to infer a state counting for the higher dimensional black holes.

A "road map" for our discussion is provided by figure 1. Our aim is to understand $4d$ or $5d$ black holes using string theory (top right part of figure 1). As a first step one constructs solutions of the low energy effective action of the appropriate string theory that upon compactification in 4 or 5 dimensions yield the desired black hole solution (middle left part of figure 1). In these lectures we will restrict ourselves to toroidal compactifications of type II string theory. The appropriate supergravity solutions can be constructed by intersecting elementary black p-brane solutions. This yields a connection with D-branes (bottom right part of figure 1). A D-brane, which in string weak coupling is a hyperplane where strings can end, has a description as a solitonic object of the low energy supergravity. The black hole solution is obtained by fully wrapping the intersecting branes over the compactification manifold (tori in our case). At weak coupling one can use D-brane techniques in order to study the configuration of intersecting branes. In particular, for supersymmetric black holes certain quantities can be calculated in weak coupling and the result can be reliably extrapolated to strong coupling. Such quantity is the number of states that make up the black hole configuration. For non-extremal black holes such an extrapolation is not justified and a new approach is needed. Such an approach is illustrated in the left part of figure 1. Either by the shift transformation or in the "near-horizon" limit, the toroidal compactifications are turned into spherical ones (middle left part of figure 1). Upon dimensional reduction over the compact part one obtains the three dimensional BTZ black hole (top left part of figure 1)! Therefore, at least part of the physics of the 4d and 5d black holes is captured by the BTZ black hole. In these lectures we start from the bottom part of figure 1 and work our way to the top.

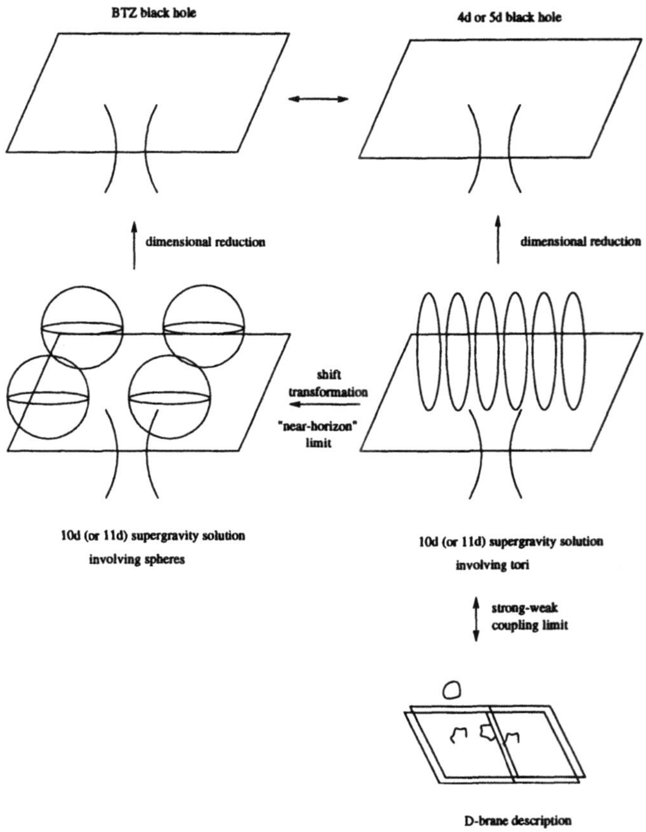

Fig. 1. "Road map": In this picture we indicate the various routes that lead to connections between 4d and 5d black holes and higher dimensional supergravity solutions, D-branes as well as the 3d BTZ black hole.

2 String theory and dualities

In this section we present some aspects of string theory. The main purpose is to set our conventions and to review certain material that will be of use in later sections.

2.1 Bosonic string and D-branes

The worldsheet action for the bosonic string is given by

$$S = \frac{1}{4\pi\alpha'}\int d\tau \int_0^\pi d\sigma \sqrt{h} h^{ab}\partial_a X^\mu \partial_b X_\mu . \tag{2}$$

where h is the worldsheet metric. The tension of the string is given by $T = 1/(2\pi\alpha')$ (α' is the square of the string length l_s). Varying the action we obtain

$$\delta S = -\frac{1}{2\pi\alpha'}\int d\tau d\sigma \sqrt{h}\delta X^\mu \Box X^\mu + \frac{1}{2\pi\alpha'}\int d\tau [\sqrt{h}\partial_\sigma X_\mu \delta X^\mu]_{\sigma=0}^{\sigma=\pi} \tag{3}$$

In order to have a well-defined variational problem the last term should vanish. This implies three different types of boundary conditions

$X^\mu(\tau,\sigma) = X^\mu(\tau,\sigma+\pi)$	closed string
$\partial_\sigma X^\mu(\sigma=0) = \partial_\sigma X^\mu(\sigma=\pi) = 0$	open string with Neumann BC
$X^\mu(\sigma=0) = const,\ X^\mu(\sigma=\pi) = const,$	open string with Dirichlet BC

The Neumann boundary conditions for the open string imply that there is no momentum flow at the end of the string. With Dirichlet boundary conditions, however, there is momentum flowing from the string to the hypersurface where the string ends ($dP^\mu = T\partial_\sigma X^\mu d\tau \neq 0$, at $\sigma = 0$ and $\sigma = \pi$). Therefore, this hypersurface, the D-brane, is a dynamical object.

One may (first) quantize the string using standard methods. The closed string consists of left and right movers. We denote the left and right level by N and $\tilde{N}$, respectively. For open strings we have only one kind of oscillators. The perturbative spectrum for the three kind of boundary conditions listed above is given by

$$\begin{aligned} M^2_{closed} &= \frac{2}{\alpha'}(N+\tilde{N}-2) \\ M^2_{open,N} &= \frac{1}{\alpha'}(N-1) \\ M^2_{open,D} &= \left(\frac{l}{2\pi\alpha'}\right)^2 + \frac{1}{\alpha'}(N-1) \end{aligned} \tag{4}$$

The term $l/2\pi\alpha' = lT$ is the energy of a string of length l stretched between two D-branes.

From (4) follows that the massless spectrum of closed strings consist of a graviton $G_{\mu\nu}$, an antisymmetric tensor $B_{\mu\nu}$ and a dilaton ϕ. The massless spectrum of open strings with Neumann boundary conditions consists of a photon A_μ. Finally, for a string that ends on a Dp-brane, i.e. the open string endpoints are confined to the $p+1$-dimensional worldvolume of the D-brane, we get a vector field A_m, $m = 0, \ldots, p$, that lives on the worldvolume of the D-brane, and $(25-p)$ scalars. The latter encode the fluctuations of the position of the D-brane.

The string coupling constant is not a new parameter but the expectation value of the dilaton field, $\langle e^\phi \rangle = g_s$. String theory perturbation theory is weighted by g_s^χ, where χ is the Euler number of the string worldsheet. A compact surface can be built by adding g handles, c cross-caps and b boundaries to the sphere. Its Euler number is given by $\chi = 2 - 2g - b - c$. Hence, the closed string coupling constant is proportional to the square of the open string coupling constant.

One may calculate the tension of D-branes[8, 14]

$$T_p \sim \frac{1}{g_s l_s^{p+1}}. \tag{5}$$

Since the tension of the D-brane depends on the inverse of the string coupling constant, D-branes are non-perturbative objects. Notice that this behavior is different from the behavior of field theory solitons whose mass goes as $1/g^2$, where g is the field theory coupling constant. The existence of such non-perturbative objects is required by string duality [7].

2.2 Superstrings

There are five consistent string theories; type IIA and IIB, type I, heterotic $SO(32)$ and heterotic $E_8 \times E_8$. All of them are related through dualities. In this review we shall concentrate on type II theories, so we briefly present some aspects of them.

The bosonic massless sector of type II theories consist of the following fields

Type IIA	$g_{\mu\nu}$	$B_{\mu\nu}$	ϕ		$C^{(1)}_\mu$	$C^{(3)}_{\mu\nu\lambda}$
Type IIB	$g_{\mu\nu}$	$B_{\mu\nu}$	ϕ	$C^{(0)}$	$C^{(2)}_{\mu\nu}$	$C^{(4)\,+}_{\kappa\lambda\mu\nu}$,

where $C^{(p)}$ are p-index antisymmetric gauge fields. The $+$ in $C^{(4)+}$ indicates that the field strength is self-dual. The graviton $g_{\mu\nu}$, the antisymmetric tensor $B_{\mu\nu}$ and the dilaton ϕ make up the NSNS sector. These fields couple to perturbative strings. The RR sector (i.e. the antisymmetric tensors $C^{(p+1)}$), however, does not couple to perturbative strings but rather to Dp-branes.

Extended objects naturally couple to antisymmetric tensors. The prototype example is the coupling of the point particle to electromagnetic field, $\int A_\mu dx^\mu$. Similarly, fundamental strings naturally couple to $B_{\mu\nu}$, and Dp-branes to $C^{(p+1)}$

$$\int_\Sigma B_{\mu\nu} dx^\mu \wedge dx^\nu$$

$$\int_{\mathcal{M}_{p+1}} C^{(p+1)}_{\mu_1 \cdots \mu_{p+1}} dx^{\mu_1} \wedge \cdots \wedge dx^{\mu_{p+1}} \tag{6}$$

where Σ and $\mathcal{M}_{p+1}$ is the string worldsheet and Dp-brane worldvolume, respectively. To each "electric" p-brane there is also a dual "magnetic" $(6-p)$-brane. (To see this notice that $*dC^{(p+1)} = d\tilde{C}^{(7-p)}$). In particular, there is a solitonic 5-brane (NS5) that is the magnetic dual of a fundamental string F1. In addition, strings can carry momentum. This corresponds in low energy to gravitational waves (W). The (Hodge) dual to waves are Kaluza-Klein monopoles (KK) (see section 3).

In summary, we have the following objects in type II theory (D(-1) are D-instanton and D9 are spacetime-filling branes)

Type IIA	W	F1	NS5	KK		D0	D2	D4	D6	D8	
Type IIB	W	F1	NS5	KK	D(-1)	D1	D3	D5	D7	D9	

We have deduced the existence of dynamical extended objects by considering perturbative string theory. These states, however, preserve half of maximal supersymmetry and therefore continue to exit at all values of the string coupling constant.

2.3 Dualities

A central element in the recent developments are the duality symmetries of string theory. The duality symmetries are believed to be exact discrete gauge symmetries spontaneously broken by scalar vev's.

The best-understood duality symmetry is T-duality. This symmetry is visible in string perturbation theory but it is non-perturbative on the worldsheet. T-duality relates compactifications on a manifold of (large) volume v to compactifications on a manifold of (small) volume $1/v$. The simplest case is compactification on a circle. Upon such compactification the two type II theories, and heterotic $E_8 \times E_8$ and heterotic $SO(32)$ theories are equivalent,

$$\begin{aligned} &[IIA]_R &&\overset{T}{\longleftrightarrow} [IIB]_{1/R} \\ &[Het\ E_8 \times E_8]_R &&\overset{T}{\longleftrightarrow} [Het\ SO(32)]_{1/R}, \end{aligned}$$

where the subscript indicates that the theory is compactified on a circle of radius R $(1/R)$.

The action of T-duality on the various objects present in II theories is given in Table 1. The T-duality may be performed along one of the worldvolume directions or along a transverse direction (for the KK monopole the transverse direction is taken to be the nut direction (see section 3)). More generally, T-duality asserts that different spacetimes with isometries may be equivalent in string theory. We shall present the argument in some detail in the next section since we will make use of these results.

A (conjectured) non-perturbative symmetry is S-duality. This is non - perturbative because it acts on the dilaton as $g_s \to 1/g_s$. Thus, S-duality relates the strong coupling regime of one theory to the weak coupling regime of another. In particular we have

$$\begin{aligned} &IIB &&\overset{S}{\longleftrightarrow} IIB \\ &Het\ SO(32) &&\overset{S}{\longleftrightarrow} Type\ I \end{aligned} \qquad (7)$$

	Parallel	transverse
Dp	D$(p-1)$	D$(p+1)$
F1	W	F1
W	F1	W
NS5	NS5	KK
KK	KK	NS5

Table 1. *T-duality along parallel and transverse directions*

Actually, IIB string theory is believed to have an exact non-perturbative $SL(2,Z)$ symmetry. In the following we shall only make use of the Z_2 subgroup that sends $\tau = C^{(0)} + ie^{-\phi}$ to $-1/\tau$, interchanges $B_{\mu\nu}$ with $C^{(2)}_{\mu\nu}$, and leaves invariant $C^{(4)+}$ (so, in terms of branes, S-duality interchanges F1 with D1, NS5 with D5, and leaves invariant the D3 brane).

S-duality allows one to get a handle to the strong coupling limit of three of the five string theories. In turns out that the strong coupling limit of IIA and heterotic $E_8 \times E_8$ theories is of a more "exotic" nature. One gets instead an 11 dimensional theory, the M-theory[7,15]. M-theory on a small circle of radius $R_{11} = g_s l_s$ yields IIA theory with string coupling constant g_s[7]. Since perturbative string theory is an expansion around $g_s = 0$, the eleventh dimension is not visible perturbatively. Likewise, M-theory on an interval gives $E_8 \times E_8$ string theory[16]. Actually, all string theories can be obtained in suitable limits from eleven dimensions.

Although we do not have a fundamental understanding of what M-theory is, we know that in low-energies M-theory reduces to 11 dimensional supergravity[17]. Eleven-dimensional supergravity compactified on a torus yields a lower dimensional Poincaré supergravity with a certain duality group. The discretized version of this duality group is conjectured[6] (and widely believed) to be an exact symmetry of M-theory. T and S duality combine to yield this bigger group, the U-duality group.

Buscher's duality Consider the sigma model

$$S = \frac{1}{4\pi\alpha'} \int d^2\sigma \sqrt{h}[(h^{ab} g_{\mu\nu} + i\frac{\epsilon^{ab}}{\sqrt{h}} B_{\mu\nu})\partial_a X^\mu \partial_b X^\nu + \alpha' R^{(2)} \phi], \tag{8}$$

where h and $R^{(2)}$ is the worldsheet metric and curvature, g is the target space metric and B is a potential for the torsion 3-form $H = dB$. This action is invariant under the transformation

$$\delta X^\mu = \epsilon k^\mu \tag{9}$$

when the vector field k^μ is a Killing vector, the Lie derivative of B is a total derivative and the dilaton is invariant,

$$\begin{aligned} \mathcal{L}_k g_{ij} &= k_{i;j} + k_{j;i} = 0, \\ \mathcal{L}_k B &= \iota_k dB + d\iota_k B = d(v + \iota_k B) \\ \mathcal{L}_k \phi &= k^\mu \partial_\mu \phi = 0 \end{aligned} \tag{10}$$

One can now choose adapted coordinates $\{X^\mu\} = \{x, x^i\}$ such that the isometry acts by translation of x, and all fields g, B and ϕ are independent of x. In adapted coordinates, the killing vector is equal to $k^\mu \partial/\partial X^\mu = \partial/\partial x$.

To obtain the dual theory we first gauge the symmetry and add a Lagrange multiplier χ that imposes that the gauge connection is flat [18]. The result (in the conformal gauge and omitting the dilaton term) is (see [19], [20] for details)

$$S_1 = \frac{1}{2\pi\alpha'} \int d^2 z [(g_{\mu\nu} + B_{\mu\nu})\partial X^\mu \bar{\partial} X^\nu + (J_k - \partial\chi)\bar{A} + (\bar{J}_k + \bar{\partial}\chi)A + k^2 A\bar{A}] \tag{11}$$

where $J_k = (k+v)_\mu \partial X^\mu$, $\bar{J}_k = (k-v)_\mu \bar{\partial} X^\mu$ are the components of the Noether current associated with the symmetry. If one integrates out the Lagrange multiplier field χ, on a topologically trivial worldsheet the gauge fields are pure gauge, $A = \partial\theta$, $\bar{A} = \bar{\partial}\theta$, and one recovers the original model (8).

If one integrates out the gauge fields $A, \bar{A}$ one finds the *dual* model. One obtains (8) but with dual background fields $\tilde{g}, \tilde{B}, \tilde{\Phi}$. In adapted coordinates $\{X^\mu\} = \{x, x^i\}$,

$$\begin{aligned}
&\tilde{g}_{xx} = \frac{1}{g_{xx}} \qquad \tilde{g}_{xi} = \frac{B_{xi}}{g_{xx}} \qquad \tilde{g}_{ij} = g_{ij} - \frac{g_{xi}g_{xj} - B_{xi}B_{xj}}{g_{xx}} \\
&\tilde{B}_{xi} = \frac{g_{xi}}{g_{xx}} \qquad \tilde{B}_{ij} = B_{ij} + \frac{g_{xi}B_{xj} - B_{xi}g_{xj}}{g_{xx}} \\
&\tilde{\phi} = \phi - \frac{1}{2}\ln g_{xx}
\end{aligned} \tag{12}$$

The dilaton shift is a quantum mechanical effect [21] (see [22] for recent careful discussion).

Another useful way to write these transformation rules is to re-write the metric as

$$ds^2 = g_{xx}(dx + A_i dx^i)^2 + \bar{g}_{ij} dx^i dx^j \tag{13}$$

where $A_i = g_{xi}/g_{xx}$. Then the duality transformations take the form[23]

$$\begin{aligned}
&\tilde{g}_{xx} = \frac{1}{g_{xx}}, \qquad \tilde{A}_i = B_{xi}, \qquad \tilde{B}_{xi} = A_i, \qquad \tilde{B}_{ij} = B_{ij} - 2A_{[i}B_{j]x} \\
&\tilde{\phi} = \phi - \frac{1}{2}\ln g_{xx} \qquad \bar{g}_{ij} \text{ invariant}
\end{aligned} \tag{14}$$

This form of the transformation rules exhibits most clearly the spacetime interpretation of the duality transformations. The form of the metric in (13) is the standard KK ansatz for reduction over x. Dimensional reduction over x leads to a $(d-1)$-dimensional theory which is invariant under the transformations in (14). These transformations act only on the matter fields and not on the pure gravitational sector.

Let us now discuss under which conditions the dual models are truly equivalent as conformal field theories.

• *Compact vs non-compact isometries*
In our discussion above we assumed that the worldsheet is trivial. Let us relax this condition. Suppose also that we deal with a compact isometry. The constraint on $A, \bar{A}$ that comes from integrating out the Lagrange multiplier χ implies $A, \bar{A}$ are flat, but in principle they still may have nontrivial holonomies around non-contractible loops. These holonomies can be constrained to vanish if χ has appropriate period[19, 20]. In summary, dualizing along a compact isometry one obtains a dual geometry which also has a compact isometry. The periods of the original and dual coordinate are reciprocal to each other. If this condition does not hold, the two models are not fully equivalent but related via an orbifold construction.

Non-compact isometries can be considered as a limiting case. Since in this case x takes any real value, the dual coordinate χ must have period zero. The dual manifold is an orbifold obtained by modding out the translations in χ.

• *Isometries with fixed points*
In our analysis we also assumed that the isometry is spacelike. If the isometry is timelike then it follows from (11) that the integration over the gauge field yields a divergent factor. If the isometry is null then the quadratic in the gauge field term in (11) vanishes. Therefore these cases require special attention. We refer to [24–26] for work concerning dualization (or the closely related issue of dimensional reduction) along timelike or null isometries.

A spacelike isometry may act freely or have fixed points. A typical example of an isometry without fixed points are the translational symmetries on tori. On the other hand, rotational isometries have fixed points. At the fixed point $k^2 = 0$. It follows from (12) (using $k^2 = g_{xx}$) that the dual geometry appears to have a singularity at the fixed point.

Taking the curvature of the spacetime to be small in string units (which is required for consistency for strings propagating in a background that only solves the lowest order beta functions) we see that we may approximate the vicinity of the fixed point by flat space. In adapted coordinates, which are just polar coordinates, the isometry direction being the angular coordinate, we have

$$ds^2 = dr^2 + r^2 d\theta^2. \tag{15}$$

Dualizing along θ we obtain

$$ds^2 = dr^2 + \frac{1}{r^2} d\theta^2, \qquad \phi = -\frac{1}{2} \ln r^2. \tag{16}$$

So indeed the fixed point of the isometry, i.e. $r = 0$, becomes a singular point after the duality transformation. Since the curvature now diverges at $r = 0$ we cannot trust the (first order in α') sigma model analysis. A more careful conformal field theory analysis[27] shows that the duality yields an exact equivalence but the operator mapping includes all orders in α'. We can read this result as follows: All order α' corrections resolve the singularity present in the spacetime described by (16) yielding an exact non-singular conformal field theory.

Studies of T-duality along a rotational isometry can be found in [28–30].

3 Brane solutions

String theory has a mass gap of order $1/l_s$. At low enough energies only the massless fields are relevant. We can decouple the massive modes by sending $\alpha' \to 0$ (so the mass of the massive modes goes to infinity). The interactions of the massless fields are described by an effective action. For IIA and IIB superstring theories the low energy theory is IIA and IIB supergravity, respectively. We have seen that in type II string theories there exist dynamical objects other than strings, namely D-branes, and solitonic branes. For each of these objects there is a corresponding solution of the low energy supergravity. The purpose of this section is to describe these solutions. For reviews see [31–33].

The relevant part of the supergravity action, in the string frame, is [2]

$$S = \frac{1}{128\pi^7 g_s^2 \alpha'^4} \int d^{10}x \sqrt{-g}\left[e^{-2\phi}\left(R + 4(\partial\phi)^2 - \frac{1}{12}|H_3|^2\right) - \frac{1}{2(p+2)!}|F_{p+2}|^2\right] \tag{17}$$

We use the convention to keep the asymptotic value of ϕ in Newton's constant ($G_N^{(10)} = 8\pi^6 g_s^2 \alpha'^4$), so the asymptotic value of e^ϕ below is equal to 1.[3]

The equations of motion of the above action have solutions that have the interpretation of describing the long range field of fundamental strings (F1), Dp-branes and solitonic fivebranes (NS5). These solutions are given by[34]

$$\begin{aligned} ds^2_{st} &= H_i^\alpha [H_i^{-1} ds^2(\mathbb{E}^{(p,1)}) + ds^2(\mathbb{E}^{(9-p)})] \\ e^\phi &= H_i^\beta \\ A^{(p+1)}_{01\cdots p} &= H_i^{-1} - 1, \text{ “electric”}, \text{ or } F_{8-p} = \star dH_i, \text{ “magnetic”} \end{aligned} \tag{18}$$

where $A^{(p+1)}$ is either the RR potential $C^{(p+1)}$, or the NSNS 2-form B, depending on the solution. $\star$ is the Hodge dual of $\mathbb{E}^{(9-p)}$. The subscript $i = \{p, F1, NS5\}$ denotes which solution (Dp-brane, fundamental string or solitonic fivebrane, respectively) we are describing. In order for (18) to be a solution H_i must be a harmonic function on $\mathbb{E}^{(9-p)}$,

$$\nabla^2 H_i = 0 \tag{19}$$

Let r be the distance from the origin of $\mathbb{E}^{(9-p)}$. The choice

$$H_i = 1 + \frac{Q_i}{r^{(7-p)}}, \qquad p < 7 \tag{20}$$

yields the long-range fields of N infinite parallel planar p-branes near the origin. The constant part was chosen equal to one in order for the solution to be asymptotically flat. The values of the parameters α and β for each solution are given in Table 2. In the same table we also give the values of the charges Q_i. The constant d_p is equal to $d_p = (2\sqrt{\pi})^{5-p}\Gamma(\frac{7-p}{2})$.

[2] There are several other bosonic terms in the action. These terms are not relevant for the solutions (18) since in these solutions there is only a single antisymmetric tensor turned on. We have also omitted all fermionic terms.

[3] The field equations are invariant under $e^\phi \to ce^\phi, C^{(p+1)} \to c^{-1}C^{(p+1)}$, where c is a constant, so one can change conventions by an appropriate choice of c.

Dp-branes	$\alpha = 1/2$	$\beta = (3-p)/4$	$Q_p = d_p N g_s l_s^{7-p}$
F1	$\alpha = 0$	$\beta = -1/2$	$Q_{F1} = d_1 N g_s^2 l_s^6$
NS5	$\alpha = 1$	$\beta = 1/2$	$Q_{NS5} = N l_s^2$

Table 2. *p-brane solutions of Type II theories.*

Apart from these solutions, there are also purely gravitational ones. There is a solution describing the long range field produced by momentum modes carried by a string. This is the gravitational wave solution,

$$ds^2 = -K^{-1}dt^2 + K(dx_1 - (K^{-1} - 1)dt)^2 + dx_2^2 + \cdots + dx_9^2 \tag{21}$$

where $K = 1 + Q_K/r^6$ is again a harmonic function and $Q_K = d_1 g_s^2 N\alpha'/R^2$. R is the radius of x_1.

Finally, there is a solution describing a Kaluza-Klein (KK) monopole (the name originates from the fact that upon dimensional reduction over ψ the KK gauge field that one gets is the monopole connection):

$$\begin{aligned} ds^2 &= ds^2(\mathbb{E}^{(6,1)}) + ds^2_{TN} \\ ds^2_{TN} &= H^{-1}(d\psi + Q_M \cos\theta d\varphi)^2 + H dx^i dx^i, \qquad i = 1,2,3 \\ H &= 1 + \frac{Q_M}{r}, \qquad r^2 = x_1^2 + x_2^2 + x_3^2 \end{aligned} \tag{22}$$

where TN stands for Taub-NUT, θ and ψ are the angular coordinates of x_1, x_2, x_3, $Q_M = NR/2$, N is the number of coincident monopoles and R is the radius of ψ.

S-duality leaves invariant the action in the Einstein frame. To reach the Einstein frame we need to do the Weyl rescaling $g_E = e^{-\phi/2}g_{st}$. Using the fact that under S-duality $\phi \to -\phi$ (and $g_s \to 1/g_s$) we get $g_{\mu\nu} \to e^{-\phi}g_{\mu\nu}$. The compactification radii are measured using the string metric. So, they change under S-duality. One can take care of this by changing the string scale, $\alpha' \to \alpha' g_s$. We therefore get the following S-duality transformation rules

$$\begin{aligned} &\phi \to -\phi \;\; (g_s \to 1/g_s), \qquad \alpha' \to \alpha' g_s \\ &g_{\mu\nu} \to e^{-\phi} g_{\mu\nu}, \qquad B_{\mu\nu} \leftrightarrow C^{(2)}_{\mu\nu} \end{aligned} \tag{23}$$

With these conventions Newton's constant, $G_N^{(10)} = 8\pi^6 g_s^2 \alpha'^4$, is invariant under S-duality.

T-duality acts as in (12) in the NSNS sector. In particular, dualization along a coordinate of radius R yields

$$R \to \frac{\alpha'}{R}, \qquad g_s \to g_s \frac{l_s}{R} \tag{24}$$

For the RR fields we get[35]

$$\begin{aligned} C_{\mu_1\cdots\mu_{p+1}} &\to C_{\mu_1\cdots\mu_{p+1}x}, \qquad x \notin \{x_{\mu_1}, \cdots, x_{\mu_{p+1}}\} \\ C_{x\mu_1\cdots\mu_{p+1}} &\to C_{\mu_1\cdots\mu_{p+1}} \end{aligned} \tag{25}$$

depending on whether we dualize along a coordinate transverse or parallel to the brane.

It is easy to see that the values of the charges Q_i are consistent with dualities. For instance, under S-duality: $Q_{NS5} = N\alpha' \leftrightarrow Ng_s\alpha' = Q_5$. Actually, dualities determine both the value of Newton's constant and the charges (including the numerical coefficients) [10]: The mass M of an object can be calculated from the deviation of the Einstein metric from the flat metric at infinity. In particular[36],

$$g_{E,00} = \frac{16\pi G_N^{(d)} M}{(d-2)\omega_{d-2}} \frac{1}{r^{d-3}} \tag{26}$$

where $\omega_d = 2\frac{\pi^{(d+1)/2}}{\Gamma(\frac{d+1}{2})}$ is the volume of the unit sphere S^d. Completely wrapping a given brane on torus and dimensionally reducing we get a spacetime metric in $d = 10 - p$ dimensions,

$$ds^2_{E,d} = -H^{-\frac{d-3}{d-2}} dt^2 + H^{\frac{1}{d-2}} ds^2(\mathbb{E}^{(d-1)}) \tag{27}$$

This result is obtained by using the dimensional reduction rules[37]

$$ds^2_{E,d} = e^{-\frac{4}{d-2}\phi_d} ds^2_{st}, \qquad e^{-2\phi_d} = e^{-2\phi}\sqrt{det g_{int}} \tag{28}$$

where g_{int} is the component of the metric in the directions we dimensionally reduce. If $H = 1 + c^{(d)}/r^{d-3}$ then,

$$c^{(d)} = \frac{16\pi G_N^{(d)} M}{(d-3)\omega_{d-2}}. \tag{29}$$

The mass M appearing in this formula is the same as the mass measured in the string frame since we used the convention to leave a factor of g_s^2 in Newton's constant. These masses can be easily obtained by U-duality. Knowledge of one of the coefficients in (29) is sufficient to determine G_N and therefore all other coefficients as well. In [10] the value of c_{NS5} was determined from the Dirac quantization condition. Perhaps the simplest way to proceed is to observe that the coefficient in the harmonic function of the KK monopole is fixed by requiring that the solution is non-singular.

All these solutions are BPS solution and preserve half of maximal supersymmetry. This implies that certain quantities do not renormalize. Let us sketch the argument. The supersymmetry algebra has the form

$$\{Q_\alpha, Q_\beta\} \sim (C\Gamma^\mu)_{\alpha\beta} P_\mu + (C\Gamma^{\mu_1\cdots\mu_p})_{\alpha\beta} Z^{(p)}_{\mu_1\cdots\mu_p} \tag{30}$$

where C is the charge conjugation matrix, Q_α are the supercharges, P_μ is the momentum generator, and $Z^{(p)}$ are central charges. These are the charges carried by p branes.

Taking the expectation value of (30) between a physical state $|A\rangle$ and going to the rest frame we get

$$\langle A|\{Q_\alpha, Q_\beta\}|A\rangle = (M^A - c|Z|)_{\alpha\beta} \geq 0 \tag{31}$$

where $M^A_{\alpha\beta}$ is the mass matrix, c is a constant, and we used the fact that $\{Q_\alpha, Q_\beta\}$ is a positive definite matrix.

If the matrix in the right hand side has no zero eigenvalues, then one can take suitable linear combinations of the supercharges so that the superalgebra takes the form of fermionic oscillator algebra. Then half of the oscillators can be regarded as creation and half as annihilation operators. This means that a supermultiplet contains 2^{16} states.

If the matrix in the right hand side of (31) has a zero eigenvalue (so the mass is proportional to the charge, $M = c|Z|$, i.e. we have a BPS state) then some of the generators annihilate the state. The remaining supercharges can again be divided into half creation and half annihilation operators. Thus, the BPS supermultiplet is a short multiplet. For 1/2 supersymmetric states, such as the branes we have been discussing, this means that we have 2^8 states instead of 2^{16}.

If we vary adiabatically the parameters of theory (i.e. no phase transition) the number of states cannot change abruptly, so the number of BPS states remains invariant and the mass/charge relation does not renormalize[38]. (Here we also assume that we do not cross curves of marginal stability).

3.1 M-branes

We briefly describe the connection of the brane solutions described in the previous section to M-theory. M-theory at low-energies is described by eleven dimensional supergravity. The bosonic field content of eleven dimensional supergravity consists of a metric, G_{MN}, and a three-form antisymmetric tensor, A_{MNP}. We therefore expect that this theory has solutions describing extended objects coupled "electrically" and "magnetically" to A_{MNP}. Indeed, one finds a 2-brane solution, M2, and a fivebrane solution, M5[39, 40]. The explicit form of the solution is as in (18), with $\alpha_{M2} = 1/3$ for the M2 and $\alpha_{M5} = 2/3$ for the M5 (there is no dilaton field in $11d$ supergravity, so $\beta = 0$). In addition, we have the purely gravitational solutions describing traveling waves and KK monopoles.

From the solutions of eleven dimensional supergravity one can obtain the solution of IIA supergravity upon dimensional reduction. The Kaluza-Klein ansatz for the bosonic fields leading to the string frame $10d$ metric is

$$\begin{aligned} ds^2_{11} &= e^{-\frac{2}{3}\phi(x)} g_{\mu\nu} dx^\mu dx^\nu + e^{\frac{4}{3}\phi(x)} (dx_{11} + C^{(1)}_\mu dx^\mu)^2 \\ A &= C^{(3)} + B \wedge dx_{11} \end{aligned} \qquad (32)$$

where B is the NSNS antisymmetric tensor and $C^{(1)}$ and $C^{(3)}$ are the RR antisymmetric tensors of IIA theory. Dimensionally reducing the M-branes along a worldvolume or a transverse direction one obtains all solution of IIA as follows:

$11d$ SUGRA		W		M2		M5		KK
	↙	↓	↙	↓	↙	↓	↙	↓
IIA SUGRA	D0	W	F1	D2	D4	NS5	D6	KK

3.2 Intersection rules

In the previous section we described brane solutions of supergravity theories. These solutions can be used as building blocks in order to construct new solutions [41–51] (for a review see [52]). The new solutions can be interpreted as intersecting (or in some cases overlapping) branes. In order to obtain a supersymmetric solution only certain intersections are allowed.

The intersection rules are as follows:
One superimposes the single brane solutions using the rule that all pairwise intersections should belong to a set of allowed intersections. If all harmonic functions are taken to depend on the overall transverse directions (i.e. the directions transverse to all branes) we are dealing with a "standard" intersection. Otherwise the intersection will be called "non-standard". In $D = 11$ there are three standard intersections, $(0|\,2 \perp 2)$[4], $(1|\,2 \perp 5)$ and $(3|\,5 \perp 5)$ [42, 43, 47], and one non-standard $(1|5 \perp 5)$[53, 43]. In the latter intersection the harmonic functions depend on the relative transverse directions (i.e. the directions which are worldvolume coordinates of the one but transverse coordinates of the other fivebrane). In addition, one can add a wave solution along a common string. The intersection rules in ten dimensions can be derived from these by dimensional reduction plus T and S-duality. We collect the standard and non-standard intersection rules in the table below. (For intersections rules involving KK monopoles see [49]). When both standard and non-standard intersection rules are used (as for instance in the solutions of [54]), one has to specify which coordinates each harmonic function depends on. This is usually clear by inspection of the intersection, but it can also be further verified by looking at the field equation(s) for the antisymmetric tensor field(s).

	standard	non-standard
$D=11$	$(0\|M2 \perp M2)$	
	$(1\|M2 \perp M5)$	
	$(3\|M5 \perp M5)$	$(1\|M5 \perp M5)$
$D=10$	$(\frac{1}{2}(p+q-4)\|Dp \perp Dq)$	$(\frac{1}{2}(p+q-8)\|Dp \perp Dq)$
	$(1\|F1 \perp NS5)$	
	$(3\|NS5 \perp NS5)$	$(1\|NS5 \perp NS5)$
	$(0\|F1 \perp Dp)$	
	$(p-1\|NS5 \perp Dp)$	$(p-3\|NS5 \perp Dp)$

Table 3. *Standard and non-standard intersections in ten and eleven dimensions.*

There is a simple algorithm which leads to a non-extreme version of a given supersymmetric solution (constructed according to standard intersection rules) [44]. We will give these rules for M-brane intersections. This is sufficient as

[4] The notation $(q|\,p_1 \perp p_2)$ denotes a p_1-brane intersecting with a p_2-brane over a q-brane.

dimensional reduction and duality produce all standard intersections of type II branes. It consists of the following steps:

(1) Make the following replacements in the d-dimensional transverse space-time part of the metric:

$$dt^2 \to f(r)dt^2 \ , \ dx_1^2 + \cdots + dx_{d-1}^2 \to f^{-1}(r)dr^2 + r^2 d\Omega_{d-2}^2 \ , \ f(r) = 1 - \frac{\mu^{d-3}}{r^{d-3}} \ , \tag{33}$$

and use the following harmonic functions,

$$\begin{aligned} H_T &= 1 + \frac{\mathcal{Q}_T}{r^{d-3}} \ , \qquad \mathcal{Q}_T = \mu^{d-3} \sinh^2 \alpha_T \ , \\ H_F &= 1 + \frac{\mathcal{Q}_F}{r^{d-3}} \ , \qquad \mathcal{Q}_F = \mu^{d-3} \sinh^2 \alpha_F \ , \end{aligned} \tag{34}$$

for the constituent two-branes and five-branes, respectively.

(2) In the expression for the field strength $\mathcal{F}_4$ of the three-form field make the following replacements:

$$\begin{aligned} H_T'^{-1} &\to H_T'^{-1} = 1 - \frac{Q_T}{r^{d-3}} H_T^{-1}, \qquad Q_T = \mu^{d-3} \sinh \alpha_T \cosh \alpha_T \ , \\ H_F &\to H_F' = 1 + \frac{Q_F}{r^{d-3}} \qquad\qquad\quad Q_F = \mu^{d-3} \sinh \alpha_F \cosh \alpha_F \ , \end{aligned} \tag{35}$$

in the "electric" (two-brane) part, and in the "magnetic" (five-brane) part, respectively. In the extreme limit $\mu \to 0$, $\alpha_F \to \infty$, and $\alpha_T \to \infty$, while the charges Q_F and Q_T are kept fixed. In this case $\mathcal{Q}_F = Q_F$ and $\mathcal{Q}_T = Q_T$, so that $H_T' = H_T$. The form of $\mathcal{F}_4$ and the actual value of its "magnetic" part does not change compared to the extreme limit.

(3) In the case there is a common string along some direction x, one can add momentum along x. Then

$$-f(r)dt^2 + dx^2 \to -K^{-1}(r) f(r) dt^2 + K(r) \left(dx - [K'^{-1}(r) - 1]dt\right)^2 \tag{36}$$

where

$$\begin{aligned} K &= 1 + \frac{\mathcal{Q}_K}{r^{d-3}} \ , \qquad\qquad \mathcal{Q}_K = \mu^{d-3} \sinh^2 \alpha_K \ , \\ K'^{-1} &= 1 - \frac{Q_K}{r^{d-3}} K^{-1} \ , \qquad Q_K = \mu^{d-3} \sinh \alpha_K \cosh \alpha_K \ . \end{aligned} \tag{37}$$

In the extreme limit $\mu \to 0$, $\alpha_K \to \infty$, the charge Q_K is held fixed, $K = K'$ and thus the metric (36) becomes $dudv + (K-1)du^2$, where $u, v = x \pm t$.

4 Black holes in string theory

Black holes arise in string theory as solutions of the corresponding low-energy supergravity theory. String theory lives in 10 dimensions (or 11 from the M-theory perspective). Suppose the theory is compactified on a compact manifold down to d spacetime dimensions. Branes wrapped in the compact dimensions will look like pointlike objects in the d-dimensional spacetime. So, the idea is to construct a configuration of intersecting wrapped branes which upon dimensional reduction yields a black hole spacetime. If the brane intersection is supersymmetric then the black hole will be an extremal supersymmetric black hole. On the other hand, non-extremal intersections yield non-extremal black holes.

In general, the regime of the parameter space in which supergravity is valid is different from the regime in which weakly coupled string theory is valid. Thus, although we know that a given brane configuration becomes a black hole when we go from weak to strong coupling, it would seem difficult to extract information about the black hole from this fact.

For supersymmetric black holes, however, the BPS property of the states allows one to learn certain things about black holes from the weakly coupled D-brane system. For example, one can count the number of states at weak coupling and extrapolate the result to the black hole phase. In this way, one derives the Bekenstein-Hawking entropy formula (including the precise numerical coefficient) for this class of black holes[55, 56]. We will review this calculation in section 4.1.

In the absence of supersymmetry, we cannot in general follow the states from weak to strong coupling. However, one could still obtain some qualitative understanding of the black hole entropy. On general grounds, one might expect that the transition from weakly coupled strings to black holes happens when the string scale becomes approximately equal to the Schwarzschild radius (or more generally to the curvature radius at the horizon). This point is called the correspondence point. Demanding that the mass and all the other charges of the two different configurations match, one obtains that the entropies also match [57]. These considerations correctly provide the dependence of the entropy on the mass and the other charges, but the numerical coefficient in the Bekenstein-Hawking entropy formula remains undetermined.

In [58] a different approach was initiated. Instead of trying to determine the physics of black holes using the fact that at weak coupling they become a set of D-branes, the symmetries of M-theory are used in order to map the black hole configuration to another black hole configuration. Since the U-duality group involves strong/weak transitions one does not, in general, have control over the microscopic states that make up a generic configuration. We will see, however, that the situation is better when it comes to black holes! U-duality maps black holes to black holes with the same thermodynamic characteristics, i.e. the entropy and the temperature remain invariant. This implies that the number of microstates that make up the black hole configuration remains the same. Notice that to reach this conclusion we did not use supersymmetry, but the fact that the area of the horizon of a black hole (divided by Newton's constant) tell us how many degrees of freedom the black hole contains. We discuss this approach in section 4.2.

The effect of the U-duality transformations described in section 4.2 is to remove the constant part from certain harmonic functions (and also change the values of some moduli). One can achieve a similar result by taking the low-energy limit $\alpha' \to 0$ while keeping fixed the masses of strings stretched between different D-branes. Considerations involving this limit lead to the adS/CFT correspondence[59]. This will be discussed in section 4.3.

4.1 Extremal black holes and the D-brane counting

We will analyze five dimensional black holes. Four dimensional ones [60] can be analyzed in a completely analogous manner[61, 62]. Rotating black holes have been discussed in [63–65].

5*d* Extremal Black Holes To study extremal charged five dimensional black holes we build a configuration of intersecting branes using the supersymmetric intersection rules. In particular, we consider the configuration of N_5 D5-brane wrapped in $x_1, \ldots, x_5$, N_1 D1-brane wrapped in x_1, with N_K momentum modes along x_1. The coordinates $x_i, i = 1, \ldots, 5$ are taken periodic with periods R_i. Explicitly, the spacetime fields are

$$\begin{aligned} ds^2 = H_1^{1/2} H_5^{1/2} \ & [H_1^{-1} H_5^{-1} \left(-K^{-1} dt^2 + K(dx_1 - (K^{-1} - 1)dt)^2\right) \\ & + H_5^{-1}(dx_2^2 + \cdots + dx_5^2) + dx_6^2 + \cdots + dx_9^2] \end{aligned} \tag{38}$$

and

$$e^{-2\phi} = H_1^{-1} H_5, \qquad C_{01}^{(2)} = H_1^{-1} - 1$$

$$H_{ijk} = \frac{1}{2} \epsilon_{ijkl} \partial_l H_5, \qquad i, j, k, l = 6, \ldots, 9 \tag{39}$$

$$r^2 = x_6^2 + \cdots + x_9^2 \tag{40}$$

The harmonic functions are equal to

$$\begin{aligned} H_1 &= 1 + \frac{Q_1}{r^2}, \qquad & Q_1 &= \frac{N_1 g_s \alpha'^3}{V} \\ H_5 &= 1 + \frac{Q_5}{r^2}, \qquad & Q_5 &= N_5 g_s \alpha' \\ K &= 1 + \frac{Q_K}{r^2}, \qquad & Q_K &= \frac{N_K g_s^2 \alpha'^4}{R_1^2 V} \end{aligned} \tag{41}$$

where $V = R_2 R_3 R_4 R_5$ and the charges have been calculated using (29).

Upon dimensional reduction over the periodic coordinates $x_1, \ldots, x_5$, using (28), we obtain

$$ds_{E,5}^2 = \lambda^{-2/3} dt^2 + \lambda^{1/3}(dr^2 + r^2 \delta \Omega_3^2) \tag{42}$$

where

$$\lambda = H_1 H_5 K = (1 + \frac{Q_1}{r^2})(1 + \frac{Q_5}{r^2})(1 + \frac{Q_K}{r^2}) \tag{43}$$

This is an extremal charged black hole. The horizon is located at $r = 0$. The area of the horizon and the five dimensional Newton's constant are equal to

$$A_5 = (r^2 \lambda^{1/3})^{3/2}\Big|_{r=0} \omega_3 = \sqrt{Q_1 Q_2 Q_K}(2\pi^2)$$
$$G_N^{(5)} = \frac{G_N^{(10)}}{(2\pi)^5 R_1 V} \tag{44}$$

Therefore, the entropy is equal to

$$S = \frac{A_5}{4G_5} = 2\pi\sqrt{N_1 N_5 N_K} \tag{45}$$

For the supergravity to be valid we need to suppress string loops and α' corrections. We suppress string loops by sending $g_s \to 0$, while keeping the charges Q_i fixed. These charges are the characteristic scales of the system. In order to suppress α' corrections they should be much larger than the string scale,

$$Q_1,\ Q_5,\ Q_K \gg \alpha' \tag{46}$$

Taking the compactification radii to be of order l_s we obtain

$$g_s N_1,\ g_s N_5,\ g_s^2 N_K \gg 1 \tag{47}$$

This means that $N_K \gg N_1 \sim N_5 \gg 1$.

D-brane counting We now turn to the weak-coupling D-brane configuration in order to compute the D-brane entropy. Counting the degeneracy of D-brane states translates into the question of counting BPS states in the D-brane world-volume theory[66, 58, 68–70]. For the system we are interested in, and taking the torus T^4 in the relative transverse directions to be small, $R_2, R_3, R_4, R_5 \ll R_1$, the relevant worldvolume theory is 1+1 dimensional. This theory is the infrared limit of the Higgs branch of the 1+1 gauge theory, and it has been argued to be a deformation of the supersymmetric $\mathcal{N} = (4,4)$ sigma model with target space $(T^4)^{N_1 N_5}/S^{N_1 N_5}$[68]. Since $(T^4)^{N_1 N_5}/S^{N_1 N_5}$ is a hyperkaehler manifold of dimension $4N_1N_5$, the sigma model has central charge equal to $6N_1N_5$. This is the central charge of $4N_1N_5$ bosonic and fermionic degrees of freedom (since scalars contribute 1 and fermions 1/2 to the central charge). Roughly, these degrees of freedom are the ones describing the motion of the D1 brane inside the D5 brane. For details we refer to [10].

In the worldvolume theory we get that the right movers are in their ground state and the left movers carry N_K momentum modes. Thus, the degeneracy of the D-brane system is given by the degeneracy of the conformal field theory of central charge $c = 6N_1N_5$ at level N_K. For a unitary conformal field theory the degeneracy is given by Cardy's formula[71]

$$d(c, N_K) \sim \exp(2\pi\sqrt{\frac{c}{6}N_K}) \tag{48}$$

Therefore, the entropy is equal to

$$S = \log d(c, N_K) = 2\pi\sqrt{N_1N_5N_K} \tag{49}$$

This is in exact agreement with (45).

Let us now inspect the regime of validity of the D-brane picture. Open string diagrams pick up a factor $g_sN_{1,5}$ because the open string coupling constant is g_s and there are $N_{1,5}$ branes where the string can end (or equivalently one should sum over the Chan-Paton factors). Processes involving momenta involve a factor $g_s^2N_K$[72]. Therefore, conventional D-brane perturbation theory is good when

$$g_sN_1,\ g_sN_5,\ g_s^2N_K \ll 1 \Rightarrow Q_1,\ Q_5,\ Q_K \ll \alpha', \tag{50}$$

which is precisely the opposite regime to (47) where the classical supergravity solution is good. The D-brane/string perturbation theory and black hole regimes are thus complementary. This feature is related to open-closed string duality. Due to supersymmetry, however, one can extrapolate results obtained in the D-brane phase to the black hole phase.

4.2 Non-extremal black holes and the BTZ black hole

In this section we review the approach of [58]. The idea is to use U-dualities in order to connect higher dimensional black holes to lower dimensional ones. Such ideas also appeared in [73]. The U-duality transformation essentially maps the initial black hole to its near-horizon region (but Schwarzschild black holes are also included as a limiting case). In particular, four and five dimensional black holes are mapped to the three dimensional BTZ black hole. The U-duality group of string (or M) theory on a torus does not change the number of non-compact dimensions. However, black hole spacetimes always contain an extra timelike isometry. This extra isometry allows for a duality transformation, the shift transformation[74], that yields trans-dimensional transformations. A thorough discussion (that includes global issues) of the shift transformation is given in section 4.2.

U-duality and entropy Let us discuss whether one can use U-duality in order to infer a state counting for a given black hole from the counting of a U-dual configuration. The U-duality group is conjectured (and widely believed) to be an exact symmetry of M-theory. This symmetry, however, is spontaneously broken by the vacuum. The vacua of M-theory (compactified on some manifold) are parametrized by a set of constants. These constants are expectation values of scalar fields arising from the compactification. U-duality acts on these scalars, so it transforms one vacuum to another. Therefore, from a state on a given vacuum one can deduce by U-duality the existence of another state in a new vacuum. Since the U-duality group contains S-duality which is strong/weak coupling duality, one cannot in general continue the new state back to the original vacuum, unless this state is protected from quantum corrections. States with this property are BPS states. Therefore, the spectrum of BPS states is invariant under U-duality transformations. This implies in particular that if we want to count the number of states that make up an extremal supersymmetric black hole, we may use any U-duality configuration. Indeed, the entropy formula for extremal black holes is U-duality invariant[75–78].

The question is whether it is justified to use U-duality in more general context. A remarkable fact about S and T duality transformations is that they leave invariant both the entropy and the temperature of black holes connected by S and T transformations. For S-duality this follows from the fact that S-duality leaves invariant the Einstein metric. For T-duality, this has been shown in [23]. We review this argument here.

Consider a black hole solution with a timelike isometry $\partial/\partial t$, a compact spacelike isometry $\partial/\partial x$, and a NSNS 2-form B turned on. Smoothness near the horizon requires[23] that the B_{tx} vanishes at the horizon. In order for the T-dual geometry to also be smooth (i.e. the dual 2-form to vanish at the horizon) we require in addition that $A_x = 0$ at the horizon (see (13)-(14)). (This can always be achieved by a coordinate transformation.) RR potentials that can be turned into B_{xt} by dualities are also required to vanish at the horizon.

Let us first discuss the entropy. In d dimensions the Einstein metric is given by (see (28)),

$$ds_E^2 = e^{-4\phi/(d-2)}[g_{xx}(dx + A_i dx^i)^2 + \bar{g}_{ij}dx^i dx^j] \tag{51}$$

The metric induced on the horizon is of the same form but with i, j taking values only over the $d-3$ angular variables. Therefore, the area is equal to

$$A_d = \int \sqrt{(e^{-4\phi/(d-2)})^{d-2} g_{xx} \det \bar{g}} = \int e^{-2\phi}\sqrt{g_{xx}}\sqrt{\det \bar{g}} \tag{52}$$

One may check that $e^{-2\phi}\sqrt{g_{xx}}$ is a T-duality invariant combination (and $\bar{g}$ was invariant to start with). Therefore, the entropy of black holes is T-duality invariant.

Let us also note that the entropy formula is invariant under dimensional reduction

$$S = \frac{A_{10}}{4G_N^{(10)}} = \frac{A_d}{4G_N^{(d)}} \tag{53}$$

since $A_d = A_{10}/V_{10-d}$ and $G_N^{(d)} = G_N^{(10)}/V_{10-d}$, where V_{10-d} is the volume of the compactification space.

We now turn to the discussion of the behavior of the Hawking temperature under duality transformations. Perhaps the simplest way to compute the

Hawking temperature is to analytically continue to Euclidean space by taking $t \to \tau = -it$. The black hole spacetime becomes then a non-singular Riemannian manifold provided that the Euclidean time is periodically identified with period equal to the inverse Hawking temperature. Suppose that the horizon is at $r = \mu$. One can calculate the temperature to be equal to (we assume that the event horizon is non-degenerate)

$$T_H = \left. \frac{\partial_r g_{\tau\tau}}{4\pi\sqrt{g_{\tau\tau} g_{rr}}} \right|_{r=\mu} \tag{54}$$

It follows by inspection that the Hawking temperature is invariant under non-singular Weyl rescaling. Hence, it does not make any difference whether we consider the Einstein or the string frame. We choose to work with the string frame. From (13) we get

$$g_{\tau\tau} = \bar{g}_{\tau\tau} + g_{xx} A_\tau A_\tau, \qquad g_{rr} = \bar{g}_{rr} + g_{xx} A_r A_r \tag{55}$$

Assuming that A_r is finite at the horizon (in all cases we will consider $A_r = 0$), and using $g_{xx}|_{r=\mu} = A_\tau|_{r=\mu} = 0$ we obtain

$$T_H = \left. \frac{\partial_r \bar{g}_{\tau\tau}}{4\pi\sqrt{\bar{g}_{\tau\tau} \bar{g}_{rr}}} \right|_{r=\mu} \tag{56}$$

which is manifestly T-duality invariant.

Therefore, an arbitrary combination of S and T transformations will lead to a black hole solution with the same entropy and temperature as the original one. This implies that black holes connected by U-duality transformations have the same number of microstates. This is somewhat surprising since for non-supersymmetric black holes we cannot follow the states during U-duality transformations. As we move from one configuration to a U-dual one, some states may disappear. However, an equal number of states has to appear, since the final configuration has the same entropy. We do not have a microscopic derivation of this fact. We believe that such derivation will be an important step towards further understanding of black holes.

A general U-duality transformation may involve strong/weak transitions. The U-duality transformations, however, that we will use below do not involve such strong/weak transitions. Actually we shall exclusively be in the black hole phase. We will only consider transformations, call them U_T, that are connected to T-dualities by a similarity transformation

$$U_T = U^{-1} T U \tag{57}$$

where U denotes a generic U-duality transformation and T a sequence of two T-duality transformations (so U_T acts within the same theory).

The shift transformation As we have discussed, we construct black holes configurations using appropriate non-extremal intersections of extremal branes. These configurations are solutions of the field equations provided the various harmonic functions H_i appearing in the solution satisfy Laplace's equations,

$$\nabla^2 H_i = 0, \tag{58}$$

where ∇ is the Laplacian in the overall transverse space. The constant part of the harmonic function is usually set to one in order for the solution to be asymptotically flat. Clearly, up to normalization, the only other choice is to set this constant to zero. This choice has the dramatic effect of changing the asymptotics of the solution. We will see, however, that there is a duality transformation, the shift transformation, that removes the one from the harmonic function. This duality transformation has appeared in the past in various contexts [79, 20, 28, 73, 74, 58, 80, 25].

Consider the solution describing a non-extremal fundamental string in $d+1$ dimensions

$$\begin{aligned} ds^2 &= H^{-1}(r)(-f(r)dt^2 + dx_1^2) + f^{-1}(r)dr^2 + r^2 d\Omega_{d-2}^2 \\ B_{tx_1} &= H'^{-1} - 1 + \tanh\alpha \\ e^{-2\phi} &= H \end{aligned} \tag{59}$$

The coordinate x_1 is periodic with period R_1. The various harmonic functions are equal to

$$H = 1 + \frac{\mu^{d-3}\sinh^2\alpha}{r^{d-3}},$$

$$H'^{-1} = 1 - \frac{\mu^{d-3}\sinh\alpha\cosh\alpha}{r^{d-3}} H^{-1},$$

$$f = 1 - \frac{\mu^{d-3}}{r^{d-3}} \tag{60}$$

The constant part of the antisymmetric tensor B_{tx_1} is fixed by the requirement that B_{tx_1} vanishes at the horizon. This is required by regularity[23], as described in the previous section. The entropy and the temperature are given by

$$S = \frac{1}{4G_N^{(d+1)}} 2\pi R_1 \cosh\alpha\mu^{d-2}\omega_{d-2}, \qquad T_H = \frac{(d-3)}{4\pi\mu\cosh\alpha}. \tag{61}$$

Notice that in order to calculate the area one first has to reach the Einstein frame.

We now perform the following sequence of T-dualities that we call the shift transformation:

$$shift = T_{\frac{\partial}{\partial t'}}\left(\frac{\partial}{\partial x_1'}\right) \circ T_{\frac{\partial}{\partial t}}\left(\frac{\partial}{\partial x_1}\right) \tag{62}$$

where

$$\begin{aligned}\frac{\partial}{\partial x_1'} &= -e^{-\alpha}\frac{\partial}{\partial t} + \frac{1}{\cosh\alpha}\frac{\partial}{\partial x_1} \\ \frac{\partial}{\partial t'} &= \cosh\alpha\frac{\partial}{\partial t}\end{aligned} \tag{63}$$

The notation $T_{k_1}(k_2)$ indicates a T-duality transformation along the killing vector k_2 keeping k_1 fixed.

Let us give the details. After the first T-duality, $T_{\partial/\partial t}(\partial/\partial x_1)$, we get a non-extremal wave solution,

$$\begin{aligned}ds^2 = &-H^{-1}(r)f(r)dt^2 + H(r)\left(dx_1 - (H'^{-1}(r) - 1 + \tanh\alpha)dt\right)^2 \\ &+f^{-1}(r)dr^2 + r^2 d\Omega_{d-2}^2\ ,\end{aligned} \tag{64}$$

The radius of x_1 is now α'/R_1. In addition, $g_s \to l_s/R_1$, so $G_N^{(d+1)} \to G_N^{(d+1)}\alpha'/R_1^2$. One can check that this solution has the same entropy and temperature as the solution in (59).

We would like now to dualize along (63). To do this we first reach adapted coordinates

$$\begin{pmatrix} t \\ x_1 \end{pmatrix} = \begin{pmatrix} \cosh\alpha & -e^{-\alpha} \\ 0 & \frac{1}{\cosh\alpha} \end{pmatrix} \begin{pmatrix} t' \\ x_1' \end{pmatrix}\ . \tag{65}$$

The metric in the new coordinates takes the form (we have dropped the primes)

$$ds^2 = -\tilde{H}^{-1}(r)f(r)dt^2 + \tilde{H}(r)\left(dx_1 - (\tilde{H}^{-1}(r) - 1)dt\right)^2 + f^{-1}(r)dr^2 + r^2 d\Omega_{d-2}^2 \tag{66}$$

where now

$$\tilde{H}(r) = \frac{\mu^{d-3}}{r^{d-3}}\ . \tag{67}$$

The radius of x_1 also changes to $\cosh\alpha/R_1$.

Now, that we have reached adapted coordinates we can use (12) to obtain,

$$\begin{aligned}ds^2 &= \tilde{H}^{-1}(r)(-f(r)dt^2 + dx_1^2) + f^{-1}(r)dr^2 + r^2 d\Omega_{d-2}^2 \\ B_{\tau x_1} &= \tilde{H}^{-1} - 1 \\ e^{-2\phi} &= \tilde{H}\end{aligned} \tag{68}$$

The radius of x_1 is now equal to $R_1/\cosh\alpha$. In addition, there is again a change in Newton's constant. One can calculate the temperature and entropy of this solution. The result for the entropy is the same in (61). The temperature is equal to $T_H = (d-3)/4\pi\mu$. This differs by a factor of $\cosh\alpha$ from (61). This is due to the fact that the timelike killing vectors $\partial/\partial t$ and $\partial/\partial t'$ differ by a factor of $\cosh\alpha$ (see (63)).

To summarize, the effect of the shift transformation (62) is to change the solution by removing the constant part of the harmonic functions. All the dependence of the metric and the antisymmetric tensor on the non-extremality angle α resides in the radius of the compact direction which after the shift transformation is equal to $R_1/\cosh\alpha$. In addition, $g_s \to g_s/\cosh\alpha$, so $G_N^{(d+1)} \to G_N^{(d+1)}/\cosh^2\alpha$.

The orbits of the killing vector $\partial/\partial x_1'$ are non-compact since the time coordinate is non-compact. This means that (59) and (68) are not equivalent. To make the duality transformation a symmetry we need to compactify the orbits of the killing vector $\partial/\partial x_1'$.[5] The fact, however, that the entropy and temperature of the one black hole can be deduced from the entropy and temperature of the other indicates that the two solutions are in the same universality class (in a loose sense).

The norm of the killing vector (63) is

$$|\partial/\partial x_1'|^2 = \frac{\mu^{d-3}}{r^{d-3}} \tag{69}$$

therefore the isometry is spacelike everywhere but it becomes null at spatial infinity. Let us examine the (r, x_1) part of the metric close to spatial infinity. From (66) we get

$$ds^2_{(r,x_1)} = dr^2 + \frac{\mu^{d-3}}{r^{d-3}} dx_1^2 \tag{70}$$

For $d = 5$, which will be the case in the next section where we discuss five dimensional black holes, this is exactly the same metric as in (16). This suggests to consider r, x_1 as polar coordinates and the isometry in x_1 as a rotational isometry with a fixed point at infinity.

Connection of 5*d* and 4*d* black holes to the BTZ black hole We are now ready to use our results to study non-extremal 5d and 4d black holes. We will explicitly work out the case of 5d black holes. The analysis of 4d black holes is completely analogous [58]. Four and five dimensional black holes can also be mapped by similar operations [73, 58, 82] to two dimensional black holes[83]. Let us also note that the manipulations we describe here cannot connect the BTZ black hole to higher than five dimensional black holes [58]. The relation between the near-horizon limit of higher-dimensional black holes and the BTZ black hole has also been investigated in [84].

[5] One way to make the orbits compact is to compactify time with appropriately chosen radius. It has been argued in [81] that a spatially wrapped brane should also be wrapped in time in order to avoid conical singularities at the horizon. The two issues may be related. The time coordinate is naturally compact in Euclidean black holes, the radius of the time coordinate being the inverse of the Hawking temperature. One may try to formulate the analysis in the Euclidean framework. The problem is then that the coordinate transformation (65) is complex.

The solution we will study is the non-extremal version of (38). Explicitly, the metric, the dilaton and the antisymmetric tensor are given by

$$\begin{aligned} ds_{10}^2 = H_1^{1/2} H_5^{1/2} \ &\Big[H_1^{-1} H_5^{-1} \left(-K^{-1} f dt^2 + K \left(dx_1 - (K'^{-1} - 1) dt \right)^2 \right) \\ &+ H_5^{-1} (dx_2^2 + \cdots + dx_5^2) + (f^{-1} dr^2 + r^2 d\Omega_3^2) \Big] \ , \end{aligned} \tag{71}$$

and

$$\begin{aligned} e^{-2\phi} &= H_1^{-1} H_5 \ , \quad C_{01}^{(2)} = H_1'^{-1} - 1 + \tanh \alpha_1 \ , \\ H_{ijk} &= \frac{1}{2} \epsilon_{ijkl} \partial_l H_5' \ , \quad i,j,k,l = 6, \ldots, 9 \ , \\ f &= 1 - \frac{\mu^2}{r^2} \ , \quad r^2 = x_6^2 + \cdots + x_9^2 \ , \end{aligned} \tag{72}$$

The coordinates x_i, $i = 1, \ldots, 5$, are assumed to be periodic, each with radius R_i.

The various harmonic function are given by

$$\begin{aligned} &K = 1 + \frac{\mathcal{Q}_K}{r^2}, \ K'^{-1} = 1 - \frac{Q_K}{r^2} K^{-1}, \ \mathcal{Q}_K = \mu^2 \sinh^2 \alpha_K, \\ &Q_K = \mu^2 \sinh \alpha_K \cosh \alpha_K \\ &H_1 = 1 + \frac{\mathcal{Q}_1}{r^2}, \ H_1'^{-1} = 1 - \frac{Q_1}{r^2} H_1^{-1}, \ \mathcal{Q}_1 = \mu^2 \sinh^2 \alpha_1, \\ &Q_1 = \mu^2 \sinh \alpha_1 \cosh \alpha_1 \\ &H_5 = 1 + \frac{\mathcal{Q}_5}{r^2}, \ H_5' = 1 + \frac{Q_5}{r^2}, \ \mathcal{Q}_5 = \mu^2 \sinh^2 \alpha_5, \\ &Q_5 = \mu^2 \sinh \alpha_5 \cosh \alpha_5 \ , \end{aligned} \tag{73}$$

Dimensionally reducing in x_1, x_2, x_3, x_4, x_5, one gets a $5d$ non-extremal black hole, whose metric in the Einstein frame is given by

$$ds_{E,5}^2 = -\lambda^{-2/3} f dt^2 + \lambda^{1/3} (f^{-1} dr^2 + r^2 d\Omega_3^2) \ , \tag{74}$$

where

$$\lambda = H_5 H_1 K = \left(1 + \frac{\mathcal{Q}_5}{r^2} \right) \left(1 + \frac{\mathcal{Q}_1}{r^2} \right) \left(1 + \frac{\mathcal{Q}_K}{r^2} \right) \ . \tag{75}$$

This black hole is charged with respect to the Kaluza-Klein gauge fields originating from the antisymmetric tensor fields and the metric. When all charges are set equal to zero one obtains the $5d$ Schwarzschild black hole. The metric (74) has an outer horizon at $r = \mu$ and an inner horizon at $r = 0$.

The Bekenstein–Hawking entropy may easily be calculated to be

$$S = \frac{A_5}{4 G_N^{(5)}} = \frac{1}{4} \frac{(2\pi)^5 R_1 V}{G_N^{(10)}} \mu^3 \omega_3 \cosh \alpha_5 \cosh \alpha_1 \cosh \alpha_K \ , \tag{76}$$

where $V = R_2R_3R_4R_5$ is the compactification volume in the relative transverse directions, ω_3 is the volume of the unit 3-sphere and $G_N^{(5)}$ and $G_N^{(10)}$ are Newton's constant in five and ten dimensions, respectively. The temperature is given by

$$T_H = \frac{1}{2\pi\mu \cosh\alpha_1 \cosh\alpha_5 \cosh\alpha_K} \tag{77}$$

We will now show that one can connect this black hole to the BTZ black hole times a 3-sphere using transformations of the form (57). A U-transformation is used to map a given brane to a fundamental string. The T transformation is the shift transformation (62).

For the case at hand we need to perform the shift transformation to the D1 and the D5 brane. The final result is given by the metric in (71), but with

$$H_1 = \frac{\mu^2}{r^2}\ , \qquad H_5 = \frac{\mu^2}{r^2}\ , \tag{78}$$

and, in addition,

$$\begin{aligned} e^{-2\phi} &= 1\ , \qquad C_{01}^{(2)} = H_1^{-1} - 1\ , \\ H_{ijk} &= \frac{1}{2}\epsilon_{ijkl}\partial_l(H_5-1)\ , \qquad i,j,k,l = 6,\dots,9\ . \end{aligned} \tag{79}$$

In addition the compactification volume becomes $V \to V/(\cosh\alpha_1 \cosh\alpha_5)$ (here, for convenience in the presentation, we assume that the U-duality transformation mapped the D1 and D5 into a fundamental string wrapped in one of the relative transverse directions). Furthermore, $G_N^{(10)} \to G_N^{(10)}/(\cosh^2\alpha_1 \cosh^2\alpha_5)$. Notice that the parameters α_1 and α_5 associated to the charges of the original D1 and D5 brane do not appear in the background fields anymore.

Dimensionally reducing along x_2, x_3, x_4, x_5 we find

$$ds^2_{E,6} = ds^2_{BTZ} + l^2 d\Omega_3^2\ , \tag{80}$$

where

$$\begin{aligned} ds^2_{BTZ} = &-\frac{(\rho^2-\rho_+^2)(\rho^2-\rho_-^2)}{l^2\rho^2}dt^2 + \rho^2(d\varphi + \frac{\rho_+\rho_-}{l\rho^2}dt)^2 + \\ &+ \frac{l^2\rho^2}{(\rho^2-\rho_+^2)(\rho^2-\rho_-^2)}d\rho^2 \end{aligned} \tag{81}$$

is the metric of the non-extremal BTZ black hole in a space with cosmological constant $\Lambda = -1/l^2$, with inner horizon at $\rho = \rho_-$ and outer horizon at $\rho = \rho_+$. The mass and the angular momentum of the BTZ black hole are equal to

$$M = \frac{\rho_+^2 + \rho_-^2}{8G_N^{(3)} l^2}, \qquad J = \frac{\rho_+\rho_-}{4G_N^{(3)} l}. \tag{82}$$

In terms of the original variables:

$$l = \mu \,, \qquad \varphi = \frac{x_1}{l} \,, \qquad \rho^2 = r^2 + l^2 \sinh^2 \alpha_K \,,$$
$$\rho_+^2 = l^2 \cosh^2 \alpha_K \,, \qquad \rho_-^2 = l^2 \sinh^2 \alpha_K \,. \tag{83}$$

In addition,

$$\phi = 0 \,, \qquad C_{t\varphi}^{(0)} = (\rho^2 - \rho_+^2)/l \,, \qquad H = l^2 \epsilon_3 \,, \tag{84}$$

where ϵ_3 is the volume form element of the unit 3-sphere. Therefore, the metric (80) describes a space that is a product of a 3-sphere of radius l and of a non-extremal BTZ black hole. Notice that the BTZ and the sphere part are completely decoupled.

We can now calculate the entropy of the resulting black hole. The area of the horizon is equal to

$$A_3 = 2\pi \frac{R_1}{\mu} \mu \cosh \alpha_K \,, \tag{85}$$

whereas Newton's constant is given by

$$G_N^{(3)} = \frac{G_N^{(10)}}{(2\pi)^4 V (\cosh \alpha_1 \cosh \alpha_5)(\mu^3 \omega_3)} \,. \tag{86}$$

It follows that $S = A_3/(4G_N^{(3)})$ equals (76), i.e. the Bekenstein–Hawking entropy of the final configuration is equal to the one of the original $5d$ black hole. Notice that the Newton constant in (86) contains the parameter α_1, α_5, i.e. carries information on the charge of the original D1 and D5 brane. The temperature of the BTZ black hole is equal to

$$T_{BTZ} = \frac{\rho_+^2 - \rho_-^2}{2\pi \rho_+ l^2} \tag{87}$$

Transforming to the original variables we get

$$T_{BTZ} = \frac{1}{2\pi \mu \cosh \alpha_K} = \cosh \alpha_1 \cosh \alpha_5 T_H \tag{88}$$

precisely as predicted by the duality transformations.

We finish this section by pointing out a remarkable fact: We have started with the solution (71) of the low-energy supergravity. This solution is expected to get α' corrections. Then we used the T-duality rules (12) which are also valid only to first orders in α'. The final result, however, is valid to all orders in α'!

The fields in (81), (84) have their canonical value, so that both the BTZ and the sphere part are separately exact classical solutions of string theory,[6] i.e. there

[6] For the D1-D5 system that we discuss we obtain a CFT describing a D-string. One gets a fundamental string from the S-dual system of F1-NS5.

is an exact CFT associated to each of them. For the BTZ black hole the CFT corresponds to an orbifold of the WZW model based on $SL(2,\mathbb{R})$ [79,85,86], whereas for S^3 and the associated antisymmetric tensor with field strength H, given in (84), the appropriate CFT description is in terms of the $SU(2)$ WZW model. The same result also holds in the case of *4d* black holes[58]. This time the black hole is mapped to $BTZ \times S^2$. Again all fields are such that there is an exact CFT associated to each factor. The one associated with S^2 is the monopole CFT of [87].

The situation seems quite similar to the case described at the end of section 2.3: There we had the singular solution (16) of the lowest order in α' beta function equation which becomes an exact CFT after dualization with respect to a killing vector whose norm vanishes at spatial infinity. However, to establish equivalence one needs all order in α'.

In the case of black holes we have:
The singular black hole spacetime (71) that solves the lowest order in α' beta functions becomes, after dualization with respect to a killing vector whose norm vanishes at spatial infinity (plus other dualities), the BTZ black hole which contains no curvature singularity and is an exact CFT. (So, one could argue that the original singularity is resolved by α' corrections).

We find these similarities quite suggestive. However, it is difficult to see how one could overcome the problem of the non-compactness of the orbits of the killing vector in (63).

4.3 Low-energy limit and the near-horizon geometry

Near-horizon limit of branes We have argued that the physical system describing a black hole in strong coupling becomes a set of intersecting branes in weak coupling. We emphasize that there is only one physical system. Its description, however, in terms of some weakly coupled theory changes as we change the parameters of the theory, and furthermore, at any given regime of the parameter space, there is only one weakly coupled description.

One may view the different descriptions as effective theories that are adequate to describe the system at specific range of the parameter space. As we go outside this range new degrees of freedom become important and a new description takes over. In some cases, however, a given theory may still be well-defined for any value of the coupling constant. In this case we get a dual description of this theory.

Let us consider N coincident Dp-branes. At weak coupling they have a description as hypersurfaces where string can end. There is worldvolume theory describing the collective coordinates of the brane. The worldvolume fields interact among themselves and with the bulk fields. We would like to consider a limit which decouples the bulk gravity but still leaves non-trivial dynamics on the worldvolume. In low energies gravity decouples. So, we consider the limit $\alpha' \to 0$, which implies that the gravitation coupling constant, i.e. Newton's constant, $G_N \sim \alpha'^4$, also goes to zero. We want to keep the worldvolume degrees of freedom and their interactions. Since the worldvolume dynamics are governed by open string ending on the D-branes, we keep fixed the masses of strings stretched between D-branes as we take the limit $\alpha' \to 0$. In addition, we keep fixed the coupling constant of the worldvolume theory, so all the worldvolume interactions remain present. For N coincident D-branes, the worldvolume theory is an $SU(N)$ super Yang-Mills theory (we ignore the center of mass part). The YM coupling constant is equal (up to numerical constants) to $g_{YM}^2 \sim g_s(\alpha')^{(p-3)/2}$. Thus we get that the following limit,

$$\alpha' \to 0, \qquad U = \frac{r}{\alpha'} = \text{fixed}, \qquad g_{YM}^2 = \text{fixed} \tag{89}$$

yields a decoupled theory on the worldvolume.

At strong coupling the Dp branes are described by the black p-brane spacetimes (18). Let us consider the limit (89) for this spacetime. One gets that the harmonic function becomes,

$$H_i \to g_{YM}^2 N(\alpha')^{-2} U^{p-7} \tag{90}$$

The limit (89) is a near-horizon limit since $r = U\alpha' \to 0$ and there is a horizon at $r = 0$. We see that the effect of the limit (89) is similar to the effect of the shift transformation, namely the one is removed from the harmonic function. Inserting (90) back in the metric one finds that the spacetime becomes conformal to $adS_{p+2} \times S^{8-p}$ [54, 88] (for M-branes, one gets $adS_4 \times S^7$ for the M2 brane and $adS_7 \times S^4$ for the M5 brane [89]).

Let us now consider the particular case of N coincident D3-branes. The worldvolume theory is $d = 4$, $\mathcal{N} = 4$ $SU(N)$ SYM theory. This is a finite unitary theory for any value of its coupling constant. On the other hand, this system has a description as a black 3-brane at strong coupling. In the limit (89) we get that the spacetime becomes $adS_5 \times S^5$. In order to suppress string loops we need to take N large. For the supergravity description to be valid 't Hooft's coupling constant[90], $g_{YM}^2 N$, must be large. We therefore get that the strong ('t Hooft) coupling limit of large $d = 4$, $\mathcal{N} = 4$ $SU(N)$ SYM is described by adS supergravity[59]!

$\mathcal{N} = 4$ $d = 4$ SYM theory is a well-defined unitary finite theory, whereas supergravity is a non-renormalizable theory. It is best to think about it as the low energy effective theory of strings. Therefore, one should really consider strings on $adS_5 \times S^5$. In this way we reach the celebrated adS/CFT duality[59][7]:

Four dimensional $\mathcal{N} = 4SU(N)$ SYM is dual to string theory on $adS_5 \times S^5$.

This conjecture was made precise in [92, 93], where a prescription for evaluation of correlation functions was proposed. Subsequently a large number of papers appeared, all of them supporting the adS/CFT duality.

Let us examine again our result. We obtained that five dimensional adS gravity is equivalent to $d = 4$, $N = 4$ SYM theory. In other words, a gravity theory in $d+1(=5)$ dimensions is described in terms of a field theory without

[7] Many of the elements leading to this conjecture appeared in [91]. In [58], the worldvolume theory of the D3 brane was argued to be mapped to the singleton of adS_5 by the shift transformation.

gravity in $d(=4)$ dimensions. This is just holography[94, 95]! One can further show that the boundary theory indeed has one degree of freedom per Planck area[96].

Similar results hold for other brane configurations as one can always consider the low energy limit. In the case of conformal worldvolume theories there is an *adS* factor on the gravity side. In these cases the worldvolume theory is valid at all energy scales, and these considerations provide a weakly coupled gravity description of a strongly coupled theory. In the non-conformal cases the worldvolume SYM theory is a theory with a cut-off. As we change the cut-off new degrees of freedom become relevant and the description in terms of a SYM theory may not be valid. In these cases one finds that as we change the parameters of the theory there is always some perturbative description[97, 98]. The black p-brane solution becomes conformal to anti-de Sitter spacetime and the gravity description is in terms of gauged supergravities which have domain-wall vacua[88].

Low-energy limit of black hole spacetimes Let us discuss the low energy limit for black hole configurations. We will discuss in detail the $5d$ case. The $4d$ case is very similar [99]. Rotating black holes have been considered in [100].

Consider the black hole configuration in (38). We go to low energies keeping fixed the masses of stretched strings, the radius of coordinate which the string is wrapped in and the radii of the relative transverse directions in string units,

$$\alpha' \to 0, \qquad U = \frac{r}{\alpha'} \text{ fixed}, \qquad R_1, r_i = \frac{R_i}{\sqrt{\alpha'}} \text{ fixed } \; i = 2, \dots, 5 \tag{91}$$

Notice that $R_1 \gg R_i, i = 2, \dots, 5$, as in section 4.1. Since the horizon is at $r = 0$ and $r = U\alpha' \to 0$ this is at the same time a near-horizon limit. Therefore, the resulting configuration has the same number of degrees of freedom as the original one (since the area of the horizon is a measure of the degrees of freedom).

In the limit (91) the harmonic functions (41) become

$$\begin{aligned} H_1 &\to \frac{1}{\alpha'}\frac{\tilde{Q}_1}{U^2}, & \tilde{Q}_1 &= \frac{g_s N_1}{v}, \\ H_5 &\to \frac{1}{\alpha'}\frac{\tilde{Q}_5}{U^2}, & \tilde{Q}_5 &= g_s N_5 \\ K &\to 1 + \frac{\tilde{Q}_K}{U^2}, & \tilde{Q}_K &= \frac{g_s^2 N_K}{R_1^2 v} \end{aligned} \tag{92}$$

where $v = r_2 r_3 r_4 r_5$. Notice that the low-energy limit removes the one from the harmonic function of the D1 and D5 brane exactly as in section 4.2. Let us define new variables

$$\begin{aligned} &\rho^2 = U^2 + \rho_0^2, \qquad \phi = x_1/R_1, \qquad t_{BTZ} = t\frac{\tilde{Q}_1\tilde{Q}_5}{R_1^2} \\ &\rho_0^2 = \tilde{Q}_K, \qquad l^2 = \frac{\tilde{Q}_1\tilde{Q}_5}{R_1^2} \end{aligned} \tag{93}$$

The metric (38) becomes

$$ds^2 = \alpha' \frac{R_1^2}{\sqrt{\tilde{Q}_1 \tilde{Q}_5}} [ds^2_{BTZ} + \frac{\tilde{Q}_1}{R_1^2}(dx_2^2 + \cdots + dx_5^2) + \frac{\tilde{Q}_1 \tilde{Q}_5}{R_1^2} d\Omega_3^2]$$
$$e^{-2\phi} = \frac{\tilde{Q}_5}{\tilde{Q}_1} \tag{94}$$

where ds^2_{BTZ} is the metric (81) with $\rho_+ = \rho_- = \rho_0$, i.e. the metric of the extremal BTZ black hole. The overall factor in (94) originates from the fact that we want to have the angle ϕ with unit radius. We move this overall factor to Newton's constant by a Weyl rescaling. The three dimensional Netwon's constant is then equal to (taking into account the dilaton, and arranging such that the $3d$ metric is the standard BTZ metric (81))

$$G_N^{(3)} = \frac{g_s^2}{4R_1 v\sqrt{\tilde{Q}_1 \tilde{Q}_5}} \tag{95}$$

Notice that all the factors of α' have cancelled out. The mass, the angular momentum and the area of the horizon of the BTZ black hole are equal to

$$M = Jl, \qquad J = \frac{\rho_0^2}{4G_N^{(3)} l} = N_K, \qquad A = 2\pi\rho_0 = 2\pi\sqrt{\tilde{Q}_K} \tag{96}$$

Therefore,

$$S = 2\pi \frac{R_1 v}{g_s^2}\sqrt{\tilde{Q}_1 \tilde{Q}_5 \tilde{Q}_K} = 2\pi\sqrt{N_1 N_5 N_K} \tag{97}$$

as in (45) (as it should since we just took the near-horizon limit). Therefore, at low energies the physics of extremal black holes is governed by the BTZ black hole.

Let us now move to non-extremal black holes. In this case, the low energy limit is supplemented by the condition[101],

$$\mu_0 = \frac{\mu}{\alpha'} \text{ fixed} \tag{98}$$

The non-extremal black hole (71) has an outer horizon at $r = \mu$ and an inner horizon at $r = 0$. Therefore, the low-energy limit (91), (98) is a near inner-horizon rather than near outer-horizon limit. As a result the entropies do not agree in general. To see this observe that the effect of the low energy limit (91), (98) is to remove the one from the harmonic functions H_1 and H_5 but leave K unchanged[101]. Since before we take the low energy limit, $H_i(r = \mu) = \cosh^2 \alpha_i$, $i = 1, 5$ and after the low energy limit $H_i(r = \mu) = \sinh^2 \alpha_i$, the entropies of the two configurations differ by a factor of $\tanh \alpha_1 \tanh \alpha_5$. Unless this factor is equal to one, the low energy configuration will contain different number of degrees of freedom. This factor is equal to one in the dilute gas approximation [102]

$$\alpha_1, \alpha_5 \gg 1, \tag{99}$$

and therefore the entropies agree in this approximation. Far from extremality the number of degrees of freedom changes as we go to low energies. In all cases the low energy regime is governed by the BTZ black hole. This result should be contrasted with the result in the previous section. There we also found that $4d$ and $5d$ black holes are connected to the BTZ black hole. All our transformations, however, were isoentropic, and there was no limit involved. We only needed that the supergravity approximation is valid.

Let us finish by presenting a microscopic derivation of the Bekenstein-Hawking entropy formula for extremal black hole (38) using the results of this section. It has been shown by Brown and Henneaux [103] that the asymptotic symmetry group of adS_3 is generated by two copies of the Virasoro algebra with central charge

$$c = \frac{3l}{2G_N^{(3)}} \tag{100}$$

This central charge was also derived through the adS/CFT correspondence in [104]. Therefore, any consistent theory of gravity on adS_3 is conformal field theory with central charge equal to (100).

The generators of the asymptotic Virasoro are related to the mass and angular momentum as

$$\begin{aligned} M &= \frac{1}{l}(L_0 + \bar{L}_0), \\ J &= L_0 - \bar{L}_0 \end{aligned} \tag{101}$$

where we have normalized L_0, L_0 such that they vanish for the massless black hole.

In the case of the $5d$ extremal black hole, and after the low-energy limit is taken, we obtain a geometry of the form $BTZ \times S^3 \times T^4$. One may dimensionally reduce over the compact spaces to obtain the BTZ black hole and a set of matter fields. The BTZ black hole is asymptotically adS_3 so quantum theory in this space is described by a CFT. We can calculate the central charge using (93), (95). The result is

$$c = 6N_1N_5 \tag{102}$$

This is the same value as the one we found in section 4.1! In addition, from (96) we obtain $L_0 = J = N_K$, $\bar{L}_0 = 0$. Thus, we get the same description as in the D-brane side. This is the same unitary CFT but we are now at strong coupling. Therefore, Cardy's formula apply and, (for large black holes, so $N_K \gg 1$) we get correctly (97).

This counting of states generalizes immediately to non-extremal BTZ black holes [105, 106][8]. (From (101) we get L_0, $\bar{L}_0$ in terms of M and J. We also know c from (100). Applying Cardy's formula we get the Bekenstein-Hawking entropy formula). A crucial point is that in order for Cardy's formula to apply we need the CFT to be unitary. The BTZ black hole, however, induces a Liouville theory at spatial infinity[108, 109]. This means that the effective central charge is equal to one[114] instead of $c = 2l/3G_N^{(3)}$, and one does not get correctly the Bekenstein-Hawking entropy formula (see [113] for further discussion). We argued that for the case we are discussing we have a unitary CFT because of the connection to D-branes. We find likely that the CFT corresponding to the BTZ is unitary only when the latter is embedded in string theory.

Acknowledgments

I would like to thank Jan de Boer for reading the manuscript and for discussions and comments. Research supported by the Netherlands Organization for Scientific Research (NWO).

References

1. S. Doplicher, K. Fredenhagen and J. Roberts, *The Quantum Structure of Spacetime at the Planck Scale and Quantum Fields*, Commun. Math. Phys. **172** (1995) 187-220.
2. D. Christodoulou, Phys. Rev. Lett. **25** (1970) 1596-1597;
   D. Christodoulou and R. Ruffini, Phys. Rev. **D4** (1971) 3552;
   R. Penrose and R. Floyd, Nature **229** (1971) 77;
   S. Hawking, Phys. Rev. Lett. **26** (1971) 1344;
   B. Carter, Nature **238** (1972) 71.
3. J.M. Bardeen, B. Carter and S.W. Hawking, *The four laws of black hole mechanics*, Commun. Math. Phys. **31** (1973) 161-170.
4. J.D. Bekenstein, *Black Holes and the Second Law*, Lett. Nuov. Cimento **4** (1972) 737; *Black Holes and Entropy*, Phys. Rev. **D7** (1973) 2333; *Generalized second law of thermodynamics in black-hole physics*, Phys. Rev. **D9** (1974) 3292.
5. S.W. Hawking, *Black hole explosions?*, Nature **248** (1974) 30; *Particle Creation by Black Holes* Commun. Math. Phys. **43** (1975) 199.
6. C.M. Hull and P.K. Townsend, *Unity of Superstring Dualities* Nucl. Phys. **B438** (1995) 109, hep-th/9410167.

[8] A different counting of the BTZ microstates was presented in [107]. There it was used the fact that three dimensional gravity is topological. The Einstein action can be rewritten as a Chern-Simons action[110, 111]. A Chern-Simons theory on a manifold with a boundary induces a WZW model in the boundary[112]. The degrees of freedom in the boundary are would-be gauge degrees of freedom that cannot be gauged away because of the boundary. Assuming that the horizon is a boundary and imposing certain boundary condition one gets that the boundary degrees of freedom can account for the black hole entropy[107]. A problem with this derivation is that some of the states counted have negative norms.

7. E. Witten, *String Theory Dynamics In Various Dimensions* Nucl. Phys. **B443** (1995) 85, hep-th/9503124.
8. J. Polchinski, *D-branes and RR-charges* Phys. Rev. Lett. **75** (1995) 4724, hep-th/9510017.
9. G. Horowitz, *The Dark Side of String Theory: Black Holes and Black Strings*, hep-th/9210119.
10. J. Maldacena, *Black holes in string theory*, hep-th/9607235.
11. A. Peet, *The Bekenstein Formula and String Theory (N-brane Theory)*, Class. Quant. Grav. **15** (1998) 3291-3338, hep-th/9712253.
12. M. Bañados, C. Teitelboim and J. Zanelli, *The Black Hole in Three Dimensional Space Time*, Phys. Rev. Lett. **69** (1992) 1849, hep-th/9204099.
13. M. Bañados, M. Henneaux, C. Teitelboim and J. Zanelli, *Geometry of the* $(2+1)$ *black hole*, Phys. Rev. **D48** (1993) 1506.
14. J. Polchinski, *TASI Lectures on D-Branes*, hep-th/9611050.
15. P.K. Townsend, *The eleven-dimensional supermembrane revisited*, Phys. Lett. **B350** (1995) 184-187, hep-th/9501068.
16. P. Horava, E. Witten, *Heterotic and Type I string dynamics from eleven dimensions* Nucl. Phys. **B460** (1996) 506, hep-th/9510209, and *Eleven-Dimensional Supergravity on a Manifold with Boundary*, Nucl. Phys. **B475** (1996) 94-114, hep-th/9603142.
17. E. Cremmer, B. Julia and J. Scherk, *Supergravity in eleven dimensions*, Phys. Lett. **B76** (1978) 409.
18. T. Buscher, *A symmetry of the string background field equations*, Phys. Lett. **B194** (1987) 59.
19. M. Roček and E. Verlinde, *Duality, quotients and currents*, Nucl. Phys. **B373** 1992 630-646, hep-th/9110053.
20. E. Alvarez, L. Alvarez-Gaume, J.L.F. Barbon and Y. Lozano, *Some global aspects of duality in string theory*, Nucl.Phys. **B415** (1994) 71-100, hep-th/9309039.
21. T. Buscher, *Path integral derivation of quantum duality in non-linear sigma models*, Phys. Lett. **B201** (1988) 466.
22. J. De Jaegher, J. Raymaekers, A. Sevrin and W. Troost, *Dilaton transformation under abelian and non-abelian T-duality in the path-integral approach*, hep-th/9812207.
23. G. Horowitz and D. Welch, *Duality Invariance of the Hawking Temperature and Entropy*, Phys. Rev. **D49** (1994) 590, hep-th/9308077.
24. G. Moore, *Finite in All Directions*, hep-th/9305139; C.M. Hull and B. Julia, *Duality and Moduli Spaces for Time-Like Reductions*, Nucl. Phys. **B534** (1998) 250-260, hep-th/9803239; C.M. Hull, *Timelike T-Duality, de Sitter Space, Large N Gauge Theories and Topological Field Theory*, J.High Energy Phys. 9807 (1998) 021, hep-th/9806146.
25. E. Cremmer, I.V. Lavrinenko, H. Lu, C.N. Pope, K.S. Stelle and T.A. Tran, *Euclidean-signature Supergravities, Dualities and Instantons*, Nucl. Phys. **B534** (1998) 40-82, hep-th/9803259.
26. B. Julia and H. Nicolai, *Null Killing Vector Dimensional Reduction and Galilean Geometrodynamics*, Nucl. Phys. **B439** (1995) 291, hep-th/9412002.
27. E. Alvarez, L. Alvarez-Gaume and I. Bakas, *T-duality and Space-time Supersymmetry*, Nucl. Phys. **B457** (1995) 3, hep-th/9507112.
28. I. Bakas, *Spacetime interpretation of S-duality and supersymmetry violations of T-duality*, Phys. Lett. **343B** (1995) 103, hep-th/9410104.

29. I. Bakas and K. Sfetsos, *T-duality and world-sheet supersymmetry*, Phys. Lett. **B349** (1995) 448-457, hep-th/9502065.
30. R. Gregory, J.A. Harvey and G. Moore, *Unwinding strings and T-duality of Kaluza-Klein and H-Monopoles*, hep-th/9708086.
31. M.J. Duff, R.R. Khuri, J.X. Lu, *String solitons*, Phys. Rept. 259 (1995) 213-326, hep-th/9412184.
32. K. S. Stelle, *Lectures on Supergravity p-branes*, hep-th/9701088.
33. D. Youm, *Black Holes and Solitons in String Theory*, hep-th/9710046.
34. G.T. Horowitz and A. Strominger, *Black strings and p-branes*, Nucl. Phys. **B360** (1991) 197.
35. E. Bergshoeff, C.M. Hull and T. Ortin, *Duality in the Type–II Superstring Effective Action*, Nucl. Phys. **B451** (1995) 547, hep-th/9504081.
36. R. Myers and M. Perry, *Black holes in higher dimensional spacetimes*, Annals Phys. **172** (1986) 304.
37. J. Maharana and J. Schwarz, *Noncompact Symmetries in String Theory*, Nucl. Phys. **B390** (1993) 3, hep-th/9207016; A. Sen, *Electric magnetic duality in string theory*, Nucl. Phys. **B404** (1993) 109, hep-th/9207053.
38. E. Witten and D. Olive, *Supersymmetry algebras that include topological charges*, Phys. Lett. **78B** (1978) 97.
39. M.J. Duff and K.S. Stelle, *Multimembrane solutions of $D = 11$ supergravity*, Phys. Lett. **B253** (1991) 113-118.
40. R. Güven, *Black p-brane solutions of $D = 11$ supergravity theory*, Phys. Lett. **B276** (1991) 49-55.
41. G. Papadopoulos and P.K. Townsend, *Intersecting M-branes*, Phys. Lett. **380B** (1996) 273, hep-th/9603087.
42. A. Tseytlin, *Harmonic superpositions of M-branes*, Nucl. Phys. **B475** (1996) 149, hep-th/9604035 and *No-force condition and BPS combinations of p-branes in 11 and 10 dimensions*, Nucl. Phys. **B487** (1997) 141, hep-th/9609212.
43. J.P. Gauntlett, D.A. Kastor and J. Traschen, *Overlapping Branes in M-Theory*, Nucl. Phys. **B478** (1996) 544, hep-th/9604179.
44. M. Cvetič and A.A. Tseytlin, *Non-extreme black holes from non-extreme intersecting M-branes* Nucl Phys. **B478** (1996) 431, hep-th/9606033.
45. E. Bergshoeff, M. de Roo, E. Eyras, B. Janssen and J. P. van der Schaar, *Multiple Intersections of D-branes and M-branes*, Nucl. Phys. **B494** (1997) 119-143, hep-th/9612095.
46. I.Ya. Aref'eva and O.A. Rytchkov, *Incidence Matrix Description of Intersecting p-brane Solutions*, hep-th/9612236; I.Ya. Arefeva, K.S. Viswanathan, A.I. Volovich and I.V. Volovich, *Composite p-branes in various dimensions*, Nucl. Phys. Proc. Suppl. **56B** (1997) 52-60, hep-th/9701092; I.Ya. Aref'eva, M.G. Ivanov and I.V. Volovich, *Non-extremal Intersecting p-branes in Various Dimensions*, Phys. Lett. **B406** (1997) 44-48, hep-th/9702079.
47. R. Argurio, F. Englert and L. Houart, *Intersection Rules for p-Branes*, Phys. Lett. **398B** (1997) 61, hep-th/9701042.
48. N. Ohta, *Intersection Rules for Non-Extreme p-Branes*, Phys. Lett. **B403** (1997) 218-224, hep-th/9702095.
49. E. Bergshoeff, M. de Roo, E. Eyras, B. Janssen and J.P. van der Schaar, *Intersections involving waves and monopoles in eleven dimensions*, Class. Quant. Grav. **14** (1997) 2757, hep-th/9704120.

50. N. Ohta and J.-G. Zhou, *Towards the Classification of Non-Marginal Bound States of M-Branes and Their Construction Rules*, Int. J. Mod. Phys. **A13** (1998) 2013-2046, hep-th/9706153.
51. J.D. Edelstein, L. Tataru and R. Tatar, *Rules for localized overlappings and intersections of p-branes*, High Energy Phys. 9806 (1998) 003, hep-th/9801049.
52. J.P. Gauntlett, *Intersecting branes*, hep-th/9705011.
53. R.R. Khuri, *A Comment on String Solitons*, Phys. Rev. **D48** (1993) 2947, hep-th/9305143.
54. H.J. Boonstra, B. Peeters and K. Skenderis, *Branes intersections, anti-de Sitter spacetimes and dual superconformal theories*, Nucl. Phys. **B533** (1998) 127-162, hep-th/9803231.
55. A. Strominger and C. Vafa, *Microscopic origin of the Bekenstein-Hawking entropy*, Phys. Lett. **B379** (1996) 99, hep-th/9601029.
56. C.G. Callan and J.M. Maldacena, *D-brane approach to black hole quantum mechanics*, Nucl. Phys. **B472** (1996) 591, hep-th/9602043.
57. G.T. Horowitz and J. Polchinski, *A Correspondence Principle for Black Holes and Strings*, Phys. Rev. **D55** (1997) 6189, hep-th/9612146.
58. K. Sfetsos and K. Skenderis, *Microscopic derivation of the Bekenstein-Hawking entropy formula for non-extremal black holes*, Nucl. Phys. **B517** (1998) 179-204, hep-th/9711138.
59. J.M. Maldacena, *The Large N Limit of Superconformal Field Theories and Supergravity*, Adv. Theor. Math. Phys. **2** (1998) 231, hep-th/9711200.
60. R. Kallosh, A. Linde, T. Ortin, A. Peet and A. Van Proeyen, *Supersymmetry as a Cosmic Censor*, Phys. Rev. **D46** (1992) 5278-5302, hep-th/9205027; M. Cvetic and D. Youm, *Dyonic BPS Saturated Black Holes of Heterotic String on a Six-Torus*, Phys. Rev. **D53** (1996) 584-588, hep-th/9507090; M. Cvetic and A.A. Tseytlin, *General class of BPS saturated dyonic black holes as exact superstring solutions*, Phys. Lett. **B366** (1996) 95, hep-th/9510097, and *Solitonic Strings and BPS Saturated Dyonic Black Holes*, Phys. Rev. **D53** (1996) 5619-5633; Erratum-ibid. **D55** (1997) 3907, hep-th/9512031.
61. J. Maldacena and A. Strominger, *Statistical Entropy of four-dimensional extremal black holes*, Phys. Rev. Lett. **77** (1996) 428-429, hep-th/9603060.
62. C. Johnson, R. Khuri and R. Myers, *Entropy of 4D extremal black holes*, Phys. Lett. **B378** (1996) 78-86, hep-th/9603061.
63. J.C. Breckenridge, R.C. Myers, A.W. Peet, C. Vafa, *D–branes and Spinning Black Holes*, Phys. Lett. **B391** (1997) 93-98, hep-th/9602065.
64. J. C. Breckenridge, D. A. Lowe, R. C. Myers, A. W. Peet, A. Strominger and C. Vafa, *Macroscopic and Microscopic Entropy of Near-Extremal Spinning Black Holes*, Phys. Lett. **B381** (1996) 423-426, hep-th/9603078.
65. M. Cvetic and D. Youm, *Entropy of Non-Extreme Charged Rotating Black Holes in String Theory*, Phys. Rev. **D54** (1996) 2612-2620, hep-th/9603147.
66. E. Witten, *Bound States Of Strings And p-Branes*, Nucl. Phys. **B460** (1996) 335, hep-th/9510135.
67. A. Sen, *A Note on Marginally Stable Bound States in Type II String Theory*, Phys. Rev. **D54** (1996) 2964-2967, hep-th/9510229; and *U-duality and Intersecting D-branes*, Phys. Rev. **D53** (1996) 2874-2894, hep-th/9511026.
68. C. Vafa, *Gas of D-Branes and Hagedorn Density of BPS States*, Nucl. Phys. **B463** (1996) 415-419, hep-th/9511088.
69. M. Bershadsky, V. Sadov and C. Vafa, *D-Branes and Topological Field Theories*, Nucl. Phys. **B463** (1996) 420-434, hep-th/9511222.

70. C. Vafa, *Instantons on D-branes*, Nucl. Phys. **B463** (1996) 435-442, hep-th/9512078.
71. J.L. Cardy, *Operator content of two-dimensional conformally invariant theories*, Nucl. Phys. **B270** (1986) 186.
72. M. Douglas, J. Polchinski and A. Strominger, *Probing Five-Dimensional Black Holes with D-Branes*, J.High Energy Phys. 9712 (1997) 003, hep-th/9703031.
73. S. Hyun, *U-duality between Three and Higher Dimensional Black Holes*, hep-th/9704005.
74. H.J. Boonstra, B. Peeters and K. Skenderis, *Duality and asymptotic geometries*, Phys. Lett. **411B** (1997) 59, hep-th/9706192, and *Branes and anti-de Sitter spacetimes*, hep-th/9801076.
75. S. Ferrara and R. Kallosh, *Supersymmetry and Attractors*, Phys. Rev. **D54** (1996) 1514-1524, hep-th/9602136.
76. G. Horowitz, J. Maldacena and A. Strominger, *Nonextremal black hole microstates and U-duality*, Phys. Lett. **B383** (1996) 151-159, hep-th/9603109.
77. M. Cvetic and C.M. Hull, *Black Holes and U-Duality*, Nucl. Phys. **B480** (1996) 296-316, hep-th/9606193.
78. S. Ferrara and J.M. Maldacena, *Branes, central charges and U-duality invariant BPS conditions*, Class. Quant. Grav. **15** (1998) 749-758, hep-th/9706097.
79. G. Horowitz and D. Welch, *Exact Three Dimensional Black Holes in String Theory*, Phys. Rev. Lett. **71** (1993) 328, hep-th/9302126.
80. E. Bergshoeff and K. Behrndt, *D-Instantons and asymptotic geometries*, Class. Quant. Grav. **15** (1998) 1801-1813, hep-th/9803090.
81. G.W. Gibbons, *Wrapping Branes in Space and Time*, hep-th/9803206.
82. E. Teo, *Statistical entropy of charged two-dimensional black holes*, Phys. Lett. **B430** (1998) 57-62, hep-th/9803064.
83. M.D. McGuigan, C.R. Nappi and S.A. Yost, *Charged Black Holes in Two-Dimensional String Theory*, Nucl. Phys. *B375* (1992) 421-452, hep-th/9111038; G.W. Gibbons and M. Perry, *The Physics of 2-d Stringy Spacetimes*, Int.J.Mod.Phys. **D1** (1992) 335-354, hep-th/9204090; C.R. Nappi and A. Pasquinucci, *Thermodynamics of Two-Dimensional Black-Holes*, Mod. Phys. Lett. **A7** (1992) 3337-3346, gr-qc/9208002.
84. Y. Satoh, *BTZ black holes and the near-horizon geometry of higher-dimensional black holes*, hep-th/9810135.
85. N. Kaloper, *Miens of The Three Dimensional Black Hole* Phys. Rev. **D48** (1993) 2598, hep-th/9303007.
86. A. Ali and A. Kumar, *$O(\tilde{d},\tilde{d})$ Transformations and 3D Black Hole* Mod. Phys. Lett. **A8** (1993) 2045-2052.
87. I. Antoniadis, C. Bachas and A. Sagnotti, *Gauged supergravity vacua in string theory*, Phys. Lett. **B235** (1990) 255; S.B. Giddings, J. Polchinski and A. Strominger, *Four-dimensional black holes in string theory*, Phys. Rev. **D48** (1993) 5784, hep-th/9305083..
88. H.J. Boonstra, K. Skenderis and P.K. Townsend, *The domain-wall/QFT correspondence*, J.High Energy Phys. 9901 (1999) 003, hep-th/9807137.
89. G.W. Gibbons and P.K. Townsend, *Vacuum interpolation in supergravity via super p-branes*, Phys. Rev. Lett. **71** (1993) 3754, hep-th/9307049.
90. G. 't Hooft, *A planar diagram theory for strong interactions*, Nucl. Phys. **B72** (1974) 461.
91. I.R. Klebanov, *World Volume Approach to Absorption by Non-dilatonic Branes*, Nucl. Phys. **B496** (1997) 231, hep-th/9702076; S.S. Gubser, I.R. Klebanov, A.A.

Tseytlin, *String Theory and Classical Absorption by Threebranes*, Nucl. Phys. **B499** (1997) 217, hep-th/9703040; S.S. Gubser, I.R. Klebanov, *Absorption by Branes and Schwinger Terms in the World Volume Theory*, Phys. Lett. **413B** (1997) 41, hep-th/9708005.

92. S.S. Gubser, I.R. Klebanov and A.M. Polyakov, *Gauge Theory Correlators from Non-Critical String Theory*, Phys. Lett. **B428** (1998) 105, hep-th/9802109.
93. E. Witten, *Anti-de Sitter space and Holography*, Adv. Theor. Math. Phys. **2** (1998) 253, hep-th/9802150.
94. G. 't Hooft, *Dimensional reduction in Quantum gravity*, Class. Quant. Grav. **11** (1994) 621, gr-qc/9310006.
95. L. Susskind, *The World as a Hologram*, J. Math. Phys. **36** (1995) 6377, hep-th/9409089.
96. L. Susskind and E. Witten, *The Holographic Bound in Anti-de Sitter Space*, hep-th/9805114.
97. N. Itzhaki, J. Maldacena, J. Sonnenschein and S. Yankielowicz, *Supergravity and The Large N Limit of Theories With Sixteen Supercharges*, Phys. Rev. **D58** (1998) 046004, hep-th/9802042.
98. A.W. Peet, J. Polchinski, *UV/IR Relations in AdS Dynamics*, hep-th/9809022.
99. V. Balasubramanian and F. Larsen, *Near Horizon Geometry and Black Holes in Four Dimensions* Nucl. Phys. **B528** (1998) 229, hep-th/9802198.
100. M. Cvetic and F. Larsen, *Near Horizon Geometry of Rotating Black Holes in Five Dimensions*, Nucl. Phys. **B531** (1998) 239-255 and *Microstates of Four-Dimensional Rotating Black Holes from Near-Horizon Geometry*, hep-th/9805146.
101. J. Maldacena and A. Strominger, *AdS_3 black holes and a stringy exclusion principle*, hep-th/9804085.
102. J. Maldacena, A. Strominger, *Black Hole Greybody Factors and D-Brane Spectroscopy*, Phys. Rev. **D55** (1997) 861-870, hep-th/9609026.
103. J.D. Brown and M. Henneaux, *Central charges in the canonical realization of asymptotic symmetries: An example from three-dimmensional gravity*, Commun. Math. Phys. **104** (1986) 207.
104. M. Henningson and K. Skenderis, *The Holographic Weyl anomaly*, J.High Energy Phys. 9807 (1998) 023, hep-th/9806087; *Holography and the Weyl anomaly*, hep-th/9812032.
105. A. Strominger, *Black Hole Entropy from Near-Horizon Microstates*, J.High Energy Phys. 9802 (1998) 009, hep-th/9712251.
106. D. Birmingham, I. Sachs and S. Sen, *Entropy of Three-Dimensional Black Holes in String Theory*, Phys. Lett. **B424** (1998) 275-280, hep-th/9801019.
107. S. Carlip, *The Statistical Mechanics of the (2+1)-Dimensional Black Hole*, Phys. Rev. **D51** (1995) 632, gr-qc/9409052 and *The Statistical Mechanics of the Three-Dimensional Euclidean Black Hole*, **D55** (1997) 878, gr-qc/9606043.
108. S. Carlip, *Inducing Liouville theory from topologically massive gravity*, Nucl. Phys. **B362** (1991) 111-124.
109. O. Coussaert, M. Henneaux and P. van Driel, *The asymptotic dynamics of three-dimensional Einstein gravity with negative cosmological constant*, Class. Quant. Grav. **12** (1995), 2961-2966.
110. A. Achúcarro and P. K. Townsend, *A Chern-Simons action for three-dimensional anti-de Sitter supergravity theories*, Phys. Lett. **B180** (1986) 89.
111. E. Witten, $(2+1)$-*dimensional gravity as an exactly soluble system*, Nucl. Phys. **B311** (1988) 46.

112. G. Moore and N. Seiberg, *Taming the conformal zoo*, Phys. Lett. **B220** (1989) 422; S. Elitzur, G. Moore, A. Schwimmer and Nathan Seiberg, *Remarks on the canonical quantization of the Chern-Simons-Witten theory*, Nucl. Phys. **B326** (1989) 108.
113. S. Carlip, *What We Don't Know about BTZ Black Hole Entropy*, Class. Quant. Grav. **15** (1998) 3609-3625, hep-th/9806026.
114. D. Kutasov and N. Seiberg, *Number of degrees of freedom, density of states and tachyons in string theory and CFT*, Nucl. Phys. **B358** (1991) 600-618.

Gravitational waves and massless particle fields

Jan Willem van Holten

NIKHEF, Amsterdam NL

Abstract. These lectures address the planar gravitational wave solutions of general relativity in empty space-time, and analyze the motion of test particles in the gravitational wave field. Next we consider related solutions of the Einstein equations for the gravitational field accompanied by long-range wave fields of scalar, spinor and vector type, corresponding to massless particles of spin $s = (0, \frac{1}{2}, 1)$. The motion of test masses in the combined gravitational and scalar-, spinor- or vector wave fields is discussed.

1 Planar gravitational waves

a. Planar wave solutions of the Einstein equations

Planar gravitational wave solutions of the Einstein equations have been known since a long time [1]-[3]. In the following I discuss unidirectional solutions of this type, propagating along a fixed light-cone direction; thus the fields depend only on one of the light-cone co-ordinates (u, v), here taken transverse to the x-y-plane:

$$u = ct - z, \qquad v = ct + z. \tag{1}$$

Such gravitational waves can be described by space-time metrics

$$g_{\mu\nu}dx^\mu dx^\nu = -\,dudv - K(u,x,y)du^2 + dx^2 + dy^2 = -\,c^2 d\tau^2, \tag{2}$$

or similar solutions with the roles of v and u interchanged. If the space-time is asymptotically minkowskian. With the metric (2), the connection co-efficients become

$$\Gamma_{uu}{}^{v} = K_{,u}, \qquad \Gamma_{uu}{}^{x} = \frac{1}{2}\Gamma_{xu}{}^{v} = \frac{1}{2}K_{,x}, \qquad \Gamma_{uu}{}^{y} = \frac{1}{2}\Gamma_{yu}{}^{v} = \frac{1}{2}K_{,y}. \tag{3}$$

All other components vanish. The corresponding Riemann tensor has non-zero components

$$\begin{aligned} R_{uxux} &= -\frac{1}{2}K_{,xx}, \qquad R_{uyuy} = -\frac{1}{2}K_{,yy}, \\ R_{uxuy} &= R_{uyux} = -\frac{1}{2}K_{,xy}. \end{aligned} \tag{4}$$

The only non-vanishing component of the Ricci tensor then is

$$R_{uu} = -\frac{1}{2}\left(K_{,xx} + K_{,yy}\right) \equiv -\frac{1}{2}\Delta_{trans}K. \tag{5}$$

Here the label *trans* refers to the transverse (x, y)-plane, with the z-axis representing the longitudinal direction. In complex notation

$$\zeta = x + iy, \qquad \bar{\zeta} = x - iy, \tag{6}$$

the Einstein equations in vacuo become

$$R_{\mu\nu} = 0 \qquad \Leftrightarrow \qquad K_{,\zeta\bar{\zeta}} = 0. \tag{7}$$

The general solution of this equation reads

$$K(u,\zeta,\bar{\zeta}) = f(u;\zeta) + \bar{f}(u;\bar{\zeta}) \sum_{n=0}^{\infty} \int_{-\infty}^{\infty} \frac{dk}{2\pi}\left(\epsilon_n(k)e^{-iku}\zeta^n + \bar{\epsilon}_n(k)e^{iku}\bar{\zeta}^n\right). \tag{8}$$

Note that the terms with $n = 0, 1$ correspond to vanishing Riemann tensor: $R_{\mu\nu\kappa\lambda} = 0$; therefore they represent flat Minkowski space-time in a non-standard choice of co-ordinates. For this reason we adopt the convention that $\epsilon_0 = \epsilon_1 = 0$, which is just a choice of gauge.

b. Geodesics of planar-wave space-times

We proceed to solve the geodesic equation in the gravity-wave space-time (2) along the lines of ref.[5]:

$$\ddot{x}^\mu = -\Gamma_{\nu\lambda}^{\ \ \mu}\dot{x}^\nu\dot{x}^\lambda. \tag{9}$$

Here the overdot denotes a proper-time derivative. The proper-time Hamiltonian satisfies a constraint imposed by eq.(2):

$$\begin{aligned} H &= g_{\mu\nu}\dot{x}^\mu\dot{x}^\nu \\ &= -\dot{u}\dot{v} - K(u,x,y)\,\dot{u}^2 + \dot{x}^2 + \dot{y}^2 = -c^2. \end{aligned} \tag{10}$$

Because the metric is covariantly constant, the hamiltonian is a constant of motion:

$$\dot{H} = 0. \tag{11}$$

This can be checked directly from the geodesic equation (9). Also, as v is a cyclic co-ordinate, its conjugate momentum is conserved:

$$\ddot{u} = 0, \tag{12}$$

with the simple solution $\dot{u} \equiv \gamma =$ constant. Again, this agrees with the geodesic equation, as there is no non-vanishing connection component in the u-direction: $\Gamma_{\nu\lambda}^{\ \ u} = 0$.

Only the equations of motion in the x-y-plane depend on the specific wave potential $K(u,x,y)$:

$$\ddot{x} = -\frac{1}{2} K_{,x} \dot{u}^2 = -\frac{\gamma^2}{2} K_{,x},$$
$$\ddot{y} = -\frac{1}{2} K_{,y} \dot{u}^2 = -\frac{\gamma^2}{2} K_{,y}, \tag{13}$$

Eqs. (10)-(13) specify completely the motion of a test particle, with the conservation of H taking the place of the equation for the acceleration in the v-direction:

$$\gamma \dot{v} + \gamma^2 K(u,x,y) = \dot{x}^2 + \dot{y}^2 + c^2. \tag{14}$$

If we now add $\dot{z}^2$ to the left- and right-hand side, and remember that

$$\gamma \dot{v} = \dot{u}\dot{v} = c^2 \dot{t}^2 - \dot{z}^2, \tag{15}$$

we can rewrite the hamiltonian conservation law as

$$c^2 \dot{t}^2 + \gamma^2 K = c^2 + \dot{\mathbf{r}}^2. \tag{16}$$

Finally, with $\mathbf{v} = d\mathbf{r}/dt = \dot{\mathbf{r}}/\dot{t}$, the equation can be cast into the form

$$\dot{t} = \frac{dt}{d\tau} = \sqrt{\frac{1 - \gamma^2 K/c^2}{1 - \mathbf{v}^2/c^2}} \tag{17}$$

This equation describes relativistic time-dilation as resulting from two effects:
(i) the usual special-relativistic time-dilation from the relative motion of observers in the rest- and laboratory frame, whose time co-ordinates are τ and t, respectively;
(ii) the gravitational redshift resulting from the non-trivial potential K.
Now from the conservation of $\gamma = \dot{u} = c\dot{t} - \dot{z}$ it follows, that

$$\gamma = c\dot{t}\left(1 - \frac{v_z}{c}\right), \tag{18}$$

with $v_z = dz/dt$. Eqs. (17), (18) can then be solved for γ:

$$\frac{\gamma^2}{c^2} = \frac{1}{K + \dfrac{1 - \mathbf{v}^2/c^2}{(1 - v_z/c)^2}}. \tag{19}$$

Thus, for a paticle starting at rest at infinity in an asymptotically minkoskian space-time, we find $\gamma = c$. At the same time we observe that

$$h = K + \frac{1 - \mathbf{v}^2/c^2}{(1 - v_z/c)^2} \tag{20}$$

is conserved. Now we recall that in our conventions K is at least quadratic in the transverse co-ordinates; hence the components $\ddot{x}$ and $\ddot{y}$ of the transverse acceleration vanish for $x = y = 0$. Furthermore $K(u, 0, 0) = 0$, with the result that the origin of the transverse plane moves at constant velocity along the z-axis:

$$\frac{\gamma^2}{c^2} = \frac{1 - v_z/c}{1 + v_z/c} \quad \Leftrightarrow \quad v_z = \frac{1 - \gamma^2/c^2}{1 + \gamma^2/c^2}. \tag{21}$$

In particular, the point at rest in the origin moves along the simple geodesic

$$x^\mu(\tau) = (c\tau, 0, 0, 0). \tag{22}$$

Taking this geodesic as our reference, the solution for the geodesic motion $\bar{x}^\mu(\tau)$ of any other test particle at the same time presents a measure for the geodesic deviation between the worldlines of the two particles.

2 Einstein-scalar waves

Having discussed the planar gravitational waves (2) in empty space we now turn to discuss similar unidirectional wave solutions of the combined system of Einstein gravity and a set of massless self-interacting scalar fields. The solutions of the inhomogeneous and non-linear Einstein equations, with the energy-momentum tensor that of the right- (or left-) moving scalar waves, nevertheless turn out to be a linear superposition of the gravitational field of the scalar waves and the free gravitational wave solutions discussed in the first paragraph.

We introduce a set of massless scalar fields $\sigma^i(x)$, $i = 1, ..., N$, taking values in a manifold with the dimensionless metric $G_{ij}[\sigma]$. In four-dimensional space-time the fields themselves have dimension $[\sigma] = \sqrt{E/l}$; thus, introducing an appropriate length scale $1/f$, in the context of quantum field theory we could write $\sigma^i = \sqrt{\hbar c}\, f\, \eta^i$, with $\eta^i(x)$ a dimensionless field.

The starting point of our analysis is given by the gravitational and σ-model field equations

$$\begin{aligned} \Box^{cov}\, \sigma^i \,+\, \Gamma_{jk}{}^i[\sigma]\, g^{\mu\nu} \partial_\mu \sigma^j \partial_\nu \sigma^k \,&=\, 0, \\ R_{\mu\nu} \,&=\, -\frac{8\pi G}{c^4} G_{ij}[\sigma]\, \partial_\mu \sigma^i \partial_\nu \sigma^j. \end{aligned} \tag{23}$$

Here the covariant d'Alembertian is defined on scalar fields in the standard fashion

$$\Box^{cov} \,=\, \frac{1}{\sqrt{-g}}\, \partial_\mu \sqrt{-g} g^{\mu\nu}\, \partial_\nu,$$

whilst $\Gamma_{ij}{}^k[\sigma]$ denotes the Riemann-Christoffel connection in the target manifold of the scalar fields. These equations can be derived straightforwardly from the combined Einstein-σ-model action, but we will skip the details of that procedure here. Our aim is to construct simultaneous traveling wave solutions of the full set of equations (23). Such solutions are actually quite easy to find. First, the scalar field equation is solved by taking right-moving fields

$$\sigma^i = \sigma^i(u), \tag{24}$$

with no dependence on any other co-ordinate. Next we substitute this solution of the scalar field into the second equation for the corresponding gravitational field. As before, only the uu-component of this equation survives, reading

$$R_{uu} = -\frac{1}{2}\Delta_{trans}K = -\frac{8\pi G}{c^4}G_{ij}[\sigma]\,\partial_u\sigma^i\partial_u\sigma^j. \tag{25}$$

As this is a linear equation, the general solution consists of a linear superposition of a particular solution and the general free gravitational wave of the previous section:

$$K(u,\zeta,\bar{\zeta}) = \frac{8\pi G}{c^4}G_{ij}[\sigma]\,\partial_u\sigma^i\partial_u\sigma^j\bar{\zeta}\zeta + f(u,\zeta) + \bar{f}(u,\bar{\zeta}). \tag{26}$$

Now any specific solution $\sigma^i(u)$ is a map from the real line into the target manifold of the scalar fields. Consider the special case that this curve in the target manifold is a geodesic:

$$\frac{d^2\sigma^i}{du^2} + \Gamma_{jk}^{\ \ i}\frac{d\sigma^j}{du}\frac{d\sigma^k}{du} = 0. \tag{27}$$

Then the quantity

$$I = G_{ij}[\sigma]\frac{d\sigma^i}{du}\frac{d\sigma^j}{du}, \tag{28}$$

generating translations in u, is constant along this curve: $dI/du = 0$. Moreover, for Euclidean manifolds with non-degenerate metric it is positive definite: $I > 0$. Observe, that for manifolds with compact directions (like spheres) the geodesics may be closed; then the corresponding scalar field configurations are periodi.

The special solution for the accompanying gravitational field now becomes

$$K_{scalar}(u,x,y) = \frac{4\pi GI}{c^4}(x^2+y^2), \tag{29}$$

to which an arbitrary free gravitational wave solution can be added. In this special case, upon inserting K_{scalar} into eqs.(13) the transversal equations of motion of a test mass take the particularly simple form:

$$\ddot{x} = -\frac{4\pi GI\gamma^2}{c^4}x, \qquad \ddot{y} = -\frac{4\pi GI\gamma^2}{c^4}y. \tag{30}$$

Thus the test mass executes a simple harmonic motion in the transverse plane, with frequency

$$\omega = \frac{\gamma}{c^2}\sqrt{4\pi GI}. \tag{31}$$

The solutions for the coupled Einstein-scalar field equations discussed here are not the only ones of interest. For example, the gravitational waves accompanying expanding domain walls in a theory with a spontaneously broken global symmetry can be calculated and have been discussed e.g. in [4, 5].

3 Einstein-Dirac waves

In this section we construct wave-solutions for massless chiral fermions coupled to Einstein gravity. As before the waves are unidirectional, and both left- and righthanded fermion solutions, associated with helicity ± 1 quantum states, exist.

To treat fermions in interaction with gravity, it is necessary to introduce the vierbein and spin connection into the formalism. With the local minkowski metric $\eta = \text{diag}(+1,+1,+1,-1)$, the vierbein is a local lorentz vector of 1-forms $E^a(x) = dx^\mu e_\mu^{\ a}(x)$ satisfying the symmetric product rule

$$\eta_{ab}\, E^a E^b \;=\; \eta_{ab}\, e_\mu^{\ a} e_\nu^{\ b} dx^\mu dx^\nu \;=\; g_{\mu\nu} dx^\mu dx^\nu. \tag{32}$$

In a convenient local lorentz gauge, the vierbein corresponding to the metric (2) takes the form

$$E^a \;=\; \left(dx, dy, \frac{1}{2}\left((K-1)\,du + dv\right), \frac{1}{2}\left((K+1)\,du + dv\right) \right). \tag{33}$$

The inverse vierbein is defined by the differential operator $\nabla_a = e_a^{\ \mu}\partial_\mu$ such that

$$E^a \nabla_a \;=\; dx^\mu \partial_\mu \tag{34}$$

In components it reads

$$\nabla_a \;=\; \left(\partial_x, \partial_y, -\partial_u + (K+1)\,\partial_v, \partial_u - (K-1)\,\partial_v\right). \tag{35}$$

Next we compute the components of the spin connection $\omega^a_{\ b} = dx^\mu \omega_\mu^{\ a}{}_b$ from the identity

$$dE^a \;=\; \omega^a_{\ b} \wedge E^b. \tag{36}$$

With the vierbein (33) the spin connection has only one component

$$\omega_u^{\ ab} \;=\; -\omega_u^{\ ba} \;=\; \frac{1}{2}\begin{pmatrix} 0 & 0 & K_{,x} & K_{,x} \\ 0 & 0 & K_{,y} & K_{,y} \\ -K_{,x} & -K_{,y} & 0 & 0 \\ -K_{,x} & -K_{,y} & 0 & 0 \end{pmatrix}. \tag{37}$$

In order to construct the dirac operator we introduce a basis for the dirac matrices satisfying $\{\gamma^a, \gamma^b\} = 2\eta^{ab}$, and define a set spinor generators for the lorentz algebra by $\sigma_{ab} = \frac{1}{4}\left[\gamma_a, \gamma_b\right]$. Then the dirac operator is

$$\gamma \cdot D \;=\; \gamma^a \left(\nabla_a - \frac{1}{2}\,\omega_a^{\ bc}\sigma_{bc}\right), \tag{38}$$

The results we need all depend on the property of the light-cone components of the dirac algebra:

$$\gamma^u \;=\; \gamma^a e_a^{\ u} \;=\; -\gamma_3 + \gamma_0. \tag{39}$$

This element of the dirac algebra is nilpotent:

$$(\gamma^u)^2 = 0. \tag{40}$$

The same is true for $\gamma_v = e_v^{\ a}\gamma_a = \frac{1}{2}\gamma^u$. Because of the form of the spin connection (37), the dirac-algebra valued form $\omega^{ab}\sigma_{ab}$ is itself proportional to γ^u; its nilpotency then guarantees that the spin-connection term in the covariant derivative (38) vanishes by itself:

$$\gamma^a \omega_a^{\ bc}\, \sigma_{bc} = \gamma^u \omega_u^{\ bc}\, \sigma_{bc} = 0. \tag{41}$$

Hence the only vestige of curved space-time left in the dirac operator is the inverse vierbein in the contraction of dirac matrices and differential operators:

$$\gamma \cdot D = \gamma^a \nabla_a = \gamma^\mu \partial_\mu$$

$$= i \begin{pmatrix} \partial_u - (K-1)\,\partial_v & \begin{array}{c} -\sigma_1\partial_x - \sigma_2\partial_y \\ -\sigma_3\left(-\partial_u + (K+1)\,\partial_v\right) \end{array} \\ \begin{array}{c} \sigma_1\partial_x + \sigma_2\partial_y \\ +\sigma_3\left(-\partial_u + (K+1)\,\partial_v\right) \end{array} & -\partial_u + (K-1)\,\partial_v \end{pmatrix}. \tag{42}$$

Here we have introduced the following basis for the dirac algebra:

$$\gamma_k = \begin{pmatrix} 0 & -i\sigma_k \\ i\sigma_k & 0 \end{pmatrix}, \; k = 1,2,3; \qquad \gamma^0 = \begin{pmatrix} i1 & 0 \\ 0 & -i1 \end{pmatrix}, \tag{43}$$

with the σ_k the standard pauli matrices. The zero modes of this operator with the property that the energy-momentum tensor only has a non-zero T_{uu} component are flat spinor fields $\psi(u)$ with the property

$$\psi(u) = i\,\gamma^u \begin{pmatrix} \chi(u) \\ 0 \end{pmatrix} = \begin{pmatrix} 1 & -\sigma_3 \\ -\sigma_3 & 1 \end{pmatrix} \begin{pmatrix} \chi(u) \\ 0 \end{pmatrix} = \begin{pmatrix} \chi(u) \\ -\sigma_3\chi(u) \end{pmatrix}, \tag{44}$$

where $\chi(u)$ is a 2-component (pauli) spinor. Indeed, first of all spinors of this type are zero-modes of the dirac operator:

$$\gamma \cdot D\psi = 0. \tag{45}$$

This follows by direct application of the expression (42) to the spinor (44), using the nilpotency of γ^u. Moreover, with this property it also follows that the energy-momentum tensor takes the form

$$T_{\mu\nu} = \frac{1}{8}\overline{\psi}\,(\gamma_\mu D_\nu + \gamma_\nu D_\mu)\,\psi = \frac{1}{4}\,\delta_\mu^u\delta_\nu^u\,\overline{\psi}\,\gamma_u\,\partial_u\,\psi. \tag{46}$$

To see this, first note that the u-component of the covariant derivative D_μ is the only one that does not vanish on $\psi(u)$ in general. We then only have to check that in all remaining cases with $\gamma_\mu \neq \gamma_u$ the spinor $\psi(u)$ (44) gets multiplied by a dirac matrix which can be factorized such as to have a right multiplicator of the form γ^u. Again, as $(\gamma^u)^2 = 0$, $T_{\mu\nu}$ necessarily is of the required form (46).

Finally we remark, that the upper- and lower component of the pauli spinor $\chi(u)$ in our conventions correspond to a negative and positive helicity state, respectively. Thus we find as solutions of the dirac operator in the metric (2) two massless spinor states, corresponding to right-moving zero-modes of the dirac operator with helicity ± 1, respectively.

This solution is self-consistent as the only non-zero component of the energy-momentum tensor is

$$T_{uu}(u) = -\frac{1}{2}\left[\chi^\dagger \chi'\right](u), \tag{47}$$

where the prime denotes a derivative w.r.t. u, and the dagger on χ indicates hermitean conjugation of the 2-component spinor. It is then straightforward to solve the Einstein equation for K in the presence of the energy momentum distribution of the spinor field:

$$K_{spinor}(u,x,y) = -\frac{2\pi G}{c^4}\left[\chi^\dagger \chi'\right](u)\left(x^2+y^2\right). \tag{48}$$

Again, to this particular solution an abitrary free gravitatonal wave can be added. It should be mentioned here, that consistency requires the spinors in the energy momentum tensor (46), (47) to be anti-commuting objects, i.e. if the spinor fields $\chi(u)$ are expanded in a fourier series of massless matter waves, the co-efficients take values in an infinite-dimensional Grassmann algebra. Thus the expression can be given an operational meaning only in the context of quantum theory, by performing some averaging procedure. For example, if the spinors form a condensate such that the kinetic energy $\Sigma \equiv -\langle[\chi^\dagger\chi']\rangle = \text{constant} > 0$, then such a condensate would generate gravitational waves in which test-masses perform harmonic motion of the type (30), (31) with frequency

$$\omega = \frac{\gamma}{c^2}\sqrt{2\pi G \Sigma}. \tag{49}$$

4 Einstein-Maxwell waves

As the last example we consider coupled Einstein-Maxwell fields. We look for solutions of wave-type, using the metric (2). In the absence of masses and charges, the field equations are:

$$R_{\mu\nu} = -\frac{8\pi\varepsilon_0 G}{c^2}\left(F_{\mu\lambda}F_{\nu}^{\ \lambda} - \frac{1}{4}g_{\mu\nu}F^2\right), \qquad D_\lambda F^{\lambda\mu} = 0. \tag{50}$$

With the same metric (2), we also find the same expressions for the components of the connection (3), and the Riemann and Ricci curvature tensors (4), (5). Therefore the left-hand side of the Einstein eqn. (50) is fixed in terms of the potential $K(u,y,z)$.

As concerns the Maxwell equations, the covariant derivative

$$D_\lambda F^{\lambda\mu} = \partial_\lambda F^{\lambda\mu} + \Gamma_{\lambda\nu}{}^{\lambda} F^{\nu\mu} + \Gamma_{\lambda\nu}{}^{\mu} F^{\lambda\nu} \tag{51}$$

reduces to the first term on the r.h.s., an ordinary four-divergence; this happens because in the last term the even connection is contracted with the odd field-strength tensor, whilst the middle term contains a trace over an upper and a lower index of the connection, which vanishes in our case.

Thus the Maxwell equation reduces to the same expression as in minkowski space-time, and it has the same wave solutions. We consider an elementary wave solution, which in terms of the co-ordinate system (2) is described by the vector potential

$$A_\mu = (\mathbf{a} \sin k(ct - z), 0, 0), \tag{52}$$

with the light-cone components vanishing, and with $\mathbf{a}$ a constant transverse vector: $a_z = 0$. Of course, arbitrary solution can be constructed from the elementary waves (52) by linear superposition. With $u = ct - z$ and $\omega = kc$ the angular frequency of the wave, the electric and magnetic fields are

$$\mathbf{E}_k(u) = \omega \mathbf{a} \cos ku, \qquad \mathbf{B}_k(u) = \mathbf{k} \times \mathbf{a} \cos ku. \tag{53}$$

As usual for e.m. waves, $|\mathbf{E}_k(0)| = c|\mathbf{B}_k(0)|$, and $\mathbf{E}_k \cdot \mathbf{B}_k = 0$. Indeed, the only non-zero components of the full field strength are

$$F_{ui} = -F_{iu} = ka_i \cos ku, \qquad i = (x, y), \tag{54}$$

all others vanishing. It is now straightforward to compute the stress-energy tensor components of the electro-magnetic field, with the result

$$T_{uu} = \varepsilon_0 c^2 k^2 \mathbf{a}^2 \cos^2 ku, \tag{55}$$

and all other components zero. The Einstein-Maxwell equations then reduce to

$$\Delta_{trans} K = \frac{16\pi\varepsilon_0 G}{c^2} k^2 \mathbf{a}^2 \cos^2 ku. \tag{56}$$

This has the special solution

$$K_{em} = \frac{4\pi\varepsilon_0 G}{c^2} k^2 \mathbf{a}^2 \cos^2 ku \left(x^2 + y^2\right) = \frac{4\pi\varepsilon_0 G}{c^4} \mathbf{E}_k^2(u) \bar{\zeta}\zeta. \tag{57}$$

In view of the linearity of eq.(56), the general solution is a superposition of such special solutions and arbitrary free gravitational waves of the type (8):

$$K(u, \zeta, \bar{\zeta}) = K_{em}(u, \zeta, \bar{\zeta}) + f(u, \zeta) + \bar{f}(u, \bar{\zeta}). \tag{58}$$

Next we turn to the motion of a test particle with mass m and charge q in the background of these gravitational and electro-magnetic fields. These equations are modified to take into account the Lorentz force on the test charge:

$$\ddot{x}^\mu + \Gamma_{\nu\lambda}{}^\mu \dot{x}^\nu \dot{x}^\lambda = \frac{q}{m} F^\mu{}_\nu \dot{x}^\nu. \tag{59}$$

With the only non-zero covariant components of $F_{\mu\nu}$ given by eq.(54), there are no contravariant components in the lightcone direction u. As a result the equation for u is not modified, and we again find

$$\dot{u} = \gamma = \text{const.} \tag{60}$$

This also follows, because the electro-magnetic forces do not change the proper-time hamiltonian:

$$\begin{aligned} H &= g_{\mu\nu} \dot{x}^\mu \dot{x}^\nu \\ &= = -\dot{u}\dot{v} - K(u, y, z)\,\dot{u}^2 + \dot{y}^2 + \dot{z}^2 = -c^2, \end{aligned} \tag{61}$$

except that $K(u, x, y)$ now is given by the modified expression (58). Therefore v is still a cyclic co-ordinate and equation (14) for $\dot{v}$ again follows from the conservation of H:

$$\gamma\dot{v} + \gamma^2 K(u, x, y) = \dot{x}^2 + \dot{y}^2 + c^2. \tag{62}$$

As a result we find in this case the same formal expressions for the solution of the equations of motion in the time-like and longitudinal directions:

$$\dot{t} = \frac{dt}{d\tau} = \sqrt{\frac{1 - \gamma^2 K/c^2}{1 - \mathbf{v}^2/c^2}}, \tag{63}$$

whilst

$$h = K + \frac{1 - \mathbf{v}^2/c^2}{(1 - v_z/c)^2} \tag{64}$$

is again a constant of motion. In both cases of course K now is the full solution (58).

Manifest changes in the equations of motion are obtained in the transverse directions:

$$\ddot{\boldsymbol{\xi}} = -\frac{\gamma^2}{2} \nabla_\xi K - \frac{q\gamma}{m} k\mathbf{a} \cos ku, \tag{65}$$

where $\boldsymbol{\xi} = (x, y)$ is a transverse vector and ∇_ξ is the gradient in the transverse plane. If we take for K the special solution (57), we find the conservation law

$$\frac{4\pi\varepsilon_0 G}{c^2} k^2 \mathbf{a}^2 \boldsymbol{\xi}^2 \cos^2 ku + \frac{1 - \mathbf{v}^2/c^2}{(1 - v_z/c)^2} = h = \text{const.} \tag{66}$$

Inserting the explicit form of $u(\tau) = \gamma\tau$, eqs.(65) then take the form

$$\ddot{\boldsymbol{\xi}} = -\frac{4\pi\varepsilon_0 G}{c^2}\,\gamma^2 k^2 \mathbf{a}^2 \cos^2(\gamma k\tau)\,\boldsymbol{\xi} - \frac{q\gamma}{m}\,\mathbf{k}\mathbf{a}\cos(\gamma k\tau). \tag{67}$$

Equivalently, we can use u instead of τ as the independent variable:

$$\frac{d^2\boldsymbol{\xi}}{du^2} = -\frac{4\pi\varepsilon_0 G}{c^2}\,k^2\mathbf{a}^2\cos^2(ku)\,\boldsymbol{\xi} - \frac{q}{m\gamma}\,\mathbf{k}\mathbf{a}\cos ku. \tag{68}$$

Clearly, it is useful to decompose $\boldsymbol{\xi}$ into components parallel and orthogonal to the electric field $\mathbf{E}_k$, which in our choice of electro-magnetic gauge is the same as that of the vector potential $\mathbf{a}$:

$$\boldsymbol{\xi} = \boldsymbol{\xi}_{\parallel} + \boldsymbol{\xi}_{\perp}, \tag{69}$$

with

$$\boldsymbol{\xi}_{\parallel} = \frac{\boldsymbol{\xi}\cdot\mathbf{a}}{|\mathbf{a}|^2}\,\mathbf{a}, \qquad \boldsymbol{\xi}_{\perp} = \frac{\boldsymbol{\xi}\times\mathbf{a}}{|\mathbf{a}|}. \tag{70}$$

It follows that

$$\begin{aligned} \frac{d^2\boldsymbol{\xi}_{\parallel}}{du^2} &= -\frac{4\pi\varepsilon_0 G}{c^2}\,k^2\mathbf{a}^2\cos^2(ku)\,\boldsymbol{\xi}_{\parallel} - \frac{q}{m\gamma}\,\mathbf{k}\mathbf{a}\cos ku, \\ \frac{d^2\boldsymbol{\xi}_{\perp}}{du^2} &= -\frac{4\pi\varepsilon_0 G}{c^2}\,k^2\mathbf{a}^2\cos^2(ku)\,\boldsymbol{\xi}_{\perp}. \end{aligned} \tag{71}$$

Transforming to the cosine of the double argument, the last equation can be seen to reduce to the standard Mathieu equation:

$$\frac{d^2\boldsymbol{\xi}_{\perp}}{du^2} + \frac{2\pi\varepsilon_0 G}{c^2}\,k^2\mathbf{a}^2\,(1+\cos 2ku)\,\boldsymbol{\xi}_{\perp} = 0, \tag{72}$$

whilst the other equation becomes an inhomogeneous Mathieu equation, with the Lorentz force representing the inhomogeneous term:

$$\frac{d^2\boldsymbol{\xi}_{\parallel}}{du^2} + \frac{2\pi\varepsilon_0 G}{c^2}\,k^2\mathbf{a}^2\,(1+\cos 2ku)\,\boldsymbol{\xi}_{\parallel} = -\frac{q}{m\gamma}\,\mathbf{k}\mathbf{a}\cos ku. \tag{73}$$

Obviously, one may try to find a particular solution to this equation by making an expansion in powers of $\cos ku$. The general solution is a superposition of this special one plus the general solution of the Mathieu equation (72).

A special case is that of *static* crossed electric and magnetic fields, obtained in the limit $k \to 0$. Then the eqs.(72) and (73) reduce to ordinary homogeneous and inhomogeneous harmonic equations:

$$\begin{aligned} \frac{d^2\boldsymbol{\xi}_{\perp}}{du^2} &+ \frac{4\pi\varepsilon_0 G}{c^4}\,\mathbf{E}_0^2\,\boldsymbol{\xi}_{\perp} = 0, \\ \frac{d^2\boldsymbol{\xi}_{\parallel}}{du^2} &+ \frac{4\pi\varepsilon_0 G}{c^4}\,\mathbf{E}_0^2\,\boldsymbol{\xi}_{\parallel} = -\frac{q}{mc\gamma}\,\mathbf{E}_0. \end{aligned} \tag{74}$$

The angular frequency of this harmonic motion is

$$\omega = \sqrt{\frac{4\pi\varepsilon_0 G}{c^2}}\, E_0 = 0.29 \times 10^{-18} E_0 \,(\mathrm{V/m}). \tag{75}$$

Clearly, the Lorentz force due to the constant electric field produces a constant proper-time acceleration of the test charge, but the harmonic gravitational component of the motion is very slow for practically realistic electric fields: periods of a year or less require a field strength of the order of 10^{10} V/m or more.

This work is part of the research program of the Foundation for Fundamental Research of Matter (FOM)

References

1. H. Bondi, Nature 179 (1957), 1072
2. J. Ehlers and W. Kundt, in: *Gravitation: an Introduction to Current Research* (Wiley, N.Y.; 1962), ed. L. Witten
3. D. Kramer, H. Stephani, M. MacCallum and E. Herlt, *Exact Solutions of Einstein's Field Equations* (Cambridge Univ. Press; 1980)
4. J.W. van Holten, Phys. Lett. B352 (1995), 220
5. J.W. van Holten, Fortschr. Phys. 45 (1997), 6

MIX
Papier aus verantwortungsvollen Quellen
Paper from responsible sources
FSC® C105338

If you have any concerns about our products,
you can contact us on
ProductSafety@springernature.com

In case Publisher is established outside the EU,
the EU authorized representative is:
Springer Nature Customer Service Center GmbH
Europaplatz 3, 69115 Heidelberg, Germany

Printed by Libri Plureos GmbH
in Hamburg, Germany